In-pack processed foods

Related titles:

Thermal technologies in food processing
(ISBN 978-1-85573-558-3)
Thermal technologies have long been at the heart of food processing. The application of heat is both an important method of preserving foods and a means of developing texture, flavour and colour. An essential issue for food manufacturers is the effective application of thermal technologies to achieve these objectives without damaging other desirable sensory and nutritional qualities in a food product. Edited by a leading authority in the field, and with a distinguished international team of contributors, *Thermal technologies in food processing* addresses this major issue. It provides food manufacturers and researchers with an authoritative review of thermal processing and food quality.

Improving the thermal processing of foods
(ISBN 978-1-85573-730-3)
Thermal technologies must ensure the safety of food without compromising its quality. This important book summarises key research both on improving particular techniques and measuring their effectiveness in preserving food and enhancing its quality. Part I examines how best to optimise thermal processes, Part II focuses on developments in technologies for sterilisation and pasteurisation, there is a group of chapters considering the validation of thermal processes, and a final group of chapters which detail methods of analysing microbial inactivation in thermal processing.

Modelling microorganisms in food
(ISBN 978-1-84569-006-9)
While predictive microbiology has made a major contribution to food safety, there remain many uncertainties. There is growing evidence that traditional microbial inactivation models do not always fit the experimental data and an awareness that bacteria of one population do not behave homogeneously, that they may interact and behave differently in different food systems. These problems are all the more important because of the growing interest in minimal processing techniques that operate closer to death, survival and growth boundaries and thus require a greater precision from models. Edited by leading authorities, this collection reviews current developments in quantitative microbiology. Part I discusses best practice in constructing quantitative models and Part II looks at specific areas in new approaches to modelling microbial behaviour.

Details of these books and a complete list of Woodhead's titles can be obtained by:

- visiting our website at www.woodheadpublishing.com
- contacting Customer Services (e-mail: sales@woodhead-publishing.com; fax: +44 (0) 1223 893694; tel.: +44 (0) 1223 891358 ext.130; address: Woodhead Publishing Ltd, Abington Hall, Granta Park, Great Abington, Cambridge CB21 6AH, England)

In-pack processed foods

Improving quality

Edited by
Philip Richardson

CRC Press
Boca Raton Boston New York Washington, DC

WOODHEAD PUBLISHING LIMITED
Cambridge England

Published by Woodhead Publishing Limited, Abington Hall, Granta Park,
Great Abington, Cambridge CB21 6AH, England
www.woodheadpublishing.com

Published in North America by CRC Press LLC, 6000 Broken Sound Parkway, NW,
Suite 300, Boca Raton, FL 33487, USA

First published 2008, Woodhead Publishing Limited and CRC Press LLC

British Library Cataloguing in Publication Data
A catalogue record for this book is available from the British Library.

Library of Congress Cataloging in Publication Data
A catalog record for this book is available from the Library of Congress.

Woodhead Publishing ISBN 978-1-84569-246-9 (book)
Woodhead Publishing ISBN 978-1-84569-469-2 (e-book)
CRC Press ISBN 978-1-4200-7433-8
CRC Press order number WP7433

The publishers' policy is to use permanent paper from mills that operate a sustainable forestry policy, and which has been manufactured from pulp which is processed using acid-free and elementary chlorine-free practices. Furthermore, the publishers ensure that the text paper and cover board used have met acceptable environmental accreditation standards.

Typeset by Ann Buchan (Typesetters), Middlesex, England
Printed by T J International Limited, Padstow, Cornwall, England

Contents

Contributor contact details

(* = main contact)

Editor

Philip Richardson
Campden and Chorleywood Food Research Association (CCFRA)
Station Road
Chipping Campden
Gloucestershire
GL55 6LD
UK

e-mail: p.richardson@campden.co.uk

Chapter 1

Graham Pape
Crown Packaging UK plc
Crown Technology
Downsview Road
Wantage
OX12 9BP
UK

e-mail: Graham.PAPE@eur.crowncork.com

Chapter 2

Lynneric Potter
Campden and Chorleywood Food Research Association (CCFRA)
Station Road
Chipping Campden
Gloucestershire
GL55 6LD
UK

e-mail: l.potter@campden.co.uk

Chapter 3

Jose M. Lagaron,* Maria J. Ocio and Avelina Fernandez
Institute of Agrochemistry and Food Technology (IATA)
CSIC
Apdo. Correos 73
Burjassot 46100
Spain

e-mail: lagaron@iata.csic.es

Chapter 4

Emma Hanby
Campden and Chorleywood Food Research Association (CCFRA)
Station Road
Chipping Campden
Gloucestershire
GL55 6LD, UK

e-mail: e.hanby@campden.co.uk

Chapter 5

Christopher Holland
Holmach Ltd
The Dingle
Pilsgate
Stamford
PE9 3HW
UK

e-mail: holmachltd@btconnect.com

Chapter 6

Richard Walden
Director, Zinetec Ltd
22 Highworth Road
Faringdon
Oxfordshire
SN7 7EE
UK

e-mail: rwalden@shakaprocess.com

Chapter 7

Mark L. Seiboth and Geoff H. Shaw*
Ellab UK Limited
3 Lodge Farm Barns
New Road
Bawburgh
Norwich, Norfolk
NR9 3LZ
UK

e-mail: gs@ellab.com

Chapter 8

John J. Sullivan
Mesa Laboratories Inc
12100 West 6th Ave
Lakewood
CO 80228
USA

e-mail: JSULLIVAN@MesaLabs.com

Chapter 9

Gary Tucker
Section manager – Process Development
Food Manufacturing Technologies Department
Campden and Chorleywood Food Research Association (CCFRA)
Station Road
Chipping Campden
Gloucestershire
GL55 6LD
UK

e-mail: g.tucker@campden.co.uk

Chapter 10

Osvaldo H. Campanella* and Guibing Chen
Faculty of Agricultural and Biological Engineering
Purdue University
West Lafayette
IN 47907
USA

e-mail: campa@purdue.edu

Chapter 11

Cuiren Chen*
Research and Development Center
World Headquarters of Campbell Soup Company
1 Campbell Pl
Camden, NJ 08086
USA

e-mail: cuiren_chen@yahoo.com

H. S. Ramaswamy
Department of Food Science and Agricultural Chemistry
McGill University
21111 Lakeshore
Ste Anne de Bellevue
Quebec
Canada, H9X 3V9

e-mail: hosahalli.ramaswamy@mcgill.ca

Chapter 12

Phil James*
School of Mathematics and Statistics
University of Plymouth
Drake Circus
Plymouth
Devon, PL4 8AA
UK

e-mail: p.james@plymouth.ac.uk

Gary Tucker
Section manager – Process Development
Food Manufacturing Technologies Department
Campden and Chorleywood Food Research Association (CCFRA)
Station Road
Chipping Campden
Gloucestershire
GL55 6LD
UK

e-mail: g.tucker@campden.co.uk

Chapter 13

Peter McClure
Safety and Environmental Assurance Centre
Unilever
Colworth Science Park
Sharnbrook
MK44 1LQ
Bedfordshire
UK

e-mail: Peter.McClure@unilever.com

Chapter 14

Sandra C. Stringer and M. W. Peck*
Institute of Food Research
Norwich Research Park
Colney Lane
Norwich
NR4 7UA
UK

e-mail: mike.peck@bbsrc.ac.uk and sandra.stringer@bbsrc.uk

Chapter 15

Carolina Perez-Locas and Varoujan A. Yaylayan*
Department of Food Science & Agricultural Chemistry
McGill University
21111 Lakeshore
Ste Anne de Bellevue
Quebec
Canada H9X 3V9

e-mail: varoujan.yaylayan@mcgill.ca

Chapter 16

Joss Delves-Broughton
Senior Application Specialist
Sales Application Food Ingredients
Danisco
6 North Street
Beaminster
Dorset
DT8 3DZ
UK

e-mail: joss.delves-broughton@danisco.com

Chapter 17

I. Oey, * T. Duvetter, D. N. Sila, D. Van Eylen, A. Van Loey and M. Hendrickx
Center for Food and Microbial Technology
Katholieke Universiteit Leuven
Kasteelpark Arenberg 22
3001 Leuven
Belgium

e-mail: indrawati.hartono@biw.kuleuven.be

Chapter 18

Paul L. Dawson
Department of Food Science and Human Nutrition
Clemson University
204 Poole Hall
Clemson SC 29634-0316
USA

e-mail: pdawson@clemson.edu

Chapter 19

Dagbjørn Skipnes*
Norconserv AS
Niels Juels gt. 50
4008 Stavanger
Norway

e-mail: Dagbjorn.Skipnes@norconserv.no

Marc Hendrickx
Center for Food and Microbial Technology
Katholieke Universiteit Leuven
Kasteelpark Arenberg 22
3001 Leuven
Belgium

e-mail: Marc.hendrickx@biw.kuleuven.be

Preface

Thermal technologies continue to be at the core of preservation strategies used by the food industry worldwide. Traditional applications of thermal treatments to render products commercially sterile in cans and glass are being supplemented through the availability of new, alternative packaging formats such as pouches. These provide fresh opportunities for product and process innovators as they strive to service an ever-demanding consumer-driven marketplace. As yet, few of the so-termed minimal processing/non-thermal technologies have had widespread application in the industry, other than in some niche areas, and it is difficult to see this situation changing in the medium term.

This is the third book in the series relating to thermal technologies in food manufacturing and follows the same format as previous volumes. The book is an edited collection of contributions from eminent practitioners, industrial and academic, addressing the key challenges and opportunities being presented by today's technologies. The focus of this volume is the emerging opportunities to improve the quality of thermally processed foods. The often cited criticism of thermal technologies is the perceived adverse effect on product quality (organoleptic and nutritional). However, there are opportunities available to minimise any such effects.

The book is divided into a number of focused sections:

The first addresses the recent developments in packaging. It discusses recent developments in cans but majors on the new and emerging plastic containers and pouches that are penetrating the market.

In the second section, attention is turned to the developments in steriliser/pasteuriser design, and also to the available techniques to model and measure the thermal process delivered to the containers during sterilisation and pasteurisation. The area of process simulation is discussed, with particular emphasis on the application of emerging techniques to the thermal processing area for optimisation and improved process effectiveness and efficiency.

The third section discusses some of the topical issues related to food safety. A prerequisite for all processed foods is that they must be safe for the consumer.

The final section of the book explores not only some of the opportunities for thermal processing in relation to meat and fish, but also the opportunities presented by high pressure in combination with heat to deliver high-quality fruit and vegetable products. This is an interesting opportunity as in many cases combination processes point the way to new avenues for development.

The pace of innovation in equipment and packaging is fast, supporting the introduction of many thousands of new products annually. Coupling this with the new insights offered from the academic communities, further innovative opportunities emerge. Thermal processing is well understood and the impact on the microbiology and chemical constituents of foods is well defined. Manufacturers have confidence in the technology and it has good consumer acceptability, albeit with some reservations about the opportunity to deliver optimised product quality. When compared with the level of knowledge about, and confidence in, alternative 'new technologies' such as ultra-high pressure or pulsed electric fields, clearly thermal processing has a bright future in mainstream food production.

Philip Richardson

Part I

Advances in packaging formats for in-pack processed foods

1

Advances in can design and the impact of sterilisation systems on container specifications

G. Pape, Crown Packaging UK plc, UK

1.1 Introduction

A significantly large majority of ambient shelf-stable food products rely on the metal can to provide a robust container that has an outstanding safety record stretching back many decades. The cans of today are much different in construction to those of the early 1900s when canning was in its infancy, but they have to provide the same functionality. They must:

- be heat processable to allow achievement of commercial sterility,
- have integrity which prevents bacterial ingress after heat processing,
- provide a total oxygen barrier to minimise degradation of the product inside,
- resist handling abuse during distribution and retailing,
- be easy to open for the consumer, and
- in today's environmentally sensitive world be recyclable as a primary material.

The metal can, whether manufactured from steel or aluminium, performs all of these functions well. However, environmental and economic factors continue to challenge the can design engineers. The market for food containers is driven by innovation, whatever the material of choice. Can design continues to rise to this challenge by progressively using lower gauge materials to achieve the same can and end performance. This ensures that the key criteria for the package outlined previously are not compromised. There is a point at which innovative design cannot go forward without the sign-on of the canners.

It has become apparent over recent years that the hurdles to introduction of innovative metal packaging lie not only in the domain of the can manufacturer, but now also include hurdles in the cannery, often focussed around can handling, but also challenged by the sterilisation systems themselves. Large canneries sterilise the bulk of their output through large continuous cooker systems, such as hydrostats, and reel and spiral cookers. These cookers provide high efficiency but they are limited in their capabilities. Economy of scale also means that they are costly to install and have a long service lifetime, often 20–30 years. The can has changed significantly in the last 20–30 years; thus can designs of today are still expected to perform in equipment installed a generation ago.

In order to understand how the can designs of today are able to perform well when sterilised through systems which are in use in the canneries today, whatever their vintage, this chapter will seek to review the following key questions:

- What influences can performance?
- How are can performance requirements determined?

A series of case histories will be employed to illustrate the challenges which face the can designers of today.

1.2 The influence of sterilisation systems on container specifications

A number of simple questions often arise in the debate to determine what container specification may be used for a new business opportunity. Classically, the first question of the marketeer is:

- What is the optimum specification for packing product X for customer Y in cannery Z?

The commercial and technical team will then ask :

- Why can we use specification A for this application at canneries X and Y, but not at cannery Z?

The manufacturing team might ask :

- Are we able to reduce the end peaking performance for diameter B?

and:

- If we optimise the can and/or end performance, what are the limitations?

Having taken these questions onboard, we need to understand what are the driving forces behind these questions.

For the *packaging manufacturer* there is a desire to supply a minimum portfolio of specifications to limit manufacturing line changeovers. A reduced portfolio will allow more cost-effective material purchasing with fewer specifications required. Also, it is more cost effective to operate with a minimum inventory using just-in-time manufacturing to limit working capital. Ideally, a ‘one specification fits all’ philosophy gives the most cost- and material-efficient solution, but will this solution meet the requirements of the canner ?

For the *canner*, the can must function on current can filling lines with heat processing achieved through existing sterilisation equipment and have cost-effective packaging specifications.

Fundamental to being able to provide a can design solution to these requirements is an understanding of the factors that might be influencial in affecting can performance:

- *Fill level* – Extremes of fill level/associated pack headspace will affect performance requirements. The headspace is the part of the can not filled with product. It is conventionally measured on an open can from the top of the can flange on an unseamed can, or from the top of the double seam on a seamed can.
- *Fill temperature* – High and low temperatures of the solid or liquid phase components of a product recipe will have an effect. A high level of a cold garnish component on a soup pack can affect the initial pack temperature significantly.
- *Can style* – Depending on materials and profiles, a two-piece can base will have more/less volume expansion capability than a classic/sanitary end used on a three-piece can. Can base and end profiles may be flexible or rigid, depending on the design brief, and may significantly influence the level of internal pressure generated in the can.
- *Steriliser system* – Batch overpressure retort systems offer significant flexibility in their temperature and pressure capability, compared with the rigidity of capability of these parameters in continuous hydrostat or reel and spiral cooker systems, or in batch steam retorts.

How are these factors understood ?

In the past it was only possible to carry out pilot-plant simulations with pressures in the packs measured using conventional wired pressure transducers. The limitations of these systems were generally governed by the ability of the electronic components and wiring to withstand sterilisation in steam at temperatures typically up to 140 °C.

Systems were available that used silicone oil-filled capilliary tubes to connect to a pressure transducer positioned outside of the steam environment; however, the nature of these systems limited their use to the pilot plant.

During the mid-1990s, autonomous data loggers with measurement transducer, power source and memory device were developed to allow measurement of temperature and pressure in the hostile steam environment. Such loggers were thus able to record data in industrial sterilisers, opening up the opportunity to understand the differential pressures experienced by the package during sterilisation. A number of suppliers, including TMI Orion, Ellab and Datatrace, developed such autonomous dataloggers, which are now extensively utilised within the food canning industry and in other medical science and industrial applications where extreme conditions of temperature, pressure and humidity are experienced.

Initially, dataloggers were relatively large – typically 100 ml or greater. However, the volume of the dataloggers reduced to 30–40 ml (e.g. TMI Orion Nanovacq, Ellab Tracksense logger) by the mid-1990s were further reduced to volumes <10 ml (e.g. TMI Orion Picovacq, Ellab Tracksense Pro Micro logger Datatrace Micropack III) within 10 years.

Differential pressure is the key parameter, which is seen by the package at each and every stage of the sterilisation cycle. Differential pressure is expressed as the difference between the internal pressure, measured by a datalogger placed inside the package, and the external pressure, measured by a datalogger placed outside of the container, measuring the retort environment. Resultant differential pressure values are therefore positive (+) when there is excess internal pressure in the package, and negative (–) when there is a net external pressure acting on the package.

Peaking/buckling of ends is a performance feature associated with positive differential pressure values, and panelling/collapse of cans is a performance feature associated with negative differential pressure values.

1.3 Case histories illustrating the influence of sterilisation systems on can performance

For petfood, the difference in can performance requirements will be illustrated by reference to product sterilised in hydrostatic cookers and in batch overpressure retorts.

For sweetcorn, the difference in can performance requirements will be illustrated by reference to product sterilised in reel and spiral cookers, and in overpressure hydrostatic retorts.

For soups or mushrooms, the can performance requirements will be illustrated by reference to product sterilised in a reel and spiral cooker.

1.3.1 Case history – petfood

The driving force behind this case history is one of reducing the body gauge/thickness on three-piece welded cans whilst maintaining panelling-free performance. (Panelling is generally defined as collapse of the can body, resulting in a permanent indentation of the can body over the majority of its height, such that the axial performance/stacking ability of the pack is compromised.) There was an additional constraint in that the existing body gauge specification gave rise to sporadic panelling during sterilisation. The underlying question raised was 'How can we utilise a reduced body gauge with existing product recipes and sterilisation regimes?'

A panelling risk is created during the come up phase of the sterilisation cycle as the pack passes through the infeed water leg of the hydrostat (Fig. 1.1). During this phase, the pressure increase outside of the can in the water column increases faster than the internal pack pressure. This is caused by the slow rate of heating of the product inside the pack, and the associated slow increase in internal pressure. At the bottom of the water column, the pack then passes into the steam chamber and rapidly heats, creating internal pressure (Fig. 1.2).

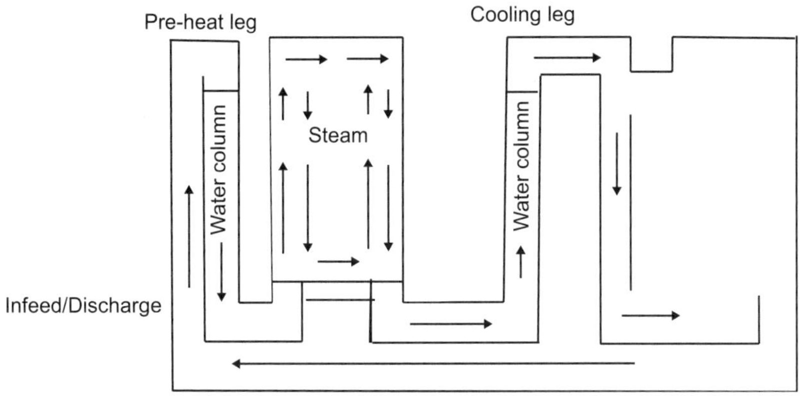

Fig. 1.1 Hydrostatic cooker – schematic diagram.

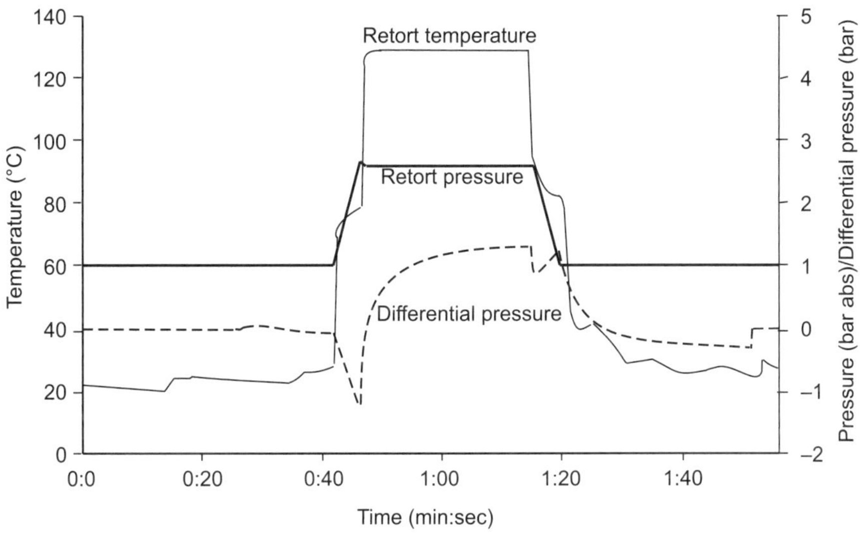

Fig. 1.2 Petfood – sterilised in hydrostat.

By increasing the water temperature in the infeed water leg, it is possible to increase the temperature gradient across the can wall and thus increase the pack temperature sufficiently fast to increase internal pack pressure (Fig. 1.3). The differential pressure therefore becomes more positive, alleviating the panelling risk. As the pressure experienced become more positive, it is possible to utilise a container specification with a reduced panelling strength, thereby allowing a reduction in body gauge to be introduced without loss of ultimate can performance – panelling as a problem should disappear.

By comparison, for a batch overpressure retort, it is possible to design the temperature and pressure regime to limit differential pressure by balancing the

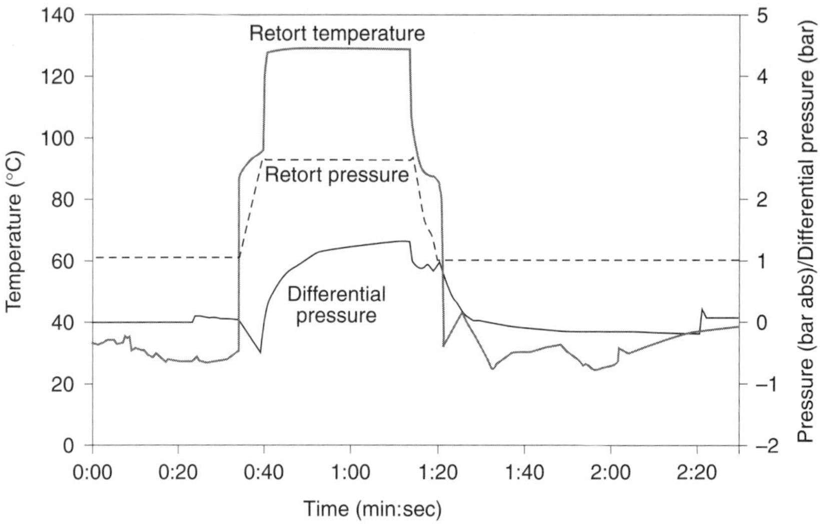

Fig. 1.3 Petfood – sterilised in hydrostat – reduced negative differential pressure/increased infeed water leg temperature.

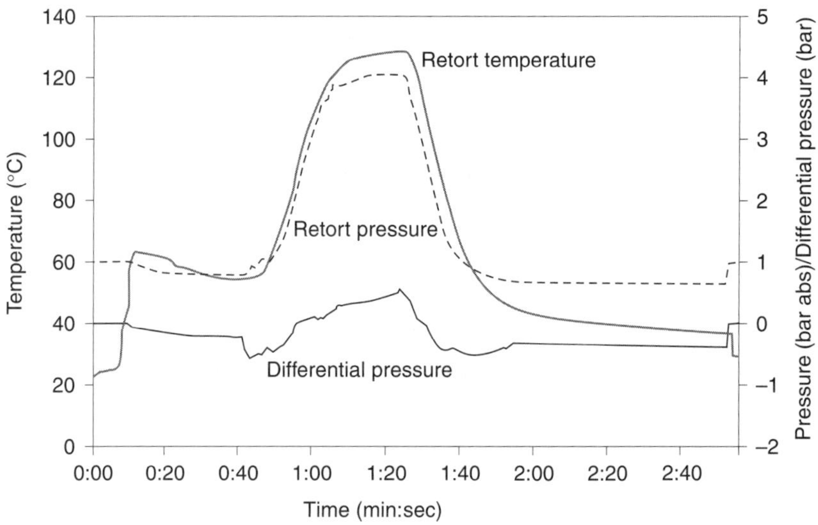

Fig. 1.4 Petfood – sterilised in batch overpressure retort.

internal pack and external retort pressures. Differential pressure at each stage of the retort cycle can be better controlled and thus the risk of panelling can be reduced, allowing a reduced gauge body (Fig. 1.4).

It should be noted that differential pressure and panelling risk are not the only factors that will play a part in the choice of specification for a particular component.

The quality of can handling is a key factor in the equation that determines the suitability of a particular specification. In this instance, damage sustained by the pack during normal can handling activities from depalletisation through filling, seaming, sterilisation, labelling and secondary packaging all play a part in determining suitability. Despite a low risk of panelling during sterilisation, a reduced gauge specification may not ultimately be introduced if routine can handling damage levels are unsustainable.

1.3.2 Case history – vacuum-packed sweetcorn

The driving force behind this case history is one of supplying a common body gauge/thickness on three-piece welded cans for sweetcorn and other vegetable packs in both 65 × 71 mm and 83 × 85 mm sizes, whilst maintaining panelling-free performance.

Sweetcorn is traditionally packed with a low level of liquor present in the pack to facilitate a high-temperature, short-time process, leading to enhanced product quality. Certain types of steriliser, by their design, can lead to sporadic problems of body panelling and reverse peaking. The question raised is why do we have problems with some types of sterilisers, or some steriliser configurations?

Sweetcorn cans are seamed in a vacuum chamber at around 600–700 mbar vacuum. Sterilisation is normally carried out in reel and spiral cookers, having three or four shells, promoting a low risk of body panelling or reverse peaking of the end. In a three-shell (two cook shells, one pressure cooling shell) or four-shell (two cook shells, one pressure cooling shell, one atmospheric cooling shell) reel and spiral cooker, the minimum differential pressure reached around –1 bar, which is well within the performance limits of cans conventionally used for vegetable packs such as peas, carrots, and green beans (Fig. 1.5).

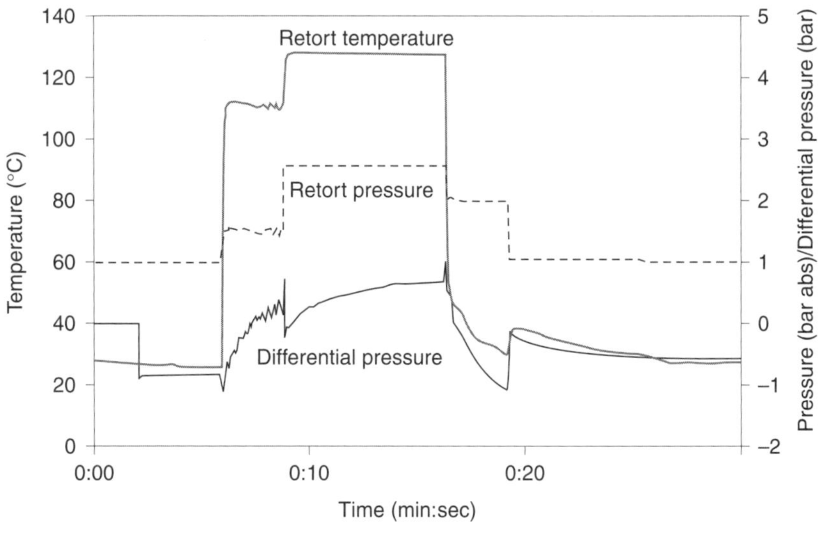

Fig. 1.5 Sweetcorn (vacuum packed) – four-shell reel and spiral cooker.

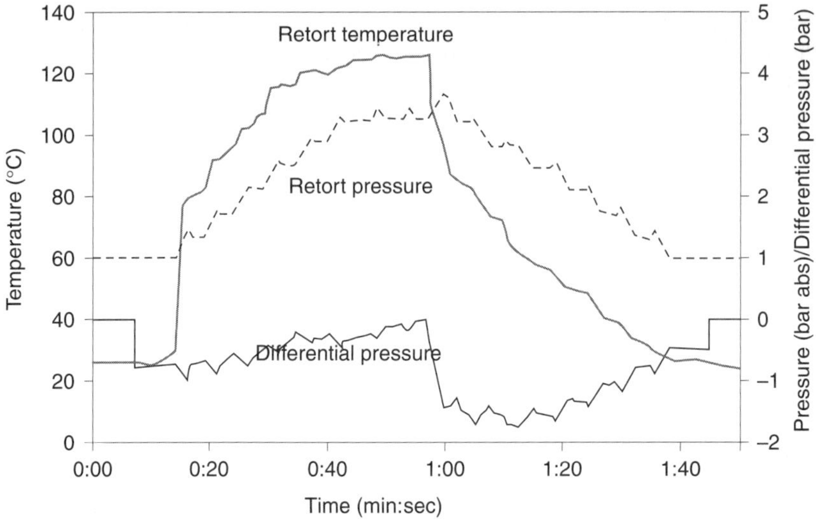

Fig. 1.6 Sweetcorn (vacuum packed) – Hunister overpressure hydrostat.

In parts of Europe, a Hunister overpressure hydrostatic cooker is used. This type of cooker can lead to a high risk of body panelling or end reverse peaking during the cooling phase of the retort cycle, necessitating the use of increased metal gauge for both can body and end. The Hunister is a six-stage overpressure hydrostat, used extensively for pasteurised or sterilised glass packs and cans mainly in Eastern Europe. Differential pressure is commonly negative (more pressure outside of the pack than inside), reaching a maximum differential pressure no more than *ca.* + 100 mbar during sterilisation (Fig. 1.6). During cooling, the combination of overpressure and the high pack vacuum, reformed by rapid condensation of the steam generated inside the can during the initial stages of cooling, results in a high negative differential pressure at a level greater than that sustainable by standard specification vegetable cans. The result is that there is a need to increase can body and end gauge to provide adequate resistance to body panelling or end reverse peaking performance. There is no simple solution to this problem through adjustment of the cooker pressure profile.

1.3.3 Case history – reel and spiral cooker for mushrooms/soups

The driving force behind this case history is to provide a single can specification for all soup products, independent of whether they are products with a vegetable garnish or are a single-phase liquid product. Similarly, there is a desire to use a common specification for mushrooms and other vegetables.

The principal can performance issue is that cans of normal body gauge for a specific can size may give rise to sporadic body panelling and end peaking during sterilisation at high temperatures (130–135 °C). This problem is equally applicable to soup products and to mushrooms. The overriding question to be answered was how to use a common body plate gauge for all products utilising existing recipes, filling conditions and sterilisation regimes.

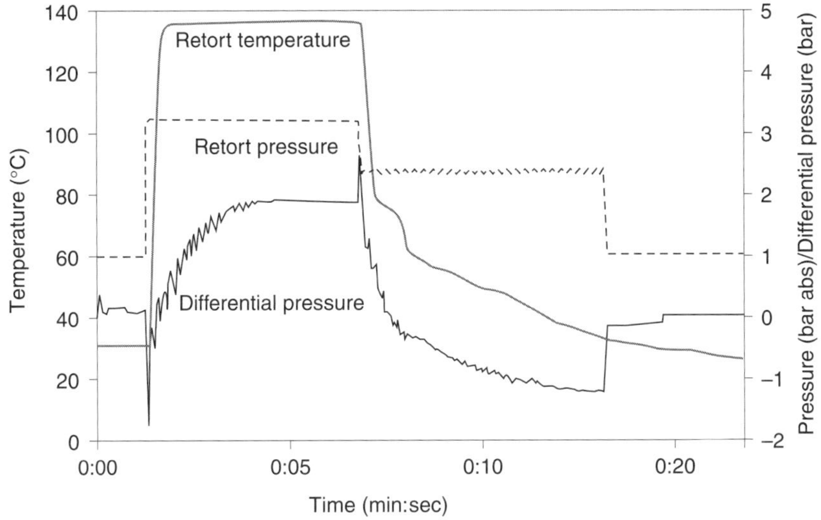

Fig. 1.7 Mushrooms in brine – two-shell reel and spiral cooker.

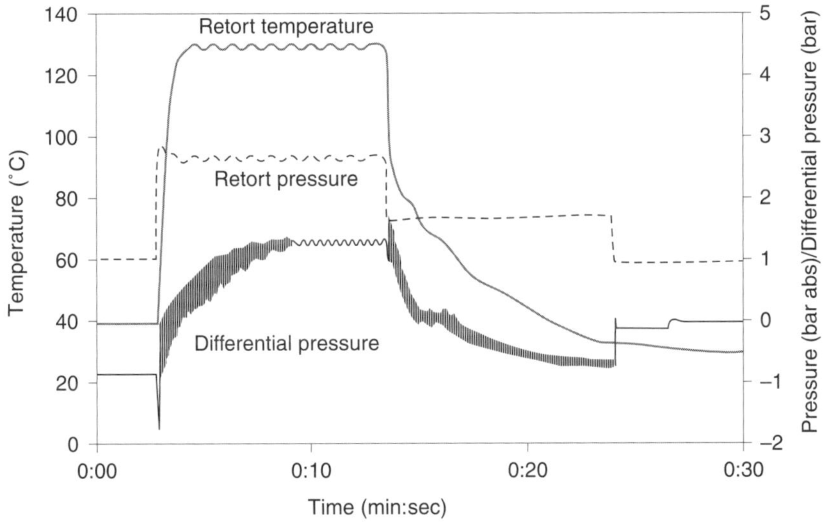

Fig. 1.8 Soup – two-shell reel and spiral cooker.

For both soups and mushrooms sterilised in a two-shell (one cook shell, one pressure cooling shell) reel and spiral cooker, the risk of can body panelling initiates as the cans enter the cook shell at full sterilisation temperature and pressure. The cans are subjected to the maximum external pressure, whilst their internal pressure is low on entry to the sterilisation shell and before the product or can headspace has opportunity to heat up (Figures 1.7 and 1.8). As the can transits through the sterilisation shell, the internal pressure builds up to a level, which is within the performance capability of the ends.

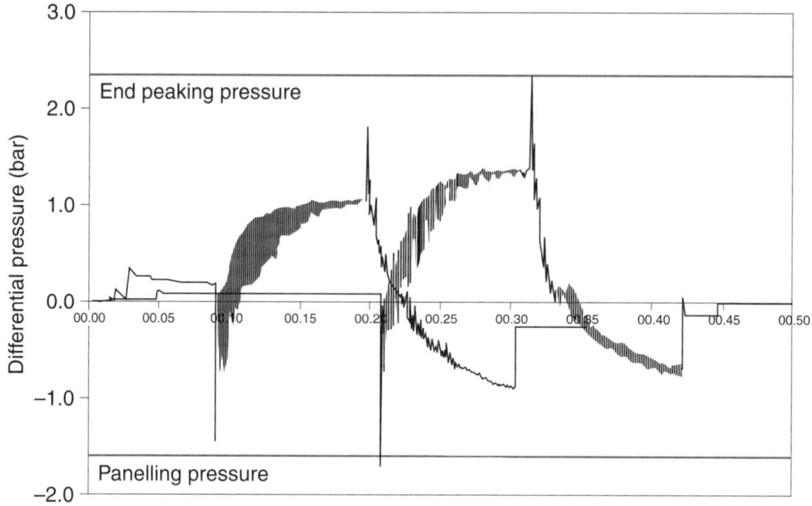

Fig. 1.9 Soup/mushrooms – two-shell reel and spiral cooker – differential pressures exceed can/end performance limit.

On transfer to the pressure cooling shell, there is an instantaneous increase in pressure difference between the internal and external pressures, caused by the lower pressure in the cooling shell. This increase is generally transient because, once the can enters the cooling water, the pressure in the can headspace collapses and the can contents start to cool, further reducing internal pressure. Finally, as the can cools, the combination of internal pressure (vacuum) present in the can and the external pressure may cause the panel, initiated on entry to the cooker, to reform. Under these circumstances, the panel reforms independent of the level of pressure present in the cooling shell.

Classically, body panelling and end peaking are caused by opposite extremes of pressure – body panelling by excess external pressure and end peaking by excess internal pressure. Pressure studies have shown that it is indeed possible for the two phenomena to occur in the same can (Fig. 1.9):

- Panelling occurs at the point of entry to the cooker (Figs 1.10 and 1.11 point 1).
- Panel formed on entry is 'pushed out' during sterilisation.
- Peaking risk occurs at transfer from sterilisation to cooling before the internal pressure of the can falls (Figs 1.10 and 1.11 point 2).
- Panelling re-occurs during cooling due to the creation of a 'pre-panel' at the entry to the cooker (Figs 1.10 and 1.11 point 3).

There are a number of potential solutions to this problem:

- Reduce sterilisation temperature and increase sterilisation time. This will have an adverse impact on plant throughput.
- Add a pre-heat shell to the cooker. This is a capital cost item and may not be

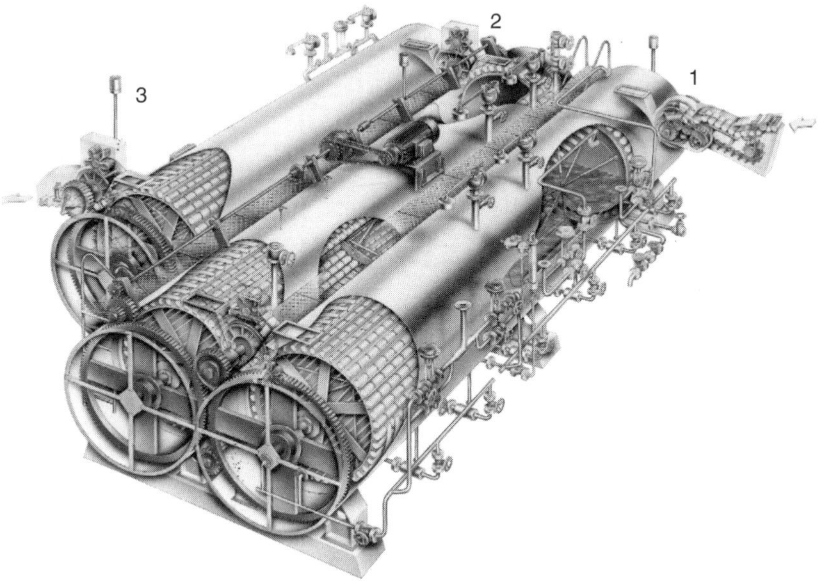

Fig. 1.10 FMC Sterilmatic Cooker. (Image kindly supplied by FMC Foodtech.).

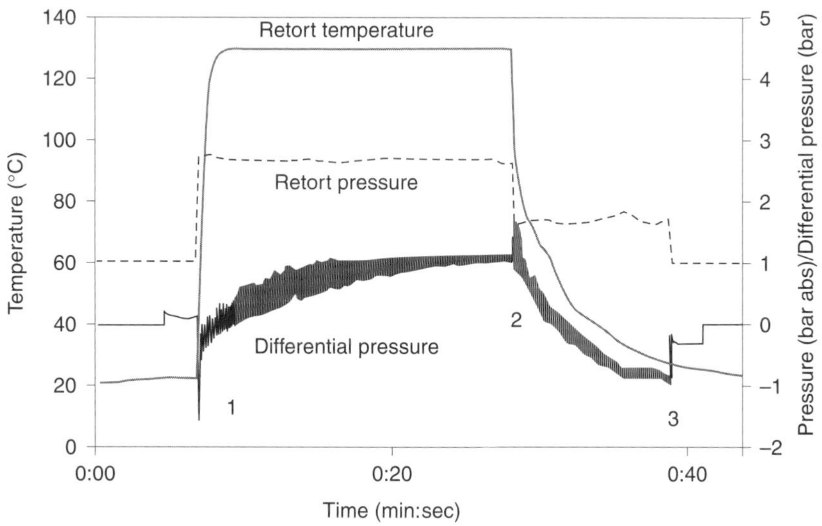

Fig. 1.11 Soup – sterilised in two-shell reel and spiral cooker – showing high levels of positive/negative differential pressures.

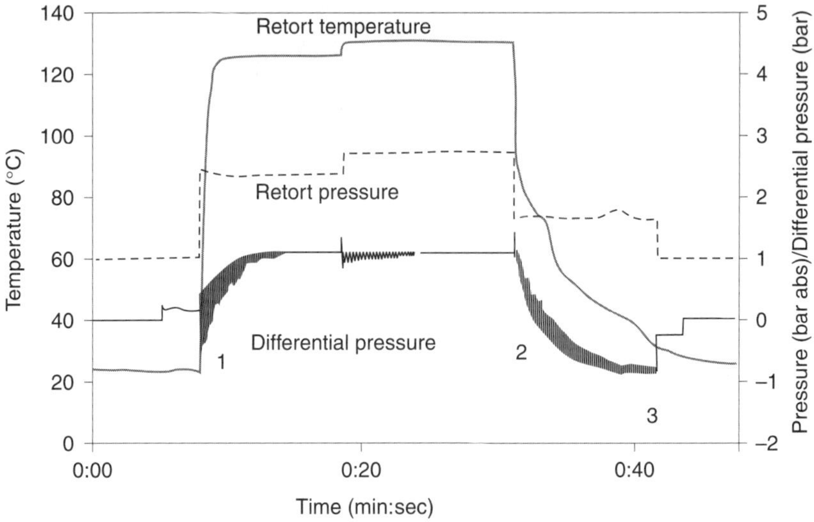

Fig. 1.12 Soup – sterilised in three-shell reel and spiral cooker – showing reduced levels of positive/negative differential pressures.

possible due to plant layout constraints. Even adding a pre-heat shell to the cooker may not always improve the situation, especially if the shell were to be operated at full sterilisation temperature (Fig. 1.12).

- Improve the radial strength of the can body and the end peaking strength. These will potentially involve increasing the metal gauge with added on-cost.

1.3.4 Case history – trouble shooting

The driving force behind this case history is a problem of end peaking on 3 kg (153 × 178 mm) cans of vegetables in brine, sterilised through a five-shell (three cook shells, one pressure cooling shell, one atmospheric cooling shell) reel and spiral cooker. The pressure performance of the ends concerned was found to meet normal specification limits. The same ends were known to be performing with the same range of products without peaking problems when used on cans sterilised through a similar cooker at a different cannery.

Differential pressure was measured and found to remain within the pressure performance limits of the end during passage through all three sterilisation shells, at the transfer from the final sterilisation shell to the pressure cooling shell, but not at the transfer from the pressure cooling shell to the atmospheric cooling shell (Fig. 1.13). On transfer from sterilisation to pressure cooling, the external pressure was normally increased in order to prevent a risk of peaking. This was contrary to accepted normal practice and resulted in the limitation of end panel expansion. Efficient heat transfer, both into and out of the can, relies on the presence of a defined minimum headspace (pack volume not filled with product) to allow mixing of the product during can rotation. The use of a high external pressure,

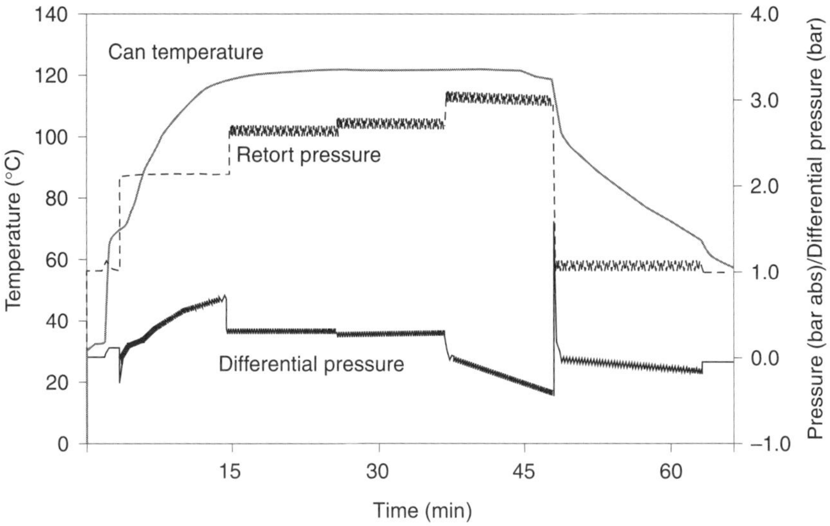

Fig. 1.13 Vegetables in brine – sterilised in five-shell reel and spiral cooker.

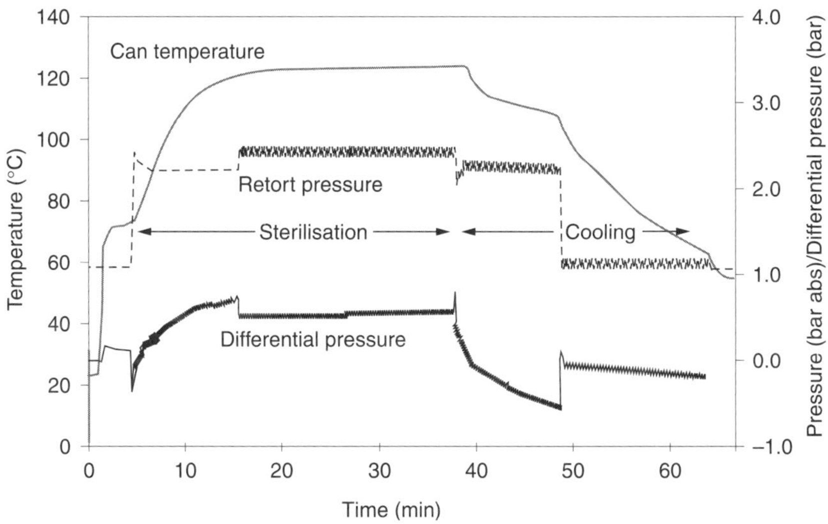

Fig. 1.14 Vegetables in brine – sterilised in five-shell reel and spiral cooker with modified pressure profile.

above normal levels, during passage of the can through the pressure cooling shell, limited the extent of cooling possible. This caused the internal pressure of the can to be maintained at an unexpected high level. On transfer from the pressure cooling shell to the atmospheric cooling shell, the differential pressure increased for a short period to a level that was higher than the end's performance limit and resulted in peaking.

The resolution to this problem was achieved through a reduction in the pressure in the pressure cooling shell, which in turn allowed the end panels to expand sufficiently to allow product mixing to occur. Once internal temperature was allowed to fall, the internal can pressure was reduced. The differential pressure on transfer to the atmospheric cooling shell was subsequently at a lower level, within the performance limits of the end (Fig 1.14).

1.4 Summary

We set out to answer a number of questions:

- What is the optimum specification for packing product X for customer Y in cannery Z ?
- Why can we use specification A for this application at canneries X and Y, but not at cannery Z ?
- Are we able to reduce the end peaking performance for diameter B ?
- If we optimise the can and/or end performance, what are the limitations ?

Advances in differential pressure measurement equipment have allowed both packaging manufacturers and canners to answer these initial questions, and to understand the factors which drive cost-effective packaging solutions.

These factors may be many and varied, being both influenced and limited by:

- Product and filling conditions,
- Packaging specifications,
- Steriliser type, configuration and operating its regime.

2

Retortable pouches

L. Potter, Campden and Chorleywood Food Research Association, UK

2.1 Introduction

Retortable pouches are flexible, laminated food packages capable of withstanding the time and temperature demands of thermal processing cycles. The shape and structure of the pouch often lead to a reduced thermal processing time compared with that of a can. Good-quality, shelf-stable products are being produced with improved texture and nutritional value, comparing favourably with those of frozen foods. They are convenient, have brand enhancement and are more environmentally friendly, with a lower weight volume and less waste. However, high capital costs for machinery, slower filling rates, and considerations such as residual air and the delicate nature of the laminate structure have previously limited their uptake. With the development of new materials and equipment, the pouch market has increased, with over 10 billion packs being sold worldwide in 2004 (Schreiber, 2005).

This chapter discusses the materials used to construct retortable pouches and the properties of these materials required to achieve a quality, shelf-stable product. There are a number of techniques that can be put in place during filling to ensure no contamination of the seal area, so reducing the likelihood of defects and the risk of microbiological growth within the product. Also discussed are processing requirements such as placing the pouches into racking systems to ensure efficient heat transfer, and methods to measure the heat distribution within the pack. Finally, outer packaging and distribution to prevent damage to the pouches are discussed. Throughout the production of retortable pouches, quality assurance checks need to be carried out to ensure the product and packaging are safe and of a good quality.

Initial research was carried out on pouches in the 1940–50s with thermal processes below 100 °C using polyvinylidene chloride (PVdC) materials with fish and pulse products, mainly for use by the military. It was during the 1960s and 1970s when the significant changes and developments to the pouch took place, with the first boilable pouch. In 1965, the first retortable pouches were produced in Italy (Canadian Food Inspection Agency, 2002). Pouches became popular in the Japanese and Chinese market in the late 1960s, with products such as meat with dumplings and curry being packed in them. The world's first commercial retortable pouch product was introduced by Otsuka Chemical, Japan, in 1969. In Japan in 1971, 160 million pouches were produced and by 1989 this had increased to one billion pouches, with products including curry sauce, pasta, meat and rice (Yoshida, 2005).

The popularity of the pouch did not reach the West, due to economical and technological reasons, until the 1990s when they flooded the retail market, replacing a number of traditional packaging methods. Soups, rice and pet food were the main products packed in retortable pouches, followed by fish, meat and sauces. Research carried out by PIRA indicated that Europe consumed almost 3.8 million stand-up pouches in 2001, compared to 3.2 billion in the US. The UK is the largest national market, accounting for 20% of the pouches consumed in Europe, followed by France with 19% and Germany with 18% (Swinden, 2002). The pouch market is expected to increase by 20%, reaching maturity, by 2010.

2.2 Materials for retortable pouches

The types of materials used for pouches have improved since the early development of the pouch. However, the costs of the raw materials have had an influence on the materials selected. The cost of materials such as plastics and aluminium are never static, and these costs need to be taken into account alongside the cost of conversion of the materials into a pouch. This is of particular importance for high volume, low cost food products. Other important cost factors include those of production, together with filling, sealing and processing.

Correct material selection for the manufacture of pouches is of critical importance. Clearly defined performance characteristics for the finished pack should be agreed and be presented as an approved packaging specification. Food processors may receive pre-formed pouches requiring only filling and sealing, or laminated film on the reel for form–fill–seal applications; the general requirements are largely similar for either application. The characteristics required by the laminated film/package will normally include:

- Heat sealability.
- Capability of being heat processed (usually 115–125 °C).
- Barrier properties to gas and water (and light, depending on product).
- Physical strength to resist penetration by the foodstuff contained, and abuse during handling and processing.

- Inertness, so as not to impart any taint or odour to the contents.
- Ability to meet the requirements of local and international food contact and other regulatory standards.

There are two main types of pouches, gussetted and pillow packs. Gussetted pouches have an insert in the bottom, allowing them to stand in an upright position. They are considered weaker than pillow style pouches, mainly because the gussetted style have to be heat sealed through four layers at the corner seal, which, when quality tested, is found to be the main failure point. However, gusseted pouches have the advantage of taking up less space on the supermarket shelf and can advertise the product clearly in this position.

Laminate film structures vary considerably, dependent upon application. Most pouches are constructed from either three- or four-ply laminates, with an inner food contact heat-seal layer (normally polypropylene), barrier layer (aluminium, ethylene vinyl alcohol (EVOH), silicon oxide (SiO_x) or aluminium oxide (AlO_x)), an optional nylon layer and an outer polyester layer as shown in Fig. 2.1. In-between the layers of film are adhesives, usually with a polyurethane base, to hold the laminates together.

Polypropylene on the inner layer of the pouch acts as a sealing layer and can provide some strength and flexibility to the pouch. Aluminium provides a complete barrier to oxygen, light, moisture and aroma, but the integrity can sometimes be challenged. The aluminium layer may not be punctured but if damaged could lose some of its barrier properties. Annealed foil can provide extra flexibility to the retortable pouch and avoids flex cracking. The optional nylon layer can increase the strength of the pouch due to its puncture and abrasion resistance. The outer polyester layer is heat resistant and provides an area for printing, often reverse printing.

For a transparent pouch (Fig. 2.2) where the product is visible and which can be used in a microwave oven, the aluminium layer can be removed. The removal of the metallised layer to produce a transparent pack has produced a number of alternatives.

Polyester (PET)
Aluminium
Polyamide (nylon)
Polypropylene

Fig. 2.1 Aluminium-based laminate retort pouch structure.

Polyester (PET)
Polyamide (nylon)
Polypropylene

Fig. 2.2 Aluminium-free laminate retort pouch structure.

Silicon and aluminium oxide coatings have trade names such as SILAMINAT®, techbarrier® and CERAMIS®. The silicate layer is a very fine layer of flexible glass that acts as a gas and moisture barrier (May, 2004) to match those of aluminium film, although they do not provide a barrier to light (Fenn-Barrabab and Prommesberger 2003). There is also the convenient advantage that they are microwaveable and are able to pass through metal detectors. Silicon oxide films are manufactured in a high vacuum, using an environmentally friendly process that involves no solvents or chemicals. They are applied to the laminate using electron beam evaporation (www.ceramis.com).

All components including inks and adhesives must also be able to withstand high temperatures during retort processing, and be suitable for water immersion, steam processing, air processing or a combination of each.

2.2.1 Quality assurance checks

Quality checks should be carried out on pouches, regardless of whether they are bought in or made in house, to ensure that they meet regulatory and specification requirements. Normally, the packaging manufacturer would carry out these tests and the results may only be available from him on request. Some of the tests that may be included are:

(i) *Visual assessment* – to check there are no defects on the pouches both decorative and that may have an adverse effect on the process and shelf-life of the product.
(ii) *Seal strength* – to measure the strength of the seal for processing and storage.
(iii) *Burst testing* – to ensure the pack has adequate seal strength and integrity.
(iv) *Oxygen and water transmission rates* – to measure the barrier properties of the pouch.
(v) *Bond strength* – to ensure the pouch has sufficient strength to prevent delamination of the layers.

Other tests may include puncture resistance, drop test, compression test and slip characteristics.

2.3 Filling

There are a number of factors that need to be taken into account during filling, which can have a fundamental effect on the processing of the product. Filling equipment needs to be matched to each of the product's characteristics. Inadequate filling operations can cause contamination of the seal, resulting in an inadequate heat seal. The heat process can be affected by the level of fill into the pack and the initial temperature of the product.

2.3.1 Tempering of pouches

In some circumstances, it may be necessary to condition the pouches/film prior to use, by holding the material at a constant slightly elevated temperature. This gives the pouch greater flexibility during the filling and processing phases. If there is a lack of flexibility with the pouch during filling, the machine may not reliably be able to open the pouch. This leads to one of the main problems associated with filling – inconsistent presentation of the pack to the filling nozzle. Tempering is usually carried out at approximately 30 °C.

2.3.2 Filling equipment

When considering a filling/sealing machine for either pre-formed pouches or reel stock, important criteria include the type of product, and ingredients and their characteristics (frozen, size, solid or liquid). For example, if the product contains sharp or hard materials such as bones, they may endanger the subsequent integrity of the pack. Other considerations are the number and order of fills (i.e. single shot or multiple shot), fill volume required, type of pouch (pillow or gusset) and whether there is a requirement to pull vacuum or gas flush.

Most modern pouch filling machines have automated pouch loading stations and it is essential to ensure that the pouches are loaded correctly to ensure an even feed into the machine. Opening the pouch correctly just prior to filling will avoid contamination, folds and creases in the seal area. Pneumatic suckers and/or cones can be used to assist with the opening of the pouch. If using gussetted pouches, the additional thickness at the bottom of the pouch needs to be considered to prevent an uneven feed. The product should never come into contact with the sealing area of the pouch. Dripping from the filling nozzle is a potential seal contamination issue. Positive cut-off pumps, or blow off or suck back systems can remove drips from the dispensing nozzles. Protective shields can also reduce fouling of the seal area, particularly when food with particles is being deposited (Canadian Food Inspection Agency, 2002).

2.3.3 Control of filled weight

The filled volume and weight related to pouch size needs to be taken into account during filling, to allow for a sufficient headspace during processing. With liquid products, the fill should not exceed the nominal volume of the pouch indicated by the supplier, so as to avoid pack creasing, contamination during sealing or straining of the seals during subsequent heat processing. Excessive pack thickness should also be avoided as this may lead to an extended or insufficient heat process.

Products may be hot or cold filled into the pouch. When hot filling, the product temperature must be maintained over 63 °C to prevent growth of microorganisms and toxins. If the temperature does fall below 63 °C for a length of time, the product should be considered an ambient fill and the retort process adjusted accordingly. An ambient fill is considered to be between 8 and 63 °C. If held for a length of time, microbiological growth may occur, so it is essential to set a holding time period to minimise this. The product may be filled chilled or frozen (below 8 °C) but this temperature must be monitored and the holding time limited as psychotrophic organisms can still grow at these temperatures.

2.3.4 Contamination of the sealing area

During filling, it is very important not to contaminate the seal area of the pouch as this will compromise the integrity of the pack and its contents. The filling technique used cannot be a guarantee that there will not be any contamination in the seal area. Close monitoring of the filling process is essential. The type of equipment used may have an influence. Drip-free dispensing nozzles, vacuum suck back systems and positive cut-off pumps can be used, along with movable protective shields – all reduce the risk of contaminating the seal area of the pouch. The type and viscosity of the product being filled into the pouch will have an effect on the levels of contamination. Once a product has contaminated the seal, the area should not be wiped as this can push the product into the corners of the seal preventing a hermetic seal being formed; instead the pouch should be discarded.

The remaining headspace within the pack will have an effect on the heat process given to the product. Gases can be trapped within the product or from the filling process. Due to the flexibility of the pouch, it is not possible to create a significant vacuum during filling and any gas remaining in the headspace following sealing will expand during processing, placing a strain on the seals of the pack. Large volumes of air or gas in the headspace will also retard the rate of heat transfer. The levels of gas in the headspace can be reduced by steam flushing prior to sealing. Excessive oxygen in the headspace can cause oxidative reactions with the product during its shelf-life, so it is important to minimise these levels. Processing and storage trials should be carried out to determine acceptable levels of oxygen within the pack. With very oxygen-sensitive foods, the pouches can be back flushed with nitrogen.

If a vacuum is applied to the pouch, care needs to be taken to ensure that the product is not pulled into the seal area, causing contamination; this can occur if the vacuum application is too short, with insufficient time allowed for air removal. Care should also be taken to prevent damage such as flex cracking to the pouch. The type of product being packed will determine whether it is suitable for vacuum packing. Solid products without any additional liquid in the pack should be vacuum sealed to draw the package around the product, minimising the headspace. Products at a low temperature allow a higher vacuum to be pulled on the pack. Headspace can also be reduced by physically flattening the pouch but care should be taken not to move the product into the seal area, causing contamination.

2.3.5 Sealing

The correct sealing of the pouch is critical to prevent post-process contamination of the product. To ensure that a hermetic seal is produced, the correct sealing equipment and techniques should be applied. Most common seal failures are due to food or water vapour contamination. Water vapour occurs during hot filling, with steam condensate forming water droplets in the seal area. These droplets can join together, increasing the area of contamination and preventing a complete fusion seal.

The main critical control points when sealing are considered to be temperature, pressure and dwell time, but these will all vary according to the material and sealant used.

The type of seal will depend on the following factors:

- Width of the seal, which should be not be less than the minimum width supplied by the manufacturer when thermal processing the pack.
- Position of the seal, which ideally should be flush with the end of the pouch to prevent contamination; the presence of a tear notch should also be taken into account.
- Sealing head design, as there are various different sealing heads available, including flat, profiled and cross-hatched. Profiled sealing heads can help to push out any contamination in the sealing area. Cross-hatch seals can make it difficult to detect defects in the sealing area.

There are a number of pieces of equipment available to form a seal, including hot bar, impulse and ultrasonic.

When setting up the parameters of the heat sealing equipment, the following factors need to be taken into account:

- Sealing temperature. If the temperature is too low, the seal will be insufficient; if the temperature is too high, it can burn the material.
- Sealing pressure. If the pressure is too low, the bonding of the sealant on each side of the package may be insufficient; if the pressure is too high, heat damage can occur.
- Dwell time.
- Further processing.
- Changes to the packaging materials.
- Throughput.

The position of the seal on the pouch is important to prevent contamination. If the seal is applied at the end of the pouch, there should be a sufficient amount of seal. However, if the seal is further down the pouch, an unclosed flap remains between the closing seal and the pouch top edge. This is potentially a trap for water, contamination and microbiological growth. To prevent this, the seal should be at the top of the pouch, or a secondary seal should be applied. Alternatively, the top of the pouch should be trimmed. If trimming, care should be taken not to compromise the integrity of the seal. Table 2.1 describes some of the defects that can occur with retortable pouches during sealing and production.

It is important in ensuring a good seal, that the seal area of the pouch and the sealing bars are free from contamination; also that there are no folds in the sealing area.

2.4 Processing requirements

There are a number of considerations to be taken into account during the loading and processing of the pouches. The design and the loading of the pouches onto the

Table 2.1 Pouch defects

Defect	Description	Trouble shooting
Abrasion	A scratch through one or more layers of the film	Damage during production by machinery, debris on sealing bar
Channel leaker	An unbonded area across the seal	Sealing parameters not correct, contamination in the seal area
Compressed seal	Separation of the laminated layers within the seal	Sealing temperature too high
Contaminated seal	Foreign matter trapped within the sealing area	Contamination in the seal area
Crooked seal	A seal that is not parallel to the cut edge	Pouch not positioned correctly during sealing
Delamination	The layers of the package separate	Sealing temperature too hot, inadequate seal strength
Flex cracking	Small breaks in one or more sealing layers of the pack	Excess flexing of the pouch during handling and processing, too high a vacuum pulled
Hot fold	Permanent bend in seal formed between sealing and cooling	Folding of the material during sealing
Incomplete seal	Seal does not cover the width of the seal	Pouch not positioned correctly during sealing
Swollen pouch	Inflation of the pouch due to microbial growth or residual air	Inadequate processing, seal defect
Wrinkling	A fold of material in the seal area producing a pleated appearance	Sealing surface not flat during processing

Adapted from CCFRA 2006.

retort racks are essential to ensure an adequate process, both for the quality and the safety of the product. It is also important to realise the effects that the holding time and temperature of the pouches prior to loading of the pouches into the retort have on both the product and process. Ultimately, any food safety risks due to microbiological growth during that period of the manufacturing process must be eliminated.

2.4.1 Racking systems

The design of the trays and racking systems to hold the pouches during processing is critical to ensure pack integrity, adequate temperature distribution and heat transfer throughout the retort. Pouch racking systems must have the structure and strength to support the weight of the containers inside and should be made stackable so that there is no pressure applied to any of the pouches held within them.

All of the pouches should be exposed to the heating medium, by allowing adequate circulation of the heating/cooling medium throughout the retort load during sterilisation/pasteurisation and cooling. The position and the orientation of the pouch are important and critical factors in heat processing; pouches are commonly held in a horizontal position within the racking system to prevent bulging or snagging at the base of the pouch, which can occur if it is held vertically. The pouches should not be overlapped or touching while in the retort as this will create container defects and affects their integrity, as the edges of the pouches are quite rigid and sharp. Large pouches should not be folded when loaded. Tight pouch stacking patterns can also retard the circulation of the heating medium.

2.4.2 Holding time

There may be a period of holding prior to processing for reconstitution and rehydration. This is desirable in order to ensure that the required textural and organoleptic quality of the product is achieved but this should be carried out under controlled conditions. The additional hold may be an essential part of the process in order that physical changes, such as rehydration, gelling and cooking of some ingredients, can occur (e.g. stabiliser systems, pasta, rice). However, overcooking during this period may also happen, causing degradation of gels, excessive softening of vegetables, undesirable cooked flavours or other quality defects. Hold-ups due to breakdown of equipment may have a detrimental effect on the product, with enzymic activity taking place during the holding time if the temperature is in a certain range. This may result in possible loss of quality. The temperature of the product will also be affected, and this needs to be taken into account when determining the process.

2.4.3 Thermal processing

The purpose of the heat process is to preserve the pack contents by destroying the microorganisms that can cause spoilage or pose a safety risk for the food contained within the pack; also, the enzymes capable of activity in the pack during subsequent and normal storage. The heat process may additionally be required to cook the food, whilst developing required characteristics such as flavour, texture and colour. When considering the thermal process required for the foodstuff, the following factors need to be considered:

- Elimination of thermophilic bacteria.
- Size and amount of solids and volume of liquid in the pack.
- Headspace.
- Arrangement of packs in the racking systems.
- Type of retort, i.e. water, steam or combination.
- Water activity, salt content and pH of the product.

2.4.4 Scheduled heat process requirements

With foods containing solids and liquids, the heat transfer of the solids can be slower or faster depending on their size (Yamaguchi 1990), so the quantity and size of solids in each pack needs to be controlled. Additionally, the viscosity of the product will affect the heat transfer, so the consistency within each pack needs to be controlled.

In order to develop an appropriate and safe retorting process, the initial temperature of the product at the coldest point within the pack (i.e. at the commencement of retorting) must be known. This initial temperature will contribute to the amount of energy required to be given to the product for the pasteurisation or sterilisation process to achieve the required degree of lethality (F_0 value or F_p value) required for commercial sterility or pasteurisation (CCFRA, 2006). The initial temperature of the product should be determined on the pouches with the highest temperature change between processing, filling and the commencement of retort (i.e. worst case scenario). In most cases this is likely to be the first product from the processing or blending vessel and the first pouches filled off for each retort loading

The temperature distribution of the retort with the selected stacking arrangement will have previously been used for process validation. To establish the scheduled heat process for foods in flexible packs, the degree of the lethal heat required to achieve commercially stable products needs to be considered. Heat penetration studies with the product under simulated production will determine the heat distribution throughout the heating vessel in order to achieve commercial sterility. This will also determine the coldest spot in the container, which is the point of measurement for heat penetration equipment.

At least the minimum *C. botulinum* heat process should be given to the sterilised ambient products. This is normally achieved through application of a heat process that has a value (F_0 value) equivalent in lethal affect to not less than 3 minutes at 121.1 °C, known in the industry as commercially sterile. If the product is to be pasteurised, the treatment will be dependent on the expected microbial load of the raw materials and the desired quality of the final product. In order to design an effective microbial treatment, the kind and number of pathogens likely to be present within the food need to be considered because the pasteurisation process cannot be expected to destroy sporeformers. Also whether the pathogens are infective or produce toxins must be taken into account.

For chilled foods (<8 °C) with a greater than 10-day shelf-life where psychrotrophic *Clostridium botulinum* is the organism of concern, the Advisory Committee on the Microbiological Safety of Food (Anon., 1992) recommend a heat process of 90 °C for 10 minutes or equivalent lethality (z value 9). Alternatively, one, or a combination, of the following controls may be imposed:

- A pH of 5 (or less) throughout the food and throughout all components of complex foods.
- A minimum salt level of 3.5% in the aqueous phase throughout the food and throughout all components of complex foods.
- An A_w of 0.97 or less throughout the food and throughout all components of complex foods.
- A combination of heat and preservative factors, which can be shown consistently to prevent growth and toxin production by psychotrophic *Clostridium botulinum*.

Where a heat process is applied, this should be followed by prompt cooling to prevent outgrowth of spore forming organisms that will survive a 90 °C for 10 minutes pasteurisation process, e.g. less than 4 hours between 63 °C and 5 °C (Anon., 1998).

2.4.5 Heat distribution and penetration tests

A heat distribution test is performed to study the uniformity in lethality throughout the retort with emphasis on the identification of the position in the retort, that potentially results in lowest lethality and the slowest arrival at the scheduled processing temperature. Heat distribution tests must be carried out on each tray configuration and should be tested on installation of the retort and repeated every 3 years to confirm that there is no change in performance of the retort (Department of Health, 1994).

A heat penetration test is conducted to determine the heating and cooling characteristics in the coldest point of a given product, in a given container, under specified process conditions, usually in the coldest zone of the retort (as determined by the heat distribution study). Particular attention is required when determining the cold spot of stand-up pouches as potentially it could migrate, depending on the type of product (Khurana *et al.*, 2004). The results for the heat penetration trials in the pilot scale and production equipment should form the basis of the scheduled heat process used in production.

Traditionally, pouches have been heat processed in batch retorts; however, continuous sterilisation can be carried out in a hydrostat. The pouch is attached to a cable-driven conveying system that moves through the hydrostat (Khurana *et al.*, 2004). There are no agreed standards for the design of equipment for heat processing food in flexible packs. However, a number of requirements are common to most designs.

Particular attention should be paid to the attainment of uniform distribution of steam and to the avoidance of the direct impingement of steam onto pouches that could result in some pouch delamination. There should be automatic controls of heat-process temperatures and pressure during the hold period at the heat process temperature. The complexity of the operation of most heat vessels used for the heat processing of flexible packs makes it preferable for the whole heat-process cycle to be automated. Pouches should be processed with an overpressure (one in which the external pressure is above that of steam at the sterilising temperature). The external pressure is used to balance the internal pressure that develops in the container during sterilisation due to expansion of the pack contents, dissolved gases being released into the headspace and expansion of the headspace (May, 2004). Failure to apply an overpressure will result in ballooning of the pack, in turn weakening the seals and possibly bursting of the pack. If, during processing, the uniformity of the temperature changes, the overpressure may be inadequate, in effect causing ballooning and bursting of the pouches. This can also affect the heat penetration and ultimately the process, resulting in under- or over-processing of the product. To establish the correct overpressure, deflection equipment can be

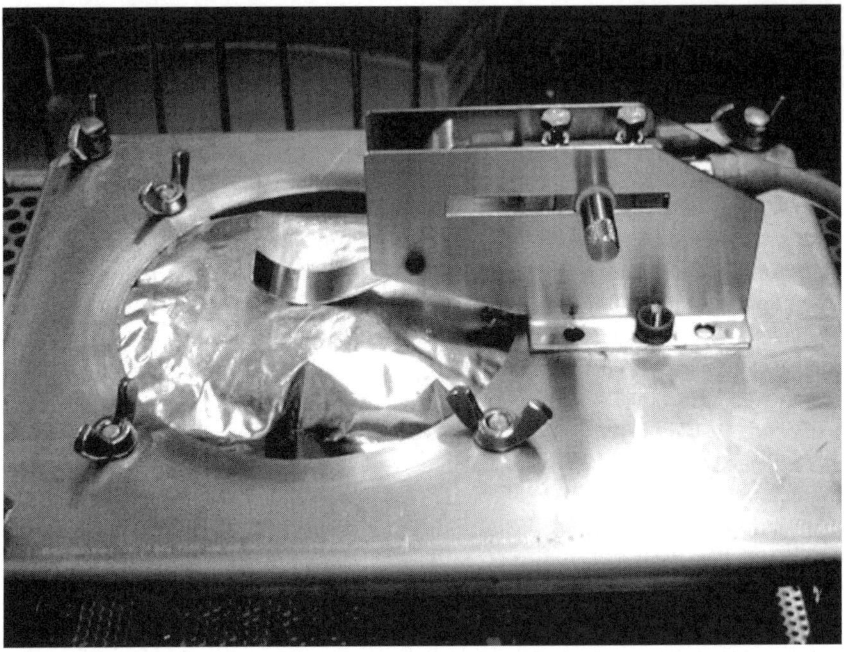

Fig. 2.3 Ellab deflection transducer for measuring movement of pack surfaces (photo courtesy of CCFRA).

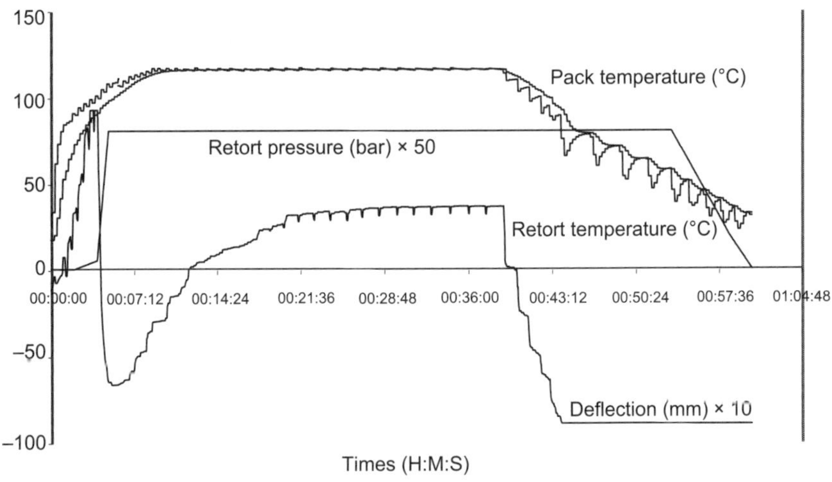

Fig. 2.4 Example of a deflection profile for a pouch processed in an over-pressure retort.

used, as shown in Fig. 2.3. Deflection equipment can detect when a pouch starts to expand excessively – then the pressure being applied to the pack can be increased until the expansion is under control (this is most easily accomplished in a retort with an element of manual control). Figure 2.4 shows a deflection profile from a pouch sterilised in a steam or air retort. During processing, the pouch expanded during the venting phase due to the rapid increase of temperature in the headspace compared to the temperature of the product, but was brought under control once the retort was pressurised. During the cooling period, the headspace will cool and contract quite rapidly, and again the overpressure can be reduced with vessel temperature.

2.4.6 Post-process handling

The objective of post process handling is to take wet, heat-processed pouches from a retort vessel and convey them ready for further packing whilst not compromising the integrity of the pouches, the product in the pouches or any secondary packaging that the pouches are packed into. A risk assessment should be carried out to confirm that the post-process handling is satisfactory when designing a process. Seals are generally weakest immediately after processing and great care should be taken to prevent post-process contamination. Studies have shown (Anema and Schram, 1980) that it is most important that the packs are dried as quickly as possible. This is because moisture in contact with the pouch acts as an effective carrier medium and may lead to increased spoilage of the contents if the pouch integrity has been compromised.

Following processing, a holding period will allow the evaporation of water from the surface of the pouch. To assist with the drying of the pouches, the internal temperature of the product should be reduced to approximately 40 °C during the cooling period of the process (May and Dobie, 1996). This will assist with the evaporation of water from the surface of the packs; a lower internal temperature will increase the drying time required. During the drying period it is also considered helpful if the retort baskets are removed from the retort immediately following a cycle and tilted to allow excess water to run off. The drying environment should have a low humidity, good airflow and an ambient temperature. Mechanical methods of drying the pouches include the use of air knives, air blowers and surfactant sprays to drive off any remaining water.

One of the main concerns with pouches is the issue of post-process contamination similar to that of canned goods. The best procedure is that the pouches are not handled post-processing, but this is impractical in many cases due to the flexible nature of the pouch, which does not suit automatic handling (CCFRA, 2006). Unlike a can, if a pouch is sealed correctly, it will remain sealed to microorganisms, and quick and efficient drying reduces the risk of damage to the pouches. A study carried out by Michels and Schram (1978) on pouches that had been punctured following processing indicated that once the pouches were dried there was very little recontamination compared to storing these pouches wet, and manual handling once they were dry had negligible effect. However, if wet handling of the pouches is carried out, it is essential to ensure the environment is clean and any contact is minimal.

2.5 Distribution

Prior to distribution of the pouches, checks need to be carried out to ensure that the integrity of the pouch is good and that the product is safe. The types of tests that are carried out in industry are visual inspection, tensile strength to determine the quality of the seal, burst testing to assess the seal strength, dye penetration to identify any minute holes within the pouch and incubation testing where the pouches are held at a temperature for a specified time to encourage the growth of microorganisms.

During distribution and retailing, it is important to ensure that there is no physical damage to the pouches and that the risk of any type of contamination is minimised. The outer packaging is required to protect the product and may be composed of paperboard, plastic or a combination of both. The outer packaging should be sized to prevent movement of the pouches but not such that it crushes or causes damage to the pouch, particularly during transportation when there is movement from the vehicle. Excessive vibration can also cause flex cracking of the pouches which may fracture the laminate and cause pinholes, compromising the safety of the product. During transportation, the packs should be stacked so as not to cause damage to the pouches or allow movement. The temperature of the mode of transport needs to be considered; high temperatures above 45 °C may pose a risk with thermophilic organisms that survived the heat process. Alternatively, low temperatures, close to 0 °C, may reduce the flex crack resistance of the pouch (CCFRA, 2006). High humidity or moisture within the vehicle may also damage the outer packaging.

2.6 Future trends

As the trend for convenience foods continues, the types of foods packaged into retortable pouches will grow. The use of retortable pouches will also penetrate other markets such as cosmetics, pharmaceuticals and household goods. The convenience of the pouch will be enhanced further with the increased use of spouts and zippers, and the ability to microwave the product in the pouch.

With techniques such as pouch filling, sealing speeds and volumes continuing to lag behind that of the canning industry due to the shape and delicate nature of the product. Challenges face the industry to improve equipment.

As with all packaging, there will be the requirement to produce recycleable versions and reduce the amount of waste produced.

2.7 Sources of further information and advice

Campden and Chorleywood Food Research Association
Chipping Campden
Gloucestershire
GL55 6LD
UK
Tel: +44 (0) 1386 842000
Fax: +44 (0) 1386 842100
e-mail info@campden.co.uk

Grocery Manufacturers Association
1350 I Street, NW Suite 300
Washington, DC 20005
USA
Tel: +202/639-5900
Fax: +202/639-5932
E-mail: info@gmaonline.org

Pira International
Randalls Road
Leatherhead
Surrey
KT22 7RU
UK
Tel: +44 (0) 1372 802000
Fax: +44 (0) 1372 802238

2.8 References

Anema PJ and Schram BL (1980), Prevention of post process contamination of semi rigid and flexible containers, *Journal of Food Protection*, 43(6) 461–464.

Anon. (1992), *Report on Vacuum packaging and associated processes*, Advisory Committee on the Microbiological Safety of Food. London, HMSO.

Anon. (1998), *Food and drink good manufacturing practice. A guide to its responsible management*, Institiute of Food Science and Technology. London.

Canadian Food Inspection Agency (2002), *Flexible Retort Pouch Defects Identification and Classification Manual.*

CCFRA (2006), Guideline on good manufacturing practice for heat processed flexible packaging, *CCFRA guideline 50.*

Department of Health (1994), *Guidelines for the safe production of heat preserved foods*, HMSO, London.

Fenn-Barrabab C and Prommesberger M (2003), Barrier materials for flexible packaging, *Packaging Technology International*, Vol 1, 21.

Khurana A, Awnah G and Charbonneau J (2004), *Thermal Processing Considerations for Retortable Pouches*, Centre for the Development Research Policy and New Technologies, National Food Processors Association, Washington DC.

May N (2004), Developments in packaging formats for retort processing. *Improving the Thermal Processing of Food*, Cambridge, Woodhead, 140–151.

May N and Dobie P (1996), Guidelines for batch retort systems: Full water immersion – raining water – steam/air, *CCFRA Guideline No 13.*

Michels MJM and Schram BJ (1978), Effects of handling procedures on the post process contamination of retort pouches, *Journal of Applied Bacteriology*, 47, 1–5.

Schreiber GM (2005), New retort pouch applications from Europe, *Retort Pouch Conference 2005*, The Packaging Group Inc, US.

Swinden P (2002), Bulging pockets, *Flexible, The Journal of Plastic Packaging Technology*, Issue 1, May/June, PIRA International, 57–64.

Yamaguchi K (1990), Retortable Packaging, *Food Packaging*, 185–211.

Yoshida K (2005), New retort pouches, cups and tubes from Japan, *Retort Pouch Conference 2005*, The Packaging Group Inc, US.

3

Improving the performance of retortable plastics

J. M. Lagaron, M. J. Ocio and A. Fernandez, CSIC, Spain

3.1 Benefits and markets for retortable plastics

When evaluating the quality and safety of food products, several parameters have to be taken into account. Food safety is usually referred to as the absence of food-borne pathogens, while food quality comprises not only microbial quality, but also sensory or organoleptic quality and, of course, nutritional quality. In packaged foods, quality also extends to the appearance and preservation of the functional properties of the package. Among the agents that can cause food spoilage, microorganisms, enzymes and chemical reactions are the most relevant. In order to control the action of these agents, diverse preservation methods are applied that seek to destroy unwanted enzymes and microorganisms, while maintaining the original texture and nutritional characteristics of the food products.

Nowadays, packaging plays a significant role in the preservation of foods, and plastic packaging has gained importance because of its balanced characteristics (transparency, flexibility, versatility, low cost, ease of processing, etc.) and the wide variety of formulations that allow the development of packaging structures for specific product requirements. Many food products are commercialized inside plastic packaging, which, in fact, constitutes a critical parameter in determining their shelf-life because, unlike other traditional packaging materials such as glass or tinplate, polymers are permeable to gases and low molecular weight substances. Today, many preservation processes are applied to the already packaged food products and, therefore, the package's characteristics must be tailored and controlled

Table 3.1 Typical requirements for retortable flexible and tray packaging

Property requirement	Flexible	Trays
Temperature	116–145 °C	121–135 °C
Oxygen permeability	< 0.05 cc/m^2 day atm	< 0.05 cc/m^2 day atm
Water permeability	< 0.003 g/m^2 day atm	< 0.003 g/m^2 day atm
Thermosealing temperature	160–260 °C	
Closure resistance in tension	> 7.5 kg/15mm	>25 N/25mm
Resistance to internal pressure	17.2 10^4 Pa	–

to withstand such processes and to guarantee the quality of products during their shelf-life. The chemistry of polymers is one of the most important factors defining structure and, therefore, end properties (Lagaron *et al.*, 2004).

Thus, due to the excellent properties of polymers as packaging materials, there is a trend in the food industry towards the replacement of classic packages manufactured with materials like glass or tinplate, with lighter, cheaper and versatile plastic packages. These polymeric structures must, however, assure the quality and safety of the packaged products without compromising food shelf-life. Typical requirements for flexible packaging structures are provided in Table 3.1.

Many food products are to be packaged with high-barrier polymeric materials because oxygen is a ubiquitous element involved in many food deterioration reactions, such as fat oxidation and vitamin loss (Ackerman *et al.*, 1995). But furthermore, several food products, such as the increasingly demanded precooked foods (ready-to-eat products), currently benefit from being processed inside the package by retorting treatment (typically 121 °C during 20 minutes in an industrial autoclave, i.e. in the presence of pressurized water vapor) before being sold. Thus, apart from the already mentioned high-barrier conditions, plastic packages must withstand such kinds of processes without suffering undesirable changes. As mentioned, in many food industrial processes, food is packaged prior to the application of the preservation technology (such as thermal treatments) in order to optimize preservation processes and minimize product manipulation. Therefore, the package is incorporated in the same production line and the preservation technologies are applied to the already packaged product (IOPP, 2002). Commonly employed heat treatments, such as pasteurization and sterilization, require airtight, high-barrier, retortable film structures containing high-barrier polymers such as polyethylene terephthalate (PET), various polyamides, poly(vinylidene chloride) (PVDC) or the most commonly used ethylene–vinyl alcohol copolymers (EVOH) (Wood, 1990). A recent development in this area is the production of retortable, multilayer, blow-molded containers that incorporate polycarbonate as the tough outer layer, with EVOH or PVDC as the barrier, and polypropylene (PP) as the food contact layer. The advantages of polycarbonate are its light weight, improved heat stability and good optics compared to the polyolefins that are normally used.

EVOH copolymers are without doubt the most widely used family of semicrystalline materials in high-barrier packaging and in retortable food packaging because, apart from being excellent barriers to oxygen and aroma compounds, they also have high chemical resistance to organic compounds (aroma components or ink solvents), excellent chemical and optical characteristics (transparency), good thermal resistance and very fast crystallization kinetics. The high oxygen barrier characteristic of EVOH materials is provided by the hydroxyl groups of their structure, which confer upon them high crystallinity and both high cohesive energy and low fractional free volume for the permeable amorphous phase, reducing the space between the polymeric chains available for gas exchange. However, these hydroxyl groups make the copolymers water sensitive and, therefore, in high relative humidity environments, their barrier characteristics are greatly impaired. For that reason, in most packaging applications, EVOH is used in multilayer structures, sandwiched between at least two layers of a hydrophobic material such as polyethylene (PE) or polypropylene. In the case of retorting applications, PP or PET are most commonly used as the structural layers in the multilayer system to provide thermal resistance, mechanical integrity and water barrier to the overall design. In the flexible area, typical retortable structures include EVOH as the barrier layer enclosed between layers of PET, oriented polyamide (OPA), copolymers of PE–PP or retortable linear low density PE (LLDPE).

Regarding markets, it is estimated that the global packaging industry will keep a 4.2% annual growth rate with a total output value of €391 billion in 2009. In this context, it is considered that, for instance, the US market for retort pouches will be a €2.6 billion business. In the US alone, the demand for pouches will climb 7% annually throughout 2008, driven by the rapidly expanding stand-up pouch segment as well as the demand for flat pouches. The incorporation of such value-added features as resealability, spouts, and retort and aseptic properties will further stimulate advances for all varieties of pouches, particularly stand-up pouches (Industrial Technology Development Institute, see web address).

3.2 Impact of retortable plastics on food quality and safety

The traditional sterilization process has been one of the most widely used methods of food preservation during the 20th century in the food industry and has contributed to extend the shelf-life of various food products. This thermal process consists of heating food containers at a specified temperature for a defined length of time. The sterilization parameters are calculated on the basis of achieving sufficient bacterial inactivation in each container to comply with public health standards and to ensure that the probability of spoilage will be below a certain threshold value. This thermal processing is nearly always associated to some undesirable degradation of heat-sensitive quality attributes. The challenge of developing advanced thermal processing for the food industry is related to the demand for enhanced food quality without compromising food safety. Loss of quality is very dependent on food type and composition, packaging and storage conditions. Quality loss can be minimized at any stage, and thus quality depends on overall control of the processing chain. Selecting appropriate packaging materials and linking the packaging operations with heat preservation are thus necessary to improve food security and the general quality of life. Nevertheless, several studies have revealed that the application of plastic-based packages in overpressure retorts has to be done with knowledge and caution to avoid adverse or unwanted effects on the product.

In most cases, the objective when designing a thermal process for low acid food is to inactivate viable spores present in the food while minimizing nutrient degradation. The heating effect on microbial spores is strongly dependent on processing temperature. Consequently, a combination of short time and high temperature is the most effective way to retain product quality parameters and to inactivate a large number of *Clostridium botulinum* spores (z-value = 10 °C), which is the basis for establishing a safe thermal process. Thus, the degradation of, for instance, a heat-labile quality attribute such as thiamin (z-value = 28 °C) is known to be minimized when shorter times at high temperatures are applied. It has been observed that heat-labile nutrients as well as sensory quality attributes including color, texture and flavor follow similar patterns for specific food products. Taking into account this fundamental concept, Tung and Smith (1980) used a computer modeling method to demonstrate that, after heating 400 g of product processed into cylindrical cans and into thin retortable plastic pouches, the curve for the retortable plastic pouches showed much lower quality destruction for any given processing temperature. The reason for this is the fact that the profile temperature needed to reach the critical point is more easily attained in the plastic pouches. The thin profile attained in retortable plastic based pouches permits a shorter heating time and thus lowers the risk of overcooking the product, while producing better color, firmer texture and lower nutrient loss. Then during the heat treatment, minimal overcooking of the product near the peripheral container areas is produced. From the modeling results, the authors concluded that using high-temperature (>125 °C) and short-time processes, clear advantages in food quality attributes could be obtained. Theoretical and experimental investigation of thermal inactivation of *Bacillus stearothermophilus* in beef–vegetable soup packaged in pouches was carried out by Abdul Ghani *et al.* (2002). The results showed that there is a good agreement between the measured values of *B. stearothermophilus* spores during heat treatment with those predicted by modeling. Using the mathematical model developed in the work, the authors concluded that it is possible to predict the heat treatment required for any given pouches in order for them to be safe. The use of *B. stearothermophilus* spores, which are extremely resistant to heat, is thought to be adequate to predict and validate the sterilization process that can guarantee the food safety of retortable pouches. Several studies have revealed that mathematical models are very good predicting tools to optimize thermal processing in terms of minimizing the quality gradient in foods heated in retortable pouches, and to determine optimum temperature profiles (Simpson *et al.*, 2004).

In view of the above, the retortable plastic pouch has been considered the most significant advance in food packaging since the metal can, and has the potential to become a feasible alternative to the metal can and the glass containers. However, more research must be done in developing suitable plastic-based packaging materials and retort processing equipment. Nevertheless, several research studies have reported an important limitation with the use of metal cans: this is the generation of off-flavours (undesirable taste) imparted to many processed foods including seafood, fish, some meat products, and some vegetables during storage. Flexible pouches can offer an efficient alternative to metal cans in this respect. The existing literature has proven the feasibility of different flexible plastic pouches for producing thermally processed shelf-stable foods. The effect of processing time on the quality attributes of pouches compared to canned product has also been widely studied. In this context, Dymit (1973) reported that shrimp in retort pouches were superior in flavor and color to canned products. Chen and George (1981) showed that green beans processed in pouches had better overall acceptability than when processed in a can, but in this case the color of the canned beans was preferred. Durance and Collins (1991) described significantly less off-flavor and greater acceptance of salmon when processed in retort pouches than when packaged in cans. Ali *et al.* (2005) showed texture quality improvement of sardines in oil when processed in pouches compared to cans. Mohan *et al.* (2006) studied the effect of thermal processing on the quality of 'Shrimp kuruma' from India in retortable pouches and aluminium cans. The authors concluded that, after reaching equal lethality in both cases, the product packaged in pouches was lighter in color, more succulent, and more desirable in firmness compared to the canned product. There are also several studies in the literature reporting that the quality characteristics of pouched packaged products are maintained after extended storage under ambient conditions. Gopal *et al.* (2001) reported that mackerel fish curry processed in retortable pouches resulted in a product with good sensory attributes which also had a shelf-life of at least 12 months at room temperature. Chandrasekar *et al.* (2004) demonstrated after a sensory evaluation of mushroom curry prepared from stored mushrooms that the product had high acceptability, and no leakage, bad odor or any other sign of spoilage during one-year storage. Bindu *et al.* (2004) demonstrated the shelf-life stability of ready-to-eat fried mussel meat packaged in retortable pouches after 12 months storage at atmosphere temperature, using a combination of vacuum packaging and heat processing.

As already indicated, the choice of materials for the manufacture of retortable pouches is very important. The packaging materials must protect the food product against light degradation, moisture changes, microbial invasion, oxygen ingress and food–package–environment interactions. The plastic material must maintain its structural integrity and be able to withstand retort temperatures as well as normal handling abuse. It must also comply with regulatory requirements. It is known that seal and burst strengths of plastic packages are much lower at thermal processing temperatures. These packaging materials must maintain adequate seal strength to assure protection of package integrity when high temperature processing is applied. Contamination of the seal area is the major problem that affects the hermetic seal of the flexible pouch. This is mainly caused by incorrect vacuum or improper pouch handling. Incorrect handling of pouches during processing and post-processing could cause physical damage to the pouch and seal, which could then weaken the seal or compromise the pouch hermeticity. There is, however, a general lack of knowledge in the literature about the potential effect of retorting conditions on the above properties and of retortable pouches how this impacts upon food quality and safety aspects (Cnadian Food Inspection Agency, 2002).

Food package interactions also play an important role in the proper selection of packaging materials for various food applications including retortable food packaging. These food/packaging interactions involve the transport of low molecular weight compounds such as gases or vapors and water from the food or the environment to the package and/or from the package into the food (IFT, 1988). These phenomena could also produce chemical changes in the food, package or both, resulting in food contamination, loss of package integrity and/or decrease in quality. These processes are relevant here because most transport phenomena are temperature activated and hence thermal processing of the packaging materials can enhance the extent to which these processes take place in packaged foods. There are, in the literature, several studies describing the changes in quality parameters in food products as affected by packaging conditions, such as their variable permeability to light, gases and vapor after application of industrial processes. For instance, changes in several physicochemical and sensory attributes and also in the acceptability of retort-packaged beefsteak or beef stew packed under different headspace levels (10–40 cc) and stored at temperatures of 4, 27 and 38 °C were evaluated over a six-month period by Sepulveda *et al.* (2003). No significant changes in microbial counts, pH, residual oxygen, color, hardness and oxidation occurred in the foods. The authors recommended increasing the headspace to 40 cc without significant detrimental effects on quality and acceptability of both products. In the case of high storage temperatures, the acceptability and some specific attributes, particularly moistness and firmness, were negatively affected.

Apart from food component oxidation caused by light degradation and/or diffusion of oxygen into the packaging, another reason for losses in quality may be the interactions of aroma compounds with the packaging material itself. An early step in this process could be the adsorption of volatile components inside the polymer, followed by permeation through the packaging material facilitated by, for instance, thermal processes. In that sense, organic polymers commonly used as inner coating films which normally are in direct contact with the food, are known to be able to adsorb volatile compounds from foods. Having this problem in mind, several researchers have compared the influence of different packaging materials on the adsorption of food volatile components. Czerny and Schieberle in 2006 carried out a study to indicate the influence of chemical structure on sorption (scalping) behavior by comparing key aroma compounds in two different UHT-milks, either packed in glass or in polyethylene bottles. The results of the study indicated that, by application of Aroma Extract Dilution Analysis (AEDA) and quantitative measurements, several lactones, aldehydes and free fatty acids were confirmed as important odor-active compounds in UHT-milk. The tendency of these compounds to adsorb into the glass bottle was weak, whereas the adsorption into a polyethylene packaging was much stronger, indicating that the quality of UHT-milk was poorer in this latter case. More investigation is needed to understand the effect of processing temperature in food retortable packaging interactions.

Due to the increasing awareness of consumers in terms of health matters and the innovative advances of packaging in food manufacturing, the importance of the migration of substances from food packaging materials to foodstuffs has attracted the interest of researchers and legislators. It is recognized that chemicals from packaging can migrate into food products and beverages during heat processing and/or the storage and be ingested by the consumer. The substances that can migrate and can thus affect the safety of the food are determined by the nature of the packaging materials. In the last few years, the introduction of novel packaging materials in the food industry has increased the number of specific hazards to which humans are potentially exposed due to migration from packaging to the food (Arvanitoyannis and Bosnea, 2004). Furthermore, the migrant species may not necessarily be the substances used in the production or conversion of the material, but unknown reaction products. Non-intentionally added substances, such as degradation products from additives or monomers, impurities and solvents can also migrate into the food under certain conditions, including heat treatments (Skjevrak *et al.*, 2005; Poças and Hogg, 2007). Nevertheless, currently many of these substances, which are not included in the positive lists, are increasingly becoming subject to regulation.

Another concern in food quality and safety regarding the use of retortable plastic-based structures is the potential employment of reusable plastic food materials. This practice has increased considerably in food industry over the last few years. In particular, some refillable packages are already on the market in some countries, such as polyethylene terephthalate (PET) refillable soft drink bottles and polycarbonate (PC) dairy bottles. These are used in some cases without specific legislation permitting their use. In this context, there was an ambitious EU project designed to develop a comprehensive package of quality assurance criteria for ensuring the quality and safety-in-use (sensory, microbiological and chemical) of reused plastics for food packaging. The results of the study were published mainly by Jetten *et al.* (1999) and Jetten and Kruijf (2002). The plastic articles investigated in both papers were bottles of PET, PC and PP. In the first study, the work was focused on evaluating the potential risk of release chemicals to the food components, inertness to chemical scouring and physical abrasion and inertness as a surface for microbial attachment, in order to establish efficient commercial washing processes. In general, it was concluded that reuse of the articles did not significantly influence any of the properties investigated. However, most of the articles will cause flavor carry-over to a new filling if they are contaminated with strongly flavored products. In the second paper, the influence of repeated use on the migration of plastic constituents, degradation products of plastic additives, barrier properties and surface characteristics were investigated. The overall conclusion of this study was that the intrinsic properties of the refillable articles were not significantly influenced by repeated use. Only the hydrophobicity of the refillable PC and PP bottles seemed to be influenced by repeated washing. Nevertheless, the effect of temperature processing or retorting was not considered in the study and seems relevant, given the case that most migration processes are known to be strongly affected by temperature. Thus, it is obvious that more investigation is required to provide a more solid basis for future legislation on refillable articles that will be thermally processed.

3.3 Improving the performance of retortable plastics

Even polymers containing polar groups (such as polyamides and the most widely implemented material in retortable plastics, the ethylene–vinyl alcohol copolymers), which are the most suitable barrier elements in retortable plastics, suffer from barrier deterioration after retorting. Thus, previous work proved that, even protected by hydrophobic layers, the oxygen permeability of the copolymers is greatly deteriorated during the sterilization processes (Lopez-Rubio *et al.*, 2003, 2005b). In fact, early work by Tsai and Wachtel (1990) already noticed the phenomenology of this effect. These authors found that, after retorting, there is a strong increase in oxygen permeability due to the ingress of water and subsequent plasticization and also due to irreversible changes in the material structure. Thus, it is thought that, even when EVOH is sandwiched between water barrier polymers such as polyolefins or polyesters, the pressurized heated water in the retort is capable of traversing the structural layers to be hosted in the EVOH layer. After the process ceases, the structural layers again enhance the water barrier performance and, therefore, as the water cannot easily evacuate from the EVOH barrier layer, there is a prolonged transient state that lasts for over six months before reaching steady-state oxygen permeation. More specifically, Tsai and Wachtel reported that the oxygen barrier of retortable packages containing an EVOH barrier layer was initially reduced by two orders of magnitude when these containers were subjected to steam or pressurized water during thermal processing, and during long-term storage (>200 days) the barrier partially recovered (by a factor of 10).

In the work by Lopez Rubio *et al.* (2003) it was found, from a fundamental viewpoint, that the application of dry heat improved the EVOH polymer crystalline morphology, leading to a higher, denser, and more stable crystallinity, i.e. to a higher gas barrier structure. On the other hand, moisture sorption was found to result in melting of ill-defined crystals, particularly for the lowest ethylene content copolymers. This water sorption-induced crystal melting process had not been reported before but it was seen to be largely suppressed by enhancing crystal stability. Combined temperature and humidity effects, such as those generated in retorting autoclaves, were found to dramatically deteriorate the polymer crystallinity (leading to gas barrier losses and to decreased layer integrity), irrespective of initial crystal robustness. By making use of simultaneous time-resolved WAXS/SAXS experiments during *in-situ* retorting of a water-saturated EVOH copolymer with 32 mol % of ethylene, it was found that heated moisture very readily weakened the polymer crystalline morphology; it melted around 80 °C below the polymer original melting point.

The previous results meant, from an applied viewpoint as later published by Lopez-Rubio *et al.* (2005C), that PP/EVOH/PP structures do have, as reported, a

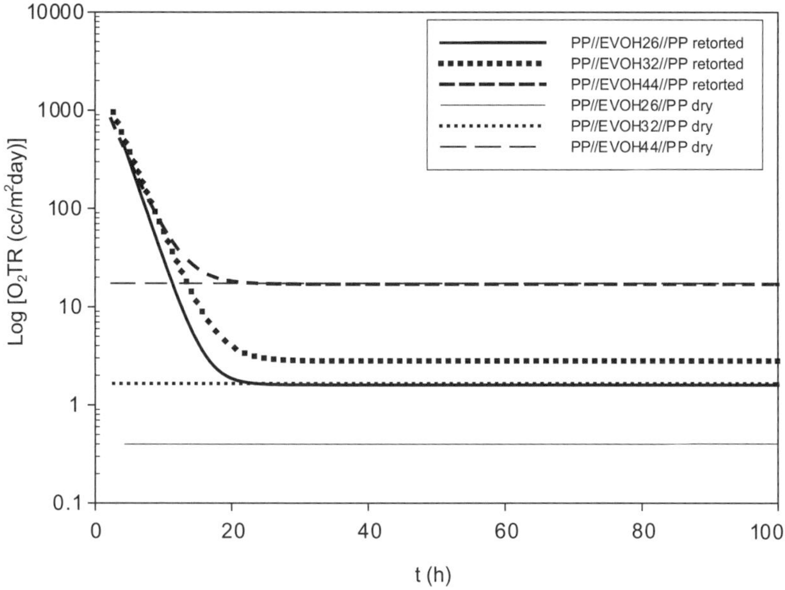

Fig. 3.1 Oxygen transmission rate of retorted multilayer structures (PP//EVOH//PP) vs. time after retorting and after a subsequent drying step.

substantial decrease in oxygen barrier properties and that the kinetics of recovery strongly depends on the copolymer ethylene fraction (see Fig. 3.1).

A morphological deterioration was also observed as a result of retorting, particularly for packaging structures composed of EVOH copolymers of low ethylene contents. This dramatic crystallinity deterioration suffered by EVOH films during retorting can be partially overcome by efficient protection between water barrier polymers such as polypropylene (PP). From our studies using synchrotron radiation, it was observed that, by appropriate shielding of the EVOH layer between polypropylene layers of a critical thickness (40 mm), the EVOH integrity was largely maintained during a typical retorting process, i.e. the gas barrier layer did not melt during the *in-situ* retorting study in the multilayer system (Lopez-Rubio *et al.*, 2005b). The experiments were carried out in a specially designed retorting cell with polypropylene layers of varying thickness, until no significant structural damage of the barrier polymer was seen. Nevertheless, even if the barrier polymer does not collapse, it still undergoes structural damage to some extent and considerable plasticization. In spite of this, it was found that polymer morphology and oxygen barrier properties can be substantially restored, if, after retorting, the materials are given a dry thermal treatment. More interestingly and more importantly, it was also found that pre-annealing of the EVOH copolymers rendered them more resistant to the retorting process by means of promoting both a more robust crystallinity and a lower water sorption capacity. Thus, a heating step before and/or after retorting of the multilayer system is thought to enhance significantly the resistance of EVOH-based multilayer systems

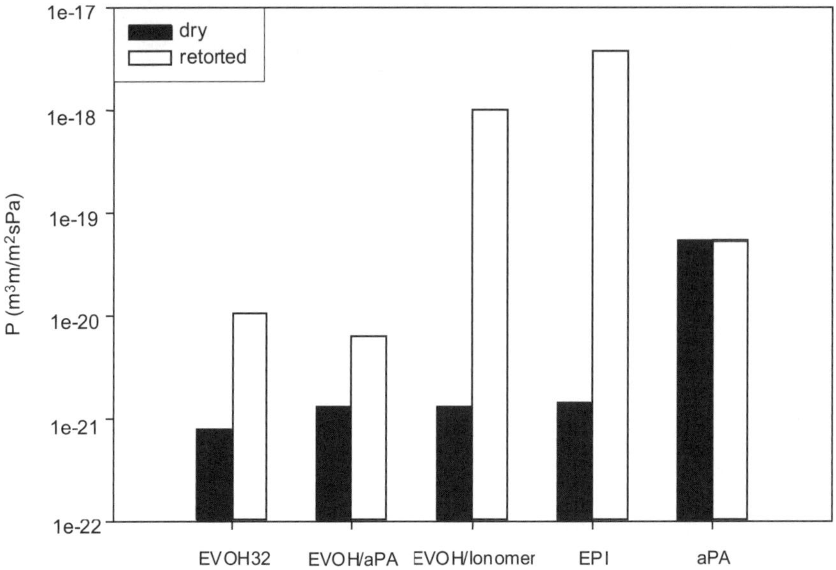

Fig. 3.2 Oxygen permeability of EVOH32, aPA, and binary and ternary (EPI) blends of these with nylon-containing ionomer measured at 0% RH and 45 °C before and after retorting. In all blends, 80 wt.% of EVOH32 was used in the blend formulation.

to the humid heat process. The industrial feasibility of these processes is being currently investigated.

Another solution reported to enhance the resistance to retorting of EVOH copolymers is by blending with other polymers. Thus, the effect of retorting on the morphology, structure and thermal characteristics of extruded films of binary and ternary blends of a 32 mol% ethylene vinyl–alcohol copolymer (EVOH) with amorphous polyamide (aPA) and a nylon-containing ionomer as blending additive was investigated (Lopez-Rubio and Lagaron, 2007). From the results, it was found that the thermal properties, the crystalline structure and the water sensitivity of the EVOH fraction in the blends were improved upon retorting compared to neat EVOH (see Fig.3.2). Nevertheless, only the binary blend with aPA showed a real enhancement in oxygen barrier properties immediately after retorting compared with neat EVOH. This unprecedented and surprising effect was additionally ascribed to the retorting-induced compatibility between EVOH and aPA components of the blend.

Significant improvements in oxygen barrier properties against relative humidity, and hence an improvement in the resistance to plasticization by water sorption, was also obtained with nanocomposites of EVOH containing layered nanoclays (Lagaron *et al.*, 2005). A reduction of at least 75% in oxygen permeability across the whole humidity range was demonstrated by means of this nanotechnology, using food contact approved specific nanoclay grades.

In spite of the fact that EVOH copolymers are unique in terms of properties for high barrier packaging applications, the plasticization suffered by water ingress during retorting has justified the extensive research done by industry and academia to re-design or modify them to enhance their resistance, and to develop alternative materials with similar barrier properties and better water resistance. Thus, other resins with high-barrier properties of potential interest, in some cases being already applied in retorting applications, are being developed by a number of companies in response to a big increase in the demand for high-barrier films. The amorphous polyamide (aPA) resins offer good oxygen and flavor/odor barrier, good clarity, good mechanical properties at elevated temperatures, and good processability. This materials show unusually improved oxygen barrier properties at high relative humidity conditions compared to dry conditions. Unfortunately, the gas barrier properties of the polymer and some other properties are not as good as those of EVOH. Nevertheless, recent work by Lopez-Rubio *et al.*, 2006A) proved that retorting of aPA does, as opposed to EVOH, improve the material barrier properties to gas and water. Films of this polymer are currently used for meat wrap, snack food bags and cereal box liners.

Other materials of interest are the copolymers of vinylidene chloride and vinyl chloride (Cassiday *et al.*, 1990). These polymers are useful packaging materials with exceptional clarity, toughness and unusually high impermeability to both water and gases. However, the use of these plastics has given rise to a number of safety and environmental concerns during disposal, regarding their chlorine chemistry. In the snack food area, metallized plastics are being increasingly used. Metallized oriented PP is particularly popular for this application.

Nylon-MXD6 is a very interesting candidate for substitution of EVOH. This polymer is a crystalline polyamide resin produced through polycondensation of *meta*-xylylene diamine (MXDA) with adipic acid. It is a unique polyamide resin which contains *m*-xylylene groups in the polymer chain. It has the best gas barrier property of all nylon resins (better than EVOH in a humid atmosphere) and retains excellent gas barrier property even after retorting or boiling treatment (Wood, 1990).

Finally, another family of materials with excellent prospects in high barrier retorting applications are the aliphatic polyketones (Lopez-Rubio *et al.*, 2006b). Analysis of the consequences of a typical humid thermal plastic food packaging sterilization (retorting) process (at 121 °C during 20 minutes in the presence of pressurized water vapour) over the crystalline morphology and gas barrier properties of a high barrier aliphatic polyketone terpolymer was reported. From a structural viewpoint, it was observed that the retorting process led to a less crystalline material: however, crystallinity was fully restored by a post-drying process, similarly to EVOH. From a barrier perspective, transport properties (P, D and S) to oxygen were measured at 21 °C (around T_g) and at 48 °C (well above T_g). The oxygen permeability at 21 °C was observed to increase by about nine times immediately after the humid treatment, i.e. much less than for EVOH copolymers, but the barrier character quickly recovered over time, as opposed to EVOH behavior. It was also suggested that a simple post-drying process at relatively moderate temperatures can easily restore the morphology and the barrier properties.

Overall, it was suggested that aliphatic polyketones can withstand far better the process of retorting in comparison to, for instance, other high barrier polymers such as the EVOH copolymers reported earlier and, therefore, can offer, even as a monolayer, a promising alternative in retortable food packaging applications.

3.4 Effects of complementary and alternative preservation technologies on plastics performance

As mentioned previously, the convenience food market is steadily growing. In Europe, this newly emerging market is expected to double the actual rates in the next ten years. Consequently, many current developments are dealing with new formulations or packaging technologies to boost the availability of ready-to-eat meals. Some advanced approaches provide solutions to stabilize the shelf-life by using retorting techniques and convenient reheating in microwave or conventional ovens. Nonetheless, the most usual options for ready-to-eat meals are usually not capable of providing innovative, stable, sterile products. On the other hand, as evidence confirms that shortening the heating and cooling steps helps to improve the quality of many food products, food technologists are introducing alternative ways (so-called emerging technologies) to minimize the thermal impact of high-temperature/long-time processing of foods, aiming at increasing the availability of shelf-stable products. Therefore, encouraging research in the field of new technologies is being undertaken, since this will be capable of assisting thermal treatments such as *high-pressure-assisted thermal sterilization* or *microwave heating*, providing a considerable reduction in processing intensity and length. Other alternatives, such as *ionizing radiation*, have the advantage of not being thermal technologies, providing excellent quality and enhanced stability of certain goods, although many aspects related to the possible formation of toxicological substances remain to be investigated before overcoming the strong regulatory constraints. The three above mentioned technologies offer in-pack solutions to prevent food recontamination, aiming at obtaining high-quality products. Therefore, an understanding of the interactions between these emerging technologies and the packaging materials is basic to developing innovative solutions and take advantage of the possible positive interactions, while maintaining high standards of food quality and safety.

3.4.1 Microwaves

Microwave reheating of foods at 2450 MHz is a technology familiar to the consumer, and many retortable pouches and other solutions to obtain shelf-stable foods aim at developing products that are both microwavable or can be reheated in conventional ovens. This is not a trivial challenge. The technology narrows the range of materials to be used: they should have good oxygen and moisture barrier properties, be able to withstand high temperatures (over 135 °C) and be sealable. When trying to achieve dual ovenable solutions, polyethylene terephthalate (PET) is probably the material of choice, since it has a use temperature of up to 205 °C, has moderate gas barrier properties and is grease resistant.

Until now, however, only very few microwave industrial applications have been exploited and pasteurization or sterilization using microwaves is still a methodology that needs to be further checked before extended implementation. Central technical constraints for microwave sterilization deal with the construction of monomode chambers at 2450 or even 915 MHz, able to withstand pressures to achieve an autoclaving effect. The requirements of plastics to be maintained under these conditions will probably have to go above the requirements of currently used retortable packaging materials, since short-time treatments will have to be more intense than conventional to achieve safe microbiological yields.

To achieve current requirements in convenience food, passive containers have been designed not to interact with microwaves. But besides helping to avoid recontamination of foods in industrial applications, in-pack microwave heating in passive containers might help in getting a homogeneous temperature distribution and foils are also recommended to be used in products to be microwaved at home. They must fulfil design requirements in shape and size to avoid, for example, the problems associated with runaway heating (Ryynämen and Ohlsson, 1996) or must have self-venting devices to control overpressure due to water vapour (Keller, 1999). Microwave challenges are also related to the development of creative solutions in obtaining crisping and browning of foods (Keefer, 1987). The end-results and the food quality during microwave processing are often conditioned by the use of active packages containing metal pieces acting as susceptors that are attached to the packaging surface (Habeger, 1997). Metallized surfaces are usually attached to a thin layer of biaxially oriented PET and an adhesive resin to bond the film to a paperboard (Schiffmann, 2005). Therefore, and due to the extreme temperatures achieved in susceptor packages or even inside the passive plastic pouches during microwave heating, most of the small amount of institutional research performed in this field is actually dealing with the migration of paper and plastic components into the food.

Migration of PET oligomers from roasting bags and susceptors was found to be higher after microwave treatment than in conventional ovens (Castle *et al.*, 1989). The parameters conditioning the amount of migrated oligomers were the temperature achieved during processing, the time of exposure, the contact with the food and the nature of the food surface. Furthermore, Begley *et al.* (1991) stated that the PET layer does not act as a sufficient barrier to avoid contact of the epoxy resin with the food during microwave reheating. And some plasticizers (acetyltributyl citrate and di(2-ethyl-hexyl) adipate were also found to migrate after microwave cooking in PVDC/PVC films (Castle *et al.*, 1988). Migration of polyisobutylene from polyethylene/polyisobutylene film into foods has been studied in domestic applications such as wrapping of foods and reheating in a microwave oven. Low molecular weight fractions of polyisobutylene could be found up to levels of 4 mg/kg in microwave reheated pizzas (Castle *et al.*, 1992). On the other hand, benzene release from different susceptor types was found to be below the detection limits of 2 mμg/kg (Jikells *et al.*, 1993) in pizza and french fries. Wax bags recommended for microwave heating were investigated and a simulated domestic processing in the microwave gave rise to the migration of waxy mineral hydrocarbons into foods (Castle *et al.*, 1994). A potential source of risk for human health is perfluorochemicals migrating from paper with fluorochemical coatings or additives. Microwaveable popcorn showed a migration of fluorotelomers hundreds of times higher than expected for the first use of a cookware paper (Begley *et al.*, 2005).

3.4.2 High hydrostatic pressure

High pressure is a well-known and readily investigated technology for food preservation when applied at mild temperatures, since it causes minimal effects on food quality (Rastogi *et al.*, 2007). Pressure alone is not able to break covalent bonds and does not affect small molecules related to the nutritional value and sensory properties of food. Thus, pressure up to 600 MPa is nowadays successfully exploited to process several in-pack food products. Containers to be used in high-pressure treatments must be flexible to be able to transmit the pressure. Indeed, in the various pressurised commercial applications, the currently available materials for other technologies are being used due to the very low pressure effects reported on plastics materials.

In spite of this, only very few references exist dealing with the effects of high-pressure and high-temperature combinations on plastic materials. High pressure application at sterilization levels (Juliano *et al.*, 2006) is a new science and is of limited application in foods, due to the high technical requirements for achieving feasible autoclaves and the problems in obtaining a homogeneous heating distribution. So far, results with pressure/temperature combinations showed fairly positive effects in, for example, pure EVOH copolymers and multilayers containing EVOH at 75 °C and approximately 800 MPa during short holding time treatments (10 min) compared with similar thermal treatments and retorting. Crystallinity was slightly enhanced in low ethylene-containing EVOH copolymers (Lopez-Rubio *et al.*, 2005), resulting in reinforced barrier properties; and migration of 1,2-propanediol through nylon/EVOH/PE pouches was considerably reduced during pressure treatments at high temperature (Schauwecker *et al.*, 2002). Furthermore, Irganox (a typical antioxidant used in flexible plastics) migration through PP pouches was only slightly accelerated by high pressure as the treatment temperature increased up to 60 °C (Caner *et al.*, 2005). Very little effects could be seen in the migration rate of food simulants (alcohol, water) through pressurized plastics (Masuda *et al.*, 1992; Mertens, 1993) and also the sorption of aroma compounds (*p*-cymene and acetophenone) was lower in pressurised films (Masuda *et al.*,1992; Kübel *et al.*,1996), probably due to an induced glassy state of the polymers under pressure. Also, LDPE showed slightly enhanced barrier properties after pressure treatments at 10° C (Le-Bail *et al.*, 2006). Parallel studies have shown that the organoleptic properties of orange and orange–carrot–lemon juices were not influenced by the type of packaging material (PP or Barex) used during pressurization at room temperatures (Fernández *et al.*, 2001).

A recurrent subject when treating multilayers by high pressure is delamination. Schauwecker *et al.* (2002) showed delamination between aluminium and PP layers of polyester/nylon/aluminium/PP pouches at relatively low pressure (200 MPa) combined with 90 °C. Additionally, pressure caused major changes in permeability and mechanical properties such as tensile strength, elongation and elasticity modulus of metallized PET (Caner *et al.*, 2000a,b); furthermore, polyamide/PE and polyamide/Surlyn films showed a 25% and 16% increase in oxygen permeability, respectively, during pressure treatments (Lambert *et al.*, 2000). Several other multilayer films investigated (PET/EVA, PE/Nylon/PE, PE/Nylon/EVOH/PE, etc.) have not shown deffects in permeability or mechanical properties (Masuda *et al.*, 1992; Caner *et al.*, 2003) due to pressure treatments at temperatures not exceeding 45 °C.

3.4.3 Ionizing radiation

The use of ionizing radiation, in the form of electron beams, gamma rays or X-rays, is of high interest in food processing because of the excellent opportunities to reduce microbial populations, control insect infestation, and stop sprouting. Irradiation is not a thermal processing so it would be a very useful tool to sanitize raw products such as meat or fruits that would, in turn, keep their freshness. But besides favourable reports from WHO, FAO, etc. (WHO 1988, 1994 and 1997) based in abundant work showing positive results for irradiated foods (CAC, 1983; Farkas, 1988; Giddings and Marcotte, 1991; Ross and Englejohn, 2000; Prakash *et al.*, 2000; Chaudry *et al.*, 2004; Grégoire *et al.*, 2003; Lee *et al.*, 2006), results commonly remark upon considerations concerning chemical changes (Vanamala *et al.*, 2005; Fan *et al.*, 2004) that might affect the organoleptic and nutritional quality of irradiated foods. Chemical changes induced by irradiation are related to the generation of free radicals and are not completely well investigated, thus not permitting this technology to have a completely safe image for the authorities or consumers. In the mean time, regulations within EU state members allow the irradiation of different foodstuffs (poultry, egg white, onion, garlic, potatoes), even though the only products with overall approval are dried herbs and spices.

Food packaging materials are exposed to ionizing radiation during the treatment of prepacked foods, or when they are sterilized for semi-aseptic packaging. Changes in polymers during irradiation are predominantly chain-scission and cross-linking. It is thought that, if irradiation takes place in the absence of oxygen, cross-linking will dominate, but if it does not occur under vacuum, chain-scission reactions will prevail (Anh and Lee, 2004). Therefore, ionizing radiation is a promising technology to enhance certain polymer properties (as reviewed by Chmielewski *et al.*, 2005), but to gain an adequate balance between degradation and induced cross-linking of polymers the processing conditions have to be carefully considered when designing in-pack food packaging. Indeed, irradiation is showing great potential to provide solutions that will enhance the generally low processing capabilities of biodegradable polymers, such as to obtain a certain improvement in thermal profile due to irradiation-induced cross-linking. This is particularly important for poly(epsilon caprolactone) (PCL) and its composites (Ikeo *et al.*, 2006), although this polymer does not show, at the moment, favourable perspectives for food packaging. Doses authorized for the most extended in-pack treatments are lower than 10 kGy (cold pasteurization) and they usually do not produce remarkably deleterious effects in the mechanical or barrier properties of plastics, or produce even positive effects (Zenkiewicz, 2004; Saito *et al.*, 2004; Fernández *et al.*, 2007). But, in general, if the dose is sufficiently high, the formation of degradation products will generate off-flavours, changes in color or modifications in the mechanical properties of the irradiated films. Therefore, the complete evaluation of food contact materials during irradiation must consequently still be performed in all food contact intended plastics materials, since for food contact applications, the production of radiolysis products at low doses is considered to be more important than the induced mechanical changes. However, if foods are kept under vacuum, less radiolysis products should be formed between the package material and the food, as was described by Anh and Lee (2004) for meat products, which would considerably discharge the safety concerns regarding irradiated foods.

The information to be gathered concerning plastics in contact with irradiated foods concerns the post-irradiation stability, the mechanical strength and the ultimate barrier properties. However, it is of particular relevance to the migration of plastics components such as additives and antioxidants. Volatiles generated during irradiation frequently remain trapped in the polymer; the formation and release rates depend on the nature of the polymer matrix and the nature of the food or food simulant. Thus, Buchalla *et al.* (1999, 2000) reported the generation of low molecular weight compounds from some common polymers treated at 25 kGy were found. In the case of PVC and PP, fragments of stabilizers and phenol-type antioxidants were detected. Irradiated PE produced only traces of hydrocarbons, aldehydes, ketones and carboxylic acids, which disappeared within a few weeks after irradiation. Irradiated polystyrene produced mainly acetophenone, benzaldehyde, phenol, 1-phenylethanol, and phenylacetaldehyde. Riganakos *et al.* (1999) also confirmed that volatiles responsible for off-flavours might be generated during the irradiation of LDPE, EVAc and PET/PE/EVOH/PE. Other works (Welle *et al.*, 2000) additionally reported the formation of off-flavours during irradiation of common packaging plastics (PVC, PP, PET, PA, etc.) and Krzymien *et al.* (2001) reported volatile products from additives in irradiated PET. Favourable reports have been published for EVOH copolymers, since amounts of non-volatile compounds identified after irradiation were comparable with non-irradiated samples (Kothapalli and Sadler, 2003); additionally, no mechanical changes and a decrease in oxygen permeability were reported at low radiation doses, although radiolysis compounds in low amounts were identified (Lopez-Rubio *et al.*, 2007; Byun *et al.*, 2007). Currently, some more elaborate approaches are trying to introduce optimal amounts of antioxidants and UV stabilizers in the plastics that might help prevent the formation of free radicals during irradiation, as they are found to delay the outcome of undesirable reactions in the polymer

(Kawamura, 2004). Some common antioxidants used to stabilize plastics films are effective in trapping free radicals generated during irradiation; among them, Irgafos 168 has been reported by Kawamura (2004) to be very suitable to stabilize polyolefins.

3.5 Future trends

Because developing new barrier materials is an ever-increasing costly operation for most chemical companies (who are becoming aware that return on investment is taken longer and longer), the most immediate future for enhancing the performance of retortable plastics is most likely going to be seen in the tuning of existing materials by improving formulations or redesigning existing polymer molecular architectures to make them more intrinsically resistant to humid heat processing. In this respect, the incorporation of nanoadditives, such as clay-based nanocomposites, or of other types of nanoreinforcing elements (nanofibers, nanobeads, etc..) can have significant advantages while retaining the good optical properties and toughness of the matrix. Blending with other polymers may also provide promising results, despite the relatively few success stories where polymer blends have made their way to the market. Development of more efficient multilayered structures with more advanced polymer formulations and/or making use of improved retortable grades from the manufacturers will also have an impact in the short term by compounders and converters. In this area, the use of active packaging technologies such as oxygen scavengers in the plastics formulations can help alleviate the gas barrier losses that could be generated by for instance retorting of EVOH-based structures.

Alternatively, as biodegradable and/or sustainable materials such as biopolyesters, polysaccharides and proteins, are entering the packaging scene very steadily, new possibilities and material combinations are becoming available (Petersen *et al.*, 2001; Weber *et al.*, 2002). Nevertheless, most of these materials bring in even weaker thermal and water resistant properties and, therefore, a strong area of research is currently devoted to enhance the properties of these new bioplastics in retortable applications (Cava *et al.*, 2006).

In parallel with all of the above developments, there it is a clear trend to reduce thermal treatments in the food packaging industry to preserve as much as possible the freshness of the products. In this perspective, emerging preservation technologies such as high hydrostatic pressure treatments, irradiation, microwaves, etc., either on their own or in combination with active technologies such as scavengers, antioxidants and antimicrobials, will be more and more implemented, hence reducing the need for heat sterilization and/or pasteurization processes that are known to reduce the quality of foods and materials' performance.

3.6 References

Abdul Ghani, A.G., Farid, M.M., Chen, X.D. (2002). Theoretical and experimental investigation of the thermal inactivation of *Bacillus stearothermophilus* in food pouches. *J. Food Eng.*, 51, 221–228.

Ackerman, P., Jagerstad, M., Ohlsson, T. (eds) (1995). *Foods and Packaging Materials: Chemical Interactions*. Royal Society of Chemistry, London, UK.

Ali, A.A., Sudhir, B., Gopal, T.K.S. (2005). Effect of heat processing on the texture profile of canned and retort pouch packed oil sardine (*Sardinella longiiceps*) in oil medium. *J. Food Sci.*, 70(5), 350–354.

Anh, D.U., Lee, E.J. (2004). Mechanisms and prevention of off-odor production and color changes in irradiated meat. Irradiation of Food and Packaging: Recent Developments, *ACS Symposium Series*, 875, 43–76.

Arvanitoyannis, I.S., Bosnea, L. (2004). Migration of substances from food packaging materials to foods. *Critical Rev. Food Sci. Nutr.*, 44, 63–76.

Begley, T.H., Biles, J.E., Hollifield, H.C. (1991). Migration of an epoxy adhesive compound into a food-simulating liquid and food from microwave susceptor packaging. J. Agric. Food Chem. 39(11), 1944–1945.

Begley, T.H., White, K., Honigfort, P., Twaroski, M.L., Neches, R., Walker, R.A. (2005). Perfluorochemicals: Potential sources of and migration from food packaging. *Food Add. Contam.* 22(10), 1023–1031.

Bindu, J., Gopal, T.K.S., Nair, T.S.U. (2004). Ready-to-eat mussel meat processed in retort pouches for the retail and export market. *Packag. Technol. Sci.*, 17, 113–117.

Buchalla, R., Boess, C., Bogl, K.W. (1999). Characterization of volatile radiolysis products in radiation-sterilized plastics by thermal desorption–gas chromatography–mass spectrometry: screening of six medical polymers. *Rad. Phys. Chem.* 56(3), 353–367.

Buchalla, R., Boess, C., Bogl, K.W. (2000). Analysis of volatile radiolysis products in gamma-irradiated LDPE and polypropylene films by thermal desorption–gas chromatography–mass spectrometry. *Appl. Rad. & Isot.* 52(2), 251–269.

Byun, Y.J., Hong, S.I., Kim, K.B., Jeon, D.M., Kim, J.M., Whiteside, W.S, Park, H.J. (2007). Physical and chemical properties of γ-irradiated EVOH film. *Rad. Phys. Chem.* 76, 974–981.

CAC (1983). *The Microbiological Safety of Irradiated Food*. Codex Alimentarius Commission, CX/FH/83/9, Rome.

Canadian Food Inspection Agency (2002). *Flexible Retort Pouch Defects: Identification and Classification*, Canada.

Caner, C., Harte, B. (2005). Effect of high-pressure processing on the migration of antioxidant Irganox 1076 from polypropylene film into a food simulant. *J. Sci. Food & Agric.* 85(1), 39–46.

Caner, C., Hernandez, R.J., Harte, B.R. (2004A). High-pressure processing effects on the mechanical, barrier and mass transfer properties of food packaging flexible structures: A critical review. *Pack. Tech. Sci*, 17(1), 23–29.

Caner, C., Hernandez, R. J., Pascall, M., *et al.* (2004B). The effect of high-pressure food processing on the sorption behaviour of selected packaging materials. *Pack. Tech. Sci.* 17(3), 139–153.

Caner, C., Hernandez, R.J., Pascall, M.A. (2000A). Effect of high-pressure processing on the permeance of selected high-barrier laminated films. *Pack. Tech. Sci.* 13(5), 183–195.

Caner, C., Hernandez, R.J., Pascall, M.A., Buchanan, J. (2000B). Effects of high pressure processing on high barrier multi-layered flexible packaging materials: Mechanical properties and permeability to oxygen, carbon dioxide and water vapor. Presented at *Ann. Mtg., Institute of Food Technologists*, Dallas, TX, 10–14 June.

Caner, C., Hernandez, R.J., Pascall, M.A., *et al.* (2003). The use of mechanical analyses, scanning electron microscopy and ultrasonic imaging to study the effects of high-pressure processing on multilayer films. *J. Sci. Food & Agric.* 83(11), 1095–1103.

Cassiday, M.D., Streu, R.J., Wence, R.L., DeLassus, P.T. (1990). Barrier Packaging: Performance of Vdc Copolymers in Commercial Rigid Containers, *Journal of Plastic Film and Sheeting*, Vol. 6, No. 4, 268–275.

Castle, L., Jickells, S.M., Sharman, M., Gramshaw, J.W., Gilbert, J. (1988). Migration of the plasticizer acetyltributyl citrate from plastic film into foods during microwave cooking and other domestic use. *J. Food Prot.* 51(12), 916–919.

Castle, L., Mayo, A., Crews, C. (1989). Migration of poly(ethyleneterephthalate) (PET) oligomers from PET plastics into foods during microwave and conventional cooking and into bottled beverages. *J. Food Protection* 52(5), 337–342.

Castle, L., Nichol, J., Gilbert, J. (1992). Migration of polyisobutylene from polyethylene/ polyisobutylene films into foods during domestic and microwave oven use. *Food Additives and Contaminants* 9(4), 315–330.

Castle, L., Nichol, J., Gilbert, J. (1994). Migration of mineral hydrocarbons into foods. 4. Waxed paper for packaging dry goods including bread, confectionery and for domestic use including microwave cooking. *Food Add. Contam.* 11(1), 79–89.

Cava, D., Gimenez, E., Gavara, R., Lagaron, J.M. (2006). Comparative performance and barrier properties of biodegradable thermoplastics and nanobiocomposites versus PET for food packaging applications, *Journal of Plastic Film & Sheeting*, 22, 265.

Chandrasekar, V., Gopal, T.K.S., Rai, R.D. (2004). Heat penetration characteristics and shelf-life studies of mushrooms in brine processed in retort pouches. *Packag. Technol. Sci.*, 17, 213–217.

Chaudry, M.A., Bibi, N., Khan, M., Khan, M., Badshah, A., Qureshi, M.J. (2004). Irradiation treatment of minimally processed carrots for ensuring microbiological safety. *Rad. Phys. Chem.* 71(1–2), 171–175.

Chen, T.S., George, W.Z. (1981). Ascorbic acid retention in retort pouched green beans. *J. Food Sci.*, 46, 6423

Chmielewski, A.G., Haji-Saeid, M., Ahmed, S. (2005). Progress in radiation processing of polymers. *Nuclear Instruments and Methods in Physics Research Section B – Beam Interactions with Materials and Atoms,* 236, 44–54.

Czerny, M., Schieberle, P. (2006). Influence of the polyethylene packaging on the adsorption of odor-active compounds from UTH-milk. *Eur. Food Res. Technol.*, 225(2), 215–223.

Durance, T.D., Collins, L.S. (1991). Quality enhancement of sexually mature chum salmon *Onchorhynchus keta* in retort pouches. *J. Food Sci.*, 56(6), 1282–1286.

Dymit, J.M. (1973). Today's food service system. *Act. Rep. Dev. Assoc. Mil. Food Package Syst.*, 25(2), 56–57.

Fan, X.T., Niemira, B.A., Thayer, D.W. (2004). *Low-dose ionizing radiation of fruit juices: Benefits and concerns. Irradiation of Food and Packaging: Recent Developments*, ACS Symposium Series, 875, 138–150.

Farkas, J. (1988). *Irradiation of Dry Food Ingredients.* CRC Press, Inc., Boca Raton, Florida.

Fernández A, Butz P, Bognar A, Tauscher B (2001). Antioxidative capacity, nutrient content and sensory quality of orange juice and an orange–lemon–carrot juice product after high pressure treatment and storage in different packaging. *Eur. Food Res. Technol.* 213, 290–296.

Fernández A., Sánchez M.D., Ankerfors M., Lagaron J.M. (2008). Effects of ionizing radiation in ethylene-vinyl alcohol copolymers and in composites containing microfibrillated cellulose. Accepted by *J. of Applied Polymer Science.*

Giddings, G.G., Marcotte, M. (1991). Poultry irradiation for hygiene/safety and market life enhancement. *Food Rev. Int.* 7(3), 259–82.

Gopal, T.K.S., Vijayan, P.K., Balachandran, K.K., Madhavan, P., Iyer, T.G.S. (2001). Traditional Keral style fish curry in indigenous retort pouch. *Food Control*, 12, 523–527.

Grégoire, O., Cleland, M.R., Mittendorfer, J., Dababneh, S., Ehlermann, D.A.E., Fan, X., Kappeler, F., Logar, J., Meissner, J., Mullier, B., Stichelbaut, F., Thayer, D.W. (2003). Radiological safety of food irradiation with high energy X-rays: Theoretical expectations and experimental evidence. *Rad. Phys. Chem.* 67(2), 169–183.

Habeger, C.C. (1997). Microwave interactive thin films. *Microwave World* 18(2), 8–22.

Ikeo, Y., Aoki, K., Kishi, H., Matsuda, S., Murakami, A. (2006). Nano clay reinforced biodegradable plastics of PCL starch blends. *Pol. Adv. Technol.* 17(11–12), 940–944.

Industrial Technology Development Institute, www.pcierd.dost.gov.ph/forum/water%20retort.pdf

Institute of Food Technologists (IFT). (1988). Migration of toxicants, flavors, and odor-active substances from flexible packaging materials to the food. Risch, S.J. (author). *Food Technol.*, 42(7), 95–102.

IOPP (Institute of Packaging Professionals). (2002). *Fundamentals of Packaging Technology*, 3rd Ed., Naperville, IL.

Jetten, J., de Kruijf, N. (2002). Quality and safety aspects of reusable plastic food packaging materials: Influence of reuse on intrinsic properties. *Food Add. Contam.*, 19(1), 76–88.

Jetten, J., de Kruijf, N., Castle, L. (1999). Quality and safety aspects of reusable plastic food packaging materials: A European study to underpin future legislation. *Food Add. Contam.*, 16(1) 25–36.

Jickells, S.M., Philo, M.R., Gilbert, J., Castle, L. (1993). Gas-chromatographic mass-spectrometric determination of benzene in nonstick cookware and microwave susceptors and its migration into foods on cooking. *J. AOAC Int.*, 76(4), 760–764.

Juliano, P., Toldra, M., Koutchma, T., Balasubramaniam, V.M., Clark, S., Mathews, W., Dunne, P., Sadlerand, G., Barbosa-Canovas, G.V. (2006). Texture and water retention improvement in high-pressure thermally treated scrambled egg patties. *J. Food Sci.*, 71(2), 52–61.

Kawamura, Y. (2004). *Effects of gamma irradiation on polyethylene, polypropylene, and polystyrene. Irradiation of Food and Packaging: Recent Developments*, ACS Sympositum Series, 875, 262–276.

Keefer, R.M. (1987). *Microwave heating package and method*, US Patent No. 4 646 325.

Keller, K. (1999). *Ventable microwave-safe food package*, European Patent No. PCT/CH98/00546 (US Patent No. 6 607 764 2003).

Kothapalli, A., Sadler, G. (2003). Determination of non-volatile radiolytic compounds in ethylene co-vinyl alcohol. *Nuclear Instruments and Methods in Physics Research, Section B: Beam interactions with Materials and Atoms*, 208(1–4), 304–344.

Krzymien, M.E., Carlsson, D.J., Deschenes, L., Mercier, M. (2001). Analyses of volatile transformation products from additives in gamma-irradiated polyethylene packaging. *Food Add. & Contam.*, 18(8), 739–749.

Kübel, J., Ludwig, H., Marx, H., Tauscher, B. (1996). Diffusion of aroma compounds into packaging films under high pressure. *Pack. Tech. Sci.*, 9, 143–152.

Lagaron, J.M., Catala, R., Gavara, R. (2004). Structural characteristics defining high barrier properties in polymers, *Materials Science and Technology*, 20, 1.

Lagaron, J.M., Cabedo, L., Cava, D., Feijoo, J.L., Gavara, R, Gimenez, E. (2005). Improving packaged food quality and safety: Nanocomposites, *Food Additives and Contaminants*, 22(10), 994–998.

Lambert, Y., Demazeau, G., Largeteau, A., Bouvier, J.M., Laborde-Croubit, S., Cabannes, M. (2000). Packaging for high pressure treatments in the food industry. *Pack. Tech. Sci.*, 13, 63–71.

Le-Bail, A., Hamadami, N., Bahuaud, S. (2006). Effect of high-pressure processing on the mechanical and barrier properties of selected packagings. *Pack. Tech. Sci.*, 19(4), 237–243

Lee, N.Y., Jo, C., Shin, D.H., Kim, W.G., Byun, M.W. (2006) Effect of gamma-irradiation on pathogens inoculated into ready-to-use vegetables. *Food Microbiol.*, 23(7), 649–656.

Lopez-Rubio, A., Lagaron, J.M., Gimenez, E., Cava, D., Hernandez-Munoz, P., Yamamoto, T., Gavara, R. (2003). *Macromolecules*, 36, 9467.

Lopez-Rubio, A., Lagaron, J.M., Hernandez-Munoz, P. (2005a). Effect of high pressure treatments on the properties of EVOH-based food packaging materials. *Innov. Food Sci. & Emerg. Technol.*, 6(1), 51–58.

Lopez Rubio, A, Hernandez-Munoz, P., Catala, R., Gavara, R., Lagaron, J.M. (2005b). *Food Additives and Contaminants*, 22(10), 988–993.
Lopez-Rubio, A., Hernandez-Munoz, P., Gimenez, E., Yamamoto, T., Gavara, R., Lagaron, J.M. (2005C). *Journal of Applied Polymer Science*, 96, 2192.
Lopez-Rubio, A., Gavara, R., Lagaron, J.M. (2006A). Unexpected Partial Crystallization of an Amorphous, Polyamide as Induced by Combined Temperature and Humidity, *Journal of Applied Polymer Science*, 102, 1516–1523.
Lopez-Rubio, A, Gimenez, E, Gavara, R, Lagaron, J.M. (2006B). Gas Barrier Changes and Structural Alterations Induced by Retorting in a High Barrier Aliphatic Polyketone Terpolymer, *J Appl. Polym. Sci.*, 101, 3348–3356.
Lopez-Rubio, A., Lagaron, J.M., Yamamoto, T, Gavara, R. (2007). Radiation-induced oxygen scavenging activity in EVOH copolymers, *J. Appl. Polym. Sci.*, 105, 2676.
Lopez-Rubio, A., Lagaron, J.M. (2007). Improving the resistance to humid heat sterilization of EVOH copolymers through blending, *J Appl. Polym. Sci.*, accepted for publication, 2007.
Masuda, M., Saito, Y., Iwanami, T., Hirai, Y. (1992). Effects of hydrostatic pressure on packaging materials for food. In: *High Pressure and Biotechnology*. Balny C, Heremans K, Masson P (eds), Colloque Inserm, 224, 545–547. John Libbey Eurotext (London).
Mertens, B. (1993). Packaging aspects of high pressure food processing technology. *Pack. Tech. Sci.*, 6, 31–36.
Mohan, C.O., Ravishankas, C.N., Bindu, J., Geethalakshmi, V., Gopal, T.K.S. (2006). Effect of thermal process time on quality of 'Shrimp Kuruma' in retortable pouches and aluminum cans. *J. Food Sci.*, 71(6), 496–500.
Petersen, K., Nielsen, P.V., Olsen, M.B. (2001). Mechanical and Physical Properties of Biobased Materials – Starch, polylactate and polyhydroxybutyrate. *Starch*, 53, 356.
Poças, M.F., Hogg, T. (2007). Exposure assessment of chemicals from packaging materials in food: A review. *Trends Food Sci. Technol.*, 18, 219–230.
Prakash, A., Guner, A.R., Caporaso, F., Foley, D.M. (2000). Effects of low-dose gamma irradiation on the shelf life and quality characteristics of cut romaine lettuce packaged under modified atmosphere
Rastogi, N.K., Raghavarao, K.S.M.S., Balasubramaniam, V.M., Niranjan, K., Knorr, D. (2007). Opportunities and challenges in high pressure processing of foods. *Crit. Rev. Food Sci. & Nut.*, 47(1), 69–112.
Riganakos, K.A., Koller, W.D., Ehlermann, D.A.E., Bauer, B., Kontominas, M.G. (1999). Effects of ionizing radiation on properties of monolayer and multilayer flexible food packaging materials. *Rad. Phys. Chem.*, 54(5), 527–540.
Ross, R.T., Engeljohn, D. (2000). Food irradiation in the United States: Irradiation as a phytosanitary treatment for fresh fruits and vegetables and for the control of microorganisms in meat and poultry. *Rad. Phys. Chem.*, 57(3–6), 211–214.
Ryynämen, S., Ohlsson, T. (1996). Microwave heating uniformity of ready meals as affected by placement, composition and geometry. *J. Food Sci.*, 61(3), 620–4.
Saito, F., Yotoriyama, T., Nagashima, Y., Suzuki, Y., Itoh, Y., Goto, A., Iwaki, M., Nishiyama, I., Hyodo, T. (2004). Study of ion irradiated poly-lactic acid using slow positron beam. Positron Annihilation, *ICPA-13, Proceedings Materials Science Forum*, 445–446, 340–432.
Schauwecker, A., Balasubramaniam, V.M., Sadler, G., Pascall, M.A., Adhikari, C. (2002). Influence of high-pressure processing on selected polymeric materials and on the migration of a pressure-transmitting fluid. *Pack. Tech. Sci.*, 15, 255–262.
Schiffmann, R. (2005). Packaging for microwave foods. In: *The microwave processing of foods*. (Ed. Schubert H. and Regier M.) Woodhead Publishing in Food Science and Technology. 192–216.
Sepulveda, D., Olivas, G., Rodriguez, J.J., Warner, H., Clark, S., Barbosa-Canovas, G.V. (2003). Storage of retort pouch beefsteak and beef stew packed under four headspace levels. *J. Food Process. Preserv.*, 27(3), 227–242.

Simpson, R., Almonacid, S., Mitchell, M. (2004). Mathematical model development, experimental validation and process optimization: Retortable pouches packed with seafood in cone frustum shape. *J. Food Eng.* 63, 153–162.

Skjevrak, I., Bred, C., Steffensen, I.L., Mikalsen, A., Alexander, J., Fjeldal, P. (2005). Non-target multi-component analytical surveillance of plastic food contact materials: Identification of substances not included in EU positive lists and their risk assessment. *Food Add. Contam.*, 22(10), 1012–1022.

Tsai, B.C., Wachtel, J.A. (1990). In: *Barrier Polymers and Structures* (W.J. Koros, ed.). American Chemical Society: Washington DC, 192.

Tung, M.A., Smith, T. (1980). Innovations in thermal processing. In *Processing 2000 Symposium*, McGill University Publications, Montreal, QC, pp. 103–121.

Vanamala, J., Cobb, G., Turner, N.D., Lupton, J.R., Yoo, K.S., Pike, L.M., Patil, B.S. (2005). Bioactive compounds of grapefruit (*Citrus paradisi* Cv. Rio red) respond differently to postharvest irradiation, storage, and freeze drying. *J. Agric. Food Chem.*, 53(10), 3980–3985.

Weber, C.J., Haugaard, V., Festersen, R., Bertelsen, G. (2002). Production and Applications of biobased packaging materials for the Food Industry. *Food Addit. Contam.*, 19, 172.

Welle, F., Haack, G., Franz, P. (2000). Investigation into migrational and sensorial changes of packaging plastics caused by ionising irradiation. *Deutsche Lebensmittel-Rundschau*, 96(11), 423–430.

WHO (1988). *Food Irradiation. A Technique for Preserving and Improving the Safety of Food.* World Health Organization, Geneva, Switzerland.

WHO (1994). *Safety and Nutritional Adequacy of Irradiated Food.* World Health Organization, Geneva, Switzerland.

WHO (1997). *Food Irradiation.* WHO Press Release. WHO/68, 19 September, World Health Organization Press Office, Geneva.

Wood, S.A. (1990). Performance polymers are finding greater use in packaging markets. *Mod. Plast.*, 67(8), 62.

Zenkiewicz, M. (2004). Effects of electron-beam irradiation on some mechanical properties of polymer films. *Rad. Phys. Chem.*, 69(5), 373–378.

4

Advances in sealing and seaming and methods to detect defects

E. Hanby, Campden and Chorleywood Food Research Association, UK

4.1 Introduction: the importance of sealing

There are a variety of pack formats used when processing in-pack, including cans, trays and pouches. Each of these formats is sealed differently. However, the one factor they all have in common is that the seal made is designed to prevent leakage of the food out and contamination in.

The types of seals used on these packs are manufactured using a variety of technologies and the knowledge behind some of these is limited to specialist manufacturers. There is also a limited understanding of the type of methods available used to assess the integrity of these seals. Seal integrity is vitally important as without it the food can spoil and lead to concern over its safety as well as having an adverse affect on the visual and organoleptic properties of the food.

Food packaging has a number of roles including containing, protecting, preserving and promoting the food inside. It is essential to ensure that the pack has a good seal integrity to prevent physical, chemical and microbiological damage. An integral seal is crucial when packing foods that are moisture and/or oxygen sensitive, to prevent the ingress of oxygen or moisture into the pack.

This chapter highlights the various methods and technologies used to seal food packages. The defects that can occur in the sealing area are highlighted, along with a range of detection methods and equipment.

4.2 Sealing methods

When selecting a sealing method, a number of factors need to be considered including the properties of the packaging material being used. Some are unable to withstand high sealing temperatures; therefore it is important to bear in mind the melting point and heat sensitivity of the material. The product being packaged will affect the type of seal used, and will also determine the strength of the seal required. For example, a fluid product that has the potential to move around within the package during transportation may require a stronger seal. The cost of the sealing equipment may also be a deciding factor; some technologies require a higher initial cost but are more cost effective in the long term. The role of the seal will determine the sealing method, i.e. whether it needs mechanical strength, hermetic integrity or whether it needs to be able to peel.

4.2.1 Adhesive seals

An adhesive is described as a substance designed to stick to things. An adhesive should be able to provide close contact between itself and the sealing surface. For these applications it should be a liquid which has a low viscosity of application but which then hardens to solid.

The ideal sealing adhesive would have the following properties:

- Suit different plastic films and trays.
- Be compatible with a variety of food products.
- Have a wide sealing temperature range to suit the thermal properties of the material with which it is being used.
- Resist separation.
- Be unaffected by low temperature during transportation and storage.
- Remain flexible throughout storage, even at different humidities.

The drawback with using adhesives is that, like all polymers, they are not hermetic in the long term and can allow oxygen and water vapour to permeate through.

Defects can occur in the seal area of a package if the adhesive is not suited to the material, or if the package has been stored incorrectly or packed or handled in an incorrect manner. The type of defects seen within the adhesive are described in Table 4.1.

4.2.2 Heat seals

Sealing using heat is the most widely used method. The heat can be applied using wires, dies, metal strips or rotating wheels. The temperature of the heat sealing and whether it is constant or fluctuating is dependent upon the packaging material. Some materials will begin to melt at 80 °C whereas others can withstand temperatures of more than 200 °C. Temperature is one of many factors to be considered when heat sealing; others include:

Table 4.1 Defects seen when using adhesives

Defect name	Defect description
Hot creep	Movement of one package surface over another before the adhesive has set
Slow seal failure	Separation of the adhesive due to low stress-resistance
Steriliser creep	Softening of the adhesive caused by temperature and/or humidity
Cold creep	The adhesive layer becoming brittle due to low temperatures during transport; this can lead to a fracture in the seal area.

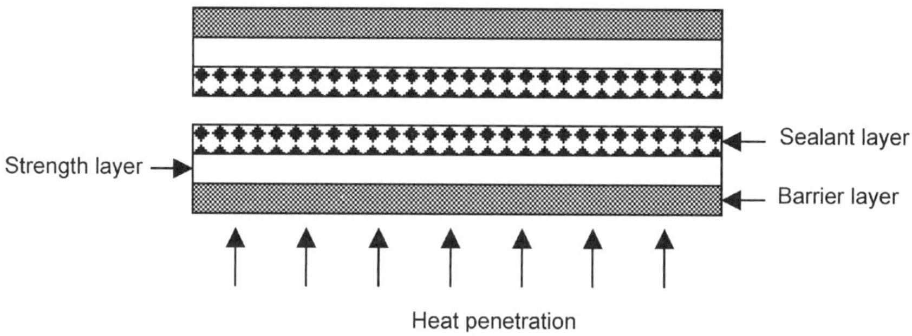

Fig. 4.1 Heat penetration during heat sealing.

- Pressure
- Dwell time
- Storage conditions after sealing
- The number of packs being sealed.

The same factors will also affect the final strength and integrity of a heat seal.

During sealing, the heat has to permeate through a number of layers from the outside, through a sealant layer, to the inside layer as shown in Fig. 4.1. Laminate films are made up of many layers; more layers mean it takes longer for the sealant to heat sufficiently to melt.

Laminated materials have a non-thermoplastic layer on the outside, allowing the heating bar to be directly applied without melting the pouch.

The temperature used for sealing is critical; if inadequate heat is applied, it will prevent a seal from forming or produce a weak seal. However, if the sealing temperature is too high, there is a risk that the film will be burnt or damaged, leading to the formation of a weak seal.

The integrity of a heat seal is also affected by the design of the sealing jaw. A crimp jaw leaves a crimped impression in the seal area after sealing. The thickness of the film impacts on the effectiveness of a crimp jaw, as the pressure is distorted along the seal, varying the quality of the seal along the length. Depending on the requirement of the final package, the seal produced can be of differing lengths and either vertical or horizontal.

Unlike some of the sealing technologies explained in this chapter, heat sealing machines have low flexibility and high costs related to machine tooling.

4.2.3 Ultrasonic sealing

Most thermoplastics can be sealed with the use of ultrasonics. This sealing method can be used for pouches and trays. During ultrasonic sealing, a converter (usually lead zirconate titanate) converts electrical energy into high frequency energy waves, which are then amplified by a booster. These high frequency waves cause a vibration; this vibration then causes friction in the molecules within the thermoplastic, which generates heat and therefore has a melting effect, which leads to sealing.

There are a number of advantages to ultrasonic sealing:

- No tooling templates are required.
- Pre-heating of the equipment is not required.
- No heat is required.
- It is suitable for small sealing areas.

'As there is no heat required, the packaging cannot be burnt outside the sealing area, although the sealing can become overheated, and a strong seal can be formed as well as making it easy opening for the consumer.' (Anon., 2006)

4.2.4 Laser sealing

This technique uses a laser beam to melt the plastic requiring to be sealed. A powerful beam of radiation (infra-red) is generated from the laser and the beam is concentrated on the area to be sealed. This stimulates a resonant frequency within the molecules, which results in the heating of the surrounding material.

No tooling is required when using a laser to seal. It is a non-contact technique and is flexible to shapes and sizes. Trays of different shapes and sizes can be sealed on the same production line without the need to change tooling plates. Once the package is sealed, it is also possible for the laser beam to cut the film. The laser has the added advantage that the cut made is very 'clean', which reduces the occurrence of misaligned seals or variations in the tension.

4.2.5 Radio-frequency sealing

Radio-frequency heat sealing works by stimulating the polar molecules contained within certain types of plastics. Radio-frequency waves activate these polar molecules and they begin to move; as the movement speeds up, heat is created. The material is heated much more quickly than when using a standard heat sealing bar; therefore production rates are improved. As this type of sealing requires less heat to seal, energy use is also reduced.

A number of different plastics can be sealed used radio-frequency sealing including; PETG and Nylon. The UK-based company Stanelco market the

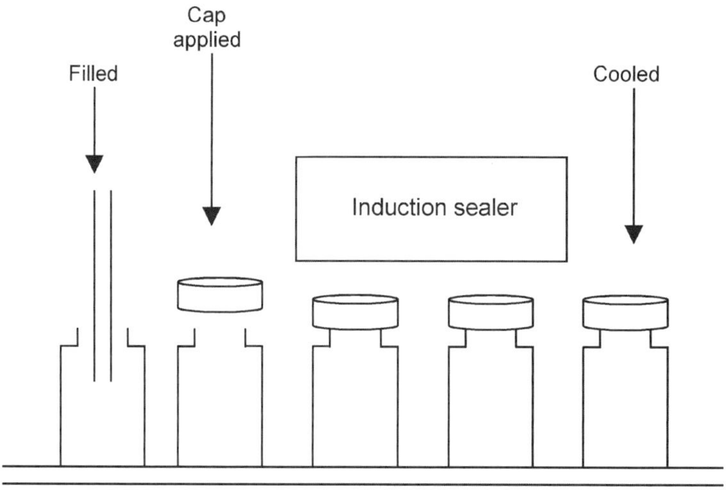

Fig. 4.2 Induction sealing of bottles.

Greenseal method of sealing plastic food trays using radio-frequency. They claim their technology is cheaper and more efficient than standard heat sealing.

4.2.6 Induction sealing

Induction sealing is a non-contact sealing technology used on plastic containers which requires a metal liner to be attached before a plastic cap. This type of seal is used to reduce leakage, lengthen shelf-life and show evidence of tampering, therefore increasing consumer confidence. An example of a food packed in this way is tomato ketchup.

Induction sealing is carried out with the use of machines that work by transmitting an electro-magnetic field, which creates currents in the metallic element within the liner. This current generates heat (temperatures of 185–350 °C) causing the metal liner to fuse to the plastic container. 'Sealing can be carried out at low or high frequency. A low frequency means that a more uniform heat is applied across the sealing area, whereas with higher frequency a greater sealing edge is given, which can assist with faster sealing rates.' (Potter *et al.*, 2006).

A typical set-up is shown in Fig. 4.2. The bottle is filled and the cap (plastic lid containing a metal liner) is applied. The liner contains a sealing material adhered to a foil layer. An induction current heats the foil liner, which in turn melts the sealant. The pressure of the cap upon the bottle provides enough pressure for the liner to seal to the bottle. The sealing layer needs to be cooled before it is fully adhered to the bottle. 'For a good seal exactly the right amount of heat must go into the foil – too little and the seal will lack strength and may leak, too much and burning or degradation of the foil will occur – and it must be evenly distributed over the foil surface.'(Line Patrolman, see web address)

As with all sealing technologies, sealing problems can occur, which lead to a

Table 4.2 Defects seen when induction sealing

Defect	Identified by	Likely cause
No seal or weak seal	Leakage Lack of seal Misalignment Foil easy to remove Cap difficult to remove	Under heating (setting too low) Conveyor speed too fast Sealing head position incorrect Insufficient cap torque Product contamination Incompatible liner material
No seal or weak seal	Possible leakage Blisters in the foil Melting of the bottle Loose cap Foil is difficult to peel	Over heating (setting too high) Conveyor speed too slow Sealing head position incorrect

number of quality issues. Table 4.2 lists some of the defects seen with induction sealing.

4.3 Seaming

Seaming is the term used when sealing cans. Canned foods have a good reputation with regard to safety and their use means food can be stored for long periods of time. However, a failure in the integrity of a can can lead to leaker infection and in turn to microbial spoilage.

Canning involves placing food inside a container and closing it with a hermetic seal. Heat is then applied; the heat process has to be sufficient as to ensure commercial sterility and to prevent recontamination. Commercial sterility is achieved by the application of heat which renders the food free from viable micro-organisms, including those of known public health significance. The heat process has three main objectives: the cooking and preparation of the food, the destruction of enzymes (which could cause chemical deterioration of the food) and the destruction of micro-organisms to a condition of commercial sterility.

There are many advantages to canning foods:

- Long shelf-life without the need for refrigeration.
- Robust form of packaging, which is beneficial in terms of distribution and storage.
- Well understood by the consumer.
- Does not require the use of additives.
- Foods prepared in this way are convenient as they are already prepared.

Cans used in food processing can be round, rectangular or irregular, but the structure of the double seam remains the same. The double seam can be defined as the interlocking of the curled edge of a can end with the flanged lip of the can. The double seam consists of three thicknesses of the end component and two thicknesses of the body component with an appropriate sealing compound trapped

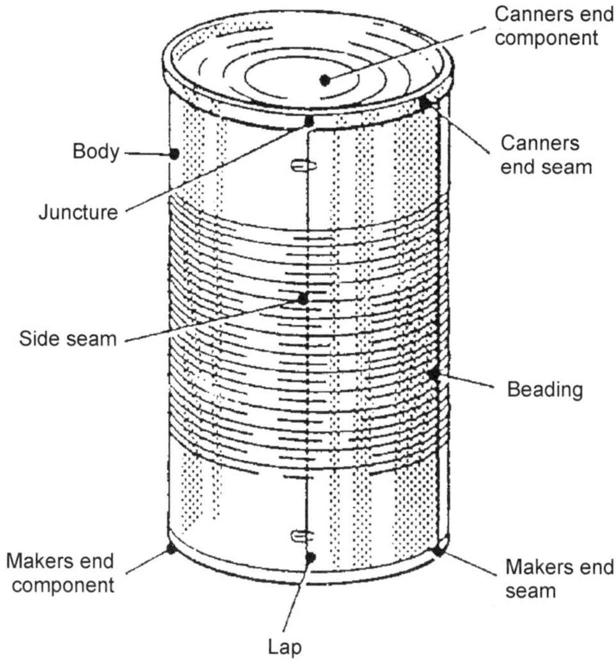

Fig. 4.3 Typical open-top processed food can.

within the folded double seam to form a hermetic seal. Figure 4.3 shows the terminology used in a typical open-top processed food can.

When producing canned foods, mechanical damage can lead to 'leakers' through defects formed during the formation of the hermetic seal. These leaks can lead to microbial spoilage and food poisoning incidents. The types of defects seen in cans are described in Table 4.3.

4.4 Defect detection methods

The main factor that affects seal integrity is contamination of the seal area with food or liquid. Others factors linked to seal integrity are:

- Heat processing, both pasteurisation and sterilisation.
- Position of packs during storage.
- Handling and transportation.

In the past, seal integrity checks on food packs have been visual checks by humans, drop tests and immersing packs under water to look for escaping air (seen as bubbles). All of these methods can be affected by the differences in the human tester. There are now numerous kinds of equipment available for testing the integrity of seals. Different types of equipment employ different technologies. The method used may be destructive or non-destructive. The disadvantages with

Table 4.3 Can defects

Defect	Description and position
Peaked	Distortion of the can end in the form of one or more permanent ridges extending into the double seam countersink region.This is likely to be caused by inadequate countersink depth at the time of seam formation.
Cut over	The seaming panel becomes flattened and the metal is forced over the seaming chuck, forming a sharp lip at the chuck wall. This can be formed in a variety of ways: Excessive solder at the side seam, wear on the seaming chuck, the seaming rolls set too tight, or incorrect seaming roll profile for the end in use.
Droop	Smooth projection of the double seam at the side seam and end seam juncture. This could be caused by: excessive solder in the side seam, excessive body hook, excessive or uneven distribution of the lining compound.
Deformed end seam	Similar to droop but formed away from the side seam. This can be caused from product entrapment in the seam, an excessive body hook or excessive or uneven distribution of the lining compound.
Spur	This is a localised irregularity characterised by a sharp 'v' protrusion at the bottom of the seam.
Knocked down curl	This occurs when the end hook is not engaged with the body hook, resulting in the end curl extending below the seam. This is seen in damaged end curls or inadequate curls.
False seam	This is a seam fault where the end and the body hooks are not engaged, although they may give the appearance of a properly formed seam. This can be caused in a variety of ways: ends not setting correctly off the seaming chuck or the use or an incorrect seaming setting. This type of defect is not always detected by external examination.
Torn back end	The damage results in a fracture during the seaming formation.
Perforation/pierced	This is seen as complete penetration of the can body or end. It is usually caused by a puncture from a sharp object.

methods that destroy the pack are that they all vary in their precision; sample selection is random and, as the product is lost, it can be expensive. 'The vast majority of package integrity test methods detect defects in packages, not the presence of microbes, and many of the current tests are destructive – both the package and its contents are destroyed during the course of testing.' (Bix *et al.*, 2004)

4.4.1 Inflation tests

Inflation tests are suitable for most pack types (pouches, bags and trays) and materials. This type of test includes burst, creep and creep-to-failure tests. These methods can suggest the site at which the pack is leaking or is weak.

Burst testing is performed to establish the strength of the seal and therefore its ability to withstand transportation and handling. It can be used on both flexile and

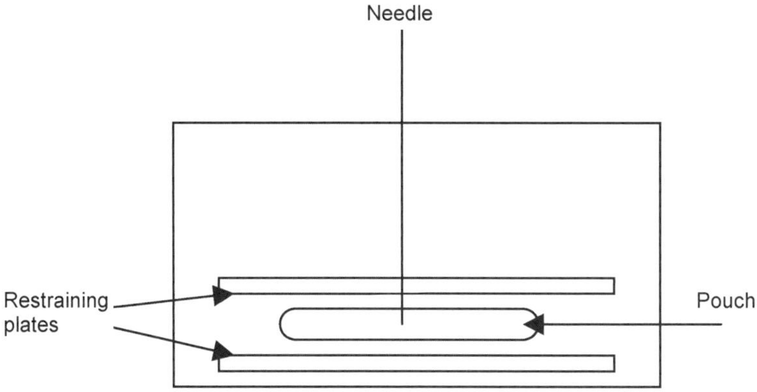

Fig. 4.4 Burst tester.

semi-rigid packaging. Figure 4.4 illustrates how a burst tester works. The pack is restrained between two plates. A needle is inserted into the pack through which compressed air flows. The restraining plates help to keep the pack flat and therefore prevent internal peeling of the pack as it inflates. The air flows into the pack until it bursts and the pressure is measured at the point at which the pack bursts. This method is used to find the weakest part of the seal. If the weak point is recorded at the same point on a number of packs when routinely carrying out a quality control check, it could be an indication of problems with the sealing or filling machinery, e.g. product contamination, incorrect sealing temperature, pressure or dwell time.

Hanby *et al.* (2006) have carried out trials to look at the effect of the profile of the sealing head on the number of failures when burst testing. There was strong statistical evidence of the effect of product when using a reduced profile and burst testing ($P = <0.001$), i.e. if the seal was contaminated and sealed using a reduced profile sealing head, the pack always burst.

A creep test can also be used to establish the strength of a seal. The pack is inflated using compressed air to a constant pressure and then held for a set time or until the pack bursts. Creep testing can be used before retort processing as a means of testing whether the pouch/pack can withstand the heat and pressure that it is subjected to during processing. This method will identify weaknesses in the corners of the pack and the seal, and leaks through pinholes in the material.

4.4.2 Leak detectors

These can be on- or off-line and all use various methods of determining whether a leak is present.

Helium leak test

Gilchrist *et al.* (1989) developed a method of determining leaks in rigid metal cans. An overpressure of helium was applied to the outside of the can and the helium leaked into the can if holes were present. The can was then pierced and the head space analysed for the presence of helium. This method was not suitable for use with flexible retort pouches as the overpressure used caused the pouches to change shape and internal pressures to equalise; therefore no helium would seep through. However, a method was developed for flexible retort pouches; one end of the pouch was cut open and emptied, dry sand was added and the pouch was resealed. The overpressure of helium could then be applied in the same way as when testing rigid cans as the sand provided a structure to prevent the pouch from flexing.

Since this method was developed, a number of mechanical helium leak detection methods have been marketed. These types of detectors measure the change in pressure within the pack under vacuum within a sealed chamber. Helium is inserted into the headspace of the pack; the pack is then placed into a chamber. A vacuum is drawn on the chamber and the helium escaping from the pack is detected and measured by a mass spectrometer. The helium level is compared with that of a predetermined maximum allowable leak rate. The pack fails if the helium level is above this rate.

The technology behind helium leak testing is complicated, which makes this method expensive.

Pressure vacuum decay test

This method measures changes in pressure between atmospheric pressure and the pressure within the pack. The pack is pressurised for a predetermined time at a predetermined pressure and then held (as with burst testing, restraining plates may be required when testing flexible packages). If a change in pressure is recorded, a leak is present. Pressure vacuum decay testing is sensitive and provides quantitative data; however. it does not identify the position of the leak. More recently, non-destructive pressure vacuum decay tests have been developed. The test package is placed in a chamber and pressurised. If a leak is present, air will enter the pack through the leak and the change in pressure is measured.

Vacuum test

A vacuum test is used to determine whether a leak is present and it may also identify the position of the leak. The package is put into a chamber of water, the chamber is closed and the air is removed from the chamber by use of vacuum. Any leaks in the pack will be seen as a stream of bubbles being released from the pack. The amount of vacuum drawn from the chamber and therefore the amount of pressure exerted on the pack depends on pack size and seal strength. This method can be used for both flexible and semi-rigid packs. Identification of leaks under water in this method is similar to past methods of immersing packs under water to look for bubbles.. This method, however, has the advantage of replicating typical pressures encountered during transportation, for example via air freight.

Integrity testing trials carried out by Hanby *et al.* (2006) found that, when vacuum testing, the profile of the sealing head has an effect on the number of leaks; with a curved profile 66% leaked, with a flat profile 37.5% leaked and with a reduced profile 16% leaked. This study also found strong statistical evidence of the effect of heat processing, when a full pack sealed using a flat seal was vacuum tested. The tray was more likely to leak if it was not processed.

Squeeze method
This method is used for pouches and for induction-sealed containers. The squeeze method test is carried out on an on-line conveyor at the end of the filling operation. The pack travels along the conveyor after filling and sealing. Above this conveyor runs another conveyor. The pack is weighed. As the pack enters the test site, the top conveyor pushes down on the pack at a set pressure. The pack is re-weighed. If the difference between the weights at the start and end are greater than a threshold limit, the pack is rejected.

Electrical conductivity
Another destructive but easy, inexpensive test is the electrical conductivity test. The test pack is cut in half and emptied; each half (separately) is filled with a 1% salt solution and held in a container also containing a 1% salt solution. An electrical multimeter is used to detect if leaks are present. One electrode is placed in the container solution and one electrode inside the pack; if a current is recorded, a leak is present. Electric conductivity testing can detect microholes 8 μm in diameter; it can also detect holes that do not fully penetrate the package. This method does not identify the point of the leak and does not give quantitative data.

Coloured and colour-change indicators
These can be used as a one-off test or can be incorporated into the pack to measure over time. The latter is useful if damage occurs during transportation or handling after the pack left the manufacturer. As a test method, penetrant dye is sprayed into a clean empty pack and is left for a time, depending on the manufacturer's instructions. At the end of the test, the pack can be examined for the presence of dye outside the pack; it leaves a trail so channels and holes can be identified. Indicators can be incorporated into packs and change colour if certain gases are detected within the pack. These strip indicators are designed to react to a specific level of a certain gas.

Scanning methods
Al-Habaibeh *et al.* (2004) developed a method to laser seal ambient ready meal trays and scan for seal defects using infra-red detection immediately after sealing. The method for sealing is non-contact. The film is stretched over the tray using rollers, and the laser beams scan the tray based on known dimensions and material type, ensuring a complete seal. An infra-red imager captures an image of the heat distribution during sealing and the first two seconds of cooling. If the image is 'normal', i.e. the thermographical behaviour is normal, the tray is moved to the visual image processing system for integrity testing.

Ultrasound can also be used as an imaging tool to detect leaks and defects in packages. This is carried out using a device that directs high frequency sound waves into the package material. The sound waves are deflected if a defect is present. This method is capable of detecting delamination, incomplete seals and channel leakers.

Bio-testing
This is the definitive integrity test. It can be used for both rigid and semi-rigid containers. The package is tested using a liquid suspension of micro-organisms.

The bio-test procedure used by Lampi (1980) to investigate hole size versus bacterial penetration is:

(i) Drill holes in package, using a laser.
(ii) Sterilise the package with nitrous oxide (to avoid heat or other gases altering the defect size).
(iii) Coat inside of the package with semisolid nutrient agar containing dextrose.
(iv) Bio-test with *Aerobacter aerogenes* by removing residual air, sealing and putting the pack in a container with bacteria, followed by incubation at 37 °C for 14 days; checking for swelling.
(v) Measure the rate of gas flow through the holes.
(vi) Scanning electron microphotograph the defects and calculate the hole diameter.

Lampi's results showed that, of the 500 pouches tested, only 30 permitted gas flow. It was also concluded that bacteria penetration was unlikely with holes less than 10 μm in diameter.

Anema *et al.* (1980) carried out a bio-test study where they handled half the batch of flexible pouches very carefully and the other half roughly. They were all evaluated for leakage using a bio-test method. Of the roughly handled pouches, 0.32% leaked, and of the carefully handled pouches, 0.08% leaked. Their study concluded that careful control during sealing, high standards of hygiene post-process and good transportation will reduce the risk of contamination, ensuring a safe product.

4.5 Future trends

As we enter a new global phase where our sustainable future is a high priority, new materials are being developed in order to reduce energy usage or enable them to be biodegraded, composted or recycled. The introduction of these new materials may require new sealing methods which, in turn, will require new methods of integrity testing or validation alongside current methods as discussed in this chapter.

Wastage and energy reduction are high on the world agenda: new sealing techniques are already being developed for use with fresh produce, such as salad leaves, to reduce the amount of packaging used. Methods have been developed to produce a hermetic seal on vertical-form fill seal bags; these methods produce a narrower seal, eliminating the crimp seal at either end of the pack, therefore reducing the amount of packaging required. New sealing techniques will follow for in-pack processed foods, e.g. retort pouches with the same aim of reducing the amount of packaging used. As these new sealing techniques are developed, the need for new integrity test methods will increase.

With packaging reduction high on the agenda, there is also an increasing need for fast, repeatable, non-destructive package integrity test methods. At present, manufacturers that use destructive package integrity testing have to discard the food and package. New methods are needed to reduce this wastage.

The food market is becoming more convenient for its consumers – food on the go, food in a hurry, for example. With regard to packaging, manufacturers are beginning to develop and market easy-to-open packaging. With current aluminium laminated pouches, the consumer struggles to tear these open without tearing too far down the pack or having to resort to using scissors or a knife. New scoring technologies using lasers are being developed to enable easier opening. With laser technology, the inconsistent score found with conventional blade scoring is eliminated. Advantages of using this technology include:

- Consistent score.
- Repeatable score/tear properties.
- No tooling required.
- Laser is always sharp.
- Produces user-friendly packaging which is especially useful for the young, the old and those with disabilities, e.g. arthritis.
- Eliminates the need for tools with blades and sharp edges.

The question must be asked: Do these packs have the same seal integrity as packs that are not marketed as easy-to-open?

This is also true of resealable packaging: Can the methods of resealing a pack provide the packs with integrity after opening? Zipper technologies are becoming increasing widespread enabling consumers to re-seal the pack after opening.

Other advances include peelable films, particularly on ready-meal trays, enabling consumers to get into the pack without cutting the film with a knife or scissors. Research is needed to establish whether these films have the strength and integrity of non-peelable films and whether the same methods of package integrity testing can be used.

4.6 Sources of further information and advice

Campden and Chorleywood Food Research Association have published a number of guideline documents and reviews over the years relating to the aspects discussed in this chapter. Some that maybe of use are listed below.

Campbell, A.J. (1990). The shelf stable packaging of thermally processed foods in semi rigid plastic barrier containers. A guideline to GMP. *CCFRA Technical Manual* 31, CCFRA, UK.

Campbell, A.J. (1995). Protocols for performance testing of shelf stable heat sealable containers. *CCFRA Guideline* 7.

Gaze, J. (1992). Safety aspects of the biotest methodology. *CCFRA Technical Manual* 36.

Thorpe, R.H. & Can Defects Working Party (1994). Guidelines on the prevention of visible can defects. *CCFRA Technical Manual* 37.

Thorpe, R.H. and Barker, P.M. (1984). Visual can defects. *CCFRA Technical Manual* 10.

ASTM International (formally known as American Society for Testing and Materials) are a standards development organisation. Technical standards are produced for materials, products, systems and services. These standards are developed by technical experts representing producers, users, consumers, government and academia in over 100 countries. ASTM standards relating to package integrity testing are:

ASTM. F2391-05 Standard test methods for measuring package and seal integrity using helium as the tracer gas.

ASTM. F2338-05 Standard test method for non-destructive detection of leaks in packages by vacuum decay method.

ASTM. F2054-00 (2005) Standard test method for burst testing of flexible package seals using internal air pressurisation with restraining plates.

The British Standards Institute (BSI) is a standards and quality services organisation, which holds all British standards, European standards and International standards. A useful standard is:

BSEN 14401:2004. Rigid plastics containers. Methods to test the effectiveness of closures.

4.7 References

Al-Habaibeh, A., Shi, F., Brown, N., Kerr, D., Jackson, M., Parkin, R.M. (2004). A novel approach for quality control system using sensor fusion of infrared and visual image processing for laser sealing of food containers. *Measurement Science and Technology*, 15, 1995–2000.

Anema, P.J., Schram, B.L. (1980). Prevention of post-process contamination of semi-rigid and flexible containers. *Journal of Food Protection*, 43, 461–464.

Anon. (2006), Guidelines on good manufacturing practice for heat processed flexible packaging. *CCFRA Guideline* 50, Campden and Chorleywood Food Research Association, UK.

Bix, L., Lwaszkiewicz, R.A., Severin, J.E. (2004). *Update from the Institute of Packaging Professional's Sterile Barrier Integrity Task Group*. Business Briefing: Global Healthcare – Advanced Medical Technologies.

Gilchrist, J.E., Shah, D.B., Randle, C., Dickerson, R.W. (1989). Leak detection in flexible retort pouches. 52, 412–415.

Hanby, E., Potter, L., Creaney, C., Campbell, A. (2006). An investigation into the integrity of seals for packaged foods. *R&D Report* No 241, CCFRA, UK.

Lampi, R.A. (1980). Retort pouch: The development of a basic packaging concept in today's high technology era. *Journal of Food Process Engineering*, 4, 1–18.

Line Patrolman – www.bottle-cap-sealing.com

Potter, L., Lloyd, E., Campbell, A. (2006). The manufacture and integrity of seals for packaged foods. *CCFRA Review* No 48, CCFRA, UK.

Part II

Advances in processing technology

5

Advances in retort equipment and control systems

C. Holland, Holmach Ltd, UK

5.1 Introduction

Autoclaves, known as retorts in the food industry, have been a major technological feature of preserved foods for over 100 years. Indeed, the first documented use of heat to preserve foods in a commercial application dates back to the Napoleonic Wars of the early 1800s and has been a reliable and straightforward methodology ever since. By the time of Waterloo, British troops had canned rations and these also assisted many of the 19th and early 20th century explorers to reach their destinations. It was a confectioner, Nicholas Appert in 1804, who sealed glass jars with pitch and discovered a way of keeping food such as mutton, vegetables and milk from spoiling. Then in 1810, Pierre Durand, was granted an English patent for the metal can, which, with a few minor differences, is still the most common container for preserved foods today. But it was not until Louis Pasteur in 1854 demonstrated the effect of micro-organism spoilage on milk that we understood why heat is such an effective and safe form of preservation.

Holmach Ltd, which was previously called Peter Holland Food Machinery, has worked with nearly all of the major retort manufacturers over the last 34 years. As a business, it can rightly claim to be at the forefront of retorting technology at a time when many of the new packaging formats that we now see on retailers shelves have been introduced. Holmach continues to provide thermal processing solutions for many of the world's largest food manufacturers Over this period we have seen the introduction of a new generation of materials such as pouches, plastic pots and trays, allowing more convenience, better product quality through reduced processing time, reduction in costs for food manufacturers and better surfaces on which to provide the increasing depth of information that consumers and marketeers demand.

Apart from retorts, which are batch operations where the product is introduced in a crate or basket, there are continuous systems such as hydrostatic, continuous rotary and Hydroloc systems, but these are universally used for metal cans and have not been widely used successfully for novel applications. As such, they are not detailed in this chapter.

All of the retort types listed are available in static mode, where the basket does not move during the process, and rotary mode where the baskets revolve. This allows the containers to be turned end-over-end during the process, with benefits in quality and reduced cycle times on those products that are suitable.

Whilst as much detail as possible has been covered here, inevitably some information will not have been available; this is usually due to commercial sensitivity.

5.2 Retort process types

There are three basic retort processes: steam, falling water and full water immersion. There are also sub-divisions within each of these categories including steam/air, steam-spray, water spray and half immersion, and a brief explanation with some features and benefits is given below. It should be emphasised that all the processes, bar simple steam, will work on all of the current container formats, so there is not necessarily a 'wrong' process for novel applications. However, it is clear that some principles of transferring heat to sealed containers do have advantages over others when specifying particular packaging media. This is particularly so when choosing between rigid, semi-rigid and flexible formats. Whilst the tin can is processed in all three heating mediums, plastic pouches and CPET (crystalline polyethylene terephthalate) trays used for ready-meal production have different requirements, so it cannot be assumed that retorts installed fifteen years ago can be utilised without challenges in control and heat distribution.

5.2.1 The steam process

Steam

This is the oldest form of autoclave and is usually a top-loaded, vertical pressure vessel with straight forward controls. Pressurised steam is admitted to the chamber, driving the air atmosphere out of the top of the vessel in a 'vent' phase lasting up to ten minutes.

This valve is then closed and the temperature raised by injecting steam, creating overpressure from the temperature increase. Thus there is no independent control of temperature and pressure and because the gaseous medium is not mixed or agitated, air must be eliminated during the vent phase to ensure there are no 'cold spots' – pockets where the temperature is significantly lower due to stratification – since air and steam do not readily mix without assistance. Cooling is achieved by flooding the chamber.

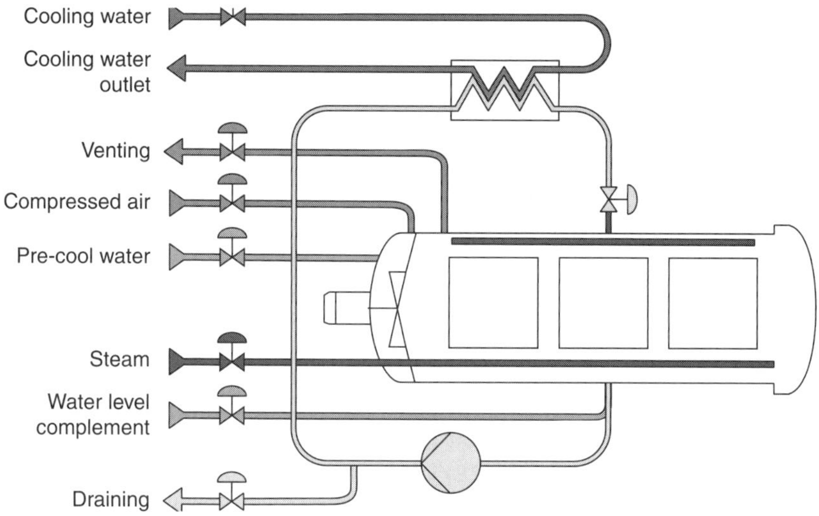

Fig. 5.1 Steam and air.

The technology used in this type of retort presents significant challenges for modern processors. The vent procedure is expensive because up to 36% of the steam required for the process is simply exhausted to atmosphere; the pressure fluctuations mean that it is virtually impossible to process pouches, semi-rigid pots and trays without distortion of the packs or the risk of cold spots occurring if compressed air is used to artificially create overpressure; and the vertical orientation of the retort means that full automation of basket loading is not achievable.

Steam/air

The steam/air process is a highly effective development of the steam process. Major differences from the steam retort are horizontal vessels with quick opening doors to facilitate basket loading and unloading, forced steam circulation and most importantly, independent control of temperature and pressure.

The steam/air process was patented by Lagarde Autoclaves in 1972. The steam is injected directly into the vessel and a reduced venting time is achieved by the use of a high velocity fan to re-circulate and mix the steam with any residual air, eliminating the occurrence of cold spots. This highly efficient process was specifically developed for flexible and semi-rigid containers, initially for military rations in aluminium foil packs, but has seen many applications on stand-up pouches and ready meals. It offers rapid heating to give the shortest process times to maximise food quality. Cooling consists of two steps – a pre-cooling step that first cools the retort chamber, gradually replacing the pressurised steam environment with compressed air, and a second step that showers the hot containers with cold water, which is re-circulated through an energy recovering heat exchanger (see Fig. 5.1). Another manufacturer of steam/air retorts is Allpax in the United States.

With the latest technology in automation, using robots, the production from these types of retorts is highly efficient, with short cycle times.

5.2.2 The water process

Steam/spray

This is a relative newcomer to batch retorting processes. In 1983, Surdry of Spain patented a process that combined steam and water in an atomised spray. The atomised environment gives a very good heat transfer on rigid containers during the come-up or heating phase, as the water transfers its heat very quickly. A fan is not utilised to mix the atmosphere. Atomising nozzles placed around the circumference of the retort take water from a pump, re-circulating the condensate, and mix it with steam directly injected into the chamber. Whilst rapid heating can be achieved, the atomising nozzles, by their design, tend to restrict the water during cooling, leading to longer processing times than conventional cascading water, immersion or water spray types. To get round this restriction, other manufacturers of this type of retort, including FMC Foodtech of Belgium, have considered and implemented separate heating and cooling circuits to maximise the efficiency of the process.

Raining or cascading water

The raining or cascading water system uses superheated water, under overpressure to achieve sterilisation temperatures. Water is heated through a heat exchanger and then pumped through a distribution plate, and showers under low pressure onto the containers below (see Fig. 5.2). This methodology is widely used in the processing of glass containers as the water can pass between the containers as it falls, transferring heat through the side walls of the container.

The cascading water process was first introduced by Barriquand of France in 1975, and is now manufactured by a number of companies including Lagarde, Lubeca, Stock Inc. and Allpax. The exchanger in the circuit allows steam that is used to heat the process water to be recovered as condensate, and it can then be returned to the boiler for re-heating. Care should be taken in the re-use of condensate because of the concentration of minerals and the resulting acidification of the steam.

Full water immersion

Another technology in widespread use is the full water immersion retort. This is comprised of a processing vessel and a pressurised water reservoir (see Fig. 5.3). At the start of the process, hot water from the reservoir floods the lower chamber and is then re-heated to sterilisation temperatures. After the cooking process, the water is returned to the reservoir, ready for the next process. A small amount of water is retained in the processing vessel and is then re-circulated and cooled through an exchanger, to be sprayed onto the products for cooling.

Manufacturers of this system include Stock Inc., FMC, Lagarde and Lubeca.

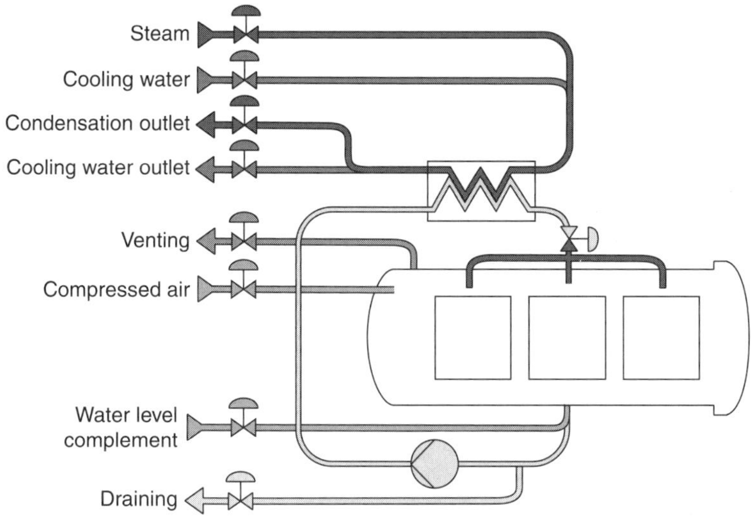

Fig. 5.2 Hot water.

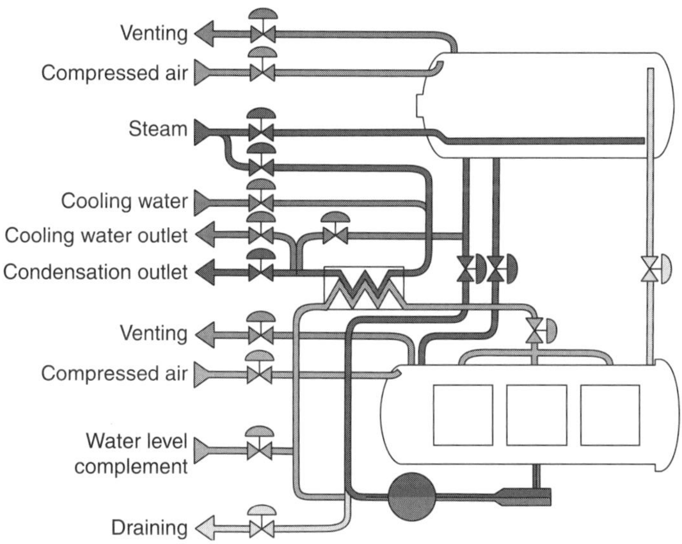

Fig. 5.3 Full water immersion.

The use of pouches and trays have tended to work against this process as the flotation of packs needs to be controlled, leading to increased costs in basket manufacture and reduced flexibility.

Half water immersion

Half immersion is where the vessel is half filled with water and part of the rotation is in water and part out of the water. This can give advantages when rotation speeds are high, as there is less turbulence created by the cage.

5.2.3 The best process?

As stated previously, it is impossible to select a single process and conclude that it is the optimal solution for every container type. However, there are distinct advantages and disadvantages when selecting a process type.

The steam/air process has been widely utilised in processing semi-rigid and pouch containers and is currently the most popular method for processing pouches, especially in the pet food sector where the highest volume of stand-up pouches is manufactured. As steam is a three-dimensional heating medium, there is virtually no difference in pouch come-up times, whatever the position of the pouch in the basket. Rapid cycle times can be achieved, as the application of energy is direct to the pack, the amount of free energy available is greater in a steam environment than in a water-based process, and there is no loss of energy through indirect heat exchangers. This leads to significant utility savings and increased output compared with a similar-sized vessel using an alternative heating method. It must be emphasized that the recirculation fan is critical to the effectiveness of the process and various safeguards to ensure fan operation are usually incorporated by the retort manufacture for both rotation and load. In addition, many novel containers have plastic laminates with gas barriers that are sensitive to water at high temperatures, so a steam process is less invasive at temperature than either cascading water or full water immersion, as the rate of condensation at the hottest part of the cycle is low.

The steam/spray process gives good distribution during come-up, but the pressure on each nozzle is critical to obtaining the correct atomisation. In addition, there are concerns with flat containers about the so-called 'umbrella effect' where, during come-up and hold, flat packs such as trays and pouches are shielded from the atomised environment due to the packs above them inhibiting the flow. Even with side sprays, good penetration cannot always be guaranteed where the layers of trays or pouches are tightly packed. Additionally, maintenance of the nozzles to ensure they remain free of scale or other particulates is critical, and the FDA recommend that a procedure is put in place to check them on a regular basis. This is somewhat difficult without entering the retort and removing the nozzles, as it is impossible to see which nozzles are blocked from the mouth of the vessel.

Falling or cascading water is highly effective at penetrating tightly packed baskets of rigid containers, such as glass or cans, as the vertical medium of water can penetrate between the containers, trickling down from top to bottom. Good thermal distribution is achieved by high flow, but the reservations are similar to those with the steam/spray process when pouches or flat pots and trays are processed. Here, the umbrella effect has to be again considered. Many of the problems with the umbrella effect can be overcome with the use of rotating retorts, where, due to the containers turning end-over-end, water can penetrate more easily. By design, the water spray process means that products take longer to heat then in a steam retort, as the water has to be heated prior to being sprayed on the containers or heated during spraying if the product is cold filled, and the heat loss throughout the exchanger should be taken into account when calculating steam requirements.

Finally, full water immersion retorts remain popular in rotary construction in the US, but have become less so in Europe where high water costs, coupled with lower process temperatures, have led to concerns about cost and thermophilic bacteria problems. In fact, the immersion process is a very good one for delicate heating for many products and containers. In heavy loads, the water immersion route does give additional support to the weight of the containers, but this is often countered by the slower process times inherent when heating such a large body of water. In addition, with the poorer nature of the seal qualities on semi-rigid and pouched containers, with contaminated water being transferred from batch to batch from 'leakers', there is an issue with colouring of white plastics, sticky residues on packs, as well as water migration through seals, that makes the application of full water immersion processes for low temperature pasteurisation more complicated.

5.3 New packaging developments and advanced control systems

With the use of thinner gauge glass and metal containers to reduce cost and with the adoption of so many plastics containers, retorts have had to change rapidly to accommodate the more accurate and shorter processes required. Control systems need to be smarter, and recently the first industrial PC-based control, SAMANTHA™, was patented and introduced by Lagarde Autoclaves. Whilst PC acquisition systems have been around to supervise PLC or digital controllers for over 15 years, the speed at which a PC can respond means that pressure and temperature control can occur to two or three decimal places. In addition, all control and data management is in real time. In practice, this means that a proper curve is achieved through the retort control rather than traditional step processes. In canning, this might be perceived as technical overkill, but with vacuum and panelled lightweight tinplate, damage can occur quite easily if there is too much fluctuation in overpressure. In addition, proper control of temperature in both the heat/hold and cooling part of the process means that process lethality during cooling can now be seriously considered.

Many of the regulatory bodies such as the FSA in the UK and FDA/USDA in the US have driven the need for more data acquisition, mainly as they are demanding better control of thermal processing in order to safeguard the consumer. Regulators are also pushing for electronic filing of this data, in a particular format, and with a PC doing the work; 'auto-filling' of a template can be completed at the end of each cycle. This document can then be emailed for filing, either directly from the machine or from a remote PC, in a readable format which does not require special software to interpret the data.

The level of control directly affects the failure rate of semi-rigid or flexible containers. Better temperature and overpressure control places less stress on the materials, ensuring the seals remain hermetic and reducing the level of wastage. This is particularly true of CPET trays, where fluctuations either lead to blown seals or deformed containers, which are then rejected by retailers. But PC control also gives additional benefits when it comes to recipe programming. Simple self-prompt menus, together with the ability to choose pre-programmed languages means that trained operators can easily see if there is a problem. Individual log-in codes, and electronic signing of all process changes, allow better traceability. Furthermore, it completely eliminates the random 'button twiddle' phenomenon so loathed by technical managers who are responsible for trying to sign-off processes that may or may not have been tampered with. There is still the backup of printed line plots against time, so there does not need to be a radical approach to the adoption of new control techniques.

Stock America have recently updated their ICON™ control system. The system is now FDA 21 CFR-Part 11 accepted. This means that food manufacturers in the US can file process data and know that it is in an accepted format. The interface also allows data and recipe control via the web.

FMC's Log-Tec™ batch retort control system also has the ability to calculate F0 values using FDA approved NumeriCAL™ Model 101. Using retort environment changes as a basis for calculation, process times can be optimised and the software built into the Log-Tec™ hardware.

Recently a number of major manufacturers have revisited the lunch-bowl format, pioneered in the 1980s. But with improved control and faster cycle times comes better perceived quality, which can now stand comparison with chilled products in a similar category, something impossible 20 years ago. And whilst the ambient sector looks with envy at the quality image of their chilled counterparts, the chilled sector, despite their large use of preservatives, natural or artificial, has been pushing to increase shelf-life to reduce the waste typical in chilled food manufacture. This has meant that many companies with no previous in-container thermal processing experience have embraced retorts to slightly improve shelf-life. The ease of control, together with accurate reporting, is critical in these environments.

The wider use of recyclable plastics in food, dairy and beverage applications also requires accurate processing, and screen-printed containers have been better served by steam rather than by water-based processing.

The stand-up or Doy-pack plastic pouch has emerged as one of the major alternative formats to glass and metal rigid containers. The most successful products have been pet food and carbohydrates, such as rice and pasta. Indeed, recent TNS data show the phenomenal growth in sales of pre-cooked rice in pouches growing at an annualised rate in excess of 25% per year. Retailer Sainsbury now allocates over 50% of its shelves for rice to the microwaveable category. Market leading Uncle Ben's expects its cooked rice sales volume to equal dried rice by the end of 2007. Stand-up pouches require careful processing as poor pressure control can lead to the failure of the gusset due to mechanical strain. As the heat penetration time to the core is so short, the process medium chosen needs to reflect this and be able to supply large amounts of energy during the heating stage very rapidly, with minimal thermal lag between top and bottom of the basket. A difference in internal temperatures of each pack will lead to varying amounts of overpressure requirement, which is impossible to resolve in one common environment. The accurate control of retort pressure is therefore crucial to success.

Other new containers such as Tetra Recart also require specific thermal processing solutions to avoid damage. A paper-based container in a wet, hot environment is not usually a good combination. However, the laminate make-up does allow full sterilisation, though care has to be taken not to damage packs and to avoid too much water at high temperature in contact with the inks behind the surface lacquer. Damage to the surface during filling can result in inks running during the process.

Most of the retort manufacturers have recognised that retort control is crucial for new-style containers and much effort is being made to use the latest electronic hardware to ensure accuracy. The ability to reduce process times to the optimum is a driver for food and consequently for autoclave manufacturers.

5.4 Advances in retort technology

Apart from control system development, retort manufacturers try to evolve their equipment continuously, so the retort available today is significantly different from the machines available 10–15 years ago. The fact that the design life of a retort is often 25–30 years also steers development. Below is a list of some of the latest retort developments.

Dual process

Most retort manufacturers are happy to argue that their process design, whether steam or water-based, is the optimum process for any type of container or product, but this can be quite easily disproved. We know in general terms that steam/air is good for plastics and pouch, water for rigid containers, such as cans and glass. But surely the optimum is a mixture of the two. Water immersion retort manufacturers such as Stock and Lubeca have offered the ability to shower or immerse products for a number of years, but recently Lagarde Autoclaves have pioneered the development and installation of dual process retorts that can use either steam or water processes at the touch of a button. The first installations have been to traditional food canners (see Fig. 5.4) with an eye to divert some of their products into more convenient or modern pack formats, such as pouch or polypropylene-based bowls. The dual process facility allows optimum processing times on rigid containers using water spray, and rapid heating to achieve short processing times for pouches. This gives the food manufacturer the ultimate retorting facility, capable of flexible, efficient manufacturing of any container shape or type.

Sterile cooling

Often fresh water is introduced to top up process water. An internal pre-heating system can now be fitted to pre-sterilise fresh water to ensure sterility in the case of sensitive packaging materials or double safety on pharmaceutical or infant food products. These exchangers can also be used as indirect cooling systems for

Fig. 5.4 Dual process retorts.

applications where water is not desired to be in contact with the container. Some vented food packs are especially sensitive, as the vent is open during cooking but needs to be closed by a fall in temperature, to ensure a hermetic seal. In traditional water cooling there would be a risk of water passing through the vent before it is fully closed, with the consequence of contaminated packs.

Rotational speed (rotary retorts)

With the requirement of shorter process times for better quality, the average speed of basket rotation is increasing. Some products are now being rotated at speeds in excess of 15rpm to obtain even re-hydration or effective starch breakdown used in sauce and soup preparation. These speeds tend to prohibit the use of water-based processes as they are subject to both centrifugal force in the case of falling water or cavitation and poor flow dynamics in the case of full water immersion, where the rotating cage acts a giant paddle. In both cases, even distribution requires the replacement of heat absorbed during the heating stage. If the water is unable to penetrate effectively to the centre of the basket, the result is poorer distribution.

Basket sizes

With pouch products, the weight of product in each basket has fallen, but because of the need to contain pouches within individual pockets to avoid damage in rotary retorts, the overall basket weight has risen. This has placed additional stresses on

Fig. 5.5 Rotary retort cage.

mechanical components within rotary retorts. Retort manufacturers have a fairly common design for the drive system of the internal cage inside rotary retorts (see Fig. 5.5). The cage shaft is powered by a rear-mounted external drive through a mechanical seal. The front of the cage is mounted on two rollers or trunnions. Traditionally, these have been mounted inside the pressure vessel. A number of manufacturers have now mounted these 'wearing' parts externally in separate pressure housings, allowing easier maintenance. The heaviest loading, however, is for processing glass jars. Line speeds continue to increase and this requires bigger basket capacities. Trunnions can now be powered by linked drives so that there is less torsion on the retort cage and less stress on the shell than with free-wheeling trunnions. Pouch basket capacities continue to grow as food manufacturers try to gain greater efficiencies from each process. The latest retorts are now capable of processing around 3600 × 400 grams of pouches per batch.

Electronic filing

As mentioned earlier in the chapter, regulatory bodies are increasingly requiring all process data be submitted electronically for filing. Photocopies of paper charts, together with handwritten batch reports, will soon be phased out, and for those who do not have modern control systems, investment in upgraded systems will need to occur. This is especially the case for companies exporting their products to a worldwide market. The onus of good manufacturing practice is now being backed up with punitive legal enforcement for breaches attributable to poor technical management of retort processes.

Shaka™ process

Whilst not a new process, the Shaka™ process from Zinetec (see Chapter 6) has been heavily promoted in the last three years. Originally sold by CarnaudMetalbox in 2001, Zinetec is now the owner of the patents concerning this novel process.

The Shaka Process is designed to reduce retorting cycle times in ambient food products by rapid oscillation of the basket to transfer heat more rapidly. Claims for better flavour, texture and colour are made and a pilot unit has been subject to in-depth trials at Campden and Chorleywood Food Research Association in the United Kingdom. Processes times are said to be cut by a factor of 10 on those products that are suitable, such as béchamel sauce, compared with traditional static retorts. Zinetec do not actually manufacture the machinery, but sell licences to the retort manufacturers. Three of these, FMC, Steriflow and Allpax, have purchased licences allowing them to industrialise the system and all have produced pilot retorts. However, at the time of going to press (early 2008), none of them had built production-ready versions. Originally designed to work with a steam process, water-based retorts are now in design.

Digital master temperature indicators

With increasing sophistication in instrumentation and control, the Food and Drug Administration (FDA) of the US has decided to review the use of digital Master Temperature Indicators (MTIs) instead of mercury-in-glass thermometers, for decades the required standard. (There has been considerable concern over the presence of both mercury and glass in food processing and production facilities with regards to food safety.) As well as the instrumentation, FDA is reviewing the protocols required for record keeping of temperature information. It is also likely that they will now include metric measurement units alongside US units for filing purposes. The FDA was due to have made a decision by the end of 2007 and the UK-based Food Safety Authority (FSA) has also begun to examine the use of digital instrumentation in place of thermometers.

5.5 Future trends

Over the last 10 years, there has been a significant change in the profile of retort users. Previously an exclusive ambient foods club, retort ownership has diversified to such an extent that many of today's major installations are on chilled pasteurised products, such as ready meals, soups and vegetables, or in component processing for use in the production of other foods, such as rice and protein cooking prior to inclusion in recipe dishes. Additionally, the food service sector has been looking at methods of extending life and making the distribution chain more efficient. Traditionally, products being delivered into the restaurant or 'pub' markets needed to be ambient or frozen, but the increasing use of microwave ovens for food preparation has increased the usage of chilled prepared meals. Hospitals and schools are also searching for more healthy foods, but with further cost savings. The reduction of salt is a major factor in this area. This has meant that the chilled food sector, with companies such as Pro-Cuisine, S & A Foods and Northern Foods Group, is investing in retort technology over traditional steam cabinet methods. The increased control and data acquisition, plus a shelf-life pick-up has meant that new markets, including export, can be contemplated. Fourteen days of shelf-life at retail outlets with no discernable reduction in quality or flavour is a major advancement. Indeed, many products benefit from a double cook or in-pack cooking, leading to a more rounded flavour.

Cheaper cuts of meat, such as shin beef or brisket, can be tenderised within the retort at low temperatures, and cooling systems mean the product can be discharged at chill temperatures across a high risk/low risk separation regime without additional handling. Because there is no down time for cleaning between processes, a retort is very flexible when it comes to handling the wide range of products found in chilled food manufacture.

The increasing use of organic and spice ingredients, some with high initial microbial counts, have led to the need to decontaminate by using pasteurisation or sterilisation, and many food ingredient manufacturers now use retorts for products such as marinades and sauces in 5 and 10 kg pouches. Hot filling at 85 °C on these types of products is not necessarily enough to eliminate yeasts, moulds and other spoilage organisms.

Many governments, due to increasing domestic health problems with obesity and cardio-vascular disease, are championing salt reduction in processed foods. Chilled products are now under intense scrutiny by officialdom and the media. Thermal processing requires no other preservatives and 'retort cooked' low-salt ready meals can now achieve long shelf-life without additional ingredients. In this way the additional cost of the thermal process can be obviated by a reduction in the cost of preservative and stabilising ingredients.

Another recent application has been the pre-cooking of ham shanks and belly pork joints. The product, once cooked and cooled in a bag, is then dispatched to stores with hot food counters, then finished off in roasting cabinets. The process not only reduces cooking time, but also ensures product safety and good quality without variability.

Seafood processors are able to use retorts for gentle controlled defrosting; the product is then prepared and finally cooked or pasteurised in the retort prior to despatch. The two largest chilled seafood producers in Europe are now using retort technology to process fish ready meals, taking shelf-life from a market-limiting 4 days to an acceptable 17 days, whilst still ensuring that the quality is the same or better compared with chilled assembly.

With the continuing consumer drive towards more convenience, vegetable growers are now reaping additional margin by processing vegetables to a ready-to-eat state in pouch and tray. Chipped, boiled, mashed and even jacket potatoes, part cooked, are flooding the chilled foods market and they require only reheating in a microwave. As well as potatoes, other root and green vegetables are processed to offer maximum convenience.

With steam/air retorts it is also possible to steam 'unpacked' product, loose in stainless retort trays, such as cauliflower, broccoli, mange-tout and leaf products including spinach. The aim is not to increase shelf-life, but to give minimal cook and then use cryogenic gas cooling to maintain colour without damaging the structure of the delicate product. Whilst more expensive than traditional cooking methods, structure, colour and nutrient loss are minimal.

5.6 Conclusions

Having covered some of the latest trends in retort design and market requirements, it should be stressed again that one retort process or design is not necessarily applicable to any and every product. Retort technology tends to evolve out of previous designs because heat processing has such a long track record of effectiveness and safety. It is straightforward to validate retort processes and it is therefore unlikely that retort manufacturers are going to adopt high-risk strategies. It is the attraction to safe processing that is drawing more and more food companies to choose to use retorts for novel packaging formats, where requirements can change rapidly, because retorts are highly flexible in coping with changed container materials, shapes and their application.

Convenience is the major driver in the food industry today. As this is unlikely to change, more and more delicate packs will emerge, to tempt consumers to associate thermally preserved foods with fresh chilled alternatives. More printed laminates will appear on processed packs as secondary packaging is eliminated, and this means retort processes must become more sophisticated in order to avoid damage to container and product.

5.7 Sources of further information and advice

Lycee Nicholas Appert
M. N. Ramesh – *A Handbook of Food Preservation*, second edition.
Florence Fabricant – February 28, 1996 – *The New York Times* Company
FDA – Guide to Inspections of Low Acid Canned Food Manufacturers, Part 2.
BPC report. Issue 37 September 2006
Zinetec Ltd – An Introduction to the Shaka Process
Federal Register/Vol 72 No. 49 – 14/3/07

Websites

Lagarde Autoclaves – www.lagarde-autoclaves.com
Stock Inc. – www.stockamerica.com
Steriflow – www.steriflow.com
Lubeca – www.lubeca.com
FMC – www.fmcfoodtech.com
Surdry – www.surdry.com.

6

The Zinetec Shaka™ retort and product quality

R. Walden, Zinetec Ltd, UK

6.1 Introduction – current retorting systems and their limitations

Current retorting systems fall into four main categories; these are batch and continuous, with or without agitation.

Batch retorts without agitation (static) capable of sterilising containers of food product at temperatures greater than 100 °C have been in existence from the middle of the 19th century. These early versions used steam as the heat transfer medium, as indeed do the majority today. Later developments included the use of water, steam/air, raining and spray water systems, all of which allowed the processing of containers that required counter-pressure so enabling them to retain their shape and integrity.

The hydrostatic retort is the most common type of continuous cooker without agitation and is generally used for cans where no counter-pressure is required, though versions providing overpressure are available.

Continuous retorts (cookers) with agitation were first introduced in the 1920s by FMC in the form of the Sterilmatic. Agitation of the contents was induced by rolling the cans, thus shortening the process compared with static.

Batch retorts with agitation were the latest of the four categories to be introduced. These generally use 'end-over-end' rotation, a system developed by Laverne Clifcorn and colleagues around 1950. They found that this method of rotation was generally better than axial rotation as utilised in the Sterilmatic retort in inducing movement of the product within the container, and hence further shortening process times. This was the last major innovation to provide a significant reduction of process times before the Shaka™ process.

Advances in retort control systems and other areas enabled processes to be optimised but this has not generally resulted in reductions in process times sufficient to allow the production of the significantly better products that the market demands.

Both the agitation methods described above suffer from the fundamental limitation that the force employed to cause the movement of the product within the container is a balance between gravity and centrifugal force. As the speed at which the container rotates increases, the process time reduces until an optimum is reached when the headspace air bubble moves through the product to cause the maximum amount of mixing. If the speed of rotation is increased further, the process time lengthens until ultimately, if the speed is sufficiently high, no mixing occurs and the process times will be virtually the same as with a static process.

The Shaka™ process is fundamentally different, using reciprocal agitation in addition to gravity. These reciprocal forces are typically two to three times that due to gravity and it is the use of the sum of these forces to cause considerably more movement of the product within the container that shortens the process. It is this much shorter process that enables the production of improved products.

6.2 The Shaka™ process

6.2.1 Concept and initial development

Considering a can of food product in a steam-filled retort, it is evident that there is plenty of thermal energy available on the inner surface of the can wall; the thermal barrier is the product. The barrier is greater for conduction than convection products. If the system is likened to a heat exchanger, it is the movement of the product through the exchanger that is critical to its performance and the only way to improve the movement of the food product within the container is to devise a better method of agitation.

In order to be able to see movement of the product in the container, a glass jar of around 400 g capacity (75 × 110 mm) was selected, filled with water containing a few chips of coloured plastic, and the jar closed leaving some headspace. Agitating the jar by hand to simulate axial and end-over-end rotation demonstrated the amount of movement these method generated, the movement of the water being highlighted by the plastic chips. The jar was then agitated in a variety of different ways, the simplest, and what appeared to be as effective as any, was to place the jar on its side and reciprocate it to and fro with a stroke of around 100 mm. The movement of the water in the jar generated by this motion was far greater than by either of the methods of rotation.

The agitation experiments were repeated using light-coloured cooking oil in the jar to simulate a thicker product, again using chips of coloured plastic to highlight the movement. The results were similar to those with water. It appeared obvious

Table 6.1 Static and rotary processing compared with reciprocation by hand

	Static	Rotary – 25 rpm, end-over-end	Reciprocating (Shaka™)
Heat-up time to 120 °C from steam on	80 min	23 min	5 min
Cooling time to 38 °C	52 min	12 min	5 min

Retort temperature – 121 °C. Can – 73 × 110 mm. Product – 5% bentonite. Headspace – 8 mm.

Table 6.2 Heat up time to 121 °C from steam on

	10 mm stroke	25 mm stroke	50 mm stroke	75 mm stroke	100 mm stroke	125 mm stroke
60 rpm				34	31	22
75 rpm				29.3	4.8	9.7
90 rpm			7.0	4.2	4.0	3.4
120 rpm		6.0	4.2	3.6	3.6	3.6
146 rpm	27	9.1	4.4	3.6	3.4	3.2

Retort temperature – 127 °C. Can – 73 × 110 mm. Product – 5% bentonite. Headspace – 8 mm. Retort come-up time – 2.9 minutes. rpm = revolutions per minute of the crank reciprocating the basket.

from the much greater movement of the product within the container that horizontal reciprocation might well increase the rate of heat transfer into the product and hence reduce the process time.

The next stage was to carry out some heat penetration experiments. To keep these as simple as possible, a table was placed in an existing small horizontal batch retort, on which a wheeled carriage holding two cans could be reciprocated. The reciprocal motion was provided by means of a rod of stainless steel, which passed into the retort via a sliding seal and which was attached to the carriage, the outer end of the rod being reciprocated by hand. The results are shown in Table 6.1. A bentonite suspension is a commonly used product simulant which, because of its inert nature, is able to be thermally processed repeatedly without its characteristics being changed. This allows the same test containers to be processed many times.

As will be seen from Table 6.1, the heat-up and cooling times were much faster than either static or rotary processing and so it appeared sensible to motorise the drive system so that reciprocation could be investigated systematically. The motorised drive system consisted of a crank disc designed to give strokes from 10 to 125 mm, with frequencies up to about 150 rpm (revolutions per minute of the crank reciprocating the basket).

Similar experiments to those above were then carried out across the range of strokes and frequencies as will be seen in Table 6.2. The results shown are the time in minutes from when steam is turned onto the retort, until a thermocouple in the centre of the can reaches 121 °C. For comparison, the time with no agitation was 62 minutes. The heat-up times using the faster agitation rates confirmed the results found with hand agitation and demonstrated that this method of agitation appeared to have significant potential. In view of these results, further experiments were carried out covering strokes up to 200 mm, various headspaces, thermocouple positions, can sizes and orientation, and concentrations of bentonite to simulate food products of varying viscosities. The results of the experiments indicated the following:

- Longer strokes appeared advantageous with higher bentonite concentrations and with smaller headspaces.
- A headspace was necessary as with other agitating retorts to allow the product to move within the container. Sensitivity to variations in headspace did not appear too great.
- Heat-up time did not appear to be affected by the position of thermocouple, though a variety of positions were used.
- Can sizes of 65 × 101 mm, 73 × 110 mm and 99 × 119 mm showed surprisingly little difference in heat-up time. With all three cans, the best orientation seemed to be for the cans to be lying on their sides and for the reciprocation to be along the longitudinal axis.
- Increasing the concentration of bentonite caused the heat-up rate to slow but these initial experiments showed that this could be largely offset by increasing agitation rates or headspace.

A starch solution plus some real products were processed to ensure the effectiveness of this type of agitation was not limited to bentonite (Table 6.3). From these results there was good evidence to indicate that reciprocal agitation could dramatically reduce the process times across a range of food products.

Based on this preliminary work, the Shaka™ system seemed to have considerable potential; in fact, it appeared almost too good to be true, so preliminary microbiological challenge experiments were undertaken to see if the lethality of the processes determined by thermocouples and expressed as F_0 appeared correct. Cans were filled with a nutrient broth, inoculated with spores of *Bacillus stearothermophilus* to three different levels, and processed through the Shaka™ retort to the appropriate F_0. The cans were then incubated and survivors recovered. The results showed good correlation between F_0 and bacterial kill, though the number of test cans was small as the pilot retort could only process two cans at a time.

Table 6.3 Comparison of process times for variety of products

Product	Process lethality F_0	Process time Shaka™	Process time static
Bentonite, 5% in water	10	3.4 (125 × 120 rpm)	47
Pureflo starch, 5.5% in water	10	4.2 (125 × 120 rpm)	62
Carbonara sauce	8	4.7 (125 × 120 rpm)	80 @ 121 °C
Chicken soup	8	5.0 (125 × 146 rpm)	85 @ 121 °C

Process time – time in minutes from when retort reaches process temperature (127 °C unless stated) until process lethality reached. Retort come-up time – 2.9 minutes. Can – 73 × 110 mm. Headspace – 12 mm.

Fig. 6.1 Crank and slider drive system for Shaka™ retort.

At the end of this initial stage it was concluded that the reciprocal agitation methods employed resulted in a large and consistent improvement in the heat transfer rates, as shown by the reduction in heat-up and process times, compared with anything previously possible for in-can sterilisation. In fact, the resultant sterilisation times were near to those achieved on some types of UHT plants. It therefore appeared that there was the likelihood of a simple in-can sterilisation system that could virtually match UHT for product quality for conduction products.

Despite the simplicity of the concept, initial literature and patent searches indicated the approach to be novel. A larger, purpose-built retort was therefore constructed.

6.2.2 The first Shaka™ retort

From the work done on the prototype retort, it was evident that the new retort should be sufficiently large to hold at least 50 73 × 110 mm cans and have a holding system that held each container separate from its neighbour. The holding system needed to be flexible so as to accommodate a variety of container sizes and types. The drive to be capable of strokes from 25 mm to 300 mm and reciprocation rates up to about 250 rpm. The retort needed to vent and come up to temperature quickly, and have capability for spray or flood cooling. From the photograph (Fig. 6.1), it can be seen that the reciprocating drive method adopted was a crank and slider mechanism, driving a basket via a drive shaft entering the retort through a boss containing seals and bearings. The basket ran on rails, guided by flanged wheels (see Fig. 6.3).

Table 6.4 Microbiological challenge experiments comparing F_0 and F_s.

	Thermal lethality (F_0)	Bacterial kill (F_s)
B. stearothermophilus	12.6–13.2	13.0–18.1
Cl. sporogenes	4.4–5.7	4.0–7.6

6.2.3 Validation and development

Mechanical, venting and heat distribution checks were carried out which, after a few modifications, proved highly satisfactory. The retort had good heat distribution, vented in around a minute, and came up to 130 °C in about 1.5 minutes, depending on thermal load.

As stated previously, the Shaka™ system seemed 'too good to be true' and, though the preliminary microbiological challenge experiment gave encouragement, it needed repeating on a much larger scale. This was now possible with the new retort.

The experiment covered two agitation conditions, two process temperatures and two test organisms (*Bacillus stearothermophilus* – TH24, and *Clostridium sporogenes* – PA3679). A total of over 1100 inoculated cans were used. As can be seen from Table 6.4, there was good correlation between the lethality in the cans, as measured by the thermocouples, and that from the survival level of the test organisms. It was evident from these results that further systematic evaluation of this method of retorting was warranted, covering the full range of conditions of which the new retort was capable.

A series of experiments were undertaken using the most widely used size of can, 73 × 110 mm, covering a range of products and headspaces. The products selected were various concentrations of bentonite (5, 7, 8, 9 and 10%) in water and headspaces (4, 8 and 12 mm, gross). All the concentrations of bentonite selected were sufficient to produce a product that, in a static retort, would heat largely by conduction. After some experimentation, the thermal treatment decided upon was to measure the time taken from when steam is turned on to the retort to when a thermocouple in the centre of the can reaches 120 °C, with the retort controller set to 130 °C. When the test cans had all reached 120 °C, cooling commenced using water sprays and the time was measured for the thermocouples to reach 40 °C. This test was carried out on the different products and headspaces across the range of strokes and frequencies available. The results were drawn up in the form of Table 6.5 which is shown as an example.

The tables of results for the different variables were similar to Table 6.5, all showing initially a rapid fall in heat-up time as the intensity of agitation increased, followed by a stable area with relatively little change. The intensity of agitation could either be in the form of a shorter stroke with higher frequency or *vice versa*. The intensity of agitation was equivalent to the maximum acceleration (equal to $\omega^2 r\,(1 + r/l)$ where ω = angular velocity in radians, r = radius of crank, half stroke, and l = length of connecting rod). This was conveniently expressed in g (acceleration due to gravity). Figure 6.2 shows a selection of strokes and frequencies from Table 6.5 calculated as g and plotted versus heat-up time to 120 °C. From the

Table 6.5 Heat-up times to 120 °C from steam on for variety of agitation conditions

	40 rpm	60 rpm	80 rpm	100 rpm	120 rpm	140 rpm	160 rpm	180 rpm	200 rpm	220 rpm	230 rpm
25 mm				37:00	7:05	7:00	7:08	6:55	6:24	5:10	5:22
50 mm			35:07	6:24	5:24	5:06	4:23	3:48	3:21	2:49	2:51
75 mm			6:56	4:35	4:15	3:23	2:54	2:44	2:38		
150 mm		8:15	3:43	3:20	2:48	2:26	2:07	2:03			
225 mm		5:46	2:58	2:44	2:15	1:59	1:55				
300 mm	27:02	4:12	2:28	2:18	1:53	1:50					

Product – 8% Bentonite. Headspace – 12 mm. Retort controller set point – 130 °C.
With no agitation (static), time to 120 °C = 50 minutes.
The top row is the revolutions per minute of the crank reciprocating the basket, the left-hand column is the movement (stroke) of the basket in millimetres and the figures in the remaining boxes are the times in minutes and seconds to reach 120 °C.

Table 6.6 Cooling time from 120 °C to 40 °C for variety of agitation conditions

	40 rpm	60 rpm	80 rpm	100 rpm	120 rpm	140 rpm	160 rpm	180 rpm	200 rpm	230 rpm
25 mm					9:33	8:09	6:58	6:51	5:23	4:20
50 mm			22:10	6:57	5:34	4:55	4:03	3:34	3:13	2:52
75 mm			7:38	5:06	4:06	3:42	3:10	2:49	2:34	2:27
150 mm		7:06	4:25	2:56	2:37	2:15	1:59	1:49		
225 mm		6:19	3:48	2:38	2:11	1:54	1:48			
300 mm	14:00	4:52	3:05	2:20	1:59	1:47				

Product – 7% bentonite. Headspace – 12 mm.
With no agitation /(static) – cooling to 40 °C = 43.25 minutes. Units as for Table 6.5.

graph, it will be seen that, with this particular product and headspace combination, there is a steep fall in the heat-up time to around 0.5 g and that, once past 1.0 g, there is very little subsequent change. This area indicates that the process is stable, meaning that small variations in either the stroke or frequency will have little effect on the rate of heating and hence the process time. With other combinations of product and headspace the form of the graph will be similar, but the position where the steep fall changes to the stable zone can change. Table 6.6 is similar to Table 6.5 but these results are times for cooling to 40 °C. As can be seen, the results show the same pattern as for heating.

This work had provided sufficient understanding of the process for a patent to be drafted. As has already been mentioned, initial searches had revealed no prior art. However, whilst the above development work was being carried out, further searches uncovered a number of patents, a US patent from 1938[1] being the most important. This described an agitating retort, which used reciprocation. However, with the new level of understanding of the process, it was evident that the intensity of agitation used would not have been sufficient to produce the reductions in heat-up and process times generated by the Shaka™ process. The patent granted on the

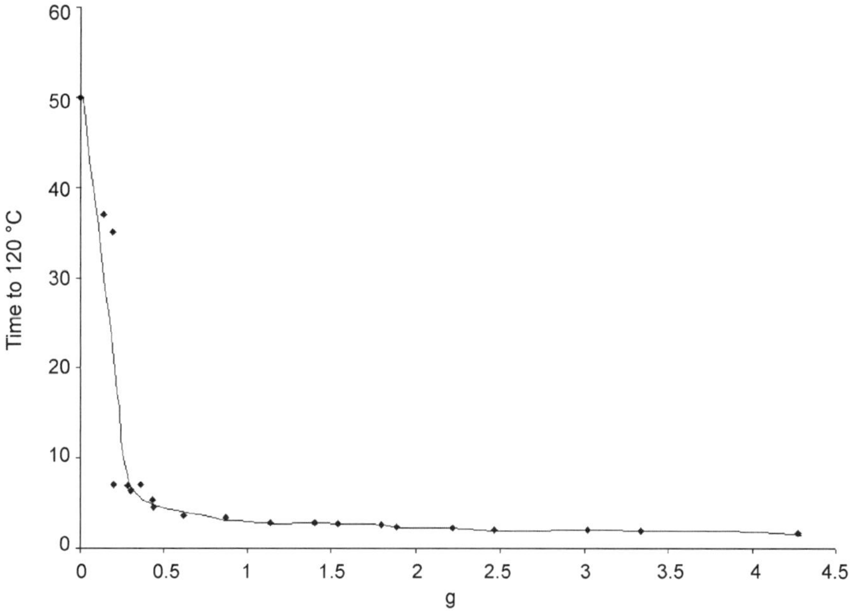

Fig. 6.2 A selection of strokes and frequencies from Table 6.5 calculated as g and plotted versus heat-up time to 120 °C.

Shaka™ process covers the reduction in heat-up time compared with static, the stability of the process and the minimum levels of maximum acceleration required.[3] Further patents on the Shaka™ process have been applied for or granted, covering further aspects of the process as well as process engineering and mechanical engineering aspects.

The work described in this section helped resolve three major issues:

(i) A larger retort using Shaka™ principles could be built and operated successfully.
(ii) In a much larger trial, the short sterilisation times demonstrated by the Shaka™ process, when measured by heat penetration methods, are matched by the bacterial kill achieved in inoculated cans.
(iii) The process works across a range of bentonite concentrations, headspaces and agitation conditions and, if the correct conditions are used, large reductions in process time can repeatably be achieved.

6.2.4 Food products

As noted, the work in the previous section was entirely based on various concentrations of a food product simulant – bentonite dispersion. The results in Table 6.3 showed that there was every indication that real food products would behave similarly. However, as the greatest benefit of the Shaka™ process would be its ability to produce much higher-quality food products (because of the shortness of

Table 6.7 Comparative sterilisation times for béchamel sauce

	Come-up	Process F_0 6	Cool to 40 °C	Total
Static	5.0	85	60	150 min
Rotary	6.0	36	36	78 min
Shaka™ process	2.0	3.5	5.8	11.3 min

the thermal process), it was necessary to confirm this by running trials with a much wider range of products. The greatest benefit was likely to be for conduction products, where processing times in conventional retorts are the longest. There would be advantages with convection products but not as great.

The results in Table 6.7 are for béchamel sauce, a conduction product, in 73 × 110 mm cans processed in static, end-over-end rotary and Shaka™ retorts. The rotary process conditions were optimised with regard to temperature and speed of rotation (121 °C and 15 rpm) by comparing the whiteness of the variables after processing. The same temperature was used for the static process. The conditions for the Shaka™ process were 130 °C and 150 mm × 120 rpm agitation. The differences in process times to $F_0$6 were dramatic, better than 20-fold compared with static and better than 10-fold compared with rotary. The resultant sauce was still white, virtually indistinguishable from the unprocessed product, whereas the rotary retorted samples were noticeably brown, and the static an even darker brown, especially against the walls of the can.

The reduction in process times shown in Table 6.7 are typical and have been seen across a wide range of products, including soups, sauces, ready meals, baby foods and desserts. The reduction is partly due to the more efficient mixing produced by the Shaka™ method of agitation and partly due to the higher temperatures that the better mixing allow to be used. If process times for 121 °C and 130 °C are compared, typically a little under half the improvement is due to the better agitation with the remainder due to the higher temperature. In the béchamel sauce example, higher temperatures (125 °C) for the rotary process were tried as part of optimisation. However, due to the more limited mixing that rotary agitation provides, more burning occurred against the wall of the container than at 121 °C, producing an inferior product even though the process time was shorter.

As was mentioned earlier, the heat treatment products receive from the Shaka™ process are similar to those from many UHT systems. Hence product quality from the two systems should also be similar. A comparative study carried out across a range of products confirmed this to be the case.

Considering products containing particles, the reduction in process times depends on the size of the particle as heat can only get to the centre of the particle by conduction – the bigger the particle the longer the process. Obviously, with solid packs, such as tuna or ham, there is no reduction in process time compared with static retorting. From work done on the Shaka™ process, particles of 3–4 mm diameter or less heat at the same rate as the carrier liquid and so make no difference to the process time, whereas larger particles (up to around 12 mm diameter) may add a minute or so. However, if particles are significantly bigger than 12 mm, it may mean that the process has to be lengthened to the extent that there is a danger of the carrier liquid (sauce) becoming over cooked, together with the outer part of the particle.

Another factor to be considered with particulate products is possible damage to the particles because of the vigorous agitation. A number of factors contribute to the likelihood of damage, two of which are:

(i) The nature of the particle and the effect of the process on this nature, in other words how tough is it before and after cooking and so how able is it to stand up to the agitation. For instance, uncooked potato is hard and quite tough but on cooking becomes fragile.
(ii) Process conditions: the time, temperature and intensity of agitation. These are interrelated but, particularly, decreasing the agitation rate and adding the small amount of extra time necessary to achieve the required lethality can be beneficial.

A programme of work is under way aimed at reducing particle damage during the process. This is giving encouraging results.

The Shaka™ process has now demonstrated significant advantages in quality across a very wide range of particulate and homogeneous products compared with the existing methods of static and rotary processing and can match UHT in some areas.

6.2.5 Container types and sizes

As will have been noted, most of the initial work on the Shaka™ system used 73 × 110 mm cans, these being the most common retail size. However, as part of the commercialisation of the process, it was necessary to be able to process a variety of other containers, including larger cans, glass jars, plastic containers and pouches. The first stage was, therefore, to convert the Shaka retort from steam-only to a system where counter pressure could be used during the heating and cooling phases. The only possible method that did not require major modifications was steam/air. Experiments showed this to work well, with only minor modifications to the air inlet system being required. The movement of the basket was found to provide sufficient mixing of the steam and air to give satisfactory thermal distribution, with no tendency for the steam and air to separate. Modifications were also carried out to the cooling system so that two-stage cooling was possible, using hot water for the first stage to prevent the cracking of glass containers due to thermal shock.

Processing large cans (153 × 178 mm or similar) proved easy, held separately and on their sides as with the smaller cans. These processed well at 130 °C using steam and standard pressure cooling. With homogeneous conduction products or products containing small particles, process times were about 10 minutes to $F_0$8–10, compared with 3–4 hours in a static retort. As can be imagined, enormous improvements in product quality resulted.

A variety of sizes of glass container have been successfully processed with both

Table 6.8 Comparative sterilization times, 99 mm diameter plastic bowl and metal can

	Come up (min:sec)	Time F_o 5 (min:sec)	Cooling (min:sec)	Total (min:sec)
Plastic bowl	1:40	6:20	3:20	11:20
73 × 110 mm can	2:00	2:20	1:50	6:10

Product – 7% bentonite. Process temperature – 130 °C. Agitation – 150 mm × 150 rpm.

push-on and screw-on lids, using, steam/air and two-stage cooling. The jars have all been significantly taller than their diameter so have all been processed on their sides and shaken along their long axis. It has been found that process times are 50–100% longer than for cans due to the lower conductivity of glass. Jars with larger diameter lids for a given volume process faster.

Plastic containers with both heat-seal closures and double-seamed ends have been processed. The heat-seal containers have all been shallow trays and have been processed with the closure uppermost, the tray being supported by its flange in a plate with suitable apertures to take the body of the container and shaken along the longer axis. Processing was carried out using steam/air and spray cooling, with selection of the agitation rate and control of the overpressure being fairly critical to produce undistorted containers. Again, due to the poor conductivity of the plastic compared with metal, the process times were somewhat longer. Most of the work on double-seamed plastic containers has been with 400 g bowls, closed with 99 mm diameter aluminium easy-open ends. By experimentation, it was found that processing was best carried out with the double-seamed end uppermost and agitation in line with the end. Again, processing was carried out with steam/air and spray water cooling. With these containers, control of overpressure was found not to be so critical.

From Table 6.8, the effect of the lower conductivity of the plastic, compared with metal, can be seen.

Both stand-up and pillow-style pouches have been processed using steam/air and spray water cooling. As would be expected, the direction of agitation is along the longer axis, with the pouches lying on their backs. To achieve undamaged pouches at the end of processing, it is vital not only to select the correct agitation rate and overpressure profile but the pouches must be held in custom-made racks in which each pouch is securely located.

6.2.6 Critical factors and determination of process conditions

For the Shaka™ process, the critical factors specifically associated with an agitating process are largely the same as for rotary retorting and include fill weight/headspace, product consistency (viscosity), solid/liquid ratio, particle size, agitation rate and container orientation. In determining the thermal process required, the variability of all these must be known in order that the worst case conditions can be used when the production process is determined. Though the Shaka™ process may possibly be more sensitive to variations in some of the critical factors, rotary processing had been carried out safely for many years before the advent of fillers of the accuracy of those available today – or of inline check weighers, viscometers and sophisticated data analysis by computer. A modern filling line should, therefore, be able to produce filled and sealed containers of sufficient consistency that not only assures safety but also gives a safety margin such that the quality of the product is not compromised.

Before determination of the thermal process, it is first necessary to decide upon the container orientation, fill weight and agitation rate. The most appropriate orientation of the container to achieve the best mixing at the lowest rates of agitation may be obvious if the container has one axis significantly longer than any other. If this is not the case, the best orientation can be determined only by experimentation.

A number of factors will come into the decision as to what fill weight to use, but the maximum fill weight (minimum headspace) will be determined by the requirements of the process. The Shaka™ process, along with all agitating processes, requires a headspace to allow the product to move within the container. The minimum required will depend on the nature of the product, type of container and agitation rate.

For homogeneous products, the agitation rate selected will depend largely on the viscosity of the product and the minimum headspace but will generally be on the high side so as to facilitate the shortest process. However, if the product contains particles, then the nature of the particle will need to be considered. Delicate particles (for instance, many vegetables when cooked) can be damaged if agitation rates are too high, so the rate chosen will have to be determined by experiment to get the best balance between the length of the process and particle damage.

Having decided upon the container orientation, fill weight and agitation rate, the presence of a 'cold spot' must be considered. With the Shaka™ process, because of the good mixing induced by the type and intensity of agitation employed, all the contents of the container heat evenly and so, generally, there is no 'cold spot'. This cannot be assumed to be the case without confirmatory data.

Finally, it is necessary to decide on the process temperature to be used. The Shaka™ process generally uses higher temperatures than more conventional retorts, around 130 °C being typical. The main arbiter will be product quality and this should be product quality determined on containers filled to worst case conditions.

The method of carrying out heat penetration determinations is fairly straight forward as conventional equipment can be used as long as the thermocouples are thin (1.5–2.0 mm diameter). The only difference is that the leads to the thermocouples must be arranged to allow for the movement of the containers and the data from the thermocouples needs to be recorded at least every 2 seconds because of the speed of the process (see Fig. 6.3).

The method used to determine process times in particulate products is similar to that used for rotary processing. To determine the rate of heat penetration, the largest particle from the product is impaled on the thermocouple. The particle will almost certainly need to be held onto the thermocouple in some way to prevent it being displaced by the agitation.

Fig. 6.3 Shaka™ retort showing wheeled basket of cans with thermocouples attached for determination of thermal process.

It has been found with products containing dry particulate matter such as pasta, that, because of the shortness of the process, it is possible they are not fully re-hydrated during the heating phase of the process. It is therefore vital to check that this is not the case or sterility will be compromised.

6.3 Product quality and the Shaka™ process

As has already been said, the major advantage of the Shaka™ process is in its ability to produce better quality food products. This has been demonstrated with soups, sauces, ready meals, spreads, dips, desserts, beverages, chopped vegetables, baby food and pet food. The step change in quality that the Shaka™ process can produce is likely to have major effects on the food market. It gives packers in the ambient shelf-stable sector the opportunity to produce a wider range of higher-value products with better flavour, colour and texture. The much lower thermal burden imposed on products by the Shaka™ process means that flavour enhancers such as salt and artificial colours can be reduced or eliminated. There are also indications that the same applies to stabilisers, emulsifiers and modified starches, leading overall to products with much 'cleaner labels'. The improvement in quality seen is often such that products produced by the Shaka™ process can compete with similar products produced by freezing, aseptic processing or for distribution under chill. This gives producers new opportunities, and the consumer increased choice, especially welcome in countries where chill and frozen distribution are not as well developed as they are, for instance, in Western Europe.

Environmental issues and the need to reduce carbon footprint have claimed the headlines in recent months. For example, Wal-Mart, Tesco, Sainsburys and Marks and Spencer have all announced their commitment to reducing carbon footprints, some aiming to be carbon neutral by 2012. Shaka™ process ambient preserved foods are considerably 'greener' than both chill and frozen – chill mainly due to the very high levels of wastage caused by the short shelf-life and frozen due to the energy needed to freeze and keep frozen. This should provide extra incentive to grow Shaka™ processed foods at the expense of chilled and frozen.

Rising levels of affluence, created, to a degree, by working longer hours and both partners in the family earning and so having less free time, have generated demand for pre-prepared premium foods, a major growth area for the supermarkets. This sector does not seem as price sensitive as the commodity area. For instance, in the UK a pack of premium soup often costs many times the supermarket's cheapest offering. Therefore, if products of appropriate quality can be produced, there will be benefits for the environment, the packer and the consumer.

6.4 Commercialisation of the Shaka™ process

As the Shaka™ process is patented, the strategy adopted by Zinetec Ltd to commercialise the system was to license both the food packers to use the process and suitable retort manufacturers to produce the equipment. It seemed evident that, in order to interest the retort manufacturers, it was first necessary to interest food packers, so major food manufacturers therefore became the initial target. Once interest was established, three retort manufacturer were approached and offered licences to manufacture Shaka™ process retorts. Three were selected to give food manufacturing customers a reasonable choice of retorts, provide a certain amount of competition and, at the same time, give each retort producer a potentially valuable market share to justify its initial design, building, testing and marketing investment. The three selected were Steriflow, France; Satori, Germany; and Allpax in the USA.

Steriflow, formerly Barriquand, makes traditional static and rotary retorts and specialises in raining water processes; it operates internationally. Satori also makes traditional retorts but specialises in rotary retorts with full water immersion; it also operates internationally. Allpax makes a wide range of retorts and handling equipment for the food industry; the company is part of the Pro Mach group and has concentrated on the American market. The three retort manufacturers introduced their first pilot machines in 2006/7, with full-scale production machines planned for 2008.

The interest of the major food manufacturers was aroused not only by the ability of the process to produce quality products but also by its simplicity and low cost, especially compared with UHT/aseptic systems. Sterilisation costs for the Shaka™

process will be similar to other batch retorts but the cost of the retorts for a given rate of production should be less, due to a Shaka™ retort being able to process many times faster. For instance, a Shaka™ retort of similar phyisical size to a rotary retort may cost somewhat more but has a much higher capacity. The ability of the Shaka™ process to produce many batches per hour means that the handling of the baskets of containers in and out of the retorts will need to be much faster than was necessary with conventional retorts to maximise throughput. The Shaka™ processes simplicity means that installation, commissioning and training costs will be little higher than for other batch retorts and, as it is only a retort, it should be compatible with the rest of existing filling lines.

The process opens up a number of marketing opportunities, including, as already mentioned, the possibility to compete with chill and frozen products in some product areas, the ability to make products previously not possible by in-container sterilisation, the chance to significantly upgrade existing products where appropriate, and perhaps for the first time enable production of a wide range of premium products in the food service sector.

A number of licences have already been signed with food companies in both Europe and the US to allow them to explore the technology, with commercial licences to produce products to follow in the near future.

6.5 Future trends

With the expected continuing growth of the premium sector, the potential market for better quality ambient preserved foods is very large. Fresher tasting, better looking, more valuable ambient foods, produced at cost levels comparable with current ambient preserved foods, is the strongest 'driver' for the technology. The process has been trialled by many of the most sophisticated food companies in the world and has performed as claimed. Shaka™ process retorts built by three reputable manufacturers have also performed as claimed. Until very recently, a major constraint was the lack of Shaka™ retorts, both pilots for product and process development, and particularly full-scale production machines that make such developments worthwhile. Pilot retorts in various sizes are now readily available, with production machines available in the near future. Zinetec sees 2007 as the year of pilot retort sales and development licences. Many more pilots should be seen in 2008, and the first production units sold, together with commercial licences. Rapid commercialisation should be seen in 2009 as more companies appreciate the benefits in the market. The overall prospects are good in both the US and Europe.

In the slightly longer term, if predictions for global warming are anything like correct, 'green' issues could become overwhelmingly important and in this respect thermally sterilised, long shelf-life, ambient stored products score well compared with chill and frozen, so forcing the market in that direction. In the two largest countries in terms of population, China and India, together with other parts of what is currently the developing world, it is doubtful whether widespread chill and frozen distribution and storage will ever exist. The reason will not only be environmental pressure but also the high cost of setting up the infrastructure and geography – the size of the countries concerned. It is noticeable that chill products make up a much smaller proportion of supermarket space in the US than in Western Europe, where the countries are small with high population densities, making the distribution of short shelf-life chill products more viable. Therefore, in the emerging countries, the food market could largely split between fresh and long shelf-life ambient storage, including thermally processed products, with the Shaka™ process having a significant role.

6.6 Sources of further information and advice

Richard Walden, Zinetec Ltd, 22 Highworth Road, Faringdon, Oxfordshire SN7 7EE, UK
+44 (0) 1365 240 650
rwalden@shakaprocess.com

John Emanuel, Utek Europe Ltd, 20, Regents Park Road, London NW1 7TX, UK
+44 (0) 20 7482 4226
john.emanual@utek.co.uk

6.7 References

1 US Patent number 2 134 817.
2 European Patent number EP 0 804 095 (US Patent number 5 857 312).

7

Optimising the processing of flexible containers

M. L. Seiboth and G. H. Shaw, Ellab UK Limited, UK

7.1 Introduction: challenges in processing flexible containers

The processing of food products by the application of heat to a sealed container offers a very simple mechanism by which the food can be preserved and stored for a significantly increased shelf-life. The two key parameters that allow this preservation technique to be successful are:

(i) The application of a scheduled process appropriate to the individual product and the target shelf-life under given storage conditions that is to be achieved. Such a process will be denoted in terms of a length of time at which the product must be held at a process temperature which, in some cases, allied with the intrinsic characteristics of product such as pH and water activity, ensures a sufficient level of microbiological kill of a target micro-organism. Whilst this aspect of thermal processing is beyond the direct scope of this chapter and will not be covered in detail, it cannot be ignored and will be considered in Section 7.6.

(ii) The integrity of the packaging, with the container being hermetically sealed to ensure that post-process contamination cannot occur. The methodology and technology of creating seals and seams is covered in Chapter 4.

In combination, the correct delivery of these two key parameters will allow the production of a safe product that will keep. However, the environment that is required for the application of (i) presents severe challenges to achieving (ii), with the processing environment being particularly challenging for plastics materials.

During the thermal process, the packaging will be exposed to an external pressure that builds up within the processing vessel, in addition to internal pressures that can be generated by components within the product itself. These pressures provide a source of stresses to the packaging that can affect the integrity of the packaging. For the most traditional format of packaging, the metal can, the design of the container has evolved to allow for pressure differentials that are likely to be experienced at typical process temperatures in saturated steam during the process. It is therefore the 'newer' packaging formats, the flexible retort pouch and the semi-rigid trays with film lids, that can be affected and lose integrity during a process.

This chapter will consider the challenges to the thermal process authority in establishing a process for a flexible or semi-rigid packaging format. The equipment that is available will be discussed and the methodology behind its application will be considered within a case study example. Whilst this will focus on the behaviour of the packaging, it is important to consider how manipulating this may impact on both the operation of the processing vessel and the heating of the product within the container, and these will be covered in Section 7.6.

7.2 Processing of flexible containers

The retortable pouch and plastic tray offer food manufacturers and retailers a visually more appealing and versatile packaging format than the traditional metal can in which to produce and sell heat-preserved foods. Each format offers greater apparent convenience to the consumer with a short cook in a microwave producing a quick meal or a component thereof.

For the food manufacturers, the retort technology, and in particular the control systems, have developed from the early saturated steam retorts. For a saturated steam retort, the pressure profile during a process would be of a fixed regime showing a correlation to the temperature within the vessel. Developments in control systems and retort designs allow pressure to be controlled independently of temperature. Retort systems known as 'overpressure' retorts, such as those manufactured by Lagarde and Steriflow SAS, permit a controlled input of air to be added to the processing vessel during a cook cycle, thus letting the user determine the pressure to which packs are exposed during a process. This ability to control pressure allows the processor to be able to limit an expansion of the packaging and, in turn, stresses on the seams of the packaging.

It is important that an understanding is made of the changes that occur with the application of heat to the product and container during the process. The major change will be an increase in internal pressure causing expansion of the packaging that can be attributed to a number of factors:

- expansion of headspace gases.
- expansion and release of gases entrapped within the food product.
- generation of water vapour.
- release of gases due to reduced solubility of dissolved gases at higher temperatures.
- thermal expansion of the solid and liquid phases of the food.

The amount of expansion that can take place within a processed food package depends on:

- the temperature of the food at the point when the pack is closed.
- the amount of vacuum drawn on the pack at the point of closing, e.g. mechanical vacuum seamers.
- the mechanical resistance of the pack.
- the external balancing pressure.

An additional problem for the processor will be that the pack will significantly soften during heating. The challenge, therefore, when choosing a pack format for a heating application or establishing a heat process for a selected pack, is to minimise the stresses upon the packaging. In the sector of the food industry that produces metal cans, the need for tailored pressure profiles has largely been avoided by careful design of the metal can. Cans are designed to withstand the pressure differentials that they are likely to experience at typical process temperatures in saturated steam. However, weaker packaging formats for heat-preserved foods, such as jars, semi-rigid plastic pots/trays, pouches or even cans with easy-open features, puts greater emphasis onto the food processors to design optimum pressure profiles during cooking that give maximum protection to the packaging.

A further hurdle arises with the desire to reduce the thickness of packaging materials. Overpressure retorts offer the capability to ramp pressures up and down during the process to optimally match the pressure changes taking place inside the packs. Such tailored pressure profiles can be used for processing relatively weak packaging formats that simply would not be possible under fixed external pressure regimes.

7.2.1 The importance of thermal stresses for packaging performance

When exposed to pressure differentials, packaging materials will be subjected to stress, and eventually to permanent deformation. Metal cans have evolved to withstand high internal pressures (possibly up to 6 or 7 bars). Can ends are designed to dome as the internal pressure increases and, in doing so, relieve some of the stress on the seals. However, if the internal pressure becomes high enough, then permanent deformation of the ends occurs and this can take the form of 'peaks'. The reverse can occur, for example during the cooling process, where the external pressure remains high after the internal can pressure drops in response to product cooling. In this case, panelling will occur, which is the inward collapse of the relatively weak sides of the can.

With semi-rigid trays and bowls, the effects of excessively high or low pressure differentials are similar to those with cans, but unlike cans there is a significant risk that the lids could burst open in the retort. It might be said that semi-rigid constructions require greater pressure control because they are highly sensitive to both low and high external pressure conditions. This is because the lidding films can be pushed inward as well as outward.

Pouches are possibly a little easier to work with when defining pressure profiles than semi-rigid formats, because they are less sensitive to high external pressures. If the product is not compressible and does not require free space for mixing in a rotary process (e.g. rice), then application of a high overpressure should control pack expansion.

7.3 Setting up an overpressure profile

In order to minimise the stresses to a packaging material, it is necessary to be able to set up an overpressure profile to run in parallel with the thermal process that is cooking the product within the container. An investigation of the behaviour of the packaging under processing conditions and the establishment of a suitable overpressure profile should be undertaken in the early stage of the development of a new product or packaging format. It will be rare for this work to be undertaken on a production retort system and, in general, a pilot retort system will be utilised that is a scaled-down version of the production retort or retorts that will eventually be used for the product. It is important that the pilot retort has a programmable control system that allows a degree of manual intervention on the retort pressure profile, so that during a trial run the pressure can be manipulated to an optimum level at each stage of the process. Should such a control system not be available, trial runs will need to be carried out with a review of the traces from the chart recorder being undertaken in between runs, to assess the changes to the retort program required for the following run. This is an iterative process with the programme nearing the optimum with each new trial. Sometimes the expected improvement does not occur at each iterative step and this often reflects a lack of consistency in the pack preparation conditions, for example, there is pack-to-pack variation in the vacuum being pulled at closure.

7.4 Equipment for establishing an overpressure profile

The preceding sections have highlighted that there are two parameters for which a physical measurement can be taken during a process cycle that can indicate how the packaging may be behaving. These are the physical movement, or deflection, of the packaging material and a value of pressure, which can be measured both externally to the pack and internally. Temperature within the retort system may also be recorded but this will not be having a direct influence on pack behaviour. Indirectly, retort temperature will be a factor as it will impact on the rate of heating of the product and so on the rate of expansion of gases held within.

The requirement to monitor and record pressure and/or deflection during a process means that the thermal process authority given the task of establishing an over pressure profile needs to have access to appropriate equipment.

7.4.1 Pressure sensors

The acquisition of pressure measurements during a process cycle can be successfully achieved using one of two types of datalogging equipment.

(i) A real-time pressure sensor from Ellab A/S, which uses an oil-filled capillary tube to transmit pressures from a pack to a strain gauge transducer located outside of the retort, which in turn is connected to an E-Val Flex datalogger. The key advantage offered by this type of sensor is the availability of real-time data, allowing adjustments to be made mid-process to the pressure within the retort system. This could reduce the number of practical runs needed to develop the correct over-pressure profile for a packaging format.

(ii) Pressure measuring wireless data acquisition units such those from TMI Orion or Ellab A/S (Fig. 7.1). These units are 'cable free', being small wireless devices that are programmed and downloaded from a reader station connected to a PC computer. The memory in such devices allows the user to collect around 14 000 individual sets of data within a working temperature range of –20 to + 140 °C. The small size (approximately 15 × 40 mm for an Ellab Tracksense Micro pressure device) gives the advantage that they can be put inside containers and have a minimal disruption to the product within. The small size is also of benefit for test containers going through continuous cookers such as reel and spiral retort systems and in batch rotary retorts. For reel and spiral applications, the time interval at which data can be gathered could be important, because faults on these cookers may occur in the can transfer valves through which cans travel in seconds. The major drawback with such devices is that they are memory dataloggers and cannot generate any real-time information. The thermal process technologist must therefore wait until the completion of a run before being able to see the impact of any adjustments that have been made to a retort profile.

Using pressure measuring devices, it is possible to measure the pressure on the inside and outside of a container in order to determine the differential and therefore the potential for pack expansion. Where there is no pressure differential, there is no driving force for pack surface movement. The bigger the differential, the more likely it is that there will be adverse stresses being applied to the packaging.

Pressure differential measurements are used for determining pressure profile requirements for metal cans, but the approach is less useful for semi-rigid or flexible containers. For containers that exhibit a high degree of flexibility, the tendency is for the internal pack pressure to equilibrate quickly with the external pressure, so pressure measured inside and outside the container will be similar. Likewise, the pressures required to cause deformations in weak plastic packs are so small that the sensitivity of the two pressure measurement sensors would need to

Fig. 7.1 Ellab Tracksense Micro Pressure Datalogger. With dimensions of only 15 × 40 mm, this device is ideal for recording pressure within a flexible container during a thermal process.

be very high to get useful results. As this method may not be suited to some flexible packaging types, alternative methodologies for developing an overpressure profile should be considered.

7.4.2 Pack deflection measurement

Where pressure differential measurements are not possible, direct measurement of the movement or deflection of the pack surface offers a good alternative and is often the preferred method within industry for establishing an overpressure profile. This approach utilises a movement transducer directly placed in contact with the pack surface so that either pack expansion or contraction can be monitored using a real-time datalogger system. A commercial system of this type from Ellab uses a spring-loaded stainless steel probe with 18mm of travel for measurement. The probe can either be placed directly in contact with the pack surface or put in mechanical contact through an appropriate holding device (Fig. 7.2). It is important that the deflection device does not adversely influence the movement of the pack surface so care must be taken to ensure that this is not the case. A key advantage of this system is that it allows real-time data to be acquired, so when used in conjunction with a real-time pressure sensor (see Section 7.4.1) and a standard thermocouple, a complete overview of the processing conditions can be acquired.

This type of system is limited to batch applications because the cables make it unsuitable for continuous cookers. It is possible to take deflection measurements from batch rotary retorts by restraining the measured container and holding the deflection sensor in a fixed position relative to it. The data can be passed out of the retort through a thermocouple slip-ring assembly. Possibly the most difficult aspect of using deflection in rotary environments is interpretation, because the

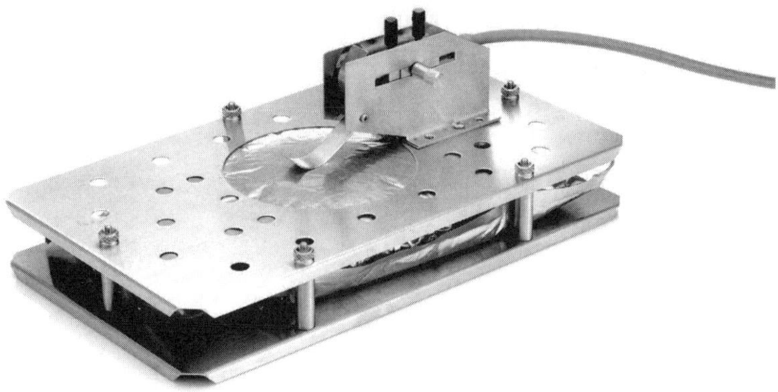

Fig. 7.2 Ellab EVal Flex Deflection Transducer. Shown here in an experimental rig for a pouch product, the deflection transducer provides a mechanism for correctly establishing a correct overpressure profile.

underlying trends in expansion/contraction of the container have to be separated from the 'noise' of the movement of the sensor and product under the effect of gravity, with each rotation of the retort basket.

An alternative to a real-time system for rotary processes would be to use a wireless deflection device such as the TMI Orion Nanovacq Deformation datalogger. As with the real-time device, a steel sensor is placed in contact with the pack surface, but for this device the movements of the sensor are recorded to a memory chip for downloading on completion of the test run.

For either of these types of sensor, the user is required to interpolate any improvements to the pressure profile that might be required manually. An alternative to this approach is to set up a feedback loop between the sensor and the retort control system. Within such a set-up, a software program determines from the output of the deflection sensor as to whether the pressure is too high or too low, and makes adjustments accordingly by either opening the retort vent valve to reduce pressure, or by opening the retort compressed air valve to increase retort pressure. Such a set up necessitates a retort that has services (vent and air valve) that can be configured in this manner. One retort manufacturer, Steritech (Saverne, France) produces a deflection sensor for use with their retort control system, whilst the Ellab CMC 92 datalogger system may be utilised on other retort systems where possible.

7.4.3 A viewing window

A viewing window in the retort system, if available, is probably the oldest method for tailoring pressure profiles for packs in retorts. The approach is simply to look through the window to judge the level of pack expansion or contraction taking place and adjust the pressure profile accordingly. Viewing is made significantly easier by having a powerful light to illuminate the retort interior. Although the approach lacks sophistication, it does offer some real benefits because the human eye can judge deformation on all pack surfaces, whereas, for example, deflection measurements are limited to those surfaces to which sensors are applied. This can be important because, in complex pack formats with surfaces of different rigidities, the control action required on one surface may be entirely wrong for another part of the pack. To take an extreme example, it is possible to measure deflection from a flexible lid on a semi-rigid pot and find that no matter how much external pressure is applied, the outward expansion of the lid cannot be restrained. Further investigation might show that the pot itself has been crushed inwards, so pushing the lid out further with every increment in pressure.

A window is a useful tool for static retorting applications but cannot be used for rotary retorts unless the rotation speed is slow or a high-speed video camera is used. Sometimes, a static test in a retort window is used as a preliminary test for a rotary process. In this case, the transfer of the pressure profile from the static to the rotary mode is facilitated by taking heat penetration measurements inside the container. With these data, it is possible to equate the pressure required at each pack internal temperature in the static mode to the pressure required at the same in-pack temperature in rotary mode, though this may not be considered an exact science.

7.5 Case study

This example considers an industrial trial that was undertaken to improve the overpressure profile and reduce the number of pack rejects that the company was experiencing. The product tested was a simple ready meal type product in a semi-rigid tray with a film lid. This was processed in a static batch retort that used rain water as the heating medium.

The first step of the trial was to acquire data for the existing conditions to identify the deflection that was taking place under the retort pressure profile then in place. A pilot retort was setup with an Ellab CMC 92 datalogging system that utilised a real-time deflection sensor to monitor the movement taking place of the film lid, and a real-time pressure sensor and a thermocouple to record the retort conditions. The data generated can be seen in Fig. 7.3. As can be seen from the profiles, the programme in place incorporated a single-step ramp up for the pressure within the retort, with this reaching a hold set value of 0.8 bar at the same time as the retort reaches the hold set point temperature of 105 °C. The ramp up time from 0.0 to 0.8 bar is approximately eight minutes. The impact that such a quick increase in retort pressure has upon the pack can be clearly seen. The deflection profile records a negative change, with the film lid being pushed in in excess of 10 mm during the ramp up phase. The deflection is actually so great that it goes beyond the recording range of the sensor, as indicated by the flat bottom that the deflection trace records for the first 12 minutes of the process. At this point in the process, the gases within the pack are expanding, which leads to the film lid being pushed out. The pressure profile is now shown to be incorrect at the end of

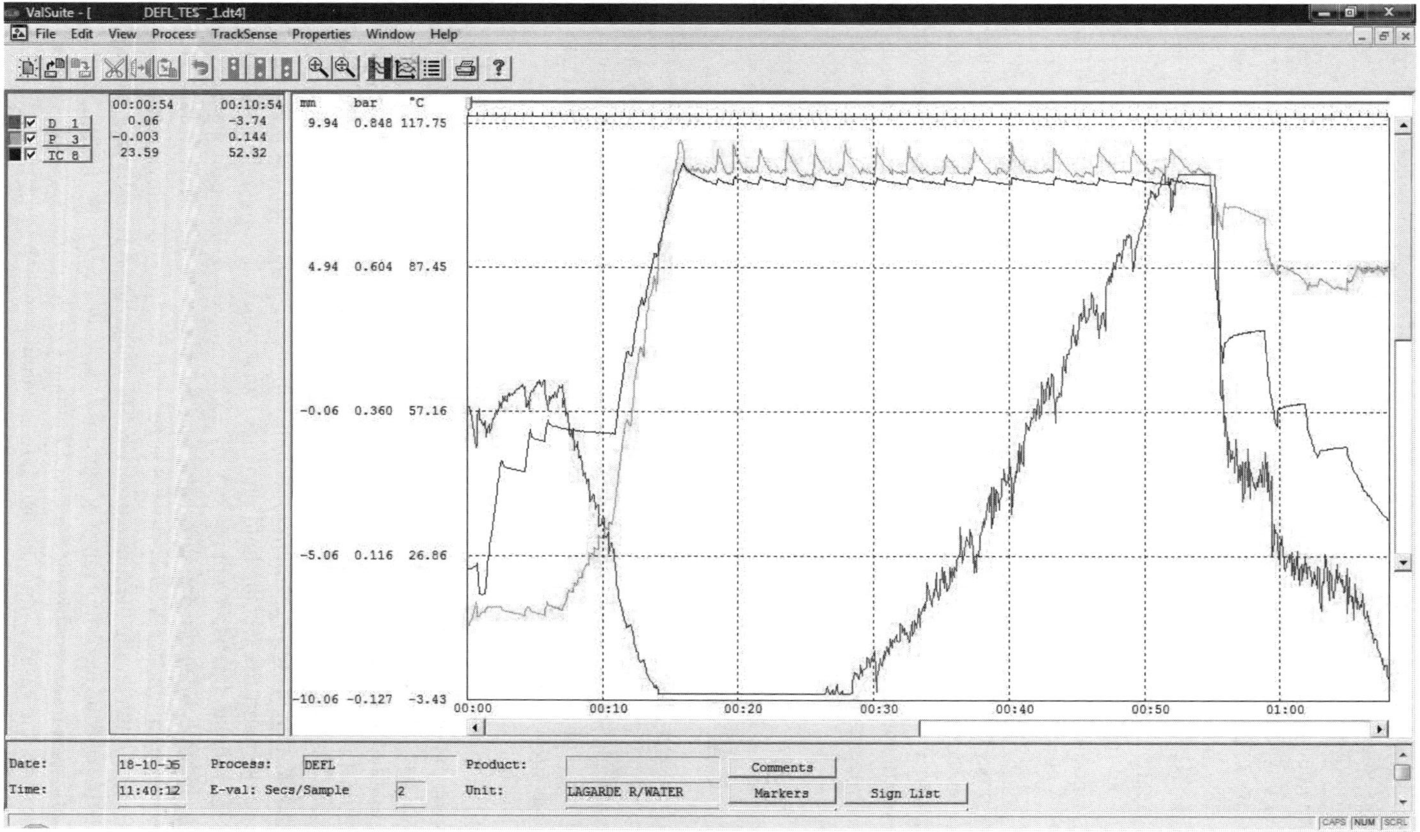

Fig. 7.3 Case study: initial retort run with incorrect retort pressure profile and excessive stresses to the container.

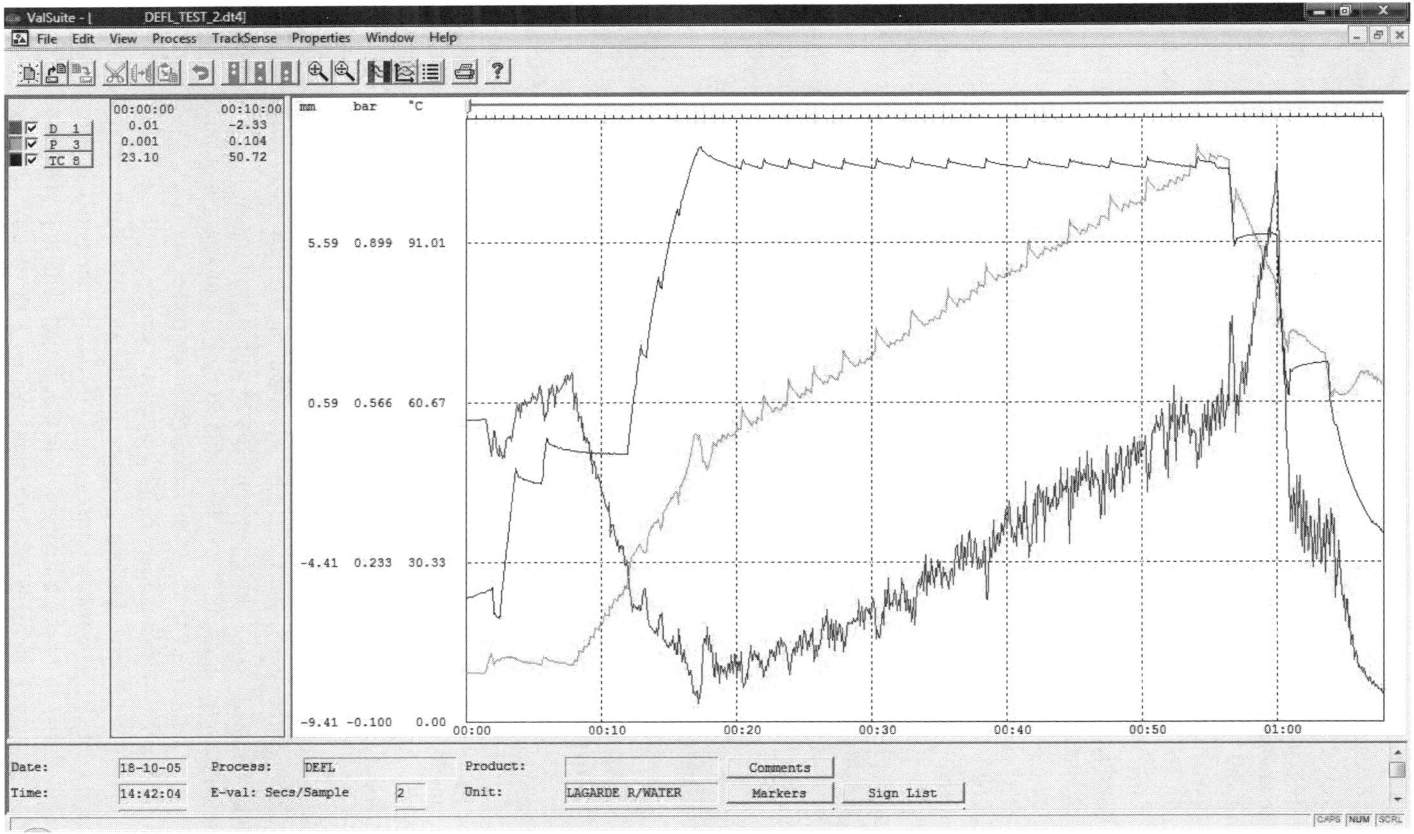

Fig. 7.4 Case study: revised retort run with adjusted retort pressure profile and reduced stresses to the container.

the hold phase as the pack expands beyond the zero start position to beyond the positive range of the sensor and again a flat plateau is recorded as the film lid deflection continues to increase at the end of the hold phase. It can therefore be seen from the first set of data that the packs processed using this retort pressure profile are being stressed both positively and negatively, with total movement of the film lid being in excess of the 18 mm range of the deflection sensor. The two key areas of the process that need adjustment to the pressure profile are the ramp up and start of the hold phase where excessive pressure is being recorded, and the end of the hold phase where the programmed pressure value is insufficient.

To reduce the deflection that the film lids are being subjected to, the company changed the retort pressure profile to incorporate a gradual ramp up of pressure for the duration of the process, with a maximum value of 1.0 bar at the end of the hold phase. A second test run was carried out and the data shown in Fig. 7.4 generated. It can be seen that, whilst the film lid is still showing some movement, it is considerably reduced from the first run and, more importantly, that changes are gradual, minimising the stresses on the seals. The range of deflection recorded during the hold phase has now been reduced to 10 mm, or half the amount seen with the original pressure profile.

This case study example shows how the use of commercially available datalogging equipment can very easily identify areas of the retort program that are leading to excessive stressing of the packaging. The data generated can be easily interpreted and improvements implemented that reduce the stresses and the volume of reject product that a company may be producing.

7.6 Implementing pressure profiles

In the event of experimental trials producing an amendment to the retort pressure profile, it is important that consideration is given as to whether or not this has an effect on other aspects of the process. One key area that should be reviewed is the performance of the retort system in the form of temperature distribution trials. Within such trials, thermocouples or wireless dataloggers will be positioned throughout a retort load of product to record the environment temperatures within the load and identify any areas of slow heating within the retort system. If the pressure profile is changed, which will in turn change the physical shape of the containers of the retort, there might be an impact on the movement of the retort heating media. This will be a particular issue within systems such as a Lagarde steam/air retort, which utilises a large fan to move a mixture of steam and air through the retort load. Once it has been ascertained that either no change in retort performance has occurred, or that a change has occurred and the variation and position within the retort has been identified, testing should also be carried within the product itself. Heat penetration testing will consider the rate of heating within a container and in turn the microbiological kill that is taking place during the thermal process. For taking measurements within a flexible or semi-rigid container, it is important to ensure that the fittings used for locating either a thermocouple

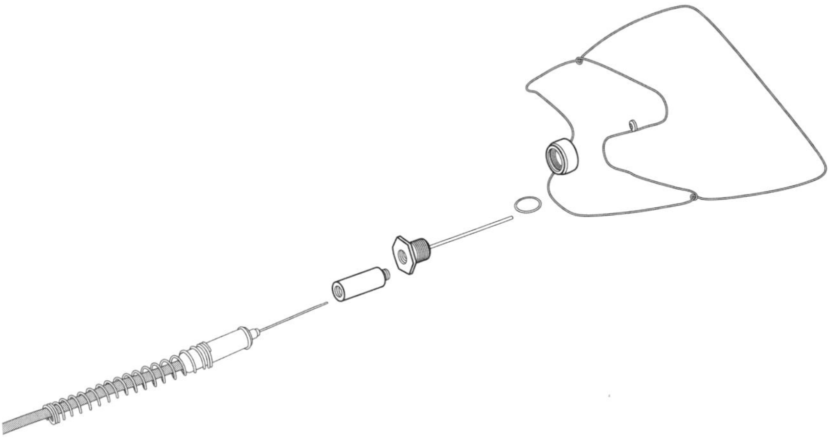

Fig. 7.5 Example of a fitting for data acquisition in a flexible packaging that ensures correct location of the temperature probe within the sample.

or a wireless datalogger within the container do not restrict any movement of that container during the process. This is of particular importance for flexible pouches, where the packaging material itself cannot support the temperature measuring device within the pouch at the cold spot of the product. In such a case, a custom-made fitting such as that produced by Ellab A/S and shown in Fig. 7.5 should be used. The positioning the test heat penetration samples is critical, and ideally these should be at the cold zone identified by the temperature distribution testing. If this is not possible, any variation of processing conditions should be allowed for within modelling software, such as the CCFRA CTemp program or the TechniCAL Calsoft program.

7.7 Future trends

Optimising the processing of flexible containers will be driven by advances and new trends in the packaging sector. As new technologies allow the development of new packaging formats and thinner materials for existing designs, there will be an on-going demand for safe processes to be established; processes that ensure stresses experienced by containers as a result of the thermal process are kept to a minimum. The technology and equipment currently commercially available from companies such as Ellab A/S and TMI Orion offer process technologists the capability and flexibility needed to test the various packaging formats. As with standard temperature datalogging, it could be suggested that the 'Holy Grail' of a retortable, wireless, transmitting device would be of benefit for the recording of both pressure and deflection measurements. Whilst developments to attain and produce such a temperature logger continue, it is questionable as to how much actual practical benefit such a pressure or deflection device would offer, with only testing of rotary processes showing clear-cut advantages in eliminating the requirement for a slip ring assembly.

7.8 Sources of further information and advice

For companies entering into the processing of flexible packaging for the first time, the input of a knowledgeable consultant could be invaluable. For companies in Europe, organisations such as the Campden and Chorleywood Food Research Association offer pilot facilities, along with the equipment and expertise to develop new processes. In addition to pilot facilities, CCFRA offer valuable information in the form of published guidelines and reports on completed research projects (see below for more detail). Datalogging equipment manufacturers such as Ellab A/S and TMI Orion can offer technical support and advice on utilising their individual products for given applications, whilst retort manufacturers or their agents in a particular country can advise on the suitability of their systems for a given process.

A valuable source of information for thermal processing as a whole is the Institute for Thermal Processing Specialists (IFTPS), who offer guidance and support to the industry through annual meetings and protocols put together by Institute members, based on their practical experience.

Details for all the companies or organisations mentioned in this chapter are as follows:

Ellab A/S, Krondalvej 9, DK – 2610 Roedovre, Denmark
Tel: +45 44 52 05 00 Website: www.ellab.com

Campden and Chorleywood Food Research Association (CCFRA), Station Road, Chipping Campden, Gloucestershire GL55 6LD, UK
Tel: +44 1386 842000 Website: www.campden.co.uk

TMI Orion, Parc Industriel et Technologique de la Pompignane, Rue de la Vieille Poste, 34055 MONTPELLIER cedex 1, France
Tel: +33 4 99 52 67 10 Website: www.tmi-orion.com/index.htm

Societe Lagarde, Z.I. Les Plaines - N° 5 bis, 26780 Malataverne, France
Tel: +33 4 75 90 58 58 Website: www.lagarde-autoclaves.com

Steriflow SAS, 32, Rue de Cambrai, 75019 Paris, France
Tel: +33 1 40 37 08 45 Website: www.steriflow.com

Steritech/ECPS SA, ZA du Kochersberg, 1, Rue de Furchhausen, 67700 Saverne, France
Tel : +33 3 88 71 04 33 Website: www.steritech.fr

Institute for Thermal Processing Specialists (IFTPS), 304 Stone Road West, Suite 301, Guelph, ON N1G 4W4, Canada
Tel: +1 519 824 6774 Website: www.iftps.org

Some useful publications of the Campden and Chorleywood Food Rearch Association are:

CCFRA Guideline 50 'Guidelines on good manufacturing practice for heat processed flexible packaging' (2006).
CCFRA Review 46 'Review of the integrity of heat processed containers through manufacturing and distribution' (2004).
CCFRA Guideline 17 'Guidelines for establishing heat distribution in batch overpressure retort systems' (1997).

8

Wireless data loggers to study heat penetration in retorted foods

J. J. Sullivan, Mesa Laboratories, Inc., USA

8.1 Introduction

From the time of harvest to the point of consumption, most foods lose a significant portion of their nutritional and organoleptic quality. Some amount of processing is required for almost every type of food product, and examples include washing, blanching, re-formulation, resizing, and retorting. The preservation of foods by in-pack heating operations is a primary means of achieving extended shelf-life. In these operations, the challenge for the industry is to achieve a balance between the desired preservation effect and maintaining acceptable quality. The general premise is that the minimal heating required to achieve preservation is best, and the focus of the industry has been to consistently achieve these processing minimums, at the same time providing an acceptable level of safety.

For heat-processed foods, it is apparent that accurate temperature measurements are required to achieve the optimum balance between adequate preservation and maintenance of high quality. Heat treatments of in-pack processed foods are designed to destroy microbial organisms; vegetative cells in the case of pasteurization and microbial spores in the case of canning. An important factor that makes it difficult to achieve the correct balance between safety and quality is the kinetics of microbial death. These processes are exponential in nature, meaning that a very small change in temperature can have a dramatic influence on the degree of microbial death. For a typical 121 °C process, a 1 °C error in process temperature can result in a 25% error in the lethality calculation (PDA, 2003). The industry has long recognized the need for accurate temperature measurements, and an accuracy of ±0.3 °C is often considered the standard (IFTPS, 1992) for both heat distribution determination in the retort and for heat penetration studies within the food product itself. This accuracy requirement dictates the type of temperature measuring equipment and calibration processes utilized. The use of highly accurate measurement instruments has become the norm in the industry and greatly facilitates achieving the balance between adequate preservation and maintenance of quality.

8.2 Introduction to temperature measurement technology for retorted foods

Temperature measurement technology has evolved from liquid-in-glass thermometers, which rely on the expansion of a fluid as it is heated, to today's modern electronic measurement and recording devices. As the technology has evolved, so has the accuracy, ease of use, reliability and safety of the devices. Although the mercury-in-glass (MIG) thermometer is still extensively used in retorts to monitor chamber temperature, and is called out in many of the standards on retorting of foods, its utility for any process development study is severely limited.

8.2.1 Wired and wireless data loggers

The primary instrument for studying heat penetration in foods is the temperature data logger. Much of the groundwork for the use of data loggers in food processing originated in the work of Ecklund (1949). This, coupled with the development of mathematical modeling of the food sterilization process (Ball and Olson, 1957), has led to the development of the industry as we know it today. All of the original work, and the development of the industry, have relied on the accurate measurement of the internal temperature of food products throughout the sterilization process using a data logger system.

Data logger systems for use in food processing are available in two primary forms, 'wired' and 'wireless'. In a wired system, temperature sensors (usually thermocouples) are hard wired directly to a recording device that has the required electronics to make the temperature measurement and record the data. In a wireless system, self-contained, battery-operated measurement and recording devices are used that have the required sensor, electronics, memory and microprocessor to record process data. Figure 8.1 illustrates the configuration of a wired and wireless data logger system. A wired data logger system continuously monitors, displays and records data as it is received. Most often, the wired data logger is connected to a PC for programming and display of data. Either during or after data collection, the data is processed, lethality calculations can be performed, and reports can be prepared. For a wireless data logger, the self-contained loggers are programmed to collect data at a specified rate and time, the loggers are deployed into the process, the loggers are retrieved following the process, and then the data is 'read' into a PC. Like the wired system, the requisite calculations and reports can be generated

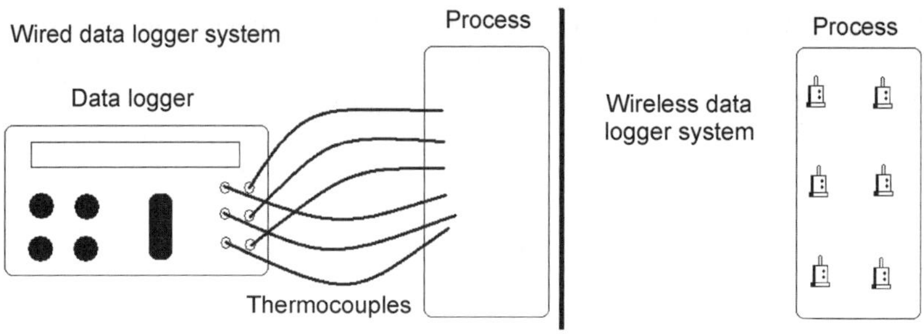

Fig. 8.1 Configuration of wired and wireless data logger systems.

following data collection. Unlike the wired system, no real-time data is available for viewing during processing.

8.2.2 Temperature measurement technology

Temperature measurement instruments are composed of two primary components, the sensor and the electronics. It is generally possible to utilize a variety of different sensors, each with its own unique characteristics and limitations, with any of the various electronics configurations. In terms of sensors, there are three primary types in use in the food industry: thermocouples, thermistors, and resistance temperature detectors (RTD). (Thermistors are technically also RTDs, but common usage in the industry has restricted the term 'RTD' to refer to those resistance temperature devices that contain either a platinum foil or wire as the primary sensing element.) Some of the key characteristics of these sensors are listed in Table 8.1. They all have their advantages and disadvantages but, in terms of temperature range, stability and accuracy, the thermistor is probably the most useful for retort applications. Thermistors can be manufactured to be extremely small, fitting into very thin probes; they have excellent stability so they do not need to be recalibrated frequently, as do thermocouples, and they are resistant to shock and vibration, unlike some RTDs. Wireless data loggers fitted with themistors have become the norm for most heat penetration studies in food processing.

Thermocouples

A thermocouple consists of two wires of dissimilar metals welded together into a junction. At the other end of the signal wires, usually as part of the input instrument, is another junction called the reference junction. A low voltage is produced by the thermocouple that is proportional to the temperature difference between the measurement and reference junctions. Figure 8.2 illustrates the relationship between temperature difference and voltage for the more common types of thermocouple. The food industry has adopted the Type T thermocouple almost universally.

There are both advantages and disadvantages of using thermocouples for temperature measurements. On the positive side, these sensors have a linear output

Table 8.1 Temperature measurement sensors

	RTD	Thermocouple	Thermistor
Range	–260 to 800 °C	–270 to 1850 °C	–80 to 200 °C
Accuracy	Best	High	High
Stability	High	Low	High
Sensitivity	Moderate	Low	Best

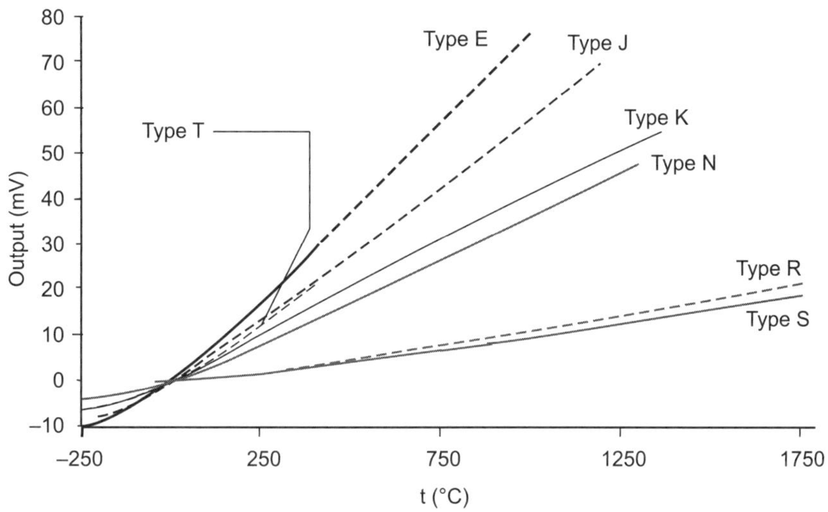

Fig. 8.2 Characteristics of various types of thermocouples.

that is well characterized, so the electronics circuitry is relatively simple. Also, thermocouples are relatively resistant to shock and vibration. There are several disadvantages, however, that limit their utility. The first of these is that thermocouples are differential sensors, which means that there must be a reference junction that is held at a constant temperature. Any drift of the reference junction will result in a measurement error. Thermocouples are prone to 'extension lead errors' which are due to inhomogeneities in the long wire between the two junctions. The requirement for a reference junction that is remote from the measurement environment also means that thermocouples are restricted to 'hard-wired' systems. Thermocouples have a tendency to drift, requiring frequent recalibration. This also limits their utility, especially compared with the more stable thermistors and RTDs. While thermocouples are inexpensive, their high drift rate and other limitations make them poorly suited for heat penetration studies in foods. Devices utilizing thermistors or RTDs are a much better choice for most studies.

Resistance temperature detectors and thermistors

Both thermistors and RTDs are sensors in which the resistance of the device varies

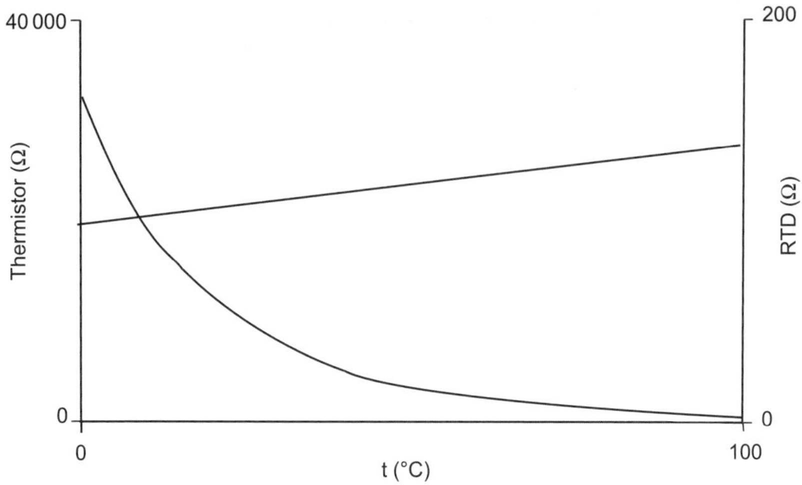

Fig. 8.3 Relationship between temperature and resistance for thermistors and RTDs.

with temperature. Thermistors are constructed of semiconductor material, most often a ceramic, while RTDs are platinum in either a wound wire or foil format. The electronics package accurately measures the resistance and translates this to temperature-based on either a known relationship or, more commonly, a calibration. Figure 8.3 illustrates the resistance change of a common thermistor and RTD in response to various temperatures.

Of note is that the RTD produces a linear response in relation to the temperature, which has implications in terms of its ease of calibration and accuracy. In contrast, the thermistor exhibits an exponential relationship between resistance and temperature, but is highly sensitive, meaning that a very large change in resistance is seen with a small change in temperature (note the different scales on the *y*-axes). It is critical to carefully calibrate a thermistor-based measurement system because of this non-linear response. Also, temperatures outside of a thermistor's calibrated range can exhibit very large errors. On the other hand, an RTD will be fairly accurate outside of its calibrated range due to its inherently linear response. Both thermistors and RTDs are available in a wide variety of different sizes, configurations and base resistivities (the resistance at a specified temperature, usually 0 °C or 25 °C).

Thermistors are widely used in industrial applications, including food processing, as they are stable, inexpensive, and if properly calibrated, extremely accurate. Additionally, they are ideal for small wireless systems because there is no need for a reference junction, as with thermocouples, and they can be fabricated extremely small, no larger than the point of a pin. It is these advantages that have made the thermistor the sensor of choice for most industrial applications.

The RTD is the sensor of choice for highly accurate temperature standards. These are used as primary standards for the calibration of other devices. The characteristics of RTDs, including their long-term stability and linearity, make them ideally suited as standards. One drawback to an RTD is its susceptibility to shock. Care must be taken to avoid either physical or thermal shocks of RTDs to maintain their long-term accuracy. Nevertheless, nearly all primary standards used in metrology are RTDs.

Electronic readouts

Beyond the temperature sensor, the electronics package or readout device is a critical component in any temperature measurement instrument. The readout can range from a very simple circuit with a digital temperature display to a highly sophisticated PC-controlled data logger. For single, immediate measurements, a simple readout device may be sufficient. A digital medical thermometer is a good example of a simple readout device. However, for time/temperature studies in food process development, a data logger is generally required. As the name implies, a data logger is a device that records sequential temperature measurements for archival and analysis. Most modern data loggers provide a connection to a PC so that data can be displayed, archived and reported. Many wired thermocouple-based data loggers also have the capability of controlling a high temperature heating block for calibration purposes.

8.2.3 Practical considerations in the selection of data loggers

When conducting either heat penetration or temperature distribution studies, the first decision is whether to use a wired thermocouple-based system or a wireless system that uses either thermistors or RTDs. The decision on which system to utilize is often dictated by the equipment on hand, but a more systematic approach would be to examine the characteristics of the two systems, as delineated in Table 8.2.

For most retort applications, a wireless data logger system is the instrument of choice because of the ease of setting up the system. There is no need to string wires into the process and into the product container, as the wireless data loggers can be placed either inside or outside of the food container, with the measuring probe tip easily positioned at various points in contact with the product. The wireless systems are also much easier to set up, as no calibration is required. Wireless data loggers come from the factory pre-calibrated and generally hold this calibration for more than one year. In contrast, wired data loggers must be calibrated in place, immediately prior to data collection. Depending on the complexity of the system, the presence of special glands, connectors, and fixtures can severely limit the calibration stability of a wired thermocouple-based system. One advantage of the wired systems is the ability to obtain real-time data, as the process is progressing. In practice, these real-time data have limited utility, as it is unlikely that a particular retort process will be altered or halted based on data observed in real-time. Most analyses of heat penetration study data are performed post-run, when data from either a wired or a wireless system are available.

Table 8.2 Characteristics of wired and wireless data loggers systems

	Wired data logger	Wireless data logger
Temperature accuracy	Depends on calibration, but generally ±0.1 to 0.3 °C.	Most are ±0.1 °C.
Calibration	Must be done on-site, immediately before and after a test. Can be very time consuming.	No user calibration required. Factory calibration generally stable for >1 year.
Ease of use	Wires must be passed through retort wall and then into container. Requires a number of seals able to withstand retort pressures.	No modification to retort required. Requires container seal if externally mounted.
Applicability	Still retorts primarily. Agitating or spiral retorts require special connector harness. Cannot be used in hydrostatic or other continuous retorts.	Can be used in any type of retort.
Real-time data	Generally available.	Not available.
Initial cost	Varies. Generally $2000 to $10,000 for an 8-channel system.	Varies. Generally $1000 to $2000 per temperature channel.
Ongoing cost	Generally high, depending on number of replacement thermocouples.	Batteries only, at $25 to $80 per year per channel.

8.3 Overview of heat penetration theory in retorted foods

One of the most critical functions of a thermal processing specialist is to conduct accurate heat penetration studies for in-pack processed foods. Developing the processing parameters that balance maintenance of the food's quality with achieving commercial sterility is an important task, with significant financial incentives. This has been recognized since the advent of the canning industry, and countless studies during the past 80 years have built a wealth of knowledge in the industry. The theory of heat penetration into packaged foods is well established and the methods used to characterize a particular product have been standardized (IFTPS, 1995).

There are many variables involved in retort processing of foods, including those associated with:

- *the container* – such as size, shape, wall thickness, and material(s) of construction.
- *the retort* – Including agitation (if any), fill medium (if any), age, temperature distribution profile, and heating source.
- *the food product* – such as viscosity, particulates, water and fat content, starting temperature, and headspace volume.

Because of the number of variables, nearly every individual food and process needs to be validated by conducting a formal heat penetration study. This has been especially critical in recent years, with the advent of newer retort packaging, such as pouches, trays and cartons. There are significant advantages in using these newer containers, as their profile allows much more rapid heat penetration, resulting in the use of shorter processes and the concomitant improvement in the food's quality. However, it is important to recognize that a properly conducted heat penetration study is even more critical because of the shorter processing times and the requirement to much more accurately locate temperature probes at the cold spot of the container. As an example, a 1 mm positioning error of a temperature probe will be much more critical in a pouch that is 20 mm thick than it will be in a can with a diameter of 60 mm.

The primary goal of heat penetration studies is to locate and model the heating characteristics of the slowest to heat position, or 'cold spot', in the container. If the cold spot experiences conditions to achieve commercial sterility, then it is assumed that the entire contents will also do so. In practice, the position of the cold spot is difficult to predict for in-pack processed foods, except for the simplest of products. Due to the large number of variables involved and the very nature of the product, it is often required to 'map' the temperature inside a container to locate the cold spot. Of course, a fundamental knowledge of heat transfer theory is helpful in designing mapping studies.

A comprehensive review of heat penetration theory in foods is beyond the scope of this chapter. However, there are a few critical concepts to keep in mind as they relate to the use of data loggers for temperature measurements. During retorting, heat is generally transferred in a food by conductive forces, convective forces, or a combination of the two. In a solid food that does not flow during processing, heat is transferred primarily by conduction and the cold spot will be near the geometric center of the container. On the other hand, in a free-flowing liquid the majority of the heat is transferred by movement of the material, and convective currents cause the cold spot to migrate to near the bottom and center of the container. These are the simplest cases, but a wide variety of other factors may influence the actual position of the cold spot. For instance, headspace may cause the cold spot to migrate up, as the air provides some degree of insulation. For containers of liquid which are agitated during heating, the cold spot may be unpredictable, as the contents are actively mixed. Also, many foods exhibit broken heating curves, which means that at some temperature the heat transfer mechanism may shift between conduction and convection due to a change in the physical condition of the contents. It is obvious that a data logger system that allows for flexibility in the positioning of the temperature probe is essential to conducting heat penetration studies.

8.4 History of wireless data loggers

The original wireless data logger system was developed in the mid-1980s by the

Ball Corporation and was named the 'DataTrace System', with the individual wireless data loggers named 'Micropack'. The DataTrace system was developed in response to an FDA 'Memorandum of Need' and was designed specifically to measure and record process temperatures during retorting of canned foods. A good overview of the system, including some initial qualification tests, was described shortly after its development (Cross and Lesley, 1985). Although many applications of wireless data logger systems have been subsequently identified, it was found to be particularly useful for retorting application, as it could be used in any type of retort, including hydrostatic, still, crateless, or agitating, without having to string thermocouple wire into the chamber or container. Beyond retorting, a wide variety of applications of wireless data loggers have been developed, including freezing, baking, storage, and transportation. Outside of the food industry, wireless data loggers are widely used in the pharmaceutical and medical device manufacturing areas for chamber validation, process control and environmental monitoring.

There are generally two types of wireless data loggers available on the market today: 'low performance' loggers, having lower accuracy (±0.5 to 1 °C) and lower price ($30 to $500 each), and 'high performance' loggers, having high accuracy (±0.1 °C) and higher price ($1000 to $2000 each). For any sterilization application, including the retorting of foods, the high performance data loggers are required due to their higher accuracy. Many of the low performance data loggers also have a limited temperature range and are not hermetically sealed, both of these factors precluding their use in retorts.

8.4.1 Characteristics of today's wireless data loggers

Significant advances in wireless data logger technology have been made in recent years, primarily impacting on their size, accuracy, and reliability. While the original wireless data loggers developed in the 1980s were found to be quite useful for hundreds of applications in many different industries, the limitations of the electronics and battery technology of the time meant that earlier systems were somewhat large and had lower performance in terms of accuracy and data storage. Table 8.3 provides a comparison of the original DataTrace Micropack system to the latest generation Micropack III. As can be seen, there are significant improvements in many aspects of the design that directly impact upon both heat penetration and temperature distribution studies in retort applications.

Table 8.3 Comparison of the original DataTrace Micropack (1985) to the Micropack III (2002)

	Micropack (1985)	Micropack III (2002)
Dimensions (diameter, height)	35 mm, 30 mm	18 mm, 20 mm
Volume	21 cm^3	4.3 cm^3
Weight	65 g	15 g
Temperature range	10 °C–150 °C	–20 °C–140 °C
Accuracy	±0.45 °C	±0.1 °C
Data storage	1 000 data points	16 000 data points

Table 8.4 Physical dimensions of various wireless data loggers currently available

	Dimensions (diameter, height)	Volume	Weight
DataTrace FRB	36 mm, 43 mm	36 cm^3	100 g
DataTrace Micropack III	18 mm, 20 mm	4.3 cm^3	15 g
Ellab Tracksense Pro	25 mm, 44 mm	22 cm^3	48 g
Ellab Tracksense Mini	20 mm, 16 mm	5.0 cm^3	16 g
Ellab Tracksense Micro	15 mm, 22 mm	3.9 cm^3	14 g
GE Kaye Valprobe	35 mm, 46 mm	44 cm^3	130 g

The current generation of data loggers from several different companies have similar performance specifications but differ widely in their size and weight, as can be seen from Table 8.4. As discussed later in this chapter, the size of a wireless data logger is important in those applications where the logger is mounted internally.

8.5 Fixtures and fittings used for heat penetration studies

No matter which type of data logger system is chosen, wired or wireless, some type of fixturing is required to accurately locate the temperature probe at a known point within the container. The type and configuration of the fixture is as diverse as the different types of data loggers, containers, and retorts. The simplest fixture is an entrance gland to allow a point of access through the container wall for a thermocouple wire or the probe of a wireless data logger, as illustrated in Fig. 8.4.

Wireless data loggers can also be mounted internally in the container for situations where external mounting would interfere with retort operation. A typical system for internal mounting of data loggers in a can is illustrated in Fig. 8.5. In this system, a hole is punched in the bottom of the can and a fitting is attached. The data logger is attached to a clip and a series of different length standoffs are utilized to position the probe tip at the desired height in the can. The data logger can be mounted either from the top or the bottom, so there is complete flexibility in the tip position at the radial center of the can. This is a good solution for relatively large cans, but it is not suitable for small cans. The same system can be used for bottles, by mounting the fitting in the cap with the probe tip protruding down into the bottle and its contents.

An alternative technique for mounting a data logger is shown in Fig. 8.6. In this system, a spring assembly holds the data logger and expands outward to the sides of the container. This positions the data logger radially in the center of the container and the logger can be mounted anywhere on the center support, to allow temperatures to be taken at various points. The primary advantage of this fixture is that no modification to the container is required. It is usable in either cans or bottles, providing a high degree of flexibility.

Mounting temperature probes in a known and consistent location inside of retort pouches can be particularly difficult because the pouches tend to flex during

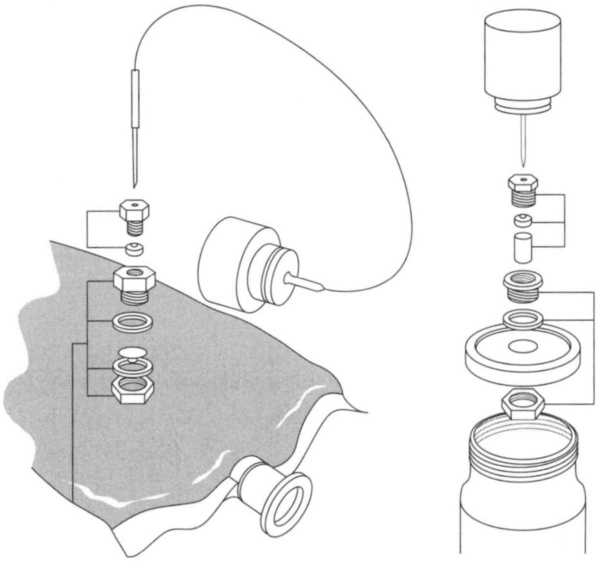

Fig. 8.4. Simple glands for positioning temperature probes in pouches and jars.

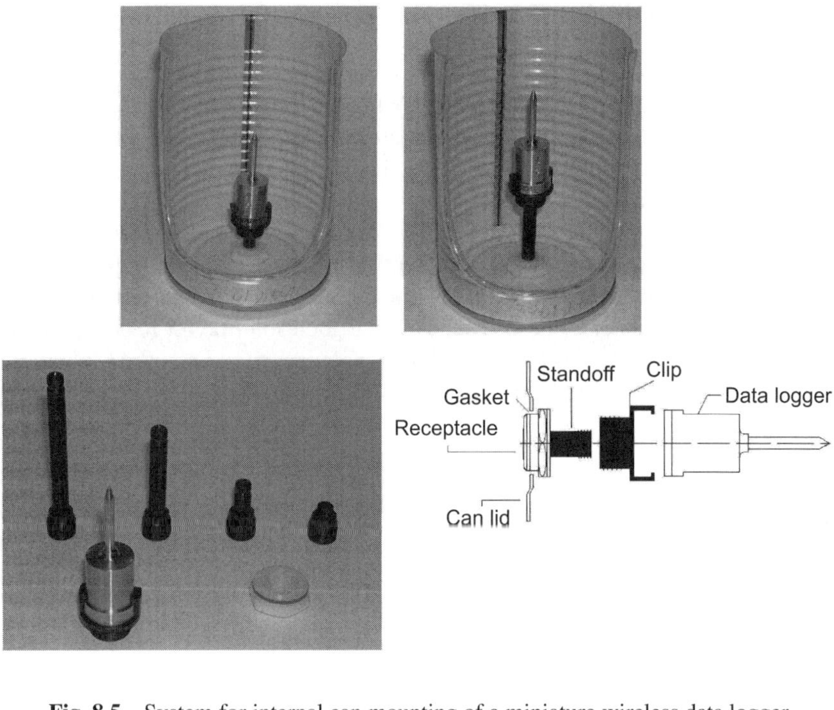

Fig. 8.5 System for internal can mounting of a miniature wireless data logger.

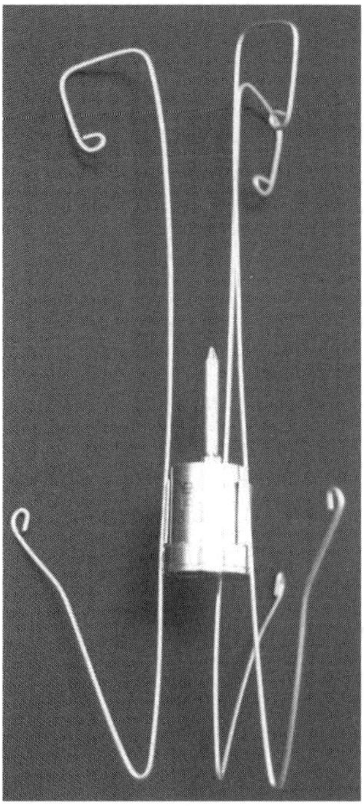

Fig. 8.6 Spring fixture assembly for internal mounting of a wireless data logger in a bottle or can.

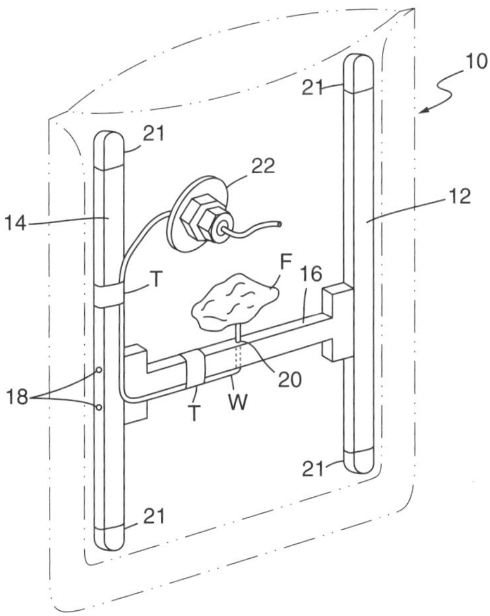

Fig. 8.7 Fixture for mounting a thermocouple inside of a retort pouch.

processing. One solution to this problem has been proposed and is the subject of a patent (Nioras, 1982). As seen in Fig. 8.7, Nioras proposed a rigid frame that is inserted into the pouch prior to processing. The frame provides support for the walls of the pouch and provides a rigid mounting location for the thermocouple. The same device could be easily used with a wireless data logger, with either internal or external mounting.

8.5.1 Minimizing errors in heat penetration studies

The goal of any heat penetration study is to use a data logger system that can measure the internal temperatures in the food product with minimal error caused by the temperature measurement system itself. It must be recognized that alterations to either the container, the food product, or to the retort can cause erroneous results, unless the errors are well characterized and quantified. Both wired and wireless data logger systems are designed to minimize errors caused by the measurement equipment. Errors can occur by a variety of mechanisms. Heat can be conducted either into or out of the container by temperature probes. Rigid temperature probes are generally made from thin-walled stainless steel tubing, a relative poor heat conductor. Sensor wires are the thinnest gauge possible to minimize heat conduction. Any fixtures used should be constructed as small as possible to also minimize heat conduction. Errors can additionally occur due to the difference in mass between the food and an internally mounted data logger. In practice, if the newer small data loggers are used, this effect is minimal. For instance, in a typical can of product, a 15 gram data logger represents less than 4% of the mass of the can and its contents. Lastly, an internally mounted data logger could have an influence on the establishment of convection currents in the container. Minimizing the size of fixtures and using the smaller data loggers should minimize these effects. Nevertheless, a prudent measure would be to characterize heat penetration in a system, using both internally and externally mounted data loggers to develop any required correction factors.

A number of studies have been conducted to characterize the influence of internally mounted data loggers on the heating characteristics of various foods. During the original development of the wireless data logger by the Ball Corporation, studies showed little or no influence on the heat penetration parameters in most situations (Lesley, 1987). However, for some products which heat mainly by conduction, significant differences were observed between internally mounted wireless data loggers and thermocouples, and a correction factor needed to be applied to obtain accurate data. In a later study, similar results were reported using internally mounted wireless data loggers and a 5% bentonite test suspension (Britt *et al.*, 1997). This study also concluded that correction factors were required to obtain results comparable to thermocouples. Additionally, in this study, it was found that, if wireless data loggers were mounted externally and only the temperature probe extended into the container, results closely matched thermocouples, which is an expected outcome. More recently, the data logger companies have introduced substantially smaller designs. For instance, the studies conducted by

Lesley and by Britt *et al.* were performed using data loggers with a volumes of approximately 20 cm^3 and weights of 48 g to 65 g. The latest generation of data loggers has volumes and weights of only about 20% of the figures (see the Micropack III and Tracksense Mini in Table 8.4). In a recent study using these smaller data loggers, no significant differences in heating characteristics were found between containers with thermocouples and those using internally mounted wireless data loggers (Awuah *et al.*, 2006). Apparently, the smaller size of the new generation of data loggers has alleviated much of the problem of the logger body influencing heating characteristics of the product.

8.6 New developments in wireless data loggers

Heat penetration studies will continue to be an important part of the development of in-pack processed foods. With increasing use of alternative packaging, such as trays and pouches, instead of traditional cans, it is required to develop processing parameters from heat penetration and temperature distribution studies for every kind of packaging in the various retort systems. The primary focus of thermal processing specialists will be developing ways to perform heat penetration studies more accurately and less expensively. In terms of accuracy, the use of the newer generation, extremely small data loggers will minimize errors due to perturbations of the normal state within an unaltered package. These extremely small data loggers are available now and any future advances to further minimize both size and weight can only provide better accuracy in the determination of heat penetration parameters.

Beyond data logger size, the most significant advance on the horizon is the development of radio frequency data loggers. The lack of real-time data from wireless data loggers has often been cited as the primary reason for using older technology wired, thermocouple-based systems. In recent years, many wireless data loggers have been developed that transmit collected data in real-time, via standard radio-frequency (RF) technology. Until now, the majority of RF data loggers have been used in low temperature applications, such as warehouse monitoring and shipping. There are not yet any commercially successful RF data loggers for use in retorts. A high temperature RF data logger would be a significant development for the food industry, as it would combine the real-time data of a wired system with the ease of use of wireless data loggers. The major hurdle to overcome in deploying RF technology for retort applications is providing the means to get an RF signal out of the process to the outside environment, where it can be picked up by a receiver and transmitted to a PC. While most plastics and glass are transparent to RF signals, any metal barrier will contain the RF signal. This includes cans, foils, and the retort wall itself. Consequently, an RF data logger cannot be mounted internally in a can as the signal will not pass through the walls. In addition, water can attenuate the RF signal so water in an immersion retort and in the food itself may represent a partial barrier. Depending on the retort, it may be required to mount an antenna on the inside of the retort to transmit the signal to the outside. The problems are not insurmountable, but creative solutions will be required to fully utilize RF technology in retort applications.

8.7 References

Awuah, G, Khurana, A, Weddig, L and Balestrini, C (2006), Heat penetration parameters: A comparative study between 'remote' temperature sensors and T-type thermocouples, *26th Annual Conference Institute for Thermal Processing Specialists,* Orlando, FL, USA.

Ball, C O and Olson, F C W (1957), *Sterilization in Food Technology: Theory, Practice, Calculations,* New York, McGraw-Hill.

Britt, I J, Zhang, Z and Tung, M A (1997), Influence of Temperature Measuring Systems on Heat Penetration Results, *17th Annual Conference Institute for Thermal Processing Specialists*, Arlington, VA, USA.

Cross, W R and Lesley, D R (1985), Self-contained microcircuity probe acquires and records food-process temperature data, *Food Technology*, 39 (12), 36–41.

Ecklund, O F (1949), Apparatus for measurement of the rate of heat penetration in canned foods, *Food Technology*, 3, 231–232.

IFTPS (1992), *Temperature Distribution Protocol for Processing in Steam–Still Retorts, Excluding Crateless Retorts,* Guelph, ON, Canada, Institute for Thermal Processing Specialists.

IFTPS (1995), *Protocol for Carrying Out Heat Penetration Studies,* Guelph, ON, Canada, Institute for Thermal Processing Specialists.

Lesley, D R (1987), Evaluation of a temperature sensing device for heat penetration data acquisition in food products, *Institute of Food Technologists Annual Meeting and Food Expo*, Anaheim, California, USA.

Nioras, R L (1982), Flexible pouch and thermocouple locator therefore, U.S. Patent #4 340 610.

PDA (2003), PDA Technical Monograph No. 1, *Industrial Moist Heat Sterilization In Autoclaves,* Bethesda, MD, USA, Peritoneal Drug Association.

9

Advances in indicators to monitor production of in-pack processed foods

G. Tucker, Campden and Chorleywood Food Research Association, UK

9.1 Introduction

One of the main objectives of thermal processing is to produce foods that are free from micro-organisms both harmful to public health and that could spoil foods. This necessitates that the process achieved at the product cold point is validated using one of several techniques available. Temperature probes, biological indicators and biochemical methods can each be used to obtain process values that are converted to log reductions of certain microbial species.

Use of biochemical methods such as time–temperature integrators (TTIs) as a technique for thermal process evaluation has received considerable attention recently. There can be advantages of using TTIs to estimate process values compared with the alternative methods. It is widely accepted in the industry that a temperature probe should be the first choice of validation method providing that it does not interfere with heat transfer. However, many situations occur whereby alternatives are required; for example, with particulates or continuous flow processes.

This chapter contains the following sections that describe the recent advances in TTI systems:

- The potential of TTIs.
- Current state of the art and limitations.
- Producing TTIs to monitor the thermal sterilisation of retorted foods.
- Future trends with pasteurisation and sterilisation TTIs.

9.2 The potential of time–temperature indicators

9.2.1 Commercial sterilisation

By definition, commercial sterility or appertisation of food is 'the condition achieved by the application of heat which renders food free from viable micro-organisms, including those of known public health significance, capable of growing in the food at temperatures at which the food is likely to be held during distribution and storage' (DoH, 1994). In practice, however, the food manufacturer makes a decision on the level of commercial risk with the applied thermal process because it is not possible to kill all of the micro-organisms and produce a saleable product. A pasteurisation process usually operates to 6 log reductions of the target organism (CCFRA, 1992, 2006), and this differs from fully sterilised foods where the intention is to achieve at least 12 log reductions in *C. botulinum* spores. The lower target log reductions for pasteurisation are because of the reduced risks associated with the target microbial species when compared with the lethal botulinum toxin, and the presence of additional preservation hurdles.

The severity of a thermal process is calculated as an integrated F-value or P-value (Ball and Olsen, 1957), using heat resistance data on the likely pathogenic or spoilage organisms present. Death of bacteria by moist heat is assumed to be almost logarithmic (Stumbo, 1965), or it follows first-order reaction kinetics in which the rate of decomposition is directly proportional to the concentration. Equation 9.1 describes the rate of change in concentration (or numbers N) of micro-organisms with time (t), where k is the proportionality constant:

$$-\frac{dN}{dt} = kN \qquad [9.1]$$

or,

$$-\frac{dN}{N} = k.dt \qquad [9.2]$$

Integrating Equation 9.2 between the limits N_0 at time zero and N after a time of heating t, results in Equation 9.3:

$$k = \frac{\ln(N_0 - N)}{t} \qquad [9.3]$$

This is usually expressed using base ten logarithms ($\log_{10}$), which are referred to in the remaining text without the subscript (log). Hence, Equation 9.3 becomes:

$$k = \frac{2.303 \log(N_0 / N)}{t} \qquad [9.4]$$

The conventional microbiological approach to quantifying thermal processing uses the decimal reduction time (D_T), which is defined as the time required to destroy 90% of the organisms by heating at a single reference temperature (T). This is calculated by the time required to traverse one log cycle on a micro-organism survivor curve, as shown in Fig. 9.1.

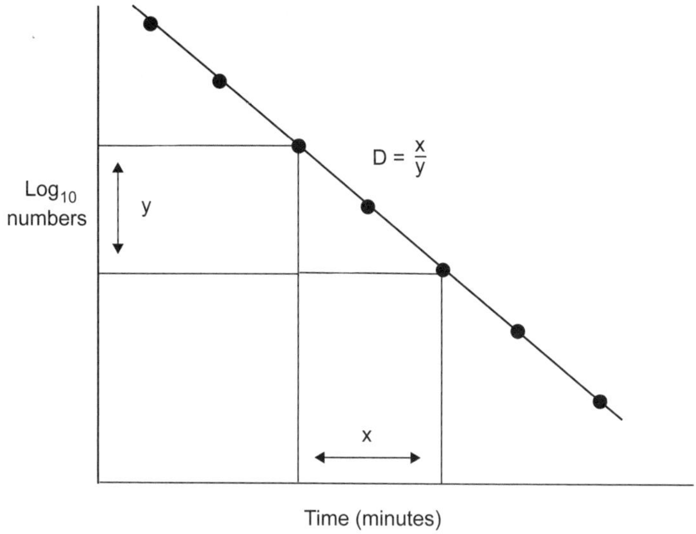

Fig. 9.1 Example of a first-order or logarithmic survivor curve.

Substituting terminology from microbiological death kinetics into the general equation for the straight line (Fig. 9.1), Equation 9.5 is obtained:

$$\log N_0 - \log N = \frac{t}{D_T} \qquad [9.5]$$

or,

$$t = D_T . \log\left(\frac{N_0}{N}\right) \qquad [9.6]$$

By comparing Equations 9.4 and 9.6, the decimal reduction time and proportionality factor can be equated. Decimal reduction time is the more convenient term used in thermal processing.

$$D_T = \frac{2.303}{k} \qquad [9.7]$$

Equation 9.6 presents the heating time (t) at a constant reference temperature in order that the number of micro-organisms are reduced from their initial population (N_0) to a final population (N). This heating time is also referred to as a sterilisation or F-value, and represents the target number of minutes at a temperature T to achieve the desired log reduction in micro-organisms (see Equation 9.8).

$$F = D_T . \log\left(\frac{N_0}{N}\right) \qquad [9.8]$$

Thus, for a sterilisation process where 12-log reductions are required, the target

F-value for an organism with *D*-value of 0.3 minutes at 121.1 °C is 3.6 minutes. The conventional approach in the UK (DoH, 1994) uses a *D*-value of 0.21 minutes at 121.1 °C for *C. botulinum* spores, which equates to a minimum *F*-value of 2.52 minutes.

9.2.2 Thermal process validation

The objective of a process validation study is to prove that the target *F*-value, as calculated by Equation 9.8, is achieved under the conditions used for the study. Conditions should be chosen to represent those that conspire to result in the lowest levels of microbiological kill, so that under normal production conditions it is not possible for the process to be less severe. This sounds simple, but requires considerable work to achieve this objective successfully.

As the number of food products and their variety increases, food companies are faced with the challenge of proving that their products are safely thermally processed. Temperature probes provide the most economical method of validating process severity with the greatest flexibility in how the data are used. This validation can sometimes be difficult if temperature probes cannot be used in the processes and other approaches need to be adopted. The main process categories that introduce these complexities include:

- Products cooked in continuous ovens or fryers, such as poultry joints, chicken nuggets, burgers, bread (Tucker *et al.*, 2005).
- Products with discrete pieces cooked in steam-jacketed agitated batch vessels, such as ready meals, soups, cook-in-sauces, fruit preparations (Tucker *et al.*, 2002).
- Particulate products processed in continuous tubular and scraped surface heat exchangers, such as cook-in-sauces, preserves, dressings (Tucker *et al.*, 2002).

If temperature probes cannot be used, a number of approaches to validating microbiological process safety are available. To prove that the thermal process has achieved the target process value or *F*-value during manufacture, it is necessary to conduct validation studies using an approved method. Various methods can be selected from the list below, and their choice depends on the costs and on the nature of the food and the process type.

- Microbiological methods can be used whereby cells or spores of a non-pathogenic organism, with similar temperature-induced death kinetics to the target pathogen, are embedded into an alginate bead (Brown *et al.*, 1984). The beads are made to mimic the food pieces in their thermal and physical behaviour and so pass through the process with the food. By adding macerated food into the calcium alginate gel, a close approximation to the physical and thermal properties of the food is obtained. Typically, 30–60 spore beads will be added to a continuous heat exchanger process in order to obtain a distribution of process values. Enumeration of the surviving organisms allows the log reduction and process values to be calculated.

- Simulated trials are carried out in a pilot plant or laboratory where the heat transfer conditions of the process are replicated (CCFRA, 1977; Bee and Park, 1978). This used to be a common approach for continuous canning systems, such as hydrostatic retorts and reel and spiral cooker–coolers, but the introduction of loggers that can travel with the cans has reduced the need for retort simulators. However, there are still some concerns over the interference with the temperature measurements that a stainless steel logger must impose. This is particularly true with cooker–coolers that rely on frictional forces to introduce can rotation in the bottom third of the reel. Changes to can density or to the centre of gravity might influence the rotational forces.
- No validation is attempted, with the process safety being inferred from temperature probing of the bulk product or the environment. Substantial over-processing is allowed, in order that the thermal process delivered to the product thermal centre is sufficient. End product testing for microbiological activity is usual. This approach is typical with the chilled foods industry; for example, with sauces cooked in steam pans and hot filled into plastic ready-meal trays (CCFRA, 1992).
- Process models can be developed that predict, for example, the temperature–time history of the critical food particulates as they travel through the heating, holding and cooling zones of the process (Heppell, 1985; McKenna and Tucker, 1991). This approach is used with continuous heat exchangers, primarily to ensure that small food particulates receive an adequate process. For larger particulates, greater than 2–3 mm, it is usual that a spore bead method is employed.
- Time–temperature integrators (TTIs) can be applied to gather similar process data to that from microbiological methods. This is a new method that originated with work by Hendrickx *et al.* (1995) in which various types of bacterial amylases were found to show kinetic properties appropriate for estimating microbiological reductions. The advantage of the amylase TTIs over many biochemical systems is that the reaction rates for amylase degradation by heat are first order, as with microbiological breakdown, and the temperature sensitivity of the reaction rates is similar to that for spore destruction.

9.3 Current state of the art and limitations

A TTI can be an enzyme, colour compound, nutrient or physical property change that breaks down in a reproducible manner during heating. Enzymes such as amylase or peroxidase are suitable for TTIs because their structure breakdown is affected by both time and temperature. Typically, with enzymes, this breakdown involves the helical structure unwinding as cross-links between molecular chains are broken. Many enzyme systems can regenerate after heating; however, an enzyme suitable for use as a TTI must exhibit a permanent denaturation.

The kinetics of the temperature-induced denaturation should match those of the death kinetics for the target microorganism. Specifically, the decimal reduction

Table 9.1 Kinetic factors for microbial destruction by wet heat. Data are selected specifically for micro-organisms relevant to full sterilisation processes. All reference temperatures are 121.1 °C

Organism	Temperature (°C)	$D_{121.1}$ (s)	z (C°)
Bacillus stearothermophilus	100–140	149	14.3
(Phosphate buffer)	100–140	170	12.3
	100–140	226	11.7
Bacillus subtilis			
(Phosphate buffer)	127–144	28.8	9.4
Clostridium botulinum			
(Phosphate buffer)	140–127	8.0	9.0
(Water)	140–127	3.1	8.5
(Pureed peas)	104–127	5.3	8.3
(Meat and vegetables)	100–113	6.6	9.8
(Sea food)	100–113	3.0	7.4
(Poultry)	100–113	3.0	7.4
(Rock lobster)	105–115.5	18.0	10.8
Clostridium sporogenes			
(Phosphate buffer)	100–120	15.0	9.1
(Strained pea)	115.5–143.3	60	9.8

time (D_T) and the kinetic factor (z) are the kinetic parameters used. The kinetic factor (z-value) is a measure of how the D-value changes with temperature, and is also calculated using a semi-logarithmic approach. Logarithms of D-values are plotted against temperature, and the temperature change required to effect a one-log change in D-value is defined as the z-value. Most bacterial spores show z-values close to 10 C°.

Thermal processes are designed to reduce microbiological populations by large numbers of log reductions, typically between six and twelve (as shown in Equation 9.8). It is unlikely that a TTI system will possess sufficient measurement sensitivity for such high log reductions in the measured parameter, whether it is a colour change or enzyme activity. Therefore, the decimal reduction time should ideally be several times as large as that for the target microbial species, otherwise there will be insufficient colour or activity left to measure from the processed TTI. As mentioned previously, the other requirement is for the z-value to be close to that for the target microbial species.

Table 9.1 presents examples of data for micro-organism death kinetics, which highlight the relatively low D_T values when compared with 'chemical' systems suitable for use as TTIs (Tables extracted from Holdsworth, 1997). Tables 9.2, 9.3 and 9.4 illustrate the wide range of D_T and z-values with vitamin, enzyme and pigment systems, respectively. Each of these systems is potentially suitable for use as a TTI.

Many of the chemical systems in Tables 9.2 to 9.4 could be used as TTI systems. However, if the TTI system is intended for estimating process values and converting these to log reductions of micro-organisms in foods, it is essential that the

Table 9.2 Kinetic factors for vitamin destruction by wet heat. Reference temperatures vary depending on the data reported

Heat-sensitive vitamin	Temperature (°C)	D_T (s)	z (C°)
Vitamin A (beta carotene)			
(beef liver)	103–127	D_{122} = 2,400	23.0
(carrot juice)	104–132	D_{104} = 23 600	25.5
Vitamin B1 (thiamin)			
(buffer)	109–150	D_{109} = 9,500	24.0
(carrots)	109–150	D_{150} = 830	22.0
(spinach)	109–150	D_{150} = 610	22.0
(pea puree)	121.1	$D_{121.1}$ = 10 000	31.3
(lamb puree)	109–150	D_{122} = 710	25.0
(pork luncheon meat)	100–127	D_{127} = 6,300	35.0
Vitamin B6 (pyridoxine)			
(cauliflower)	106–138	D_{121} = 24 000	43.0
Pantothenic acid			
(beef puree pH 5.4)	118–143	$D_{121.1}$ = 138 000	35.8
(beef puree pH 7.0)	118–143	$D_{121.1}$ = 135 000	19.3
Folic acid			
(apple juice)	100–140	D_{140} = 100 000	31.0
Vitamin C (ascorbic acid)			
(peas)	110–132	$D_{121.1}$ = 50 000	18.0
(spinach)	70–100	D_{100} = 25 900	74.4

Table 9.3 Kinetic factors for enzyme destruction by wet heat. Much of these data were taken by the frozen foods industry for the purposes of estimating enzyme breakdown during blanching, hence the lower reference temperatures

Heat-sensitive vitamin	Temperature (°C)	D_T (s)	z (C°)
Peroxidase			
(horseradish)	60–160	D_{120} = 830	27.8
(potato puree)	100–140	D_{120} = 70	35.0
Catalase			
(spinach)	60	D_{60} = 60	8.3
Lipoxygenase			
(pea/soya)	50–80	D_{77} = 720	3.4
Pectinesterase			
(guava syrup pH 4.0)	74–95	D_{96} = 35	16.5
Polyphenol oxidase			
(potato)	80–110	D_{89} = 100	7.8

z-value of TTI and micro-organism are similar. In addition, the D_T-value should be sufficiently high that changes in the measured property during a thermal process are within the measurement range of highest accuracy. This limits the choice of chemical marker, and for these reasons, the TTI systems used and developed for

Table 9.4 Kinetic factors for pigment destruction by wet heat. Reference temperatures vary depending on the data reported.

Heat-sensitive pigment	Temperature (°C)	D_T(s)	z (C°)
Green (chlorophylls)			
(green beans)	80–148	$D_{121.1}$ = 1,260	38.8
(peas)	80–148	$D_{121.1}$ = 1,500	39.4
Red			
(raspberry juice)	78–108	D_{108} = 7,000	30.4
(grapes)	76.7–121	D_{121} = 7,600	54.7
Browning reactions			
(chestnut paste darkening)	105–128	$D_{121.1}$ = 141 000	24.6
(milk, hydromethyl furfural formation)	105–160	D_{130} = 12	26.7

work on thermally processed foods are based on α-amylases. These represent a significant advance in thermal processing because it can be seen clearly that very few of the chemical systems in Tables 9.2 to 9.4 show suitable combinations of *D*- and *z*-value for use in estimating micro-organism reductions.

9.3.1 Calculation of process values with TTIs

Enzyme, or specifically, amylase breakdown by heat shows first-order reaction kinetics, as highlighted in the example in Fig. 9.2. This example is for an amylase from *Bacillus amyloliquefaciens*, showing a *D*-value of 10.1 minutes at 85 °C (Tucker *et al.*, 2002). Analysis of amylase concentration uses spectrophotometric techniques to measure the amount of amylase activity present after various heating times. These tests are known as assays.

Process values or *F*-values estimated with TTIs are calculated from the initial and final amylase activities using a similar equation to 9.8. Instead of using the initial and final number of surviving micro-organisms, as in Equation 9.8, the *F*-value equation uses amylase activities before and after heat treatment (see Equation 9.9). Activity is measured as a rate of colour development when the amylase reacts with a commercially available amylase reagent (Randox Laboratories Ltd). This is a relatively simple chemical test to perform.

$$F = D_T \cdot \log\left(\frac{A_{initial}}{A_{final}}\right) \qquad [9.9]$$

where,

A_{final} is the final amylase activity after a specific time–temperature history, minutes^{-1},

$A_{initial}$ is the initial amylase activity, minutes^{-1},

D_T is the amylase decimal reduction time at a fixed temperature (T), minutes.

Equation 9.9 presents the method for calculating a sterilisation (*F*-value) or pasteurisation (*P*-value) based on measurements of enzyme activity. Decimal

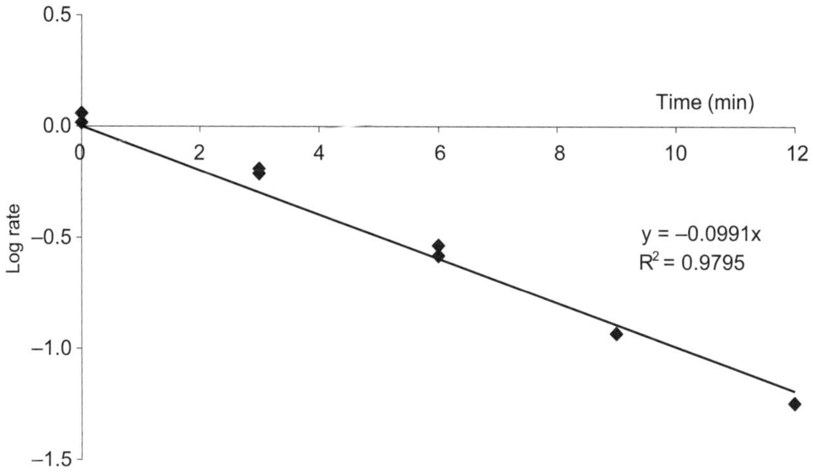

Fig. 9.2 Example of a first-order reaction curve for an amylase from *Bacillus amyloliquefaciens*, showing *D*-value of 10.1 minutes at 85 °C.

reduction time (D_T value) is used with the calculation of *P*-value from amylase activity measurements (see Equation 9.9), because this equation estimates log reductions in activity, or in other words, the number of decimal reductions. However, if a *P*-value is calculated from temperature measurements, it is the *z*-value that is used in the calculation (Bigelow *et al.*, 1920). In order that an amylase TTI system can be applied to estimate microbiological log reductions, it is essential that the *z*-value for microbiological destruction and for amylase structure breakdown are similar. Equation 9.10 presents the lethal rate equation used to calculate a *P*-value, which integrates the time and temperature effect of a thermal process, as measured from a temperature probe.

$$P = \int_0^t 10^{\frac{T(t)-Tref}{z}} \cdot dt \qquad [9.10]$$

In Equation 9.10, the reference temperature is shown as T_{ref}, which must be the same as that used in calculating the D_T-value. *T*(*t*) is the measured food product temperature that changes with time (*t*). Equations 9.9 and 9.10 should provide the same measured *P*-value or *F*-value providing that the D_T-value is quoted at the reference temperature (T_{ref}), and the *z*-value is appropriate to the amylase TTI system (see Equation 9.11).

$$P = \int_0^t 10^{\frac{T(t)-Tref}{z}} \cdot dt = D_T \cdot \log\left(\frac{A_{initial}}{A_{final}}\right) \qquad [9.11]$$

where,

$T(t)$ is the product temperature, which is a function of time, °C,
T_{ref} is the reference temperature for the D_T value, °C,
t is the process time, minutes,
z, the kinetic factor, is the temperature change required to effect a ten-fold change in the D_T value, C°.

9.3.2 Previous TTI work

Use of amylase TTIs as an alternative means of process evaluation to either temperature or microbial systems has received considerable attention in recent years (De Cordt *et al.*, 1992, 1994; Maesmans *et al.*, 1994; Hendrickx *et al.*, 1995; Van Loey *et al.*, 1996; Tucker, 1999, 2000). Reasons for the interest lie with the unique properties that bacterial amylases appear to exhibit, most importantly theirs being one of the very few chemical systems that can be characterised with a z-value close to that for micro-organism destruction.

Typically, a spore-forming microorganism will exhibit z-values in the range 9–11 C°, with vegetative cells showing slightly lower values in the range 6–8 C° (CCFRA, 1992). Table 9.1 presents examples of various spore-forming micro-organisms that are important to the sterilised food sector. Values for micro-organisms critical to pasteurised foods show similar ranges for spores and vegetative cells. Amylase TTIs are the TTI systems chosen for use as mimics for destruction of micro-organisms because they are reported to exhibit measured z-values in the range 9–10 C°.

With several TTI systems in regular use at CCFRA, a nomenclature system is used to help differentiate the TTI types. The first three letters referred to the sources of amylase, for example, *Bacillus licheniformis* amylase is BLA. This is followed by numbers that refer to the reference temperature of the micro-organism that this TTI is designed to mimic, for example, 90 °C for psychotrophic strains of *C. botulinum*. Hence this TTI is BLA90. Table 9.5 presents the full range of amylase TTIs discussed in this chapter.

Within the TTIs listed in Table 9.5, one of the TTIs is used for two different pasteurisation processes; this is the amylase from *Bacillus licheniformis*. The two pasteurisation treatments appropriate to this TTI are for acidic foods stored in ambient conditions and for low-acid foods stored chilled for extended periods of 10–40 days. These treatments target different micro-organisms but the integrated thermal effect is similar. The amylase TTI for retort temperature distribution measurements is also from *Bacillus licheniformis* but at a different concentration, which has the effect of increasing its heat stability.

9.3.3 Measurement of amylase *D*-values

Knowledge of the D-value is critical for the amylase TTI systems because this is the key variable in the calculation step for P-value estimation (see Equation 9.9). Obtaining the highest accuracy in measuring D-values requires considerable care. In order to minimise changes in D-value with different batches of amylase

Table 9.5 Range of amylase TTIs used for measuring pasteurisation and sterilisation processes

TTI code	TTI description	Organism of amylase origin	Target process	*D*-value at T_{ref} (minutes)	*z*-value (C°)	Range (minutes at T_{ref})
BAA70	Cook-chill	*Bacillus amyloliquefaciens*	2 minutes at 70 °C	8–10	8.0–9.0	2–25
BAA85	High acid	*Bacillus amyloliquefaciens*	5 minutes at 85 °C	8–10	9.0–9.5	4–30
BLA90	REPFEDS or sous-vide	*Bacillus licheniformis*	10 minutes at 90 °C	15–25	9.0–9.5	5–50
BLA93	Acid foods	*Bacillus licheniformis*	5 or 10 minutes at 93.3 °C	8–12	9.0–9.5	4–30
BLA100	Retort distribution	*Bacillus licheniformis*	–	8–10	5.5–7.5	5–30
PFA121	Sterilisation	*Pyrococcus furiosus*	3 minutes at 121.1 °C	21–24	9.0–11.0	3–40

solutions, the solutions can be prepared in sufficient quantity for use over many months or years. This relies on frozen storage of the solutions in vials that contain sufficient volume for a typical industrial TTI trial. At the time of writing this chapter, CCFRA have a stock of BAA85 solution that is over 24 months old, which has retained its heat stability properties as measured by the *D*-value at 85 °C.

Calculation of the *D*-value for either type of TTI follows a similar method to that used for obtaining the same data with micro-organisms (Stumbo, 1965). TTIs are immersed in well-stirred water baths at 85, 90 or 93.3 °C, depending on the type of amylase solution, for a series of heating times. For example, if historic data suggest that the BAA85 solution has a *D*-value of 8 minutes at 85 °C, suitable heating times would be 0, 4, 8 and 12 minutes at 85 °C. These times are chosen so that reductions in amylase activity around 1-log are measured.

Figure 9.3 illustrates an example of a BAA85 calibration test conducted at 84.3 °C. The decimal reduction time at 84.3 °C ($D_{84.3}$) is 8.4 minutes, which gave a D_{85} of 7.1 minutes when calculated with a *z* of 9.4 C° (Lambourne and Tucker, 2001; Tucker and Wolf, 2003).

Adjustment of the *D*-value is often required because the test (calibration) temperature usually differs slightly from the reference temperature. In the example presented, the test temperature is 84.3 °C and the reference temperature is 85 °C. A further calculation is used to convert the *D*-value, as shown in Equation 9.12.

$$D(T_{ref}) = D(TestTemp) \times 10^{((TestTemp - T_{ref})/z)} \qquad [9.12]$$

A *D*-value calculated from this calibration test should be close to historic values for the same commercial batch of amylase powder. Small differences, up to 15%, can be measured when making up amylase solutions, and are thought to be caused by variation in weighed quantities of amylase powder, concentration of buffer

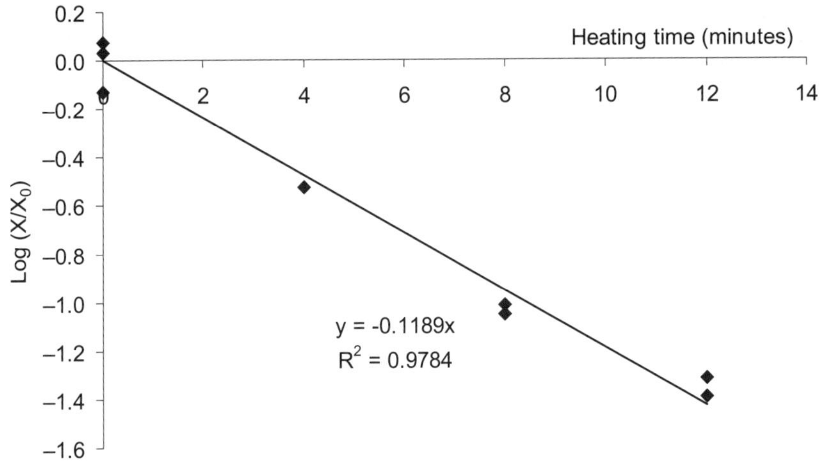

Fig. 9.3 Data from of a BAA85 calibration test conducted at 84.3 °C. $D_{84.3}$ was 8.4 minutes which gave a D_{85} of 7.1 minutes when calculated with a *z* of 9.4 C°.

components and Randox solutions, and microlitre volumes of amylase solution extracted from TTI tubes.

9.4 Producing time–temperature indicators to monitor the thermal sterilisation of retorted foods

Attempts to manufacture an enzyme-based TTI to survive a full sterilisation process at 121.1 °C for several minutes have had limited success when applied to industrial situations. Extension of the usable range upwards into sterilisation temperatures has been demonstrated by drying commercial grade amylases to precise moisture levels (Van Loey *et al.*, 1997b, Guiavarc'h *et al.*, 2002, Guiavarc'h, 2003). Results achieved in the laboratory were encouraging and showed that different levels of moisture content gave a range of heat stabilities.

One approach immobilised a mixture of *Bacillus licheniformis* α-amylase (BLA), sucrose and salts at the surface of non-porous glass beads (inert filler), and under several low-moisture conditions (a_w in the range 0–0.63 after equilibrium at 4 °C (Guiavarc'h *et al.*, 2004a,b). These systems showed potentially useful thermal stabilities in the range 100–132 °C under isothermal and non-isothermal conditions. Thermal processes up to 30 minutes at 121.1 °C could be monitored.

A similar approach was taken by Samborska *et al.* (2005), who reduced moisture content to enhance thermal stability of *Aspergillus oryzae* α-amylase. The enzyme was mixed with maltodextrin and freeze dried after equilibration above saturated salt solutions to achieve moisture contents from 3.5 to 0.029 g H_2O/g dry wt. Decreasing the moisture content from 3.5 to 0.029 g H_2O/g dry wt. led to an increase in the temperature at which inactivation of the amylase occurred, increasing it from 70–75 to 100–115 °C. The activation energy (related to *z*-value) of thermal inactivation was also affected by moisture content. Despite reducing the moisture content, the levels of heat stability for *Aspergillus oryzae* α-amylase were insufficient for use as a sterilisation TTI.

The approach involving reduced moisture amylase by Guiavarc'h *et al.* (2002 and 2004b) again used *Bacillus licheniformis* α-amylase, but equilibrated at 81% equilibrium relative humidity at 4 °C (BLA81). Due to its low water content, BLA81 denaturation could be studied in the range of 118–124 °C. Two batches of BLA81 were successfully validated under non-isothermal conditions allowing the determinations of process values (reference temperature of 121.1 °C) in the range of 1–15 minutes. BLA81 was used as a TTI to investigate potential differences of process values received by freely moving spherical particles as compared to a centrally fixed particle (single-position impact) inside cans containing water as brine. Interesting results showed the process value received by freely moving particles to be from 5.6% (4 rpm) to 19.7% (8 rpm) smaller than those with centrally fixed spheres.

Other researchers (Tucker and Wolf, 2003) have encountered experimental difficulties in controlling the sterilisation TTI based on the dried amylase approaches. Drying the amylase in a steel differential scanning calorimeter (DSC) capsule increased heat stability, and the same capsule was used for measuring process values. This had the advantage that, once sealed, the capsule did not have to be opened during the heating tests or for analysis. However, the encapsulation method did not provide adequate isolation from its environment when used in industrial sterilisation processes because the rubber O-ring gasket was unable to withstand the swings in pressure experienced within a food container during a sterilisation process. It was possible to encapsulate the DSC capsule in a silicone compound that prevented moisture ingress but this resulted in dimension changes to the TTI particulate. For some products with large particulates this was acceptable, but for most it negated the purposes of conducting process validation studies using TTIs. One further issue with the steel capsules was their high density, which prevented their use in flowing foods.

Thus, a different method is required for a sterilisation TTI, with ideally a TTI system that allows the TTI to be used as a solution within the silicone TTI tubes already proven in industrial pasteurisation processes. This is a substantial challenge that requires a novel solution.

9.4.1 A novel sterilisation TTI

One idea developed by Tucker *et al.* (2007) is to locate a micro-organism that has evolved in hostile conditions of temperature, yet is reported to be an amylase producer. Of the millions of types of micro-organisms found in nature, this narrows the search to just a few of extreme durability. The challenge is to grow this micro-organism, extract the amylase, and apply it in its native form to industrial sterilisation processes. The logic is that the amylases produced by these extreme micro-organisms must be able to withstand high temperature conditions, otherwise their structure would break down before they started work in breaking down complex starches and carbohydrates to sugars.

Micro-organisms are known to exist in hostile environments, such as volcanic pools, where they have adapted to high temperature conditions and to chemical environments considerably different from those favoured by the micro-organisms we are more familiar with (Segerer *et al.*, 1993; Stetter,1996). These 'hyper-thermophilic' micro-organisms represent a relatively new area for microbiological research and one with enormous potential for supply of heat-stable enzymes (Sterner and Liebl, 2001). A number of bacteria capable of growing at or above 100 °C have been isolated from several geothermic terrestrial and marine environments (Vieille and Zeikus, 2001). Among the many interesting features associated with these bacteria are their ability to grow and carry out biological functions at normally protein-denaturing temperatures. The enzymes that are formed by these extremely thermophilic and hyperthermophilic micro-organisms are of great interest due to their thermostability and optimal activity at high temperatures.

Several of the most promising micro-organisms referred to later are known as archaea. These are defined in the *American Heritage Dictionary* (2003) as a group of bacteria-like micro-organisms comprising a division of the Prokaryotae and usually thriving in extreme environments. Prokaryotae are unicellular organisms lacking a true nucleus and nuclear membrane, with genetic material composed of a single loop of naked double-stranded DNA. Archaea are often classified as a separate domain in taxonomic systems based on similarities of DNA sequences. However, for the purposes here of locating an amylase-producing micro-organism, archaea are bacteria that can survive in extreme environments such as high temperatures, and can produce thermostable enzymes.

Micro-organisms that produce the starch-hydrolysing enzyme α-amylase include *Pyrococcus woesei, Pyrococcus furiosus, Thermococcus celer, Fervidobacterium pennavorans, Desulfurcoccus mucosus* and *Termotoga maritima* (Leuschner and Antranikian, 1995). All these micro-organisms offer potential as sources of thermostable α-amylase; however, the one chosen by Tucker *et al.* (2007) was *Pyrococcus furiosus.*

Pyrococcus furiosus is an obligate anaerobic, hypothermophilic archaebacterium, which has been isolated by Fiala and Stetter (1986) from shallow thermal waters near Vulcano island, Italy. The motile coccus-shaped microbe, with about 50 flagella at one end, is capable of growth on complex media with or without elemental sulphur.

P. furiosus amylase is active over a broad temperature (40–140 °C) and pH range (3.5–8.0), with optimum activity at 100 °C and pH 5 (Koch *et al.*, 1990). In terms of the amylase stability to heat, no loss of activity was detected after 6 hours of incubation at 90 °C (Koch *et al.*, 1990). At 120 °C, about 10% of the initial activity was measured after 6 hours. This equated to a decimal reduction time at 120 °C of 6 hours (D_{120} = 6 h). In order to inactivate the enzyme completely, incubation had to be performed at 130 °C for at least 1 h. These levels of heat stability are higher than those required to measure a thermal sterilisation process, where the target is to exceed at least 3 minutes at 121.1 °C ($F_0$3), but the process can sometimes be as high as F_0 50.

The primary objective of the work reported by Tucker *et al.* (2007) was to determine if amylase from *Pyrococcus furiosus* was suitable for use as a sterilisation TTI. In order for this to be possible, the kinetics of the amylase destruction by heat were limited by the following two constraints:

(i) It must show sufficient heat stability for some of the active amylase structure to remain after several minutes heating at 121.1 °C. The commercial requirement is for sterilisation processes to achieve at least an equivalent process of 3 minutes at 121.1 °C. However, this is often increased to allow for product and process variability, and to target spoilage microorganisms of higher heat resistance (Stumbo, 1965). Decimal reduction time (D_T-value) for the amylase at 121.1 °C was the parameter that characterised the heat stability.
(ii) It must exhibit a temperature sensitivity of breakdown that is characterised by a z-value close to 10 °C. This is the value used to represent the destruction of *Clostridium botulinum* spores (Stumbo, 1965).

Finding a TTI material with thermal behaviour within these constraints is extremely difficult, otherwise a liquid sterilisation TTI would already have been discovered and be available for use.

9.4.2 Obtaining a supply of thermostable amylase

Several stages were involved in achieving the above objectives:

- The first was to determine whether the *Pyrococcus furiosus* micro-organism could be grown in conditions that were favourable to amylase production and in the quantities suitable for use as a TTI. This proved more difficult than expected because of the extreme nature of the fermentation conditions and doubts over the viability of the micro-organisms supplied.
- Purification of the amylase was the next issue since it is reported that up to 80 other enzymes are produced in the fermentation, as well as numerous by-products of the fermentation that may affect the amylase performance (Adams *et al.*, 2001). This work focussed on finding a suitable candidate material for the sterilisation TTI, and conducting sufficient tests to confirm that the D- and z-values were in the correct range.
- Encapsulation of the sterilisation TTI was achieved using the same method as with the pasteurisation TTIs; that was within silicone tubes capped with a silicone elastomer compound. These TTI tubes gave the greatest flexibility for applications to industrial processes. Integrity of these TTI tubes at sterilisation process conditions was unknown, i.e. temperatures of 115 to 135 °C, pressures up to 4 bar, and very rapid pressure changes.
- There was also the need to determine which assay method was appropriate to an amylase with optimal activity close to 100 °C.

Certain definitions are important at this stage. The new thermostable amylase from *Pyrococcus furiosus* is referred to as PFA121. However, when the PFA121 TTI is in the form of freeze-dried powder (FDP) from the *Pyrococcus furiosus*

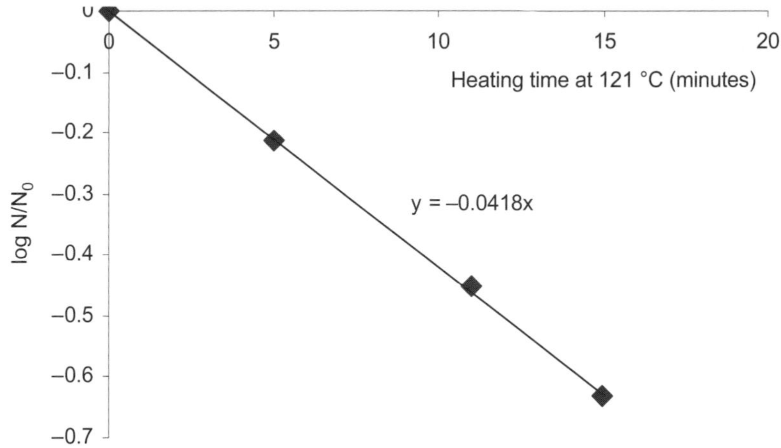

Fig. 9.4 Graph of heating time at 121 °C against log of final activity (N) divided by initial activity (N_0). D-value was 23.9 minutes at 121 °C.

fermentation, it is referred to as FDP. When the FDP is dissolved in buffer solution and encapsulated within a silicone TTI tube it is referred to as a sterilisation TTI.

Collaboration with the University of Georgia (Department of Biochemistry and Molecular Biology) was set up because of their track record with *Pyrococcus furiosus* fermentation and enthusiasm to work together (Adams *et al.*, 2001). Amylase activity was measured at high levels from a fermentation broth using peptides as the carbon source. This was a level that could be used in a dissolved form for TTI work.

9.4.3 Determination of *D*- and *z*-values

Isothermal methods for *D*- and *z*-value measurement are the industry standards because of their relative simplicity. Immersion of sealed glass capillary tubes in an oil bath at 121 °C was used to obtain the first data on the D_{121}-value for the sterilisation TTI. However, much of the thermal characterisation used non-isothermal methods because they were more effective in their use of the small quantities of FDP available.

Figure 9.4 shows the traditional approach of plotting the logarithm of ratio of final activity (N) divided by initial activity (N_0) as a function of heating time (Stumbo, 1965). They represent data using FDP at a concentration of 15 mg/ml of acetate buffer with isothermal conditions of 121 °C. A D-value at 121 °C of 23.9 minutes is suitable for estimating F_0-values for sterilised foods (Ball and Olson, 1957). The important aspect of this is that *P. furiosus* amylase as a candidate material for producing a sterilisation TTI in solution appears to work. Thermal processes with commercial F_0-values between 6 and 20 minutes can be measured using this TTI.

There are two main limitations of the traditional isothermal kinetic method for measuring D- and z-values:

- The experiments need to be repeated at several different temperatures, usually at least four, in order to calculate the z-value, which is time-consuming and uses relatively large quantities of enzyme.
- Isothermal death kinetics do not represent the nature of heating experienced by a food product, in which the particulate or pack temperatures increase gradually during heating before decreasing gradually during cooling.

Non-isothermal methods for kinetic data determination have been used by various research groups (De Cordt *et al.*, 1992; Miles and Swartzel, 1995; Van Loey *et al.*, 1997a). The method requires the sterilisation value or F-value to be calculated simultaneously (as shown in Equation 9.13) using amylase activity measured with a TTI and from temperature measurements. This will result in two F-value measurements that will equate, providing that the D-value and z-value used in either side of Equation 9.13 are correct for the TTI system.

$$F = \int_0^t 10^{\frac{T(t)-T_{ref}}{z}} \cdot dt = D_{T_{ref}} \cdot \log\left(\frac{N_0}{N}\right) \qquad [9.13]$$

In Equation 9.13, N is the final amylase activity after a specific time–temperature history, and N_0 is the initial activity. $D_{T_{ref}}$ is the decimal reduction time at the reference temperature (T_{ref}) appropriate to destruction of thermostable amylase, and in this case it was taken at 121 °C. $T(t)$ is the measured product temperature, which is a function of time (t). z is the kinetic factor for the FDP, which is the temperature change required to effect a ten-fold change in the $D_{T_{ref}}$ value.

From Equation 4.2, the integration of temperature over time (left side of the equation) will result in the same F-value as that calculated from the sterilisation TTI activities (right side of the equation), provided that first-order kinetics have been followed for the amylase destruction throughout the heat process. Hence, by applying the correct z-value to the temperature measurements and the correct D-value for the amylase activities, both sides of Equation 9.13 will be equal.

Effort was put into non-isothermal calibration because of the limited quantity of FDP. A series of experiments were carried out over varying time–temperature conditions with TTIs located at or close by the temperature sensors so that both systems measured the same process. All F-values must be measured at the end of cooling because this represents the measurement obtained from a TTI system when used for measuring processes in packs of food or in a continuous heat process (Tucker *et al.*, 2002).

One unique pair of D_{121} and z-values is appropriate for all sets of time–temperature data. To achieve a range of F-values, the thermal processing data sets utilised different product heating rates as well as different process temperatures between 121 and 131 °C. These sets of time and temperature data provided a range of different rates at which the lethal rates accumulated.

Two sets of industrial trials were set up in order to provide a wide range of *F*-values to challenge the measurement range of the TTI and thus estimate D_T and z. The objectives of these trials were two-fold:

(i) To obtain data for use in a non-isothermal measurement of *D*- and *z*-values for the FDP in acetate buffer. Several very different time–temperature and time–*F*-value measurements were experimentally set up so that *D*- and *z*-values were appropriate over a wide range.
(ii) The industrial trials were also to evaluate whether the sterilisation TTI tubes would withstand the rigours of a full thermal treatment where pressure swings of up to ±2 bar can occur almost instantaneously.

Temperature and TTI measurements were taken from the same position within the food product. This was achieved by taping at least one sterilisation TTI tube to the measuring junction of a temperature sensor. A common measuring position was assured within a few millimetres for each matching pair of TTI and probe. Sterilisation TTI tubes were approximately 8 mm in length and 2.5 mm outside diameter.

This non-isothermal method used 25 mg/ml FDP solution within silicone TTI tubes and exposed them to various thermal processes where the *F*-values accumulated at different rates. Kinetic data (i.e. *D*- and *z*-values) were evaluated with a series of coupled equations within an Excel workbook. Time–temperatures were converted to *F*-values using the left side of Equation 9.13, with *z*-value as the kinetic parameter, and amylase activity measurements using the right side of Equation 9.13, with *D*-value as the kinetic parameter.

Within the Excel worksheet, the parameters used to determine values for $D_{121.1}$ and *z* were the sterilisation values (*F*-values) calculated from the time–temperature data (referred to as *F(t-T)*) and from the TTI data (referred to as *F(TTI)*). By comparing *F(t–T)* and *F(TTI)* calculations from paired values it was possible to estimate values for $D_{121.1}$ and *z* that minimised the sum of the absolute difference between all of the TTI and probe measurements. This was done through an Excel macro using a $D_{121.1}$ of 24 minutes and a *z* of 10 C° as the starting point. The macro stepped through increments in *D*- and *z*-value of 0.05 respective units to locate the lowest sum of the absolute difference.

For Trial 1, a $D_{121.1}$ of 21.45 minutes and a *z* of 9.95 C° were estimated. Figure 9.5 shows the paired values of *F(t–T)* and *F(TTI)* for Trial 1. For Trial 2, calculated values were $D_{121.1}$ of 25.00 minutes and *z* of 11.5 C°, which were quite close to those calculated from Trial 1. Figure 9.6 shows the paired values of *F(t–T)* and *F(TTI)* for Trial 2.

Calibration of any measurement system is an essential requirement in order to provide confidence that the values are correct and within a defined error band. Estimated error bars displayed in Figures 9.5 and 9.6 were ±10% on time–temperature *F*-values and ±12.5% on *TTI F*-values. These errors were calculated from known inaccuracies with the measurement systems and variability with the relative experiments.

F-values predicted using the calculated $D_{121.1}$-value for the sterilisation TTIs

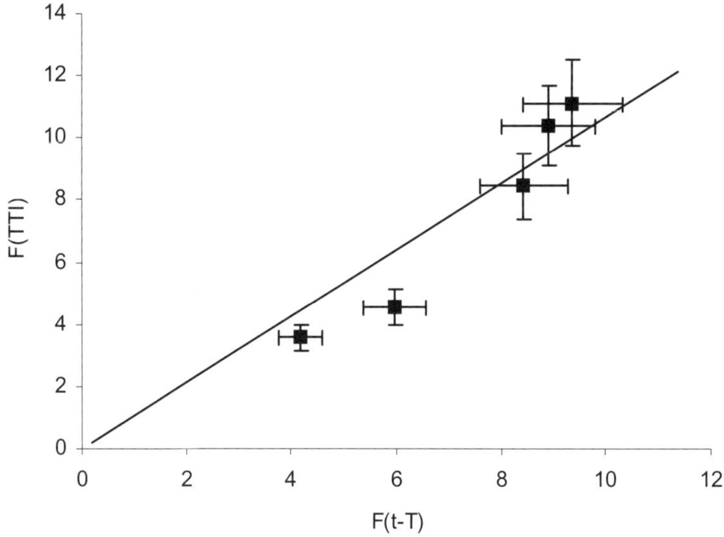

Fig. 9.5 Graphical illustration of F(t-T) and F(TTI) for Trial 1. $D_{121.1}$ was 21.45 minutes and z was 9.95 C°.

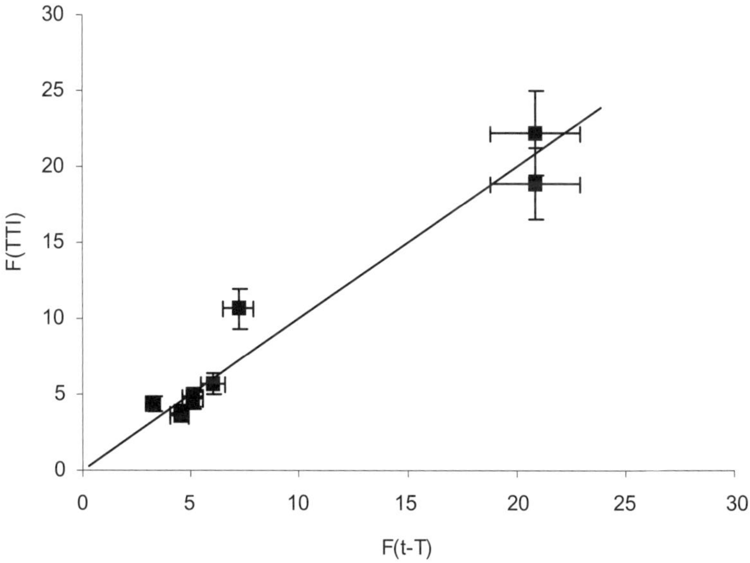

Fig. 9.6 Graphical illustration of F(t-T) and F(TTI) for Trial 2. $D_{121.1}$ was 25.00 minutes and z was 11.5 C°.

were consistently within 1.5 F-value units of the F-values from the time–temperature data. With most in-pack thermal processes operating at around F_0-values of 6 to 12 at end of heating, this was an acceptable level of inaccuracy. Continuous thermal processes with particulates usually operate to substantially higher F_0-values because of the uncertainty involved with their measurement. Thus, an error of ±1.5 minutes on a measured F-value in the region of 20–30 minutes would not be an issue.

The novel concept of an amylase produced by a micro-organism of extreme heat stability was demonstrated in this work (Tucker *et al.*, 2007). These results were the culmination of two years of effort with the sterilisation TTI in solution. Much of this time was involved with obtaining amylase from the hyperthermophilic micro-organism *Pyrococcus furiosus*, which proved difficult, but having obtained a working sample, the results for its heat stability were shown to be suitable for use as a sterilisation TTI.

9.5 Future trends with pasteurisation and sterilisation time–temperature indicators

TTI systems for measuring process values with food pasteurisation treatments are developed to a level that they can be used with commercial processes. Table 9.5 presents the range of TTIs developed for pasteurisation. Improvements are always possible with any measurement technique, and it is likely that the amylase-based TTIs will improve over the next few years. One of the main areas is to enable food companies to carry out the measurements without reliance on a service provider such as CCFRA. This can be achieved only with simplification of the methods for assaying amylase activity that currently require a laboratory spectrophotometer. Efforts were made to use a simpler colorimeter (Tucker *et al.*, 2005), but issues arose with the lack of sample temperature control. There may be a way to overcome this by consideration as to how the amylase assay can be carried out. It may not be necessary to measure the complete reaction rate but a colour change after a defined incubation time. Apart from possible simplification in the assay procedure, the pasteurisation TTIs are suitable tools for thermal process validation.

This is not yet the situation for sterilisation. Further experimental work is required in a number of areas to address the questions that arose during the sterilisation TTI research:

- It will be necessary to obtain larger quantities of FDP to enable further testing. Conditions used in the batch *Pyrococcus furiosus* fermentation may not have been optimised for amylase production and may have resulted in detrimental by-products (e.g. proteases). Continuous fermentation could be used for greater yields and consistency.
- The best conditions need to be determined for storing the FDP and the filled sterilisation TTI tubes. This is important to prevent loss in activity during transportation to/from industry trials. It was assumed that storing the TTI tubes

in acetate buffer, which were then put into a freezer, was the best method since this had been used with other TTI systems.

- Purification – how far to go? The end point for work reported here was FDP with minimal purification. Reduction in activity was found when the sterilisation TTIs were stored chilled, which was thought to be caused by proteases acting on the amylase. The FDP was not a pure amylase and so other by-products of the fermentation will be present. Some of these may be detrimental to the amylase.
- What variability should be expected for the sterilisation TTI? It has many applications to industrial thermal processes and so it will be necessary to understand the accuracy of *F*-values estimated from the TTIs. Initial work suggests that it has an accuracy that will make it suitable for estimating *F*-values.
- How to guarantee long-term supply of the FDP with reproducible heat stability properties. *Pyrococcus furiosus* fermentation may not be the best method to produce heat-stable amylase. There are many reports of the gene being expressed in bacteria such as *E. coli* or in yeast and mould. This method is being investigated by the same project team (Tucker *et al.*, 2007) using the *Saccharomyces cerevisiae* yeast as the vehicle for expressing a thermostable amylase. Initial results suggest that the heat stability is very high and it will be necessary to adjust the chemical balance of the buffer solutions in order to reduce this to levels suitable for measuring commercial sterilisation processes. This method does have the advantage that the yeast has been modified to produce thermostable amylase and so the cocktail of other enzymes produced by *Pyrococcus furiosus* should not be present. This route forwards appears to offer the best solution.

9.6 References

Adams, M.W.W., Holden, J.F., Menon, A.L., Schut, G.J., Grunden, A.M., Hou, C., Hutchins, A.M., Jenney, F.E. Jr., Kim, C., Ma, K., Pan, G., Roy, R., Sapra, R., Story, S.V. and Verhagen, M.F.J.M. (2001). Key role for sulfur in peptide metabolism and in regulation of three hydrogenases in the hyperthermophilic archaeon *Pyrococcus furiosus*. *Journal of Bacteriology*, 183, 716–724.

American Heritage Dictionary (2003). *American Heritage Dictionary of the English Language*, Fourth Edition copyright 2000 by Houghton Mifflin Company. Updated in 2003. Published by Houghton Mifflin Company.

Ball, C.O. and Olsen, F.C.W. (1957). *Sterilization in Food Technology*. McGraw-Hill Book Co., New York.

Bee, G.R. and Park, D.K. (1978). Heat penetration measurement for thermal process design. Food *Technology*, 32(6), 56–58.

Bigelow, W.D., Bohart, G.S., Richardson, A.C and Ball, C.O. (1920). Heat penetration in processing canned foods. *National Canners Association Bulletin* 161.

Brown, K.L., Ayres, C.A., Gaze, J.E. and Newman, M.E. (1984). Thermal destruction of bacterial spores immobilised in food/alginate particulates. *Food Microbiology*, 1, 187–198.

CCFRA (1977). Guidelines to the establishment of scheduled heat processes for low-acid foods, *CCFRA Technical Manual* No.3, Campden & Chorleywood Food Research Association.

CCFRA. (1992). Food Pasteurisation Treatments, *Technical Manual No.27*, Edited by J. Gaze. Campden and Chorleywood Food Research Association.

CCFRA. (2006). Pasteurisation: A food industry practical guide (second edition). *Guideline No.51*. Edited by J. Gaze. Campden and Chorleywood Food Research Association.

De Cordt, S., Hendrickx, M., Maesmans, G., and Tobback, P. (1992). Immobilised α-amylase from Bacillus licheniformis: a potential enzymic time-temperature integrator for thermal processing. *International Journal of Food Science and Technology*, 27, 661–673.

De Cordt, S., Avila, I., Hendrickx, M. and Tobback, P. (1994). DSC and protein-based time–temperature integrators: Case study of α-amylase stabilised by polyols and/or sugar. *Biotechnology & Bioengineering*, 44, 859–865.

DoH (1994). *Guidelines for the safe production of heat preserved foods*. London: The Stationery Office. ISBN 0 11 321801X.

Fiala, G. and Stetter, K.O. (1986). *Pyrococcus furiosus* sp. Nov. represents a novel genus of marine heterotophic archaebacteria growing optimally at 100°C. *Archives of Microbiology*, 145, 56–61.

Guiavarc'h, Y., Deli, V., Van Loey, A., Zuber, F. and Hendrickx, M. (2002). Development of an enzymic time–temperature integrator for sterilization processes based on *Bacillus licheniformis* alpha-amylase at reduced water content. *Journal of Food Science*, 67, 285–291.

Guiavarc'h, Y. (2003). *Development and use of enzymic time-temperature integrators for the assessment of thermal processes in terms of food safety*. PhD Thesis No.570, Katholieke Universiteit Leuven, Belgium.

Guiavarc'h, Y., Van Loey, A., Zuber, F. and Hendrickx, M. (2004a). Development characterization and use of a high-performance enzymatic time–temperature integrator for the control of sterilization process' impacts. *Biotechnology and Bioengineering*, 88, 15–25.

Guiavarc'h, Y., Van Loey, A., Zuber, F. and Hendrickx, M. (2004b). *Bacillus licheniformis* alpha-amylase immobilized on glass beads and equilibrated at low moisture content: Potentials as a time–temperature integrator for sterilisation processes. *Innovative-Food Science and Emerging Technologies*, 5, 317–325.

Hendrickx, M., Maesmans, G., De Cordt, S., Noronha, J., Van Loey, A. & Tobback, P. (1995). Evaluation of the integrated time–temperature effect in thermal processing of foods. *Critical Reviews in Food Science & Nutrition*, 35(3), 231–262.

Heppell, N.J. (1985). Measurement of the liquid–solid heat transfer coefficient during continuous sterilisation of liquids containing particles. In: *Proceedings of IUFoST Symposium, Aseptic Processing and Packaging of Foods*, Tylosand, Sweden. p.108.

Holdsworth, S.D. (1997). *Thermal processing of packaged foods*. Blackie Academic & Professional, London.

Koch, R., Zablowski, A., Spreinat, A. and Antranikian, G. (1990). Extremely thermostable amylolytic enzyme from the archaebacterium *Pyrococcus furiosus*. *FEMS Microbiology Letters*, 71, 21–26.

Lambourne, T. and Tucker, G.S. (2001). *Time–temperature integrators for validation of thermal processes*. R&D Report No.132. CCFRA, Chipping Campden, Glos., GL55 6LD.

Leuschner, C. and Antranikian, G. (1995). Heat-stable enzymes from extremely thermophilic and hyperthermophilic microorganisms. *World Journal of Microbiology and Biotechnology*, 11, 95–114.

Maesmans, G., Hendrickx, M., De Cordt, S., Van Loey, A., Noronha, J., and Tobback, P. (1994). Evaluation of process value distribution with time–temperature integrators. *Food Research International*, 27, 413–423.

McKenna, A.B. and Tucker, G.S. (1991). Computer modelling for the control of particle sterilization under dynamic flow conditions. *Food Control*, 2, 224–233.

Miles, J.J. and Swartzel, K.R. (1995). Evaluation of continuous thermal processes using thermocouple data and calibrating reactions. *Journal of Food Process Engineering*, 18, 99–113.

Samborska, K., Guiavarc'h, Y., Van Loey, A., Hendrickx, M. (2005). The influence of moisture content on the thermostability of *Aspergillus oryzae* alpha-amylase. *Enzyme and Microbial Technology*, 37, 167–174.

Segerer, A.H, Burggraf, S., Fiala, G., Huber, G., Huber, R., Pley, U. and Stetter, K.O. (1993). Life in hot springs and hydrothermal vents. Origins of Life and Evolution of the Biosphere. *Journal of the International Society for the Study of the Origin of Life*, 23(1), 77–90.

Sterner, R. and Liebl, W. (2001). Thermophilic adaptation of proteins. *Critical Reviews in Biochemistry and Molecular Biology*, 36(1), 39–106.

Stetter, K.O. (1996). Hyperthermophiles in the history of life. *Ciba Foundation Symposium*, 202: 1–10, discussion 11–18.

Stumbo, C.R. (1965). *Thermobacteriology in food processing*. Academic Press, 111 Fifth Avenue, New York.

Tucker, G.S. (1999). A novel validation method: Application of time–temperature integrators to food pasteurization treatments. *Transactions of the IChemE, Food and Bioproducts Processing*, 77, Part C, 223–231.

Tucker, G.S. (2000). Estimation of pasteurisation values using an enzymic time–temperature integrator. *Food Australia*, 52(4), 131–136.

Tucker, G.S., Lambourne, T., Adams, J.B. and Lach, A. (2002). Application of biochemical time–temperature integrators to estimate pasteurisation values in continuous food processes. *Innovative Food Science & Emerging Technologies*, 3, 165–174.

Tucker, G.S. and Wolf, D. (2003). Time–temperature integrators for food process analysis, modelling and control. *R&D Report No.177*. Campden and Chorleywood Food Research Association.

Tucker, G., Cronje, M. and Lloyd, E. (2005). Evaluation of a time–temperature integrator for mild pasteurisation processes. *R&D Report No.215*. CCFRA, Chipping Campden, Glos., GL55 6LD.

Tucker, G.S., Brown, H.M., Fryer, P.J., Cox, P.W., Poole, F.L., Lee H.-S. and Adams, M.W.W. (2007). A sterilisation time–temperature integrator based on amylase from the hyperthermophilic organism *Pyrococcus furiosus*. *Innovative Food Science and Emerging Technologies*, 8(1), 63–72.

Van Loey, A.M., Hendrickx, M. E., De Cordt, S., Haentjens, T.H. and Tobback, P.P. (1996). Quantitative evaluation of thermal processes using time–temperature integrators. *Trends in Food Science & Technology*, 7, 16–26.

Van Loey, A.M., Arthawan, A., Hendrickx, M. E., Haentjens, T.H., & Tobback, P. P. (1997a). The development and use of an α-amylase based time–temperature integrator to evaluate in-pack pasteurisation processes. *Lebensmittel-Wissenschaft und-Technologie*, 30, 94–100.

Van Loey, A.M., Haentjens, T.H., Hendrickx, M. E. and Tobback, P. P. (1997b). The development of an enzymic time–temperature integrator to assess the thermal efficacy of sterilization of low-acid canned foods. *Food Biotechnology*, 11(2), 147–168.

Vieille, C and Zeikus, G. (2001). Hyperthermophilic enzymes: Sources, uses and molecular mechanisms for thermostability. *Microbiology and Molecular Biology Reviews*, 65(1), 1–43.

10

On-line correction of in-pack processing of foods and validation of automated processes to improve product quality

O. H. Campanella and G. Chen, Purdue University, USA

10.1 Introduction: process temperature deviations during sterilization

Thermal processing is a well-established technology to protect foods against microbial spoilage and to preserve nutrients and other food quality factors over long periods of time. Thermal processes of canned foods are typically performed in pressurized continuous or batch type retorts, where the food containers are heated by a heating medium (usually steam) following prescribed process temperature–time protocols aimed to deliver nutritious and texturally acceptable foods that are commercially sterile. However, deviations in temperatures and pressures of the steam supply during thermal processes are often unavoidable. When these deviations are significant, unexpected drops of pressure and temperature in the sterilizer may occur. Any deviation from the temperature–time prescribed process schedule that is not handled properly may compromise the sterility of the canned food and lead to subsequent spoilage. Such spoilage not only results in commercial losses but also poses a health risk to consumers if the deviation is severe enough. During the past two decades, considerable efforts have been directed toward the development of 'intelligent' on-line retort control systems, capable of rapid evaluation, on-line correction, and printed documentation of any process deviations occurring during a thermal process, delivered while the process is still underway (Teixeira and Tucker, 1997; Simpson *et al.*, 2007a,b).

Computer-based algorithms for automatically correcting process temperature deviations in batch retorts have been developed (Teixeira and Manson, 1982; Datta *et al.*, 1986; Simpson *et al.*, 1993; Akterian, 1999; Kumar *et al.*, 2001, Simpson *et al.*, 2006). However, these algorithms have not been applied to continuous retorts due to the lack of a suitable method capable of providing accurate and fast evaluations of product temperature and lethality in response to process deviations. One of the main difficulties of using algorithms developed for batch retorts is that each can inside a continuous retort has a different residence time and thus is affected differently by a temperature deviation. Processors currently either stop the retort conveying mechanism and issue a complete reprocess for the containers affected by the temperature deviation, or take the more drastic resolution of discarding the product. If the duration of the temperature deviation is short, some manufacturers simply stop the conveyor chain and re-start it immediately after the retort is brought back to the processing temperature, while others have a standard operation procedure (SOP) in place that informs the retort operator to wait for a period of time closely related to the duration of temperature deviation before re-starting the process. These actions can compromise production efficiency as well as product quality due to overprocessing (the latter in the case that the temperature deviation is a temperature drop). An alternative to stopping the conveying mechanism in the event of a temperature deviation is to adjust the conveyor speed. The adjustment in this speed is required to ensure that all food containers remain in the retort long enough to achieve commercial sterility while keeping overprocessing to a minimum so that the quality (organoleptic and nutritional) of the final product is not significantly affected.

In this chapter the following topics are covered:

- On-line correction of in-pack processing of foods, including a short review of work done on batch retorts and description of new developments on continuous retorts.
- Future trends and validation of automated processes.
- Sources of further information and advice.

10.2 On-line correction of in-pack processing of foods

The primary purpose of on-line correction of process temperature deviations is to automatically vary the process time to ensure that all containers receive the heat treatment necessary to achieve the required sterilization at the cold spot (the slowest heating point) of the food product. The critical part of the control system, specifically for batch retort operations, is to implement a fast method for accurately determining the new process time necessary to compensate for the process temperature deviations. The new time is often calculated from a heat transfer model describing the changes in the temperature of the product contained in the can and real-time measurement of the retort temperature, which act as an external condition (boundary condition) to the heat transfer model. It is important to note that in batch retorts all the containers are treated equally because they have the same residence time. Conversely, for continuous retort operations, the container conveyor's speed must be varied to adjust for residence time but, as described previously, the development of a correction algorithm has the additional complexity that each can inside the retort has a different residence time.

In an on-line correction algorithm, the target process lethality is the most important constraint utilized to calculate the required new processing time. The traditional approach to evaluate the efficacy of a thermal process is based on the assumption that inactivation of microbial cells and spores follows a first-order kinetic model. However, there is a growing concern about its validity because curvilinear isothermal survival curves are frequently observed (e.g. Anderson *et al.*, 1996; Sapru *et al.*, 1992, 1993; Cole *et al.*, 1993; Hills and Mackey, 1995; Linton *et al.*, 1995; Peleg and Cole, 1998). The Weibull model has been shown to be able to provide a more accurate estimation of microbial inactivation by heat (Peleg and Cole, 1998; Peleg, 1999; Peleg and Penchina, 2000) and also by other inactivation sources, e.g. pressure and chemical agents (Peleg, 2006). This model is of special interest to thermal processes of low-acid canned foods because it has been shown that the isothermal survival curve of spores of *C. botulinum*, which is the target microorganism for these processes, can be described adequately by the Weibull model (Peleg and Cole, 2000; Campanella and Peleg, 2001; Mafart *et al.*, 2002).

Given the importance of non-linear kinetics in the evaluation of thermal processes, on-line correction strategies for continuous retorts by using both first-order kinetics and the Weibull model will be discussed and compared.

10.2.1 Batch retorts

Teixeira and Tucker (1997) give a comprehensive review of on-line control in batch retorts. There are three commonly used methods:

(i) The cold spot temperature profile is obtained directly from a real-time data acquisition system in which temperature sensors are installed in selected food containers. Based on the obtained temperature profile, the accumulated sterilization lethality is evaluated, and heating is terminated once the target lethality is attained. Representative methods were developed by Lappo and Povey (1986), Wojciechowski and Ryniecki (1989), Ryniecki and Jays (1993), and Kumar *et al.* (2001). These control systems can be very effective and reliable. However, in most large highly automated cook room operations typical in the modern food canning industry, this approach is considered impractical and cost-prohibitive (Teixeira and Tucker, 1997).

(ii) The 'table' or 'correction factor' method has been used to estimate extended process times. For a range of different constant retort temperatures, process times can be calculated in advance and held in a database table for quick reference if necessary (Teixeira and Tucker, 1997). In the case of a temperature deviation, it is assumed that the retort temperature will remain for the rest of the process and a new process time, estimated from the table, is based on this temperature. Some retort control systems (NFPA, 1980; LOG-TEC, 1984) have used this method. The major disadvantage of this method is that, once a new temperature is accepted, and by assuming that the deviation results in a lower temperature, an extended process schedule is adopted, usually resulting in overprocessing because in many instances the retort temperature quickly recovers its normal operating value (Teixeira and Tucker, 1997). To minimize this error, Giannoni-Succar and Hayakawa (1982) proposed a procedure to calculate a 'correction factor' that allows the process time to be extended to just that necessary to compensate for the deviation. However, the major disadvantage of the approach is that patterns of temperature deviations can be unlimited, and consequently it is not possible to generate in advance all the possible 'correction factors'.

(iii) This approach, which is named 'intelligent control' and is used with a heat transfer model (Teixeira and Tucker, 1997) does not require measuring product temperatures. Instead, a numerical model is used to calculate the product temperature profile at the slowest heating point of the food, and the associated lethality, as a function of the retort temperature history. In this method, the actual retort temperature is read at prescribed time intervals by temperature sensors located in the retort. When a process deviation occurs, the control system calculates the new heating process time by assuming that the momentary retort temperature does not change during the rest of the process. The process time is updated, based on the newest reading of the retort temperature. Teixeira and Manson (1982) were the first to use such a model for online correction of process deviations in batch retorts. Their method was improved by Datta *et al.* (1986). Other methods were also reported by Bown *et al.* (1986); Kelly and Richardson (1987); Tucker and Clark (1989). In all these control systems, the numerical models were developed by assuming that external surface heat transfer coefficients were large. That assumption is a good approximation of the reality, specifically in conduction-heated products, because the major resistance to heat transfer is inside the can. For cases in which the approximation is no longer valid, for example, for convection heated products, further work on models that take into account finite surface coefficients was undertaken (Tucker and Clark, 1990; Silva *et al.*, 1992; Akterian, 1999). It is important to note that all these control systems are limited to pure conduction-heated foods processed in finite cylindrical containers. This limitation was eliminated in subsequent work of a number of researchers (Teixeira *et al.*, 1992; Akterian and Fikiin, 1994; Bichier *et al.*, 1995; Noronha *et al.*, 1995).

A number of recent publications (Simpson *et al.*, 2006, 2007a,b,c) summarize the work carried out up to date. They also introduce models that include pure convection and purely agitated containers. The approach, however, is concerned with changes in the processing temperature instead of the processing time, rightly speculating that the use of high temperatures will have less detrimental effects than prolonged processing time on the quality (nutritional and sensorial) properties of the final product. Models used include ramping up (heating) and ramping down (cooling) temperatures in the retort, as well as arbitrary temperature deviations of the processing temperature.

Software packages (e.g. CAN_CALC©) for on-line correction of process deviations in batch retorts that are based on the above discussed work have been developed and applied to food canning processes.

10.2.2 Continuous retorts

Equipment

The most widely used non-agitating and agitating continuous retorts are hydrostatic, e.g. FMC Hydrostat™, and rotary retorts, e.g. FMC Sterilmatic™ (Gavin and Weddig, 1995). The hydrostatic retort is a versatile food sterilizer that can be operated over a wide range of process temperatures and pressures. The main components or stages of a hydrostatic retorting system are illustrated in Fig. 10.1. At the feed station, the containers are placed into a row (stack). The stack is fed into a carrier that holds the containers on the conveyor during their trip through: (i) a temperature-controlled preheat water leg, (ii) a controlled steam chamber set to the desired process temperature, (iii) a temperature-controlled precool water leg (discharge leg), and (iv) a water spray cooling system. Traveling from the bottom to the top in the steam chamber and *vice versa* is referred to as one pass. The number of passes as well as the speed of the container conveyor determines the residence time of the containers in the steam chamber. The speed of the conveyor must be set by the process time, which is determined *a priori* by the processing authority to ensure that each container receives the targeted microbial log-reduction.

A rotary retort uses a rotating spiral reel to transport containers through a steam-pressurized processing shell. The containers enter the shell through a transfer valve where they are deposited onto the rotating spiral. The spiral reel has steps to hold the containers as they are transported through the shell (http://files.asme.org/ASMEORG/Communities/History/Landmarks/5491.pdf). The rotating reel induces agitation to the food within the containers, which improves heat transfer and minimizes process time. The use of a pure conductive heat transfer model for this situation therefore could be problematic if the food in the container is not a solid. A typical configuration may include a preheat shell, pressure and atmospheric cooling shells, in addition to the process shell. The residence time in the processing shell depends on the rotational speed of the spiral reel.

Methods and procedures

For convenience, the steam chamber is defined as the place where food containers receive heat treatment through steam heating, i.e. the steam chamber in a hydrostatic retort and the processing shell in a rotary retort. The conveyor is used to transport the containers, i.e. the container conveyor is in a hydrostatic retort and the spiral reel is in a rotary retort. Carriers are positions on the conveyor where containers are held, i.e. container carriers in a hydrostatic retort and container steps

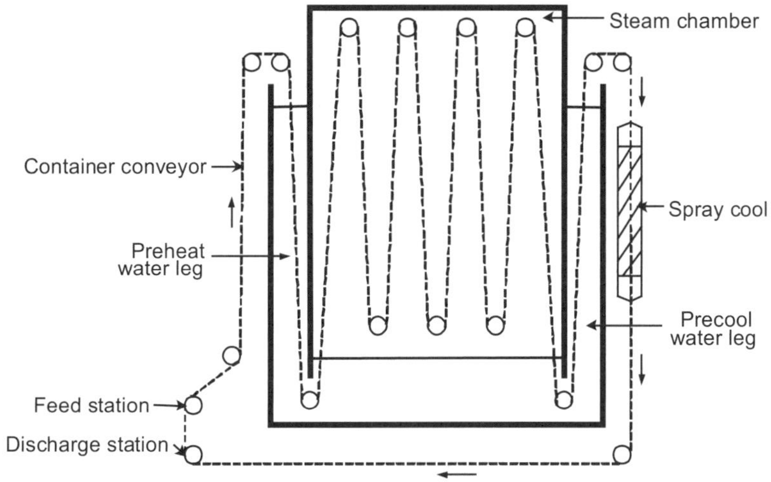

Fig. 10.1 Schematic cut-away view of a hydrostatic retort

in a rotary retort. It is assumed that the food containers in the same carrier receive identical heat treatment and thus the same microbial inactivation. Henceforth microbial inactivation in a container has the same meaning and is the same as that in a carrier.

The effects of processing temperature deviations on the product temperature profiles can be determined by temperature measurements on the product at different locations or calculated by a suitable heat-transfer model. The former approach has practical implications that have been discussed in the previous section, whereas prediction of the product temperature by using an accurate heat-transfer model offers advantages if these calculations are integrated to an on-line correction algorithm. As discussed, many of the heat-transfer models developed have focused on conductive-heat transfer that may not be suitable for all conditions used in thermal processing. When convection was incorporated, many of the models focused on perfectly mixed containers, which have simple solutions to predict the temperature in the product, albeit a uniform temperature (Simpson *et al.*, 2006).

The Apparent Position Numerical Solution method (Noronha *et al.*, 1995; Chen *et al.*, 2005) has been also used to predict temperature profiles of products including convection, although this model does not assume perfectly agitated products. Nevertheless, once a model is assumed, the product temperature profile can be estimated for given external conditions, and the accumulated lethality for non-isothermal conditions can be calculated. As discussed, for these calculations it is important to select a suitable model, able to describe the thermal death kinetics of the pertinent microorganisms. Currently, lethality is calculated by assuming first-order kinetics and an integration approach proposed by Stumbo (1965). However, the use of first-order kinetics and the F_0-value have been recently challenged (Peleg, 2006) and thus it is worth applying online corrections algorithms for situations that differ from a presumed linear relationship, e.g. first-order kinetics or Arrhenius-type dependence of microbiological parameters with temperature.

As discussed, on-line corrections to temperature deviations in a continuous retort represent a challenge to food engineers because, at a given time, each carrier has a different residence time in the retort. Thus, it is important to have a record of the temperature profiles and accumulated lethalities for each carrier inside the retort at the time that a temperature deviation occurs. The accountability of these temperature profiles and lethalities must continue during the entire period that a can is inside the retort. Given the large number of carriers processed in commercial retorts, these calculations require prohibitively long computation times that make unviable their implementation in on-line corrections algorithms for use in this type of retort. Up to date there are only two patents (Weng, 2003a,b) dealing with this issue. A recent publication (Chen *et al.*, 2008) describes methods to optimally record container temperature profiles and lethality data so that computation times are minimized and on-line algorithms aimed to adjust for process temperature deviations can be implemented.

Two algorithms, the 'Fixed Point' and the 'Worst Case', recently developed by Chen *et al.* (2008), are described in this chapter. The algorithms focus on handling retort temperature data and prediction of product temperatures with a suitable heat-transfer model; thus they are independent of the approach used to estimate microbial inactivation during the process. However, estimation of survival rates and lethalities due to changes in product residence time varies with the assumed kinetics. Therefore, two microbial inactivation approaches, first-order kinetics and the Weibull model, along with their respective lethality calculations, are considered in the description and evaluation of the developed on-line control algorithms.

Heat-transfer model: the Apparent Position Numerical Solution (APNS) method

The APNS method, a semi-empirical approach, is based on combining empirical heat penetration studies (Ball and Olson, 1957) and the analytical solution of the heat conduction equation for a sphere (Noronha *et al.*, 1995). Due to its accuracy and high calculation speed, it can be used as the heat-transfer model that calculates product temperatures. It offers advantages regarding computation times when compared with traditional methods used for cylindrical and other geometries and based on either finite differences or finite element methods. The method, however, requires the heat penetration parameters f_h (heating rate factor) and j_h (heating lag factor) obtained from experimental heat penetration studies to be incorporated into the following relationships (Ball and Olson, 1957):

$$f_h = 0.233(R^2/\alpha) \qquad [10.1]$$

$$j_h(r) = 0.63662\,(R/r)\sin(\pi r/R), \quad \text{where } 0 \le j_h \le 2 \qquad [10.2]$$

Equations 10.1 and 10.2 are used to obtain an apparent thermal diffusivity α and an apparent position r for a 'phantom' solid sphere of radius R, so that the apparent position r experiences the same heating rate and heat lag factors as those experienced by the actual product at the location used for the heat penetration studies, usually the coldest point in the container. Thus, for any product location with experimentally obtained heat penetration parameters, the temperature history can be predicted at 'its apparent position *r*' by solving the heat conduction equation in a sphere, which can be assumed to be a one-dimensional heat transfer process.

During processing and due to deviations in the retort temperature, the external conditions (boundary conditions) for the heat conduction inside the container are time dependent. Therefore, it is necessary to use a numerical scheme, in this case a finite difference method, to be able to predict the resulting product temperatures. In addition to the variable boundary conditions, the heat-transfer model assumes negligible external resistances to the heat transfer (i.e. infinite surface heat-transfer coefficient). The procedure is equally applicable to the cooling process, simply by replacing the heating factors f_h and j_h by the cooling factors f_c and j_c in Eqs. 10.1 and 10.2, respectively. Although the spherical geometry model remains that of a purely conductive heating solid, it performs well in predicting the temperature evolution even if both conduction and convection are present (Teixeira *et al.*, 1999).

This approach is adopted because it significantly decreases computation times by reducing the heat transfer to only one dimension (the radial dimension). Improvements to the APNS method to correct inaccuracies during the cooling phase have been developed by Chen *et al.* (2005). The use of the APNS method is essential for the successful development of on-line control algorithms because computation times are significantly lower than those obtained with other numerical methods, e.g. finite element or dinite differences, utilized to predict product temperatures from either 2D or 3D heat-transfer models.

Survival kinetics and lethality calculations

- First-order kinetic model for microbial inactivation and standard lethality calculations. From the product temperature profile calculated by the heat-transfer model, and the processing temperature history measured during the thermal process, the processing lethality, F_0-value, can be calculated. Due to temperature gradients existing in the product, the coldest point in the container, T_c, is generally used in the evaluation of the lethality by the following expression (Stumbo, 1965):

$$F_0 = \int_0^t 10^{\frac{T_c - T_{ref}}{z}} \, dt \qquad [10.3]$$

where t is time, T_c is product temperature at the coldest point of a container, T_{ref} is a reference temperature, usually 121.1 °C, and z is parameter that would represent the temperature required to decrease the microbial decimal reduction by a factor of ten. Then, the microbial log-reduction can be calculated as:

$$\log_{10} S = -F_0/D_{ref} \qquad [10.4]$$

where D_{ref} is the microbial decimal reduction time at the reference temperature.

- Weibull model for microbial inactivation and lethality calculation. Under isothermal conditions, the Weibull model can be expressed as (Peleg and Cole, 1998):

$$\log_{10} S(t) = -b(T)\,t^{n(T)} \qquad [10.5]$$

where $S(t) = N(t)/N_0$ is the momentary survival ratio, N_0 and N are the number of surviving microbial cells or spores initially and after an exposure time t, respectively. Coefficients $b(T)$ and $n(T)$ are temperature-dependent model parameters. The temperature dependence of $b(T)$ has been described by the log-logistic model (Campanella and Peleg, 2001; Peleg *et al.*, 2002; Peleg, 2003), whereas the parameter $n(T)$ usually has a very weak or nil dependence upon temperature (van Boekel, 2002). For a specific microorganism strain, the relationship between the coefficients and temperature can be obtained by fitting Eq. 10.5 to experimental survival data obtained under isothermal conditions. An approach to determine these parameters from non-isothermal survival curves is presented in Peleg *et al.* (2003).

Real thermal processing conditions, however, are non-isothermal; in particular, if deviations of retort temperatures are being considered. Peleg and Penchina (2000) developed an equation to calculate lethality, expressed as the decimal log reduction in the number of microorganisms under non-isothermal conditions, which consists of the solution of the following differential equation:

$$\frac{\mathrm{d}(\log S(t))}{\mathrm{d}t} = -\,b(T(t))n(T(t))\left(\frac{-\log S(t)}{b\,(T(t))}\right)^{(n(T(t))-1)/n(T(t))} \qquad [10.6]$$

There exists specialized software (e.g. MathCad, Mathematica, Matlab) that can solve the above equation with reasonable simplicity provided the dependence of the model parameters $b(T)$ and $n(T)$ with temperature and the process temperature history are known. Solution of that equation has been also implemented in the Spreadsheet Excel(T) and can be freely downloaded from Professor Peleg's website (http://www-unix.oit.umass.edu/~aew2000/GrowthAndSurvival.html).

The solution of a differential equation, however, can be time demanding if an on-line correction is implemented. To decrease computation times, a simplified method was developed by Chen *et al.* (2007).

Recording temperature histories in the retort and the carriers during thermal processing

For continuous retorts, the product temperature profile for each carrier depends on its relative position within the retort. As discussed, recording the temperature–time profile in each individual carrier of a continuous retort is highly time consuming and may require significant computer space to properly implement this information in an on-line control system. Thus, suitable algorithms must rely on the development of relationships between processing histories of carriers inside the retort. Figure 10.2 shows a schematic of the path from the entrance to the exit of a typical steam chamber in a continuous retort, which has been divided evenly into

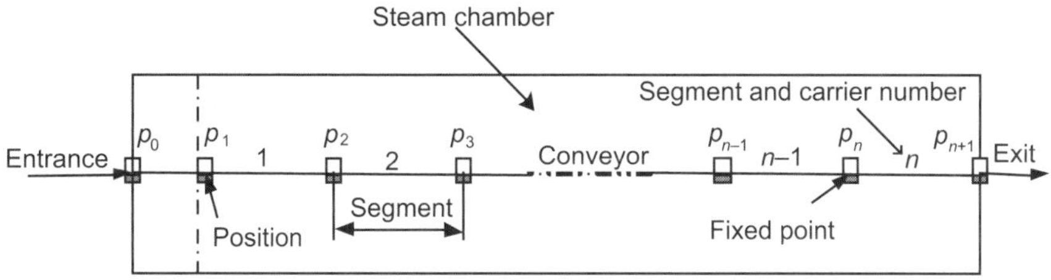

Fig. 10.2 Schematic of a hydrostatic retort, including a description of relevant components utilized in the on-line correction algorithms.

n segments. Accordingly, the carriers inside the chamber are divided into n groups and each group corresponds to one chamber segment. Because the chamber is evenly divided, there may be a small segment remainder which is represented by the distance between the carrier P_0 and the carrier P_1. However, the lethality accumulated in the small segment can be safely neglected (in this case the lethality would be slightly overestimated).

In the current scheme, the processing temperature history $T_{P_{n+1}}$ experienced by the carrier P_{n+2}, which as illustrated in the figure is located at the exit, can be represented by the vector:

$$T_{P_{n+1}} = \left[T_1 \; T_2 \ldots\ldots T_j \ldots\ldots\ldots T_n \right] \qquad [10.7]$$

where each j-th element of the vector, T_j, represents the processing temperature history experienced by the carrier during its transit along the j-th chamber segment. In addition, the processing temperature history T_j in a segment can be represented by another vector:

$$T_j = \left[T_{j,0} \; T_{j,1} \ldots\ldots T_{j,i} \ldots\ldots\ldots T_{j,k_j} \right] \qquad [10.8]$$

where now $T_{j,i}$ is the i-th temperature experienced by the carrier P_{n+1} in its transit through the segment j. Thus, at any given time, the processing temperature history experienced by each carrier in the chamber as well as the relationship among them can be represented by the following matrix where Φ indicates no heat treatment:

$$\begin{bmatrix} T_{P_{n+1}} \\ T_{P_n} \\ T_{P_{n-1}} \\ \cdot \\ \cdot \\ \cdot \\ T_{P_2} \\ T_{P_1} \end{bmatrix} = \begin{bmatrix} T_1 \; T_2 \; T_3 \; T_4 \ldots\ldots T_j \ldots\ldots T_n \\ T_2 \; T_3 \; T_4 \ldots\ldots T_j \ldots\ldots T_n \\ T_3 \; T_4 \ldots\ldots T_j \ldots\ldots T_n \\ \ldots\ldots \\ \ldots\ldots \\ \ldots\ldots \\ T_n \\ \Phi \end{bmatrix} \qquad [10.9]$$

Equation 10.9 provides a record of the processing temperatures experienced for each carrier inside the sterilizer at a given time and can be obtained from a suitable temperature sensor installed in the retort. Equation 10.9 also indicates that the processing temperature history of any carrier can be determined if the temperature history of the carrier located at the location P_{n+1}, $T_{P_{n+1}}$, is known; thus, only the temperature history of the can at that location needs to be recorded. By using this concept, a reduction in computation time and computer storage space required to estimate a new processing time from the control algorithm could be achieved in the event of temperature deviation. Once an effective method of recording retort temperature data is set, suitable on-line corrections algorithms could be implemented. Two on-line corrections have been developed (Chen *et al.*, 2008) and are briefly described below:

On-line correction algorithms

- The Fixed Point (FP) algorithm. In the FP algorithm, the starting point (P_n) of the last segment is selected as the fixed point (see Fig. 10.2). If a process temperature deviation occurs, the new processing time and new conveyor speed (*CS*) are calculated from the time needed for the carrier located at the FP, called carrier FP henceforth, to achieve the targeted log-reduction of the process under the existing processing temperature. The success of this algorithm relies on selecting a suitable number of segments. That number of segments *n* is selected by satisfying two competing conditions: the first condition is to choose *n* large enough so that the accumulated lethality of the carriers within each segment can be considered approximately identical. The second condition requires that the time needed for the FP carrier to move from its location to the chamber exit be long enough to implement a corrective action in the event of a process temperature deviation. The length of the segment where these corrections are implemented is named *d*. Thus, after selecting the appropriate number of segments, the method proceeds as follows:

  (i) Prior to the moment at which a process temperature deviation (e.g. a drop) occurs, the carrier FP has experienced a processing temperature history given by the second row of the matrix of the type given by Eq. (10.9). In a typical thermal process, the elements of this row are temperatures measured by a temperature sensor and recorded by the control system. In the event of a process temperature deviation, the time required to achieve the target lethality, t_{res}, is estimated and is closely related to the approach used to estimate process lethality. As discussed, two different approaches are considered here. One approach is based on the standard calculation method used in thermal processing (Stumbo, 1965). The other approach is based on the recognition of non-linear survival kinetics and the evaluation of process lethality for non-isothermal processes proposed by Peleg (2006). Calculations of new processing times based on these approaches are discussed in the following section.

  (ii) For calculation of the new processing time, t_{res}, it is assumed that the momentary processing temperature, which is denoted by $T^{P}_{1,0}$ (see Eq. 10.11),

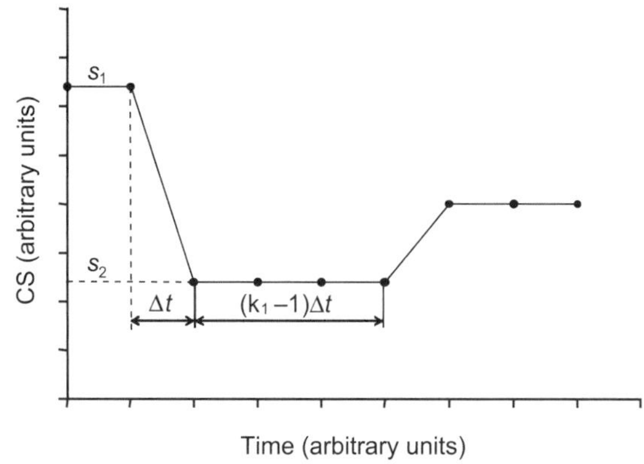

Fig. 10.3 Schematic showing the conveyor speed adjustment in response to process temperature deviations in continuous retorts.

is maintained for the remainder of the process. For practical reasons, the residence time, t_{res}, is estimated as $k_l\Delta t$ where Δt and k_l are a unit time increment and an integer providing the number of time increments necessary to achieve lethality, respectively.

(iii) Based on the calculated residence time, i.e. $k_l\Delta t$, the conveyor speed can be set. In principle, any conveyor speed setting can be used as long as the carrier FP transits the last segment, of length d, in the required time $k_l\Delta t$. But, for practical reasons, the conveyor speed setting is changed linearly from the speed s_1, speed before the temperature deviation, to a newly estimated value s_2, speed after the temperature deviation. That change is effective during the first time increment Δt. Thereafter, the conveyor speed is maintained constant for the rest of the processing time, i.e. for the time period $(k_1 - 1)\Delta t$. The new conveyor speed s_2 is calculated by Eq. 10.10. Figure 10.3 illustrates a schematic of the conveyor speed adjustment after a deviation in the processing temperature has occurred.

$$s_2 = \frac{2d - s_1\Delta t}{(2k_1 - 1)\Delta t} \qquad [10.10]$$

Because the actual processing temperatures (measured values), in the period of time $k_l\Delta t$, are recorded by a temperature sensor, they can be stored in a vector as illustrated below:

$$T_l^P = \left[T_{l,0}^P \; T_{l,1}^P \; T_{l,2}^P \; ..... \; T_{l,k_l}^P \right] \qquad [10.11]$$

where now $T_{l,j}^P$ is the j-th temperature experienced by the carrier located at the FP at the moment of the temperature deviation during its transit through the last segment. After the time $k_l\Delta t$ has elapsed, the carrier previously at P_n reaches P_{n+1} and a new vector $T_{P_{n+1}}$ defining the temperature history of the carrier currently located in FP can be expressed by the following equation:

$$T_{P_{n+1}} = \left[T_2 \ T_3 \ .... \ T_n \ T_1^P \right] \qquad [10.12]$$

(iv) The process is continued considering each carrier that consecutively occupies the FP location until a condition to stop the correction algorithm is reached. A computer flowchart describing the FP algorithm is described in Chen *et al.* (2008).

- The Worst Case (WC) algorithm. The WC algorithm is based on the fast and accurate identification of the most affected carrier, WC, as a consequence of a process temperature deviation. Thus, when a process temperature deviation occurs, the conveyor speed is adjusted to allow the WC carrier to achieve the target process lethality value. It must be noted that, by doing this, all the other carriers in the retort will exceed the targeted lethality. The algorithm is summarized as follows:

  (i) At the moment when a process temperature deviation (e.g. drop) occurs, $T_{P_{n+1}}$ can be described by Eq. 10.7. Processing temperature histories experienced by all the other carriers can be obtained from Eq. 10.9.

  (ii) The algorithm identifies the WC carrier group that requires the lowest conveyor speed to achieve the prescribed lethality under the assumption that the momentary processing temperature, $T^P_{1,0}$ (see Eq. 10.11) is maintained for the remainder of the process. The conveyor speeds for the carrier groups are calculated as the distance left to move through the steam chamber of a particular carrier group divided by the heating time required for that group. Once the conveyor speeds are calculated, the algorithm finds a global minimum among all these conveyor speeds. Because the presence of local minima cannot be ruled out, all the local minima should be found first and, from them, the global minimum speed can be determined. The approach, however, requires significant computation time as each carrier group has to be evaluated. Simulations using practical process temperature deviations (Chen *et al.*, 2008) showed that the distribution of conveyor speeds is fairly smooth so, rather than examining each carrier group individually, the local minima of a group of carrier groups are estimated. After the interval, the group of containers wherein exists the global minimum speed is determined; the required conveyor speed for each carrier group in that interval can be determined and the lowest one selected to adjust the conveyor speed. It has been shown that this approach was able to reduce significantly computational time because only a limited number of carriers are evaluated to locate the WC carrier. Since the algorithm works on discrete points, there would be chances that some local minima could be missed, particularly for cases in which sharp changes in process temperatures could be present. Chen *et al.* (2008) showed that the problem did not occur in simulations performed under practical and realistic hydrostatic retort conditions reported by Weng *et al.* (1995).

  (iii) Once the conveyor speed required by the WC carrier is determined, the transit time ($k_1\Delta t$) through one chamber segment is determined and used to set the conveyor speed. The same speed-setting procedure as that used in the FP method can be employed. The actual processing temperatures (measured values) in this period of time are recorded and represented by Eq. 10.11. After time $k_1\Delta t$, the carrier previously at P_n reaches P_{n+1}, then the new $T_{P_{n+1}}$ can be expressed by Eq. 10.12. The process is then continued until both the processing temperature and the conveyor speed reach their specified values. A computer flow chart for the WC algorithm is described in Chen *et al.* (2008).

It is important to note that the validity of the two proposed control algorithms is based on the assumption that the steam chamber reaches the new temperature instantaneously after a temperature deviation. For a hydrostatic retort, the steam pressure in the chamber is counterbalanced by water columns (legs) in the preheating and the precooling sections. When the steam pressure or temperature drops, the water levels in the two legs may decrease accordingly. This would result in less preheating of the containers before they enter the steam chamber and less precooling before the spray cooling begins. At the same time, the water level in the steam chamber increases and underprocessing may occur in over-immersed carriers. Therefore, two additional assumptions must be considered in the implementation of these algorithms:

(i) Process temperature deviations do not significantly influence the preheat and precool treatments and,
(ii) The water level in the steam chamber is correctly controlled to a level which assures that no containers are immersed in the water.

10.3 Simulation of on-line correction methods for continuous retorts

The algorithms presented for on-line correction of temperature deviations in continuous retorts deal with the development of efficient methods to store temperature histories experienced by the retort during the process and by the affected containers, the latter calculated with accurate and fast heat-transfer models. These algorithms, however, need to be controlled using process lethality as the key control parameter. As discussed, process lethality depends on the microbial survival kinetics and the method utilized for its estimation. Thus, simulations performed to test these algorithms include first-order survival kinetics and the traditional lethality calculation proposed by Stumbo (1965), which are reported in by Chen *et al.* (2008). Applications of the proposed methods are also tested for non-linear survival kinetics (Weibull) and the method proposed by Peleg and Penchina (2000) for calculating microbial inactivation during non-isothermal processes.

10.3.1 First-order kinetics and the traditional method

As discussed, after a process temperature deviation, and having records of the temperature histories in the retort and on the containers, the time required to achieve the target lethality, t_{res}, can be calculated as the root of the following:

$$\left(\log_{10} S\right)_{\text{Target}} - \log_{10} S(t) = 0 \qquad [10.13a]$$

where

$$\log_{10} S\,(t) = -\frac{1}{D_{T=T_{ref}}} \int_0^{t_{res}} 10^{\frac{T_c - T_{ref}}{z}} \qquad [10.13b]$$

and $(\log_{10} S)_{\text{Target}}$ is the target inactivation in the process. t_{res} can be obtained from a numerical method utilized for solving non-linear equations, such as the bisection method (http://mathworld.Wolfram.com/Bisection.html). In Eq. 10.13b T_c is the temperature of the product at its slowest heating point, which is estimated by the heat transfer model, the APNS method in this case. Other variables in the equation were already defined in Eqs 10.3 and 10.4. Estimation of the new processing time, for given existing external/boundary conditions, requires the solution of Eqs 10.13a and 10.13b as the solution of the heat-transfer model. These calculations are computationally time demanding and thus pose a serious challenge when they are implemented in fast on-line correction algorithms.

10.3.2 Simulation with the Weibull model for microbial inactivation

When the Weibull model is used to estimate the required processing time to achieve the targeted microbial inactivation, Eq. 10.13a also applies, but in this case the microbial inactivation, measured as the log reduction $\log_{10} S(t)$, is estimated by solution of the differential equation given by Eq. 10.6. Solution of that equation implemented in an on-line correction algorithm could be computing time demanding. Thus, in order to get a fast solution of the microbial log decimal reduction by applying the Weibull model, the numerical method developed by Chen *et al.* (2007) is another alternative to use.

10.3.3 Application of algorithms

In the simulations, spores of *C. botulinum* were considered because they are the target organisms in low acid foods heat sterilization.

The FP algorithm – first-order kinetics and classical lethality calculations

The FP algorithm was tested to adjust the conveyor speed that compensates for process temperature deviations. Details of the processing parameters and conditions used in the simulations are listed in Table 10.1 for both first-order and Weibull kinetics, whereas Table 10.2 shows the thermal inactivation parameters for *C. botulinum* spores. Temperature deviations used to test the algorithm are illustrated in Fig.10.4a, whereas Fig. 10.4b shows the corresponding adjustment of conveyor speed (*CS*) to reach a targeted log decimal reduction of 12. Additionally, Fig.10.4c shows the microbial log-reduction experienced by each carrier group after the conveyor speed adjustment. These calculations are based on the traditional method

Table 10.1 Hydrostatic retort characteristics and typical processing conditions

Preheat water leg Carrier numbers	Steam chamber Carrier numbers	Precool water leg Carrier numbers	Scheduled Carrier/min
161	1263	165	25.4 (1st order) 23.8 (Weibull)
Top/Bottom temp (°C)	Retort temp (°C)	Bottom/Top temp (°C)	Spray water temp/time (°C/min)
83.3/83.3	121.1	82.2/82.2	10.0/20.0

Table 10.2 Thermal inactivation parameters for *C. botulinum* spores

	Log linear model		Weibullian model	
T_{ref} (°C)	D_{ref} (min)	z (°C)	b	n
121.1	0.3	10	$\ln(1 + e^{(T-102.3)/3.33})$	$0.325 + \frac{0.425}{1 + e^{(T-101)/10.057}}$
	USDA (2005)		Campanella and Peleg (2001)	

proposed by Stumbo (1965) and described by Eqs 10.3 and 10.4. The labels on the x-axis of Fig.10.4c refer to the carrier groups, which are numbered starting the count from carriers leaving the steam chamber. Thus, the carrier groups can be easily translated to their locations in the steam chamber. For simulation conditions used in this study, Group 1 was located at the exit of the steam dome and group 127 (total 126 segments) at the entrance of the steam dome. Then in the next cycle, group 128 was located at the exit and 254 at the entrance of the steam chamber, and so on. As illustrated in Fig.10.4c, the log-reduction in each carrier group can be controlled in a very narrow range, which is very close to the target inactivation value of 12. A drawback of this algorithm, illustrated in Fig. 10.4b, is that the conveyor speed does not return to its normal value after the process temperature returns to the standard processing temperature (121.1 °C in this case). Figure 10.4b shows that the conveyor speed adjustment continues even after the process temperature deviation disappears. This is happening because the carriers inside the retort, which were affected by the temperature deviations (a drop in this case), still need to be processed at a speed lower than the normal speed to achieve the target log-reduction, thus resulting in overprocessing of the new carriers entering the sterilizer. Once all the carriers affected by the temperature deviation have left the sterilizer, a conveyor speed greater than that specified by the process would be needed to reduce the residence time of these new carriers and avoid overprocessing. In turn, this will result in underprocessing of carriers just entering the retort. In other words, it is not possible to achieve a stable conveyor speed once a deviation occurs.

An alternative approach that could overcome this unstable situation would be setting back the conveyor speed to its normal value once the last carrier group affected by the process deviation leaves the steam chamber. The sacrifice in doing

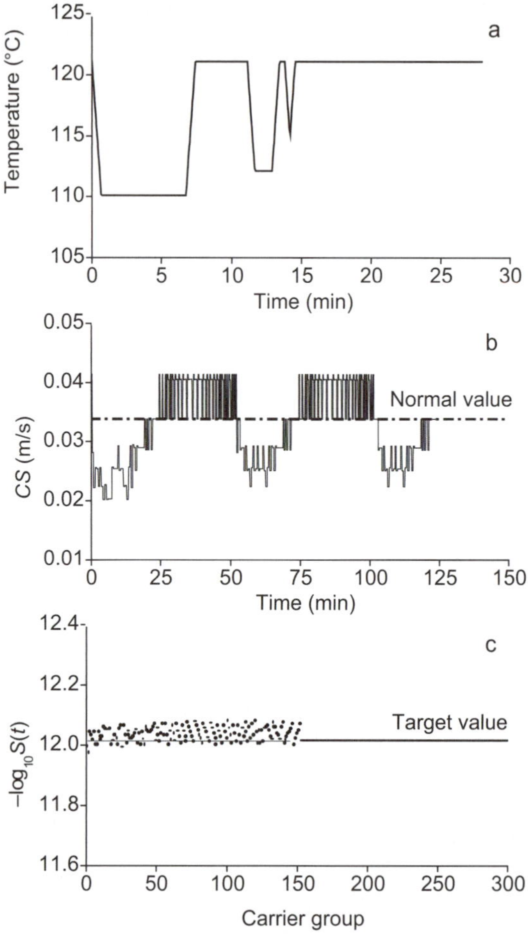

Fig. 10.4 Results showing conveyor speed and accumulated lethality in the products after imposing processing temperature deviations and applying the Fixed Point (FP) method as the correction algorithm. (a) Imposed temperature deviations in the retort. (b) Conveyor speed adjustment. (c) Microbial inactivation of the products after on-line correction. Microbial inactivation was estimated using first-order kinetics and Stumbo's method.

this is that it will result in overprocessing some of the carriers that coexisted in the steam chamber with the affected carriers. To illustrate these concepts, simulations were performed with the same temperature deviations shown in Fig.10.4a. Figures10.5a, 10.5b and 10.5c show the temperature deviations, actual values of the conveyor speed, and the microbial log-reduction in each carrier group, respectively. Figure10.5c illustrates significant overprocessing of some carrier groups that were affected by the temperature deviations.

In order to investigate the effect of a temperature deviation on the extent of the overprocessing, two types of temperature deviations were selected for further simulations. Type 1 deviation consisted of maintaining the same level of temperature

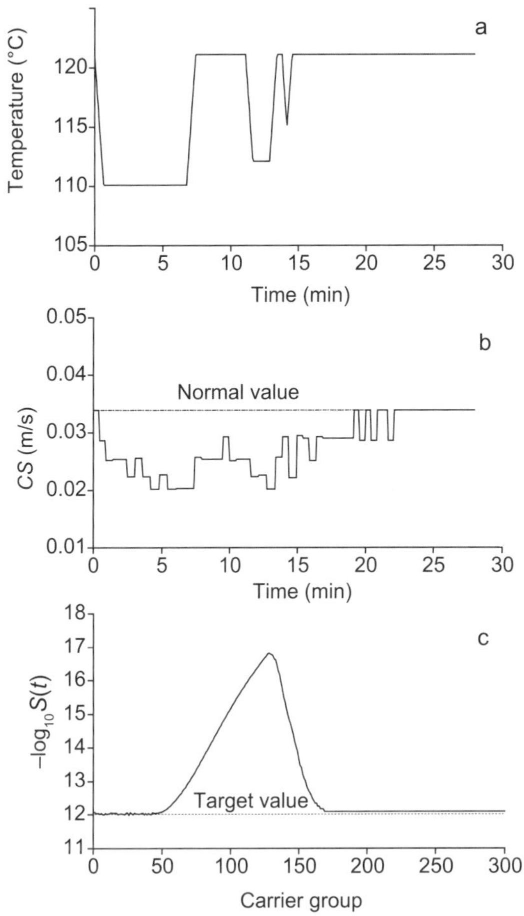

Fig. 10.5 Simulation results showing (a) imposed temperature deviations, (b) conveyor speed adjustment in response to the simulated temperature deviations and (c) microbial inactivation when conveyor speed is set to its normal value once the deviation disappears using the Fixed Point algorithm. Microbial inactivation was estimated using first-order kinetics and Stumbo's method.

drop while varying the duration of the deviation. The temperature deviations were suddenly stepped down from 121.1 °C to 110 °C and the deviation lasted for (i) 3 minutes, (ii) 5 minutes, and (iii) 10 minutes, respectively. Type 2 deviation consisted of maintaining the duration of the temperature deviations while varying the magnitude of the temperature drops. The chosen temperature deviations were stepped down from 121.1 °C to (i) 117 °C, (ii) 114 °C, (iii) 111 °C, respectively, with durations of 5 minutes. The corresponding log-reduction values after the adjustment of the conveyor speed are illustrated in Figures 10.6a and 10.6b, respectively. It is clearly shown that, although the algorithm was able to adjust the conveyor speed to keep lethalities that result in safe products, the extent of overprocessing is

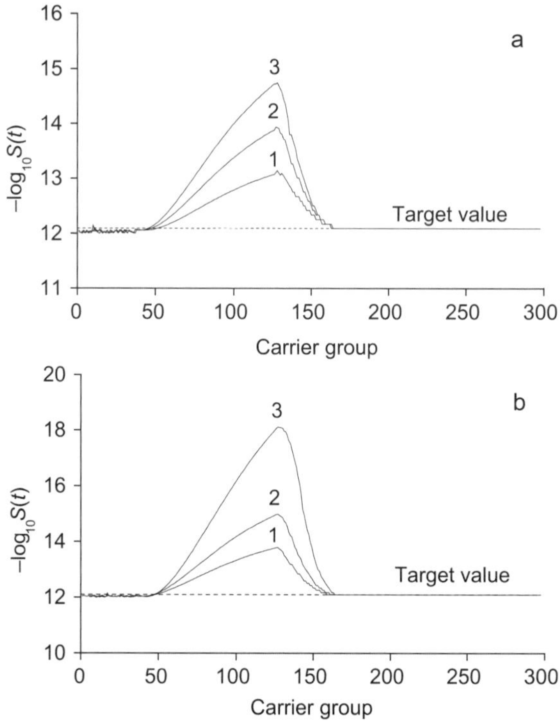

Fig. 10.6 Simulations illustrating the behavior of the Fixed Point algorithm after Type 1 and Type 2 temperature deviations (defined in the text) are imposed. Values of microbial inactivation in the different carrier groups were calculated by assuming first-order kinetics and Stumbo's method for (a) Type 1 and (b) Type 2 temperature deviations. Numbers in the figure indicate the extent of the deviation, 1 being low, 2 moderate and 3 high.

closely related to the temperature deviation's severity. Overprocessing, as expected, can be determined by either the level of the temperature drop or the duration of the deviation. Overprocessing can be acceptable for a process temperature deviation with low severity, but it can be significant if a severe temperature deviation occurs. A severe process temperature deviation could be defined *a priori*, specifically related to the process, as either a processing temperature drop that is higher than some specified value or as a deviation lasting longer than a prescribed time. In either case, these conditions can be set by the processor and readily implemented in the control algorithm.

As discussed, the two proposed alternatives to implement the FP algorithm have drawbacks. The first alternative does not set back the conveyor speed to pre-deviation conditions, whereas the second alternative may lead to significant overprocessing. Therefore, a third possibility is proposed. It consists of stopping the entry of containers once a temperature deviation occurs. After all the affected carriers have left the steam chamber and the process temperature deviations have disappeared, the conveyor speed is set back to the normal value and the containers' entry to the retort is resumed. This processing alternative does not cause product overprocessing but it may reduce the throughput significantly.

The FP algorithm – Weibull kinetics and new method to estimate log reduction calculations

The feasibility of the FP method for on-line correction of process temperature deviations in hydrostatic retorts was also tested by performing simulations that include the Weibull model to describe isothermal microbial inactivation, and the approach proposed by Peleg and Penchina (2000) to calculate microbial survival under non-isothermal processing conditions. As reported in Table 10.1, the scheduled conveyor speed at 121.1 °C was 23.8 (carrier/min), which provides a 12-log reduction. Temperature-dependent survival parameters (b and n) for *C. botulinum* spores are given in Table 10.2 as equations that are functions of temperature, rewritten below as Eqs 10.14 and 10.15. Other conditions were the same as those used in the simulations with first-order kinetics.

$$b(T) = \ln(1 + e^{(T-102.3)/3.33}) \qquad [10.14]$$

$$n(T) = 0.325 + \frac{0.425}{1 + e^{(T-101)/10.057}} \qquad [10.15]$$

Figures 10.7a,b,c show simulation results using the same temperature deviations as those used with the first-order kinetics. A comparison between these results and those obtained from the first-order kinetics indicates similar trends in the conveyor speed adjustment. The same is true for the effect of process temperature deviations on accumulated microbial log-reductions. However, the approach using non-linear kinetics provides a slower conveyor speed than that provided by the first-order kinetic approach. The approach of setting back the conveyor speed to its normal value after the deviation has disappeared is illustrated in Fig. 10.8. Figure 10.8a illustrates the temperature deviations, which are identical to those used in Fig. 10.4, and the resulting conveyor speeds are shown in Fig. 10.8b. Overprocessing resulting from this approach is illustrated in Fig. 10.8c. By comparing Figs 10.4c and 10.8c it can be observed that the extent of overprocessing is significantly smaller when using the non-linear model. Reasons for these differences can be attributed to findings reported by Campanella and Peleg (2001), who showed that the thermal death of *C. botulinum* is lower when non-linear kinetics (Weibull) are used. The effect of the type of deviation on product overprocessing using the approach of setting back the conveyor speed to it normal value when the Weibull model is used is illustrated in Fig. 10.9. Deviations Type 1 and 2, as defined above, were used in the simulations. By comparing Figures 10.5 and 10.9 it can be seen that the extent of overprocessing is lower when the Weibull model is used.

The WC algorithm – first-order kinetics and classical lethality calculations

To start the simulation, groups of 10 carriers were selected; thus the path from the entrance to the exit of the steam chamber was divided into 126 segments (see Fig. 10.3) because the total number of carriers in the steam chamber is 1263

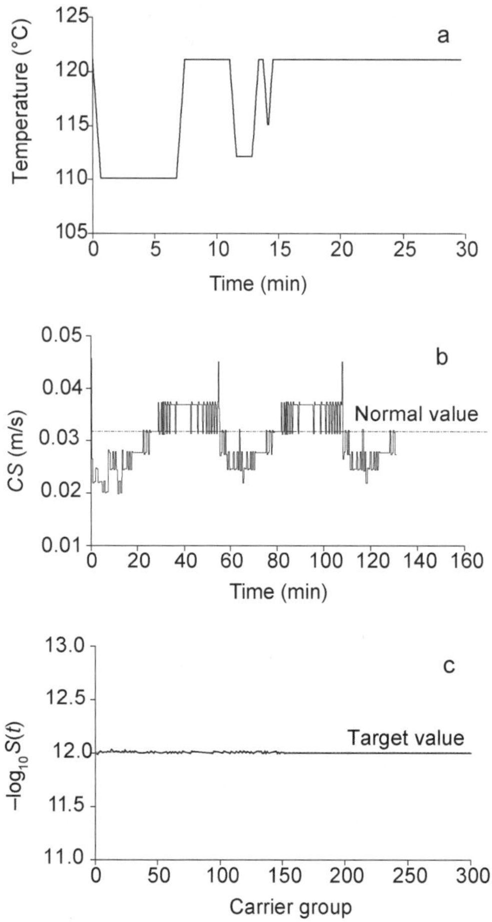

Fig. 10.7 Results showing conveyor speed and microbial inactivation in the products after imposing processing temperature deviations and applying the Fixed Point correction algorithm. (a) Simulated temperature deviations in the retort. (b) Conveyor speed adjustment in response to process temperature deviations. (c) Microbial inactivation of the products after on-line correction. Microbial inactivation was estimated using Weibull kinetics and the approach used for non-uniform temperature processes.

(Table 10.1). Simulations were conducted for a retort temperature profile with arbitrary process temperature deviations as those illustrated in Fig. 10.10a. When a temperature deviation occurs, the conveyor speed required by each carrier group corresponding to each chamber segment can be estimated from t_{res}, which is calculated from Eq.10.13, following the procedure described in Section 10.3.1. The conveyor speed *CS* can be calculated as d/t_{res}. A curve which shows the required conveyor speed for each segment and for arbitrary times, denoted by t_1, t_2, t_3 and t_4 in Fig. 10.10a, as a function of the segment number, is illustrated in Fig. 10.10b. This plot, named the conveyor speed curve, clearly shows that conveyor

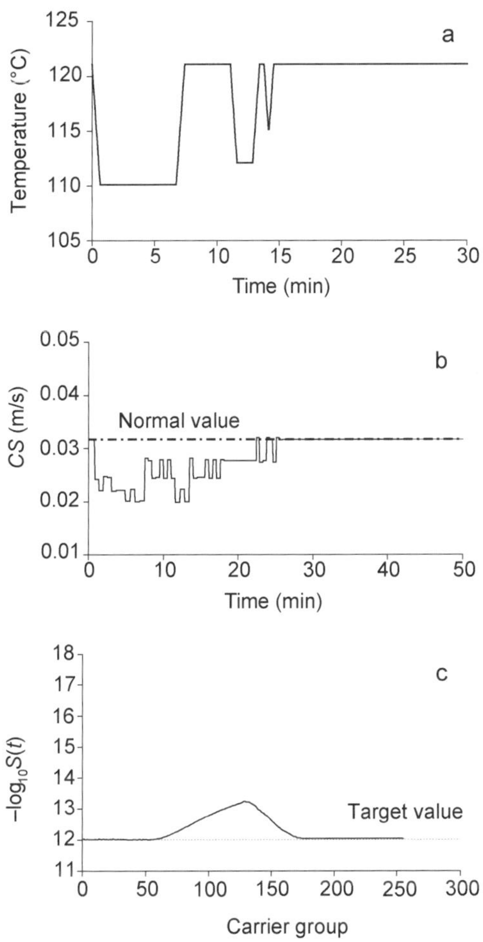

Fig. 10.8 Simulation results showing (a) Imposed temperature deviations, (b) conveyor speed adjustment in response to the simulated temperature deviations, and (c) microbial inactivation when conveyor speed is set to its normal value once the deviation disappears using the Fixed Point algorithm. Microbial inactivation was estimated using Weibull kinetics and the approach used for non-uniform temperature processes.

speed curves can have different patterns, which depend on the type of temperature deviations and the time considered. The figure also illustrates the existence of multiple local minima. A global minimum, which would correspond to the WC carrier group, can be easily identified from the conveyor speed curves themselves, or by a using a suitable optimization technique to search for that minimum, which is more suitable for a control algorithm. Thus, by applying the control algorithm, the WC carrier group is identified and the processing time necessary for that carrier is used for the process. Figure 10.10c illustrates the conveyor speed adjustment in response to the process temperature deviations. The figure clearly shows that the conveyor speed is slower than its normal value during the process temperature

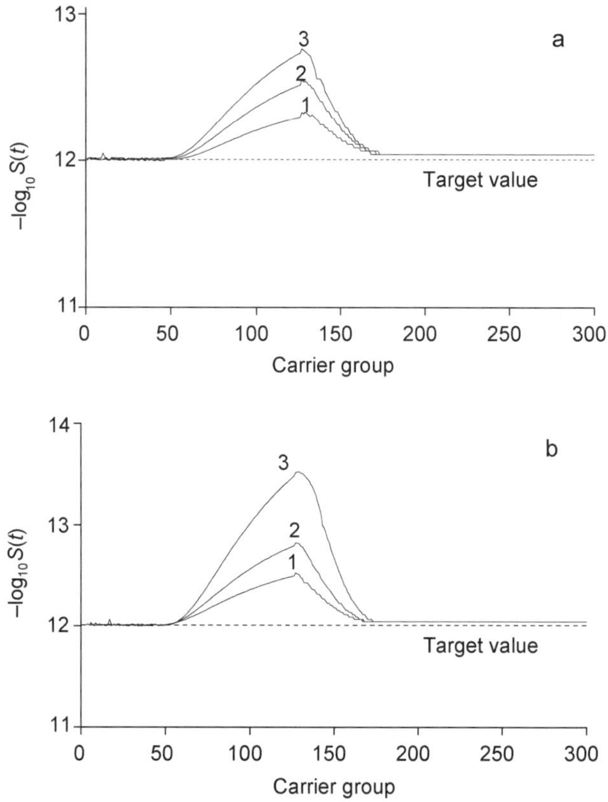

Fig. 10.9 Simulations illustrating the behavior of the Fixed Point algorithm after Type 1 and Type 2 temperature deviations (defined in the text) are imposed. Microbial inactivation in the different carrier groups was calculated assuming Weibull kinetics and the approach used for non-uniform temperature processes for (a) Type 1 and (b) Type 2 temperature deviations. Numbers in the figure indicate the extent of the deviation, 1 being low, 2 moderate and 3 high.

deviations (a drop in this case) and reaches a constant value very quickly when the steam temperature changes to a new value. Once the process temperature comes back to the normal process temperature and the carrier group existing in the Segment Number 1 (exit segment, see Fig.10.3) becomes the WC carrier group, all the carrier groups existing in the steam chamber will require a conveyor speed faster (less residence time) than the normal conveyor speed. At that time, setting back the conveyor speed to its normal value will not affect the safety of the process. This, however, results in overprocessing of all the carrier groups currently inside the steam chamber. The microbial inactivation in all affected carrier groups corresponding to the conveyor speed adjustment is shown in Fig. 10.10d. From the simulations and for the assumed retort temperature profile, it can be estimated that a total of 219 carrier groups were affected. As expected, all the carrier groups affected by the process temperature deviations exhibited a microbial inactivation

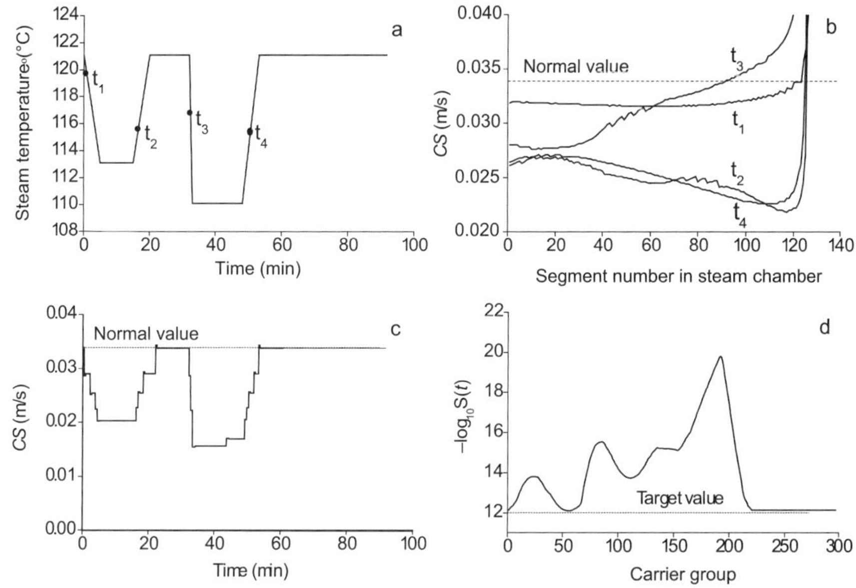

Fig. 10.10 Results showing conveyor speed and microbial inactivation in the products after imposing processing temperature deviations and applying the Worst Case (WC) method as the correction algorithm. (a) Simulated temperature deviations in the retort. (b) Conveyor speed curve for the times indicated in (a). (c) Conveyor speed adjustment in response to process temperature deviations. (d) Estimated microbial inactivation of the products after on-line correction. Microbial inactivation was estimated using first-order kinetics and the Stumbo's method.

that is higher than the targeted inactivation. This is based on the assumption that the entry of containers to the retort is not halted during the temperature deviations. As a matter of fact, overprocessed carrier groups could be reduced if the entry of containers is halted. From simulations using this algorithm it was found that the number of affected carrier groups would be 158 (based on the total number of carriers in the steam chamber plus carriers between feed station and entrance of the steam chamber) if the entry of containers is stopped. That number is significantly lower when compared to the 219 carrier groups affected in the case that the entry of containers is not stopped. The simulations show that application of the WC control algorithm may lead to the overprocessing of a number of carriers. To demonstrate the effect of different process temperature deviations on the extent of overprocessing, simulations were performed with Type 1 and Type 2 deviations, as defined previously, using the FP control algorithm. Microbial inactivation in processed carrier groups corresponding to Type 1 and Type 2 deviations are illustrated in Figures 10.11a and 10.11b, respectively. The figures clearly show that the extent of overprocessing depends on the duration of the deviation for Type 1 deviations and the magnitude of temperature drop for Type 2 deviations. As expected, for process temperature deviations of long duration and high level of temperature variation (drop in this case), overprocessing may be significant and go

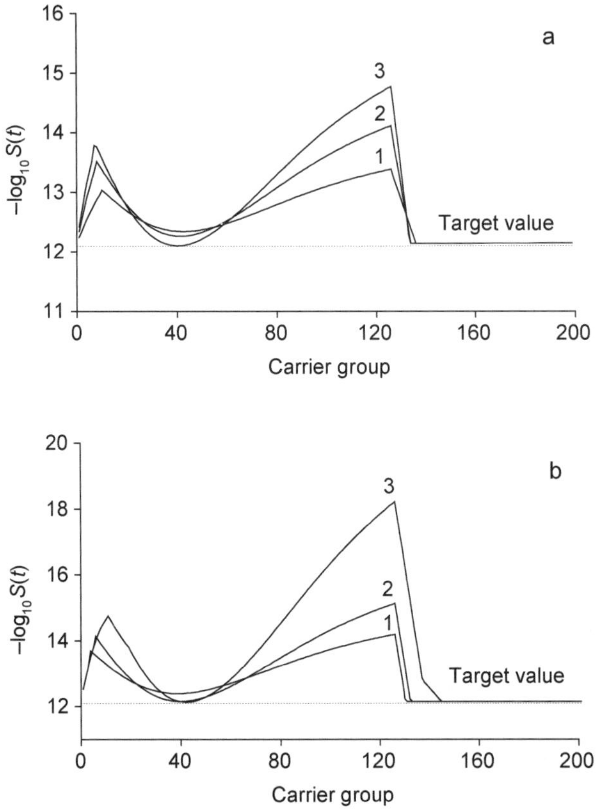

Fig. 10.11 Simulations illustrating the behavior of the Worst Case algorithm after Type 1 and Type 2 temperature deviations (defined in the text) are imposed. Microbial inactivation in the different carrier groups was calculated by assuming first-order kinetics and Stumbo's method for (a) Type 1 and (b) Type 2 temperature deviations. Numbers in the figure indicate the extent of the deviation, 1 being low, 2 moderate and 3 high.

beyond a prescribed quality criterion, and therefore the product may have to be located and discarded. One of the advantages of the WC algorithm, as shown in this simulation, is that it allows for the location and an estimation of the extent of overprocessing of the affected carriers.

The WC algorithm – Weibull kinetics and new method to estimate log reduction calculations

Simulations were also performed for the same retort temperature profile as that illustrated in Fig. 10.10a but applying Weibull kinetics and the new method to estimate microbial inactivation for non-isothermal conditions. The required heating time for a carrier group was determined by applying the bisection method (http://mathworld.Wolfram.com/ Bisection.html) to Eq. 10.13a, as explained previously, whereas the microbial inactivation was calculated using the method proposed by Chen *et al.* (2007). Figure 10.12 shows results of the simulation,

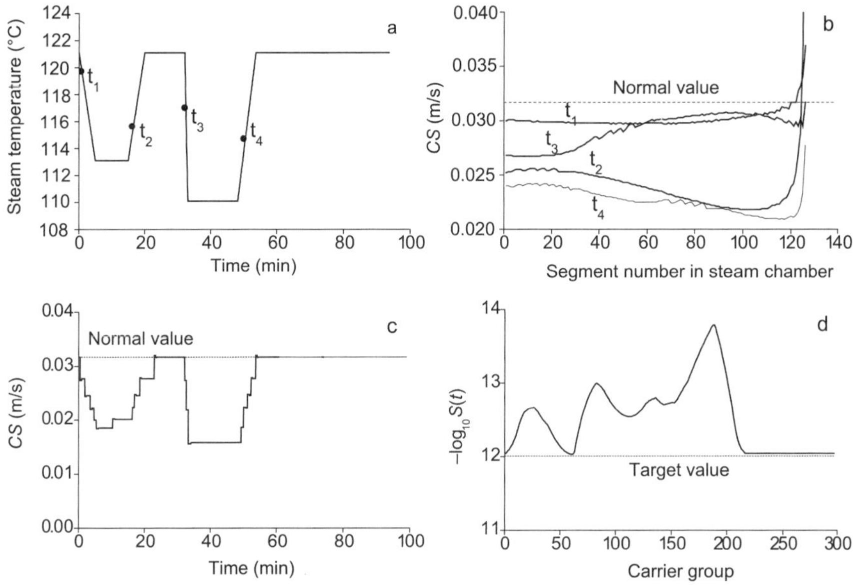

Fig. 10.12 Results showing conveyor speed and microbial inactivation in the products after imposing processing temperature deviations and applying the Worst Case (WC) method as the correction algorithm. (a) Simulated temperature deviations in the retort. (b) Conveyor speed curve for the times indicated in (a). (c) Conveyor speed adjustment in response to process temperature deviations. (d) Estimated microbial inactivation of the products after on-line correction. Microbial inactivation was estimated using Weibull kinetics and the approach used for non-uniform temperature processes.

whereas Fig 10.13 illustrates the extent of overprocessing for deviations of Type 1 and Type 2, which were defined previously. A comparison between these results and those obtained using first-order kinetics indicates a similar trend in the conveyor speed adjustment (see Fig. 10.10c and Fig 10.12c). The same is true for the extent of overprocessing (compare Figs 10.10d and 10.12d) and the effect of process temperature deviations on microbial inactivation (compare Figs 10.11a,b and 10.13a,b). However, as illustrated in these figures, the approach using non-linear kinetics provides a slower conveyor speed than that provided by the first-order kinetic approach. These comparisons also show that the calculated extent of overprocessing is significantly smaller when the non-linear model and the new approach to calculating microbial inactivation are used. The reasons for these differences are shown above and have been discussed by Campanella and Peleg (2001).

10.4 Future trends and validation of automated processes

Food thermal sterilization processes must meet required microbiological safety standards that consist of a stipulated reduction in the initial microbial count of target microbial spores at the slowest heating point of the food container. Thus, in

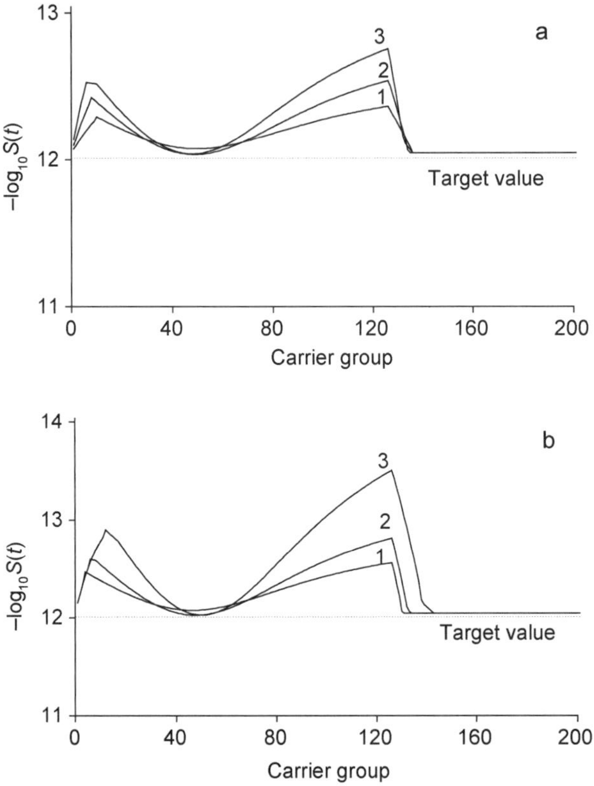

Fig. 10.13 Simulations illustrating the behavior of the Worst Case algorithm after Type 1 and Type 2 temperature deviations (defined in the text) are imposed. Microbial inactivation in the different carrier groups was calculated by assuming Weibull kinetics and the approach used for non-uniform temperature processes for (a) Type 1 and (b) Type 2 temperature deviations. Numbers in the Figure indicate the extent of the deviation, 1 being low, 2 moderate and 3 high.

order to accurately estimate the efficacy of a thermal process, information on two aspects of the food and the process must be known. The first aspect is related to the microbiology of the system and primarily concerns the inactivation kinetics of the target spores. Linear and non-linear models, such as the Weibull model, have been used to describe kinetics of spore inactivation; however, the food industry uses only first-order kinetics as a process validation criterion. New findings and reports (van Boekel, 2002) show that the presence of non-linear kinetics is more a rule than an exception. Thus, new methods are necessary to validate processes considering spores whose inactivation kinetics significantly differ from linear behavior. An attempt to show the differences between both approaches as applied to a proposed control algorithm is illustrated in this chapter. With the advent of more powerful and faster computer systems, and by using algorithms such as those described in this chapter, it is possible to perform fast calculations that can be implemented in on-line correction methods. The apparent complexity in the use of non-linear kinetics and complicated non-isothermal processes is nowadays easily overcome by available software and methods (Peleg, 2006; Chen *et al.*, 2007, 2008). In addition, some of the programs dealing with these issues are now available, as freeware, from the Internet (http://www-unix.oit.umass.edu/~aew2000/GrowthAndSurvival.html).

The second characteristic intimately related to the efficacy of the thermal process concerns the temperature–time history experienced by the product at the slowest heating point. Tucker (2002) discusses in detail various techniques that might be used to measure product temperature history at the slowest heating point. For validation of a control system implemented in a batch retort operation, a thermocouple is placed in the slowest heating point and the temperature at that location is measured and recorded under different processing conditions. Some of those conditions may include potential temperature deviations such as those used in the simulations performed in this chapter. On-line correction methods for continuous retorts have not been reported except those of the two patents mentioned (Weng, 2003a,b) and recent work carried in our lab (Chen *et al.*, 2008), so industry has not yet had the opportunity of using these methods and validating on-line correction methods for continuous retorts. However, the challenge with continuous retorts lies in the difficulty of setting suitable measuring and recording temperature systems of practical feasibility. A suitable system for product temperature measurement and recording in continuous retorts would use a wireless temperature probe. In the past few years, significant progress has been made in improving this technology. One of the available products in the market today is the ValProbe™, which satisfies FDA Regulation 21 CFR Part 11 requirements for electronic signatures, and records and complies with EN 554 for saturated steam sterilization. In order to test an online control system, process temperature deviations could be imposed in the thermal process and the control system could be used to automatically correct for these deviations. During those tests, the product temperature history at the slowest heating point could be recorded by the wireless temperature probes and sent to the control computer. With information on the product temperature at the lowest heating point, and other locations if necessary, the microbial reduction could be calculated using the methods described. The system would also incorporate kinetics discussed in this chapter which would be known *a priori* and obtained from previous microbiological tests performed on the target microorganism.

10.5 Sources of further information and advice

Useful publications as well as commercial products regarding online correction of process temperature deviations in batch retort processes are briefly discussed in this section. For continuous retort processes, information about online control strategies is more scarce. Related publications and patents are given in the list of references, including a recent publication from our research group. Government regulations on handling process temperature deviations in continuous retort processes can be found in FDA Regulation 21 CFR Part 110. As an indispensable component of an online control system, microbial inactivation kinetics must be experimentally determined and incorporated into the control system. A lot of work concerning validation purposes has focused on first-order kinetics and the traditional method of calculating the process efficacy through the lethality parameter, F_0. In recent years, work on modeling nonlinear behavior of microbial inactivation has appeared. In addition, new methodologies to estimate process efficacy for non-uniform thermal treatment without using preconceived models, which may be unsuitable for biological systems (e.g. the Arrhenius model) have been developed. Among the various nonlinear models utilized in these publications, the Weibull model has been preferred because it is simple to use and can be utilized to describe most of the harmful microorganisms present in food systems. Van Boekel (2002) provides a comprehensive list of inactivation kinetics that can be described by the Weibull model. With the advent of new computers and methods, the complexities of using non-linear kinetics and complex temperature histories to estimate, quickly and accurately, the efficacy of a process to be implemented in a control algorithm have been drastically reduced. New publications in the public domain (Chen *et al.*, 2007, 2008) will catalyze more research work in this area in the near future concerning online temperature deviation controls. The development of new wireless technology and wireless base-temperature probes will facilitate the validation of these control systems.

10.6 References

Akterian, S.G. (1999). On-line control strategy for compensating for arbitrary deviations in heating-medium temperature during batch thermal sterilization processes. *Journal of Food Engineering*, 39, 1–7.

Akterian, S.G. and Fikiin, K.A. (1994). Numerical simulation of unsteady heat conduction in arbitrary shaped canned foods during sterilization processes. *Journal of Food Engineering*, 21, 343–354.

Anderson, W.F., McClure, P.J., Baird-Parker, A.C. and Cole, M.B. (1996). The application of log-logistic model to describe the thermal inactivation of *Clostridium botulinum* 213B at temperatures below 121.1 °C. *Journal of Applied Bacteriology,* 80, 283–290.

Ball, C. O. and Olson, F. C. W. (1957). *Sterilization in Food Technology, Theory, Practice and Calculations* (McGraw-Hill, New York, USA).

Bichier, J.G., Teixeira, A.A., Balaban, M.O. and Heyliger, T.L. (1995). Thermal process simulation of canned foods under mechanical agitation. *Journal of Food Process Engineering*, 18(1), 17–40.

Bown, G., Nesaratnam, R. and Peralta-Rodriguez, R.D. (1986). *Computer Modeling for the Control of Sterilization Processes*, Technical Memorandum No. 442, CFDRA. Chipping Campden, Glos, GL55 6LD.

Campanella, O.H. and Peleg, M. (2001). Theoretical comparison of a new and the traditional method to calculate *Clostridium botulinum* survival during thermal inactivation. *Journal of the Science of Food and Agriculture*, 81, 1069–1076.

Chen, G., Campanella, O.H., Corvalan, C.M. and Haley, T.A. (2008). On-line correction of process temperature deviations in continuous retorts. *Journal of Food Engineering*, 84, 258–269.

Chen, G., Campanella, O. H., and Corvalan, C. (2007). A numerical method for evaluation of process sterility during non-isothermal processing. *Food Research International*, 40(1), 203–208.

Chen, G., Corvalan, C., Campanella, O. H. and Haley, T. A. (2005). An improved method to estimate temperatures during the cooling stage of sterilized cylindrical cans. *Trans IChemE Part C*, 83(C1), 36–42.

Cole, M.B., Davies, K.W., Munro, G., Holyoak, C.D. and Kilsby, D.C. (1993). A vitalistic model to describe thermal inactivation of *L. monocytogenes*. *Journal of Industrial Microbiology*, 12, 232–239.

Datta, A.K., Teixeira, A.A. and Manson, J.E. (1986). Computer-based retort control logic for on-line correction of process deviations. *Journal of Food Science*, 51, 480–483,507.

Gavin, A. and Weddig, L. (1995). *Canned foods: Principles of thermal process control, acidification and container closure evaluation*, 6th Ed. (The Food Processors Institute, Washington, DC, USA).

Giannoni-Succar, E.B. and Hayakawa, K.I. (1982). Correction factor of deviant thermal processes applied to packaged heat conduction food. *Journal of Food Science*, 47(2), 642–646.

Hills, B.P. and Mackey, B.M. (1995). Multi-compartment kinetic-models for injury, resuscitation induced lag and growth in bacterial cell populations. *Food Microbiology*, 12(4), 333–346.

Kelly, P.T. and Richardson, P.S. (1987). *Computer Modeling for the Control of Sterilization Processes*, Technical Memorandum No. 459, CFDRA. Chipping Campden Glos, GL55 6LD.

Kumar, M.A., Ramesh, M.N. and Rao, S.N. (2001). Retrofitting of a vertical retort for on-line control of the sterilization process. *Journal of Food Engineering*, 47, 89–96.

Lappo, B.P. and Povey, M.J.W. (1986). Microprocessor control system for thermal sterilization operations. *Journal of Food Engineering*, 5, 31–53.

Linton, R.H., Carter, W.H., Pierson, M.D. and Hackney, C.R. (1995). Use of a modified Compertz equation to model nonlinear survival curves for *Listeria monocytogenes Scott, A. Journal of Food Protection*, 58, 946–954.

LOG-TEC CCS-8 Retort Management System (1984). Central Analytical Laboratories, Inc, 2600 Marietta Street, Kenner, Louisiana.

Mafart, P., Couvert, O., Gaillard, S. and Leguerinel, I. (2002). On calculating sterility in thermal preservation methods: Application of the Weibull frequency distribution model. *International Journal of Food Microbiology*, 72, 107–113.

National Food Processors Association (NFPA) (1980). *Laboratory Manual for Food Canners and Processors*, Vol. 1, AVI, Westport, CT.

Noronha, J., Hendrickx, M., Van Loey, A., and Tobback, P. (1995) New Semi-empirical Approach to Handle Time-variable Boundary Conditions during Sterilization of Non-conductive Heating Foods. *Journal of Food Engineering*, 24, 249–268.

Peleg, M. (2006). *Advanced Quantitative Microbiology for Foods and Biosystems*. CRC Taylor & Francis, New York.

Peleg, M. (2003). Microbial survival curves: Interpretation, mathematical modeling and utilization. *Comments on Theoretical Biology*, 8, 357–387.

Peleg, M., Normand, M.D. and Campanella, O.H. (2003). Estimating microbial inactivation parameters from a single survival curve obtained under varying conditions – The linear case. *Bulletin of Mathematical Biology*, 65(2), 219–234.

Peleg, M., Engel, R., Gonzalez-Martinez, C. and Corradini, M.G. (2002). Non *Arrhenius* and non WLF kinetics in food systems. *Journal of the Science of Food and Agriculture*, 82, 1346–1355.

Peleg, M. and Penchina, C.M. (2000). Modeling microbial survival during exposure to a lethal agent with varying intensity. *Critical Reviews in Food Science and Nutrition*, 40, 159–172.

Peleg, M. and Cole, M.B. (2000). Estimating the survival of *Clostridium botulinum* spores during heat treatments. *Journal of Food Protection*, 63, 190–195

Peleg, M. (1999). On calculating sterility in thermal and non-thermal preservation methods. *Food Research International,* 32(4), 271–278.

Peleg, M. and Cole, M.B. (1998). Reinterpretation of microbial survival curves. *Critical Reviews in Food Science and Nutrition*, 38, 353–380.

Ryniecki, A. and Jays, D.S. (1993). Automatic determination of model parameters for computer control of canned food sterilization. *Journal of Food Engineering*, 15(1), 75–94.

Sapru, V., Texeira, A.A., Smerage, G.H. and Lindsay, J.A. (1992). Predicting thermophilic spore population dynamics for VHT sterilization processes. *Journal of Food Science,* 57(5), 1248–1252, 1257.

Sapru, V., Texeira, A.A., Smerage, G.H. and Lindsay, J.A. (1993). Comparison of predictive models for bacterial spore population resources to sterilization temperatures. *Journal of Food Science,* 58(1), 223–228.

Silva, C., Hendrickx, M., Oliveira, F. and Tobback, P. (1992). Optimal sterilization temperature for conduction heating foods considering finite surface heat transfer coefficients. *Journal of Food Science*, 57(3), 743–748.

Simpson, R., Almonacid-Merino, S.F. and Torres, J.A. (1993). Mathematical models and logic for computer control of batch retorts: Conduction-heated foods. *Journal of Food Engineering*, 20, 283–295.

Simpson. R., Figueroa, I. and Teixeira, A. (2006). Optimum on-line correction of process deviations in batch retorts through simulations. *Food Control*, 117, 665–675.

Simpson. R., Teixeira, A. and Almonacid, S. (2007a). Advances with intelligent on-line retort control and automation in thermal processing of canned foods. *Food Control*, 18, 821–833.

Simpson. R., Figueroa, I. and Teixeira, A. (2007b). Simple, practical and efficient on-line correction of process deviations in batch retorts though simulations. *Food Control*, 18, 458–465.

Simpson. R., Figueroa, I, Llanos, D. and Teixeira, A. (2007c). Preliminary validation of on-line correction of process deviations without extending process time in batch retorting: Any low-acid canned foods. *Food Control*, 18, 983–987.

Stumbo, C.R. (1965). *Thermobacteriology in food processing* (Academic Press, New York, USA).

Teixeira, A.A., Tucker, G.S., Balaban, M.O. and bichier, J. (1992). Innovations in conducting-heating models for on-line retort control of canned foods with any j-value. In *Advances in Food Engineering*, ed. R.P. Singh and A. Wirakartakusumah. CRC Press, Boca Raton, FL, USA.

Teixeira, A.A., Balaban, M.O., Germer, S.P.M., Sadahira, M.S., Teixeira-Neto, R.O., and Vitali, A.A., (1999). Heat Transfer Model Performance in Simulation of Process Deviation. *Journal of Food Science,* 64(3), 488–493.

Teixeira, A.A. and Manson, J.E. (1982). Computer control of batch retort operations with on-line correction of process deviations. *Food Technology*, 36, 85–90.

Teixeira, A.A. and Tucker, G.S. (1997). On-line retort control in thermal sterilization of canned foods. *Food Control*, 8(1), 13–20.

Tucker, G.S. and Clark, P. (1989). *Computer Modeling for the Control of Sterilization Processes*. Technical Memorandum No. 529, CFDRA. Chipping Campden Glos, GL55 6LD

Tucker, G.S. and Clark, P. (1990). Modeling the cooling phase of heat sterilization processes, using heat transfer coefficients. *International Journal of Food Science and Technology*, 25(6), 668–681.

Tucker, G.S. (2002). Validation of heat processes. In *Thermal Technologies in Food Processing* (Edited by Richardson, P.), Woodhead publishing, England, pp. 75–90.

USDA (2005). Principles of Thermal Processing, available at www.fsis.usda.gov/ PDF/ FSRE _ SS_ 3PrinciplesThermal.pdf

van Boekel, M.A.J.S. (2002). On the use of the Weibull model to describe thermal inactivation of microbial vegetative cells. *International Journal of Food Microbiology*, 74, 139–159.

Weng, Z., Park, D.K., and Heyliger, T.L. (1995). Process deviation analysis of conduction-heating canned foods processed in a hydrostatic sterilizer using a mathematical model. In: *Food Processing Automation* IV, FPEI, ASAE 368–379.

Weng, Z. (2003a). *Controller and Method for Administering and Providing On-line Handling of Deviations in a Hydrostatic Sterilization Process.* U.S. Patent 6440361.

Weng, Z. (2003b). *Controller and Method for Administering and Providing On-line Handling of Deviations in a Rotary Sterilization Process.* U.S. Patent 6416711.

Wojciechowski, J. and Ryniecki, A. (1989). Computer control of sterilization of canned meat products. *Fleischwirtschaft,* 69(2), 268–270 (in German).

11

Neural network method of modeling heat penetration during retorting

C. Chen, Campbell Soup Company, USA, and H. S. Ramaswamy, McGill University, Canada

11.1 Introduction

Thermal technologies have long been at the heart of food processing, involving the production, transformation, and preservation of foods. The application of heat is both an important method of preserving foods and a means of developing texture, flavor and color. An important issue for food manufacturers is determining effective application of thermal technologies to achieve these objectives without damaging other desirable sensory and nutritional qualities in a food product.

The necessity for developing advanced thermal processing for the food industry is increasing in line with the demand for enhanced food safety and quality; this is because associated with thermal processing is always some undesirable degradation of heat-sensitive quality attributes.[1] Sterilization and pasteurization are heating processes to inactivate or destroy enzyme and microbiological activity in foods. Conventional thermal processing can be divided into two types: retort processing and aseptic processing. The retort processing method is one of the most mature processing technologies, and is widely used in North American food industries, specifically for solid or liquid foods containing big particulates, although aseptic processing technologies have certain useful advantages, such as higher production efficiency and better quality. In retort thermal processing, the heat is transferred by conduction and/or convection from the heating medium to the food, depending on the type of foods being processed. The temperature inside the food during heating will be determined by a variety of processing conditions, including the type of heating medium and its temperature, initial product temperature, thermal properties of food being heated, and rheological properties for liquid foods. Theoretically, it is possible to apply a mathematical modeling method combined with modern computation techniques for the simulation of thermal processing of solid or particulate liquid foods, provided all the processing conditions can be discovered and all the thermo-physical properties of the food obtained by independent experiments. However, the biggest challenge that food modeling researchers are facing is that, unlike other engineering materials, food materials have variable thermal and/or physical properties, most of which are temperature and processing time dependent. This means that it is very difficult to discover the properties and their changes with processing temperature and time under conditions simulating the real processes. Normally, the property parameters used for modeling inputs are effective values instead of real values, which are determined by comparing the agreement level between real experimental outputs and model predicted outputs.[2] Application of effective property values instead of real values in model development is feasible for a specific case; the challenge, however, is to extend it to practical applications or to scale it up to a large size application.

In recent years, artificial neural networks (ANNs) have opened alternative pathways for modeling of complex and nonlinear processes. The advantages of ANNs over conventional mathematical methods in modeling performance have been recognized and confirmed by many research reports.[3] These advantages include:

- *Learning ability.* Neural networks have learning ability similar to the human brain; they can construct the cause-and-effect relationships through repeated training without any prior knowledge of the system being investigated. Therefore, neural networks are suitable for cases with multiple variables and complicated internal relationships, which are often difficult to describe by mathematical equations.
- *Robustness and fault tolerance.* The ANN is tolerant of noisy and incomplete data, because the information is distributed in massive processing nodes and connections. Minor damage to parameters in the network will not degrade overall performance significantly.
- *High computational speed.* The ANN is an inherently parallel architecture. The result comes from the collective behavior of a large number of simple parallel processing units. Therefore, once trained, neural networks can calculate results from a given input very quickly. This has given neural network models greater potential than conventional modeling methods used in online control systems.

This chapter focuses on an introduction to the basic principles of neural networks, development of neural network models, and their application advances in food thermal processing areas.

11.2 Principles of neural networks

Neural networks are information processing prototypes that are inspired by biological nervous systems. A typical biological neuron contains neuronal cell bodies (soma), dendrites and axons. Each neuron receives electrochemical inputs from other neurons at the dendrites. If the sum of these electrical inputs is sufficiently powerful to activate the neuron, it transmits an electrochemical signal along the axon and passes this signal to the other neurons whose dendrites are attached at the axon terminals. These attached neurons may then fire. It is important to note that a neuron fires only if the total signal received at the cell body exceeds a certain level. The entire brain is composed of these interconnected electro-chemical transmitting neurons. From a very large number of extremely simple processing units, each performing a weighted sum of its inputs and then firing a binary signal if the total input exceeds a certain level, the brain manages to perform extremely complex tasks. However, it should be noted that artificial neural networks represent only very simplified formal models of biological neurons and their interconnections, without making any attempt to model the biological system itself. Their importance lies in the fact that artificial networks are brain-inspired computational tools for solving complex problems.

11.2.1 Neural network architecture

There are two types of neural networks, feed-forward and feedback. Feed-forward ANNs allow signals to travel one way only, from input to output, while feedback networks can have signals traveling in both directions by introducing loops in the network. Figure 11.1 shows a typical feed-forward neural network with multiple layers formed by an interconnection of nodes. This neural network has an input layer, two hidden layers, and one output layer. Each layer is essential for the operation of the network. A neural network can be viewed as a 'black box' into which a specific input to each node in the input layer is sent from an external source. The network processes this information through the interconnections between nodes, although this entire processing step is hidden. After processing, the network gives an output to an external receptor from the nodes on the output layer.

When the input layer receives information from an external source, it becomes 'activated' and emits signals to its neighbors. The neighbors receive excitations from the input layer, and in turn emit signals to their neighbors. Depending on the strength of the interconnections, i.e. the magnitude of the so-called 'weight factor' that adjusts the strength of the input signal, these signals can excite or inhibit the nodes. What results is a pattern of activation that eventually manifests itself in the output layer. Finally, the values from the output layer will be compared with the desired values. If the difference between output and desired values is larger than the set error range, then the weight factors are adjusted through repeated training until the error is within the set error range, or the number of learning runs exceeds a pre-set limit.

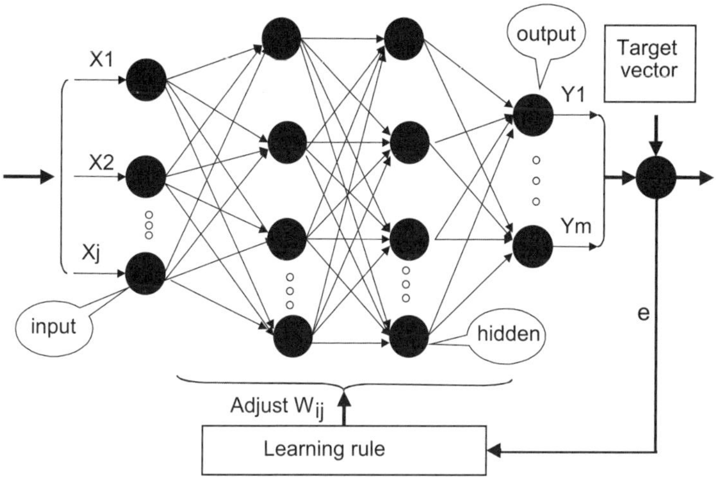

Fig. 11.1 A typical multi-layer neural network with one hidden layer (from Chen, 2001, McGill University, Canada, with permission).[13]

11.2.2 Artificial neurons

Artificial neurons are simple processing units similar to biological neurons; they receive multiple inputs from other neurons but generate only one output. This output may be propagated to several other neurons. Each neuron has two basic functions: gathering information from the other neurons in the preceding layer and sending the signals to the neurons in the next layers. The first artificial neuron model was proposed in 1943 by McCulloch and Pitts,[4] as shown in Fig. 11.2, and is based on a simplified consideration of the biological model. The elementary computing neuron functions as an arithmetic logic computing element. The binary inputs of the neurons are $x_1,x_2,\ldots, x_n$. Zero represents absence, and one represents existence. The weight of connection between the i-th input x_i and the neuron is represented by w_i. When $w_i>1$, the input is excitatory. When $w_i<0$, it is inhibitory. The net summation of inputs weighted by the synaptic strength w_i at connection i is

$$net = \sum_{i-1}^{n} w_i x_i \qquad [11.1]$$

The net value is then mapped through an activation function f of neuron output. The activation function used in the model is a threshold function,

$$y = f(net) \qquad [11.2]$$

$$f(x) = \begin{cases} 1, & x > \theta \\ 0, & \text{otherwise} \end{cases} \qquad [11.3]$$

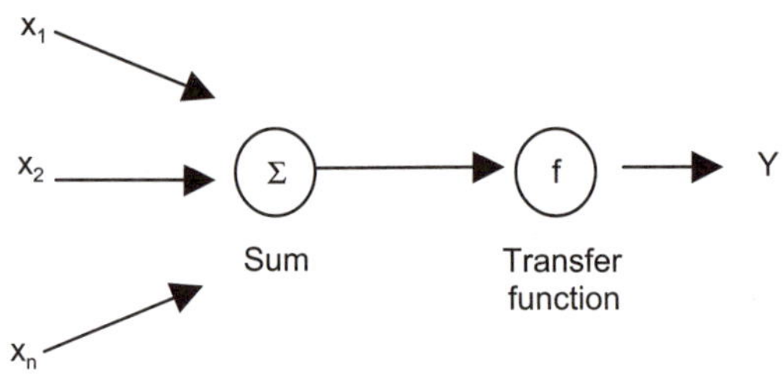

Fig. 11.2 Single artificial neuron (From McCulloch and Pitt, *Bull. Math. Biophys.*, with permission).[14]

where θ is the threshold value. Neuron models used in current neural networks are constructed in a more general way. The input and output signals are not limited to binary data, and the activation function can be any continuous function other than the threshold function used in the earlier model. The activation function is typically a monotonic non-decreasing nonlinear function. Some of the often used activation functions are:

Sigmoid function: $$f(x) = \frac{1}{1 + e^{-\alpha x}} \qquad [11.4]$$

Hyperbolic function: $$f(x) = \tanh(\alpha x) = \frac{e^{\alpha x} - e^{-\alpha x}}{e^{\alpha x} + e^{-\alpha x}} \qquad [11.5]$$

Linear threshold: $$f(x) = \begin{cases} 1 & x \geq \theta \\ x/\theta & 0 \prec x \prec \theta \\ 0 & x \leq \theta \end{cases} \qquad [11.6]$$

Gaussian function: $$f(x) = e^{-\alpha x^2} \qquad [11.7]$$

where α denotes the parameter, and θ denotes the threshold value.

11.2.3 Learning rules

There are two learning modes available for networks, supervised and unsupervised. In the supervised mode, training a neural network involves feeding the network a set of known input–output patterns, and adjusting the network weights until each input produces the appropriate output. Thus, training the neural network means the adjustment of the weight factors until the output pattern (response) calculated from the given input reflects the desired relationship. In unsupervised mode, the neural networks are presented with only a series of input patterns, without any information about their desired outputs. Thus, there is no defined criterion to adjust the weights based on the specific or target outputs. During training, the network attempts to group input patterns that are similar to each other and adapts according to a particular organization scheme.

The primary training method for supervised learning is Error-correction Learning, which uses the back-propagation of error to adjust the network weights and thresholds so as to minimize the error in its prediction on the training set; this is also known as the delta rule.

It is based on the gradient descent method to minimize the squares of differences between the actual and desired outputs, by adjusting values of the connecting weights. Mathematically, the difference between the actual and desired outputs is given as:

$$\varepsilon_i = d_i - c_i \qquad [11.8]$$

where ε_i is the output error, d_i is the desired output, and c_i is the calculated output, for the i-th neuron on the output layer only. If there are n outputs in the output layer, the total square error on the output layer can be calculated as:

$$E = \sum_{i=1}^{n} \varepsilon_i^{\,2} = \sum_{i=1}^{n} (d_i - c_i)^2 \qquad [11.9]$$

The target of network training is to minimize the total square error (value E) by adjusting node connection coefficients or weights (W). Generalized delta rule (or simply delta rule) is the most commonly used learning mechanism for multi-layer feed-forward networks with nonlinear node function such as a sigmoid. By use of this rule, the weight change can be calculated by the following equation:

$$w_{new} = w_{old} + \eta a \delta + \lambda \Delta w_{old} \qquad [11.10]$$

where

$$\delta = (d - c) f(net)(1 - f(net)) \qquad [11.11]$$

and

$$f(net) = \frac{1}{1 - \exp(-net)} \qquad [11.12]$$

where η is a linear proportionality constant for node j, called the learning rate (typically, $0<\eta<<1$), and a is the input to the neuron; λ is a constant known as momentum.

11.2.4 Development of neural network models

Neural networks can be developed either based on theory using a specific computer language or by use of commercial software. Developing neural network codes, which involves applying the theory of a particular network model into the design for a computer simulation implementation, can be a challenging task for most applied scientists and engineers who do not have the required programming skills and knowledge of neural networks. As a result, the use of commercial software has been the most popular method for developing an ANN model. With the rapid development of computer software, several ANN software packages have been developed, which can be used for developing ANN models for specific purposes. Some examples include: Neural-Ware Professional (NeuralWare, Pittsburgh, PA), Neural-Shell, Neuro-solution (Neuro-Dimension, Inc., Gainesville, FL), Matlab Neural network toolbox (MathWorks, Inc., Natick, MA), Statistica Neural Networks (StatSoft, Inc., Tulsa, OK) and Neuro-Genetic Optimizer (BioComp Systems, Inc, Bloomington, MN).

Developing a neural network by using commercial neural networks software consists of the following steps:

(i) Selection of inputs and outputs
(ii) Data collection
(iii) Optimization of configurations
(iv) Training or learning
(v) Testing or generation

The number of outputs can be determined from the problem being investigated, while the number of inputs should be all those that have a significant influence on the outputs; these can be decided by comparison of results from experiments with different inputs. The size of data required for neural network training is dependent on the complexity of the underlying function that the network is trying to model, as well as the variance of the additive noise. Normally, neural networks can process only numeric data in a limited range; it is, however, possible for neural networks to handle different types of data, such as an unusual range, missing data or non-numeric data. For example, numeric data can be scaled into an appropriate range for the network, missing values can be substituted by using mean values, and non-numeric data can be represented by a set of numeric values.

For each specific problem, in order to develop a neural network model with the best performance, the configuration parameters of the neural network must be determined by trial and error. These parameters include transfer functions, learning rules, learning rate, momentum coefficient, number of hidden layers, number of neurons in each hidden layer, and learning runs.

In the training or learning step, a set of known input–output data is repeatedly presented to train the network. During this repetition process, the weight factors between nodes are adjusted until the specified input yields the desired output. Through these adjustments, the neural network 'learns' the correct input–output response behavior. In neural network development, this phase is typically the longest and most time-consuming, and is critical to the success of the network.

After the training step, the recall and generalization step is carried out. In the recall step, the network is subjected to a wide array of input patterns used in training, and adjustments are introduced to make the system more reliable and robust. During the generalization step, the network is subjected to input patterns that it has not seen before, but whose outputs are known, and the system performance

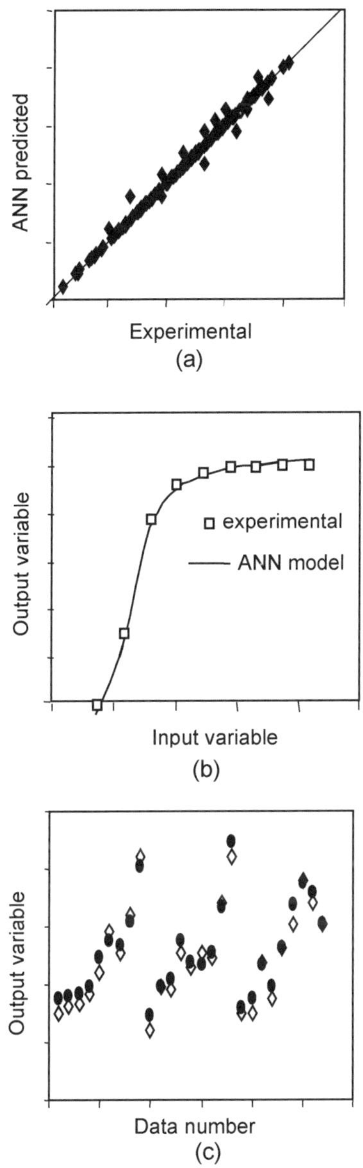

Fig. 11.3 Graphs used for evaluation of NN modeling performance (modified from Chen and Ramswamy, 2006).[5]

is monitored. The performance of neural networks can be evaluated visually by graphs or in quantity by different statistical measures. Figure 11.3 shows three kinds of graphs used for evaluating the performance of neural networks modeling.[5]

In Fig. 11.3a, *x* and *y* axes represent desired and predicted results, respectively. The diagonal line, called the ideal line, can be easily used for evaluation of the modeling performance. If the neural network predicted values are equal or close to desired results, then the points should be located on or close to the ideal line. In Fig. 11.3b, the *x* axis is one of the input variables, while the *y* axis is used for outputs, including desired and predicted results. This graph can be used to determine the correlation between the neural network model and the specific input variables. In Fig. 11.3c, the *x* axis is representing only the number or order of data used for training or testing, while the *y* axis is used for outputs as in Fig.11.3b; its emphasis is on comparison of the agreement between each pair of predicted and desired values.

Often-used statistical parameters include the coefficient of relationship between predicted and experimental results, R^2, and average relative error (E_r), which are given as:

$$R^2 = 1 - \frac{\sum_{i=1}^{n} (y_i - y_{di})^2}{\sum_{i=1}^{n} (y_i - y_m)^2} \qquad [11.13]$$

$$E_r = \frac{\sum_{i=1}^{n} |(y_i - y_{di})| / n}{y_{\max} - y_{\min}} \qquad [11.14]$$

where y_i is predicted by ANN model, y_{di} is the actual desired value (actual values), n is the number of data, and y_m is the average of actual values.

11.3 Application of neural networks in food thermal processing

11.3.1 Thermal process calculation using artificial neural network models

The basic objective of thermal processing calculations is to determine the process time for obtaining desired process lethality or evaluating the lethality of a given process. Currently, the methods used for thermal processing calculation can be divided into two types, general and formula. The general method is the most accurate method for situations where processing conditions are given, as it makes use of real-time–temperature data for process calculations. However, the application of this method could be tedious and time consuming because, for each variation in processing conditions, food product, or can size, a new set of time–temperature data is required. Several formula methods, on the other hand, have been developed, including Ball, Stumbo and Pham, which are based on linking characterized heat penetration parameters such as heating rate index (f_h) and heating/cooling lag factors (j_{ch}/j_{cc}) to destruction kinetics. Compared with the general method, formula methods are more convenient and flexible for practical applications; there are limitations, however, as these methods invariably involve the use of large tables and interpolation techniques that make the computerization of thermal process calculations complicated. As a result, studies have been carried out regarding the feasibility and potential advantages in the application of neural networks for thermal processing calculation.[5,6]

Afaghi *et al.*[6] developed two ANN models (Model A and Model B) for prediction of process time (*F*) and lethality value (*B*), respectively. The input variables for Model A included the retort temperature, initial product temperature, can size, heating behavior indexes (f_h, f_c, j_{ch} and j_{cc}), and the given lethality value, while Model B used the given process time. Both training and testing data used for the ANN modeling were generated by a confirmed finite difference computer model. The prediction performance of the trained ANN models was compared with those from conventional formula methods such as the Ball, Stumbo, or Pham methods, under the same processing conditions. The results showed that Model A, with 2.70 min mean absolute error over the entire range (reduced to 1.98 min up to a process time of 120 min), was comparable with the Pham method in the process time prediction. Model B predicted the process lethality with 2.74% of relative error, which was very close to Stumbo's method. The results indicated that ANN models, characterized by their accuracy, simplicity and on-line compatibility, can become alternative methods to the Stumbo and Pham methods, which are considered the most accurate methods of process calculations.

Sablani and Shayya[7] have carried out a similar study, using an ANN approach to predict the *g* and *fh/U* in Stumbo's formula method of thermal process calculation. The relative prediction errors, obtained through the use of developed ANN models for estimation of the process time under a given lethality value, as well as the lethality under a given process time, were less than 2% on average. The results demonstrated that using ANN models can avoid the drawbacks associated with use of Stumbo's method, while obtaining similar results.

11.3.2 Prediction of thermal inactivation of bacteria using artificial neural networks

The prediction of thermal inactivation of bacteria related to the combined effect of various processing conditions such as temperature, pH and water activity, while challenging, is useful for food processing industries. In this area, most models are developed by use of response surface methodology (RSM) and the method of Cerf. The major drawbacks of RSM are (i) it is usually quadratic, thus being inapplicable in complicated non-linear cases; (ii) co-linearity problems between factors may exist; and (iii) sensitivity analysis of input variables is difficult to perform because of the presence of cross-interactions. In Cerf's method, determining how to predict the thermal inactivation rate is difficult when the water activity differs from the modeled level values. Lou and Nakai[8] have developed an ANN model as a new predictive method for the combined effect of three environmental factors, temperature, pH and water activity, on inactivation of *E. coli*. A comparison of prediction accuracy between the ANN-based model, RSM and Cerf's model is listed in Table 11.1. It can be seen that the ANN-based approach was more accurate

Table 11.1 Accuracy comparison between ANN-based model, RSM and Cerf's models (From Lou and Nakai, 2001, Food Research International, with permission)[8]

Criterion	Model		
	ANN-based	RSM	Cerf's method
AAE	0.1169	0.1838	0.2317
AAPE	3.59	5.73	7.23
RMSE	0.1439	0.2315	0.2338
R2	0.9489	0.8684	0.8149

ANN, artificial neural networks; RSM, response surface methodology; AAE, average absolute error; AAPE, average absolute percentage error; RMSE, root-mean-squares error; R2, determination coefficient.

Table 11.2 Sensitivity analysis of *T*, pH and a_w (From Lou, and Nakai, 2001, Food Research International, with permission)[8]

		Variables		
		a_w	pH	*T*
	Rank	3	2	1
Training	VSE	0.367	0.498	0.741
	VSR	2.716	3.690	5.489
Verification	VSE	0.364	0.530	1.064
	VSR	1.982	2.885	5.793

a_w, water activity; *T*, temperature; VSE, variable sensitivity error; VSR, variable sensitivity ratio.

than either the RSM or Cerf's models. The developed ANN models were further used for sensitivity analysis as shown in Table 11.2, and generating 2D and 3D response surfaces which are useful in the optimization of the environmental conditions to control the growth and inactivation of *E. coli*. It revealed that temperature was the most sensitive factor affecting the thermal inactivation rate of *E. coli*, followed by pH and water activity.

11.3.3 Modeling and optimization of variable retort temperature thermal processing

Variable retort temperature (VRT) thermal processing is an effective technique used for improving the quality of canned foods and reducing the process time. The key for designing a VRT thermal process is to choose a reasonable (or optimal) VRT function for a given food product to be packaged in a given container and thermally processed. Since the often used VRT functions such as sine, exponential, step line, and step pulse all involve several parameters, searching out an optimal VRT function is a multi-variable optimization problem. Although many conventional methods have been used for solving a variety of optimization problems, they have several limitations, such as lack of robustness for providing optimal results

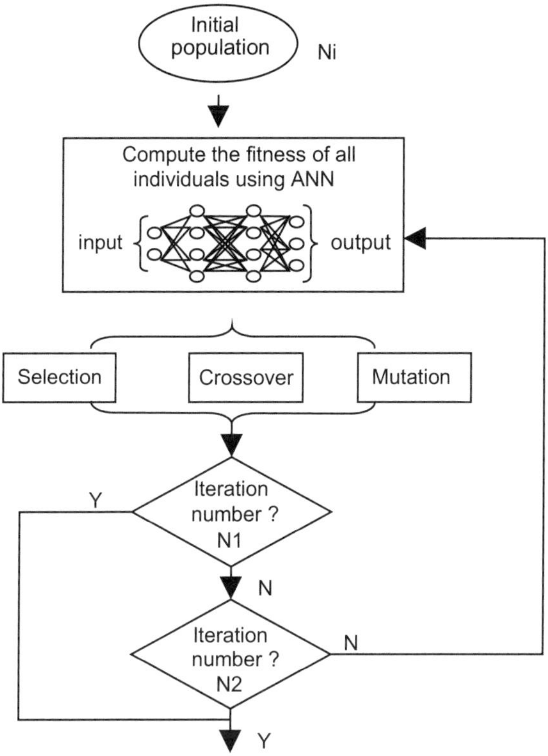

Fig. 11.4 The procedure of hybrid optimization method using ANNs and GAs (from Chen and Ramaswamy, 2002, *Journal of Food Engineering*, with permission).[9]

and low calculation speed, both of which are important for practical applications. Artificial intelligence technologies have opened new avenues to deal with such problems. Chen and Ramaswamy[9] have explored the use of ANNs and genetic algorithms to develop various prediction models of VRT processing, and aid in the search for the optimal VRT function parameters. The hybrid optimization method (shown in Fig.11.4) linking GAs with ANNs was employed for searching the optimal variable retort temperature profiles. The research objective was focused on (i) analyzing the effects of VRT function parameters on the main response variables, namely, process time, average quality retention, and surface cook value, (ii) determining the search ranges for optimization of VRT processing, (iii) developing ANN prediction models for each main response variable related to VRT function parameters, and (iv) searching the optimal processing temperature profiles using the coupled ANN–GA models.

ANN models were implemented to develop prediction models for the VRT output variables, which are average quality retention (Q_v), process time (*PT*) and surface cook value (F_s). Genetic algorithms were coupled with trained neural network models to meet the different optimization objectives of minimum *PT* and F_s under given constraints. The searching range of each independent variable was

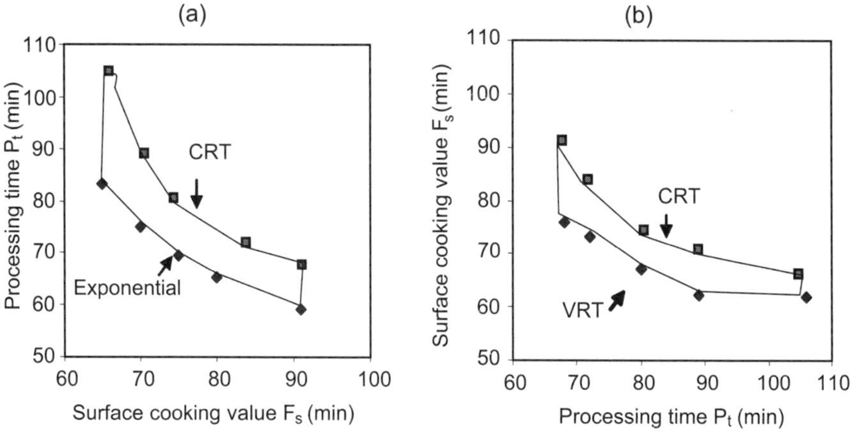

Fig. 11.5 Comparison of VRT vs CRT processing on optimal results of processing time and surface cooking value (from Chen and Ramaswamy, 2002, *Journal of Food Engineering*, with permission).[9]

based on a sensitivity analysis of the effects of function parameters on response variables. The best results for Q_v, *PT* and F_s under constant retort temperature (CRT) processing conditions were used as constraints. Test results indicated that coupled ANN–GA models could be effectively used for describing the relationships between the operating variables and VRT function parameters, and for identifying optimal processing conditions. VRT processes reduced the process time by more than 20% and surface cook value by about 7–10% as compared to the best CRT process. The benefits on both process time and surface quality by using the VRT method were predicted by use of ANN models, as shown in Figures 11.5a and b. In Fig.11.5a for processing time *PT*, surface cook value was used as a constraint condition, while in Fig. 11.5b for surface cook value, the processing time was used as the constraint condition.

11.3.4 Analysis of critical control points in deviant retort temperature thermal processing

The basic objective of thermal processing is to meet safety requirements while keeping quality degradation to a minimum. Theoretically, it is possible to design an optimal processing protocol for any food product, but in practice it is difficult to obtain truly optimal results since considerable deviations exist in process parameters. In cases where deviations go beyond a certain critical level, there can be under-processing or over-processing. The former indicates that processed products cannot meet the sterility requirements for safety and consumption, while the latter means that quality destruction is more than acceptable. Therefore, it is important to identify critical factors, assess the effect of their deviations on the process calculations, and establish control actions during thermal processing to avoid excessive process deviations.

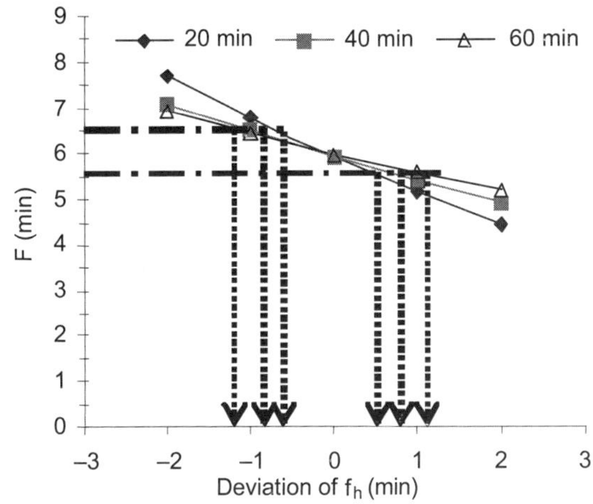

Fig. 11.6 Acceptable deviation ranges predicted by ANN models for heating rate index, f_h (from Chen and Ramaswamy, 2002, *Journal of Food Engineering*, with permission).[10]

Thermal processing is a complex system, and standard processes are established based on achieving target process lethality (*F* value) at a critical point, usually the package center. The required process time (*PT*), product initial temperature (T_i), cooling water temperature (T_w), and several product–related properties, such as heating rate index (f_h), heating lag factor (j_h), and cooling lag factor (j_c). It is necessary to understand and quantify the influence of these process parameters and process deviations on the required process time. Chen and Ramswamy[10] have developed ANN models for evaluating the relative order of importance of different critical control variables with respect to process calculations and developing predictive models to compensate for their deviations. The critical variables studied were retort temperature (*RT*), initial temperature, cooling water temperature, heating rate index, heating lag factor, and cooling lag factor. Their ranges of deviation from a set point were selected as –2 to 2 °C for *RT*, –5 to 5 °C for both T_w and T_i, –2 to 2 min for f_h, and –0.2 to 0.2 for both j_c and j_h. ANN models were developed and used for analysis of different critical variables with respect to their importance on the accumulated lethality, process time, cooling time (*CT*), and total time (*TT*) under the given processing conditions. By use of ANN models, relative orders of importance of critical variables within the deviation ranges were discovered, as follows: for *F*, $RT > f_h > j_h > T_i{*}T_w > T_i > T_i{*}f_h > RT{*}T_i > j_c > RT{*}j_h$; for *RT*, $PT > f_h >> j_h > T_i > j_c > T_i{*}T_w$; for *CT*, $j_c > T_w > f_h$; and for *TT*, $RT > f_h > j_h > j_c > T_w > T_i > T_i{*}j_c > T_i{*}T_w$. The accepted deviation ranges for various input variables under given control ranges were predicted by ANN models, one of which is shown in Fig. 11.6. Based on these graphs, it can be easily determined that, when the desired *F* values were set at 6 ± 0.5 min, the maximum acceptable deviation ranges of different variables were ± 0.3 °C for *RT*, ± 4 °C for T_i, 0.1 °C for j_h, ± 0.8, ± 1, ± 2.1 min for f_h at f_h = 20, 40 and 60 min, respectively, and 0.4 for j_c. Neural network

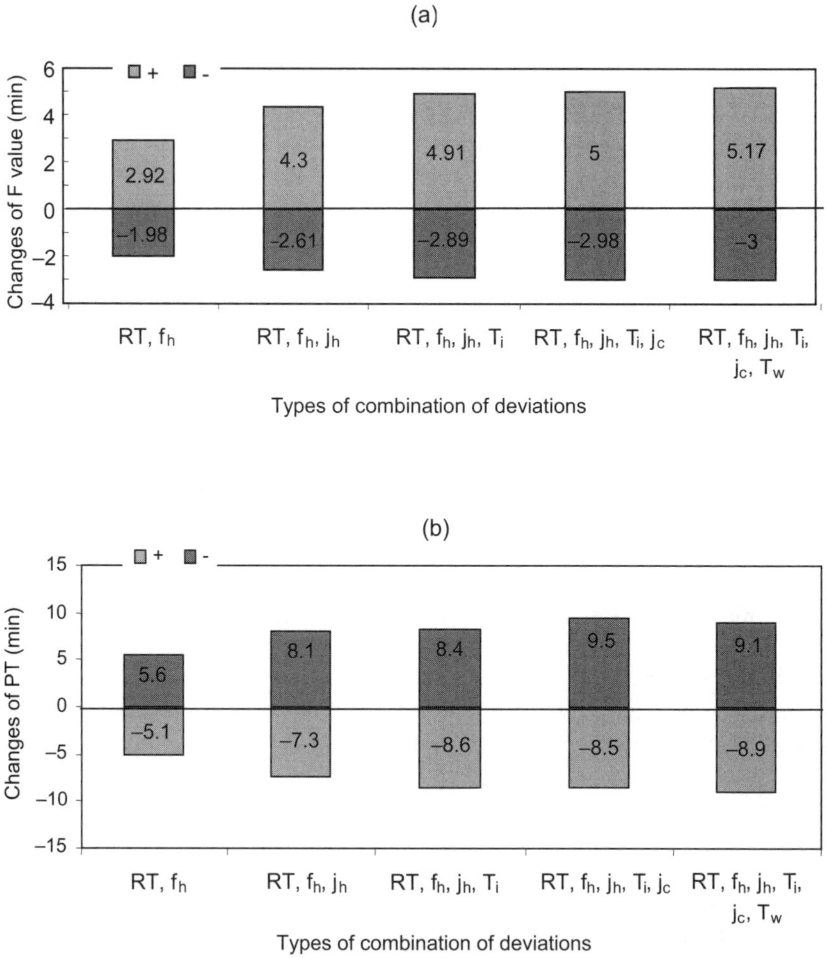

Fig. 11.7 The comprehensive effects of multiple deviations predicted by ANN models: (a) lethality value and (b) heating time (from Chen and Ramaswamy, 2002, *Journal of Food Engineering*, with permission).[10]

models were also used for analysis of the combination effect of multiple deviations on F, PT, and CT (shown in Fig. 11.7). By use of this graph, the maximum changes in F and PT for different deviation combinations could be easily determined.

11.3.5 Dynamic modeling of retort thermal processing

Dynamic modeling of thermal processing is complex, especially under real processing conditions, and difficult to predict using conventional mathematical modeling since a large number of variable parameters need to be taken into account. Chen and Ramaswamy[11] applied ANNs for dynamic modeling of retort thermal processing to predict lethality value F and quality retention Q_v, while considering different

Table 11.3 Main factors* and levels for the inputs of training ANN models (from Chen and Ramaswamy, 2002, *Journal of Food Processing and Preservation*, with permission)[9]

Code	*RT* (°C)	α ($10^{-7}m^2/s$)	*R* (mm)	*H* (mm)	*D* (min)	*z* (°C)
–2	116	1.5	38	38	150	15
–1	120	1.7	44.5	44.5	175	21
0	124	1.9	51	51	200	27
1	128	2.1	57.5	57.5	225	33
2	132	2.3	64	64	250	39

* *RT*: retort temperature; α: thermal diffusivity; *R*: radius of can; *H*: height of can; *D*: time decimal deduction of quality factor; *z*: temperature sensitivity of quality factor.

processing conditions, including the retort temperature (*RT*), thermal diffusivity (α), can radius (*R*), can height (*H*), and quality kinetic parameters (*D* and *z* values). Their ranges and levels are shown in Table 11.3.

Two modeling methods were used, a moving-window network (MV–ANN) and a hybrid network (H–ANN); both were used for modeling *F* and Q_v dynamic functions. The MV–ANN is a special hierarchical network used to model dynamic systems and unsteady-state processes.[4] A moving window provides a means for creating multiple training examples from continuous raw data. In this case, the moving window included four values, one at current time, *t*, and three other past values at $t\text{-}3\Delta t$, $t\text{-}2\Delta t$, and $t\text{-}\Delta t$, respectively. The time step Δt was set to 2 minutes. The H–ANN method consisted of two steps. The first step involved the use of polynomial regression models relating *F* and Q_v to operating variables, while the second step used neural networks to predict the regression coefficients. The modeling performance for both models is listed in Table 11.4.

In order to show the application of both types of ANN models developed, the predicted values for *F* and Q_v under a typical processing condition were plotted (Figures 11.8a and b). In these figures, only one parameter was varied at a time, while others were kept at their base values. For example, in Fig. 11.8a, the effect of retort temperature varying between 116 and 132 °C is shown, while other parameters were maintained at their base values. Graphical results confirmed that both ANN models could accurately predict the cumulative lethality value and average quality retention changes with the process time. But, by comparison of both ANN methods, it was found that the predicted values for both *F* and Q_v with the MV–ANN models were much closer to the computer experimental values than

Table 11.4 The modeling performance of both ANN models (from Chen and Ramaswamy, 2002, *Journal of Food Processing and Preservation*, with permission)[9]

	MV–ANN		H–ANN	
	F	*Qv*	*F*	*Qv*
Er	0.68%	0.81%	2.7%	4.2%
R	0.995	0.998	0.98	0.988

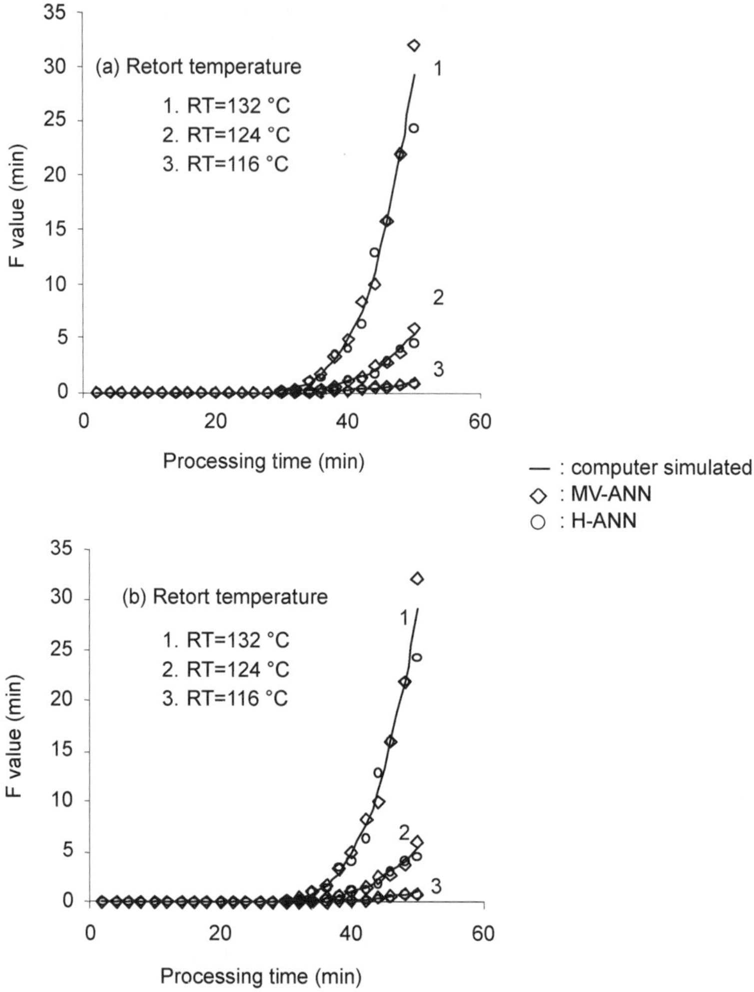

Fig. 11.8 Comparisons of prediction ability of both MV and hybrid ANNs for F dynamic functions under different retort temperatures. (Symbols: – computer simulated; ◊ MV–ANN; o H–ANN) (from Chen and Ramaswamy, 2002, *Journal of Food Processing and Preservation*, with permission)[11]

those with the H–ANN models, demonstrating that the MV–ANN models had better prediction capabilities than the H–ANN models.

11.3.6 Modeling of thermal processing with irregular shape packages

Foods with irregular shape packages, such as retortable pouches or semirigid containers, have become increasingly popular in modern food markets due to their low cost and superior package design. In addition, compared with cylindrical containers, the pouch transfers heat faster to the critical point due to its characteristic thinner profile. This fast heat transfer permits the required amount of heat for sterilization to be applied to the critical point with minimal overcooking of the product bulk near the peripheral container areas. Thus, thin profile containers can potentially provide higher quality retention for solid foods than conventional cylindrical containers.[12] The difficulty in developing a thermal process for foods with irregular shape packages is in determination of the cold spot and prediction of the heat penetration characteristic indexes. Due to the irregular shape, it is not only hard to develop a model by use of conventional modeling methods to predict the transient temperature profile at the cold spot, but also difficult to collect the heat penetration data by use of experimental methods such as thermal couples. Due to the advantages of ANNs over conventional modeling methods described in previous sections, it is credible that ANNs could be used as a tool for modeling of thermal processing of foods with irregular shape packages; there are, however, no reports concerning applications of ANNs for this area. The moving-window method used for dynamic modeling of thermal processing[11] has the best potential to be applied for modeling of processing irregular shape foods. This is because this method can continuously use the history data in the past steps to predict the value in the next steps, as demonstrated in an example,[11] which will be useful for modeling of thermal processing with irregular shapes.

11.3.7 Other examples of artificial neural network applications for thermal processing

Applications of ANNs for thermal processing have been widely reported in the last 20 years. In order to be more convenient for readers to get related information, other application reports are listed in the references section 11.6, Bibliography.

11.4 Future trends

The capability of ANN modeling is no longer in question as its potential has been confirmed through a variety of applications.[5] However, there exist some limitations for the use of ANN methods. First, neural networks work as a black box, meaning that they can provide only results but not give any reasonable interpretations between input and output variables. In this aspect, neural networks should be considered as a tool for exploring the applied rather than theoretical aspects. Second, if little input–output data exist on a problem or process, the use of neural networks should not be considered since they rely heavily on such data. Consequently, neural networks are best suited for problems with a large amount of historical data, or those that allow training the neural network with a separate simulator. In addition, there may also be situations where there exist large databases, but all training data are similar, causing the same problems as having small training data sets. Thus, a broad-based data set and experimental design are essential. Third, most training techniques are capable of 'tuning' the network, but they do not guarantee that the network will operate properly. The training may 'bias' the network, making it accurate in some operating regions, but inaccurate in others. In addition, for the application purposes, it is more important that developed ANN models be used for different aspects such as optimization, on-line control, and identification. This often combines ANN models with other techniques; for instance, fuzzy logic, expert systems, and genetic algorithms or other search techniques. Therefore, future application of neural networks should focus on developing hybrid methods by using neural networks with other techniques, which will be of more use for industrial purposes.

11.5 References

1 Sun, Da-Wen (2006). *Thermal Food Processing*. CRC Press, Inc. Boca Raton, FL, USA.

2 Chen, C.R. & Ramaswamy, H.S. (2006). Visual Basics Computer Simulation Package for Thermal Process Calculations, *Transactions of I Chem E Part C* (FBP) 83(C1), 65–79.

3 Baughman, D. R. and Liu, Y. A. (1995). *Neural Networks in Bioprocessing and Chemical Engineering*, Academic Press Inc, San Diego, CA.

4 McCulloch, W.S. and Pitts, W. (1943). A logical calculus of the ideas imminent nervous activity. *Bulletin of Mathematical Biophysics*, 5, 115–133.

5 Chen, C.R. & Ramaswamy, H.S. (2006). 'Modeling thermal processing using artificial neural networks' in *Thermal Food Processing*, edited by Da-Wen Sun. CRC Press Inc., Boca Raton, FL, USA.

6 Afaghi, M., Ramaswamy, H.S. and Prasher, S.O. (2001). Thermal process calculations using artificial neural network models. *Food Research International*, 34, 55–65.

7 Sablani, S. S. and Shayya, W.H. (2001). Computerization of Stumbo's method of thermal process calculations using neural networks. *Journal of Food Engineering*, 47, 233–240.

8 Lou, W. and Nakai, S. (2001). Application of artificial neural networks for predicting the thermal inactivation of bacteria: A combined effect of temperature, pH and water activity. *Food Research International*, 34, 573–579.

9 Chen, C.R. and Ramaswamy, H.S. (2002). Prediction and optimization of variable retort temperature (VRT) processing using neural network and genetic algorithms. *Journal of Food Engineering*, 53, 209–220.

10 Chen, C.R. and Ramaswamy, H.S. (2002). Analysis of critical control points for deviant thermal processing using artificial neural networks. *Journal of Food Engineering*, 57, 225–235.

11 Chen, C.R. and Ramaswamy, H.S. (2002). Dynamic modeling of retort thermal processing using neural networks. *Journal of Food Processing and Preservation*, 26(2), 91–112.

12 Ramaswamy, H.S. and Chen, C.R. (2005). *Novel Processing Technologies for Food Preservation in Processing Fruits*, edited by Barrett, D.M., Somogyi, L. and Ramaswamy, H.S., CRC Press, 201–219.

13 Chen, C.R. (2001). *Application of Computer Simulation and Artificial Intelligence Technologies for Modeling and Optimization of Food Thermal Processing*, Ph.D thesis, McGill University, Montreal, Canada.

14 Cerf, O., Davey, K.R. and Sadoudi, A.K. (1996). Thermal inactivation of bacteria – a new predicitive model for the combined effect of three environmental factors: Temperature, pH and water activity. *Food Research International*, 29, 219–226.

11.6 Bibliography

Afaghi, M. (1999). *Application of artificial neural network modeling in thermal process calculations of canned foods*. M.Sc. thesis, Food Science Department, McGill University, Montreal, Canada.

Chen, C.R. and Ramaswamy, H.S. (2000). Neural computing approach for modeling of residence time distribution (RTD) of carrot cubes in a vertical scraped surface heat exchanger (SSHE). *Food Research International*, 33(7), 549–556

Chen, C.R. and Ramaswamy, H.S. (2004). Multiple Ramp-Variable (MRV) Retort Temperature Control for Optimization of Thermal Processing, *Transactions of I Chem E Part C* (FBP) 82(C1), 1–11..

Hussain, M.A. and Rahman, M.S. (1999). Thermal conductivity prediction of fruits and vegetables using neural networks. *International Journal of Food Properties*, 2(2), 121–137.

Kaminski, W., Strumillo, P. and Tomczak, E. (1996). Genetic algorithms and artificial neural networks for description of thermal processes. *Drying Technology*, V14, 2117–2133.

Mittal G.S. and Zhang, J.X. (2000). Use of artificial neural network to predict temperature, moisture, and fat in slab-shaped foods with edible coatings during deep-fat frying. *Journal of Food Science*, 65(6), 978–983.

Sablani, S.S. , Ramaswamy, H.S. and Prasher, S.O. (1995). A neural network approach for thermal processing applications. *Journal of Food Processing and Preservation*, 19(4), 283–301.

Sablani, S.S., Ramaswamy, H.S., Sreekanth, S. and Prasher, S.O., (1997). Neural network modeling of heat transfer to liquid particle mixtures in cans subjected to end-over-end processing. *Food Research International*, 30(2), 105–116.

12

The role of computational fluid dynamics in the improvement of rotary thermal processes

P. James, University of Plymouth, UK, and G. Tucker, Campden and Chorleywood Food Research Association, UK

12.1 Introduction

12.1.1 An overview of the use of computational fluid dynamics (CFD) in the food processing industry

In 1999, Scott and Richardson wrote a review of the extent to which CFD was then being applied to problems in food processing. Earlier applications, to clean room design, refrigerated transport, static mixers and pipe flow, had been described by Quarini (1995), and Scott and Richardson extended this list to include the performance of baking ovens, chillers and display cabinets, all of which involve the prediction of single-phase air flow. Applications to liquid flows, in which the variation of material properties was taken into account, also appeared at around this time. Scott and Richardson concluded, amongst other things that, even though the usage of CFD in the food processing sector was relatively small, it was evident that 'CFD should be considered as an engineering tool whose application can assist in the efficient operation of a wide range of food processes'. Since that time, the scope and number of applications of CFD to food processing problems have increased substantially. For example, the recent comprehensive review by Norton and Sun (2006) cites applications in ventilation, drying, sterilisation, refrigeration and mixing, and contains around 100 references to applications of CFD in the food processing sector. It can therefore be safely concluded that CFD has arrived as a tool for the analysis and design of many of the controlling mechanisms in the food processing industry. There are, of course, technical limitations to what CFD can achieve. For example, if the flow is turbulent, then the question of how to model turbulence in the appropriate way arises, and there is a vast literature on this topic (see, for example, Wilcox, 2000). Similarly, if the fluid properties change with the flow or with temperature, then their accurate characterisation is needed if the output from a CFD simulation is to be more than just qualitative. Further complications that arise in the food processing context are the occurrence of two-phase flows, in which the foodstuff itself may be liquid or solid, and in which different phases of other products arise. Fluid elasticity may also be a significant factor and has been taken into account in a study of the wetting and peeling of doughs by Sujatha *et al.* (2003). (There is a vast research literature on the numerical solution of flows of elastic liquids – see, for example, Owens and Phillips, 2002.) Flows with free surfaces or flows in which there are discrete droplets are other examples that challenge the designers of flow simulation software, and CFD cannot yet provide answers to all the design and analysis questions that arise in food processing. A significant factor limiting the increased application of CFD is the supply of experienced practitioners. Successful application requires the modeller to have a good grasp of the physics of the problem and an understanding of what a solution means and what its limitations are. In addition, some form of validation of the output from any CFD simulation is essential before it can be used with confidence as a design tool.

In summary, while there are still areas where better physical models are required and problems for which more powerful computers are needed, it can safely be concluded that there is much to be gained from the judicious application of validated CFD to problems in the food processing industry. The major part of this chapter represents just such an application.

12.1.2 A brief review of work carried out to date on rotary thermal processing

Rotary thermal processing takes place in two main ways, namely axial rotation, which is most often found in continuous processing systems such as reel and spiral cookers, and end-over-end rotation (EOE), which normally takes place in batch retorts. In the first case, food containers, usually cylindrical cans, are placed with their axes of symmetry horizontal and are transported along a spiral path, typically of diameter between one and two metres and of varying pitch. The cans undergo relatively slow rotations as they move along the spiral path but may also experience much faster axial rotations for a part of their motion as they roll along the lower part of the reel. Further information on the nature of the rotational motion in a reel and spiral setting can be found in Vandenberghe (2001). Axial rotation is not, however, the main concern of this chapter, although it will be referred to frequently when discussing validation. The main focus of this chapter is on the EOE type of rotation, in which cans are usually fixed in place in a crate, which is then rotated as a solid body inside a retort.

Many factors influence the rate of heating of liquid foodstuffs inside cans undergoing EOE agitation. These include the rotation rate, the position of the cans within the crate, the orientation of the cans, the variation of physical properties of the foodstuff with temperature and shear rate, the heating mechanism and profile, and the headspace bubble. An article by Tucker (2004) contains a review of work, some dating back to 1950 (Clifcorn *et al.*), on the effect on convective heat transfer of agitating cans, and points out that several studies have shown that the headspace bubble can have a significant effect on mixing, and consequently heat transfer, within the cans (see, for example, Parchomchuk, 1977, Naveh and Kopelmann, 1980, and Berry *et al.*,1979).

Attempts at numerically simulating various aspects of in-container sterilisation processes have been made by several authors in recent years. Yang and Rao (1998) used the FIDAP finite element software to calculate the temperature field, due to natural convection, in a starch solution contained in a heated can. The temperature and shear rate dependence of apparent viscosity was allowed for and their predictions of the can centre temperature were in good agreement with experimental measurements. A series of papers by Abdul Ghani, Farid and co-workers (Abdul Ghani *et al.*,1999, 2001, 2002, 2003; Farid and Abdul Ghani, 2004; Abdul Ghani and Farid, 2006) describes a systematic study of the use of CFD to investigate the role of natural convection in canned food sterilisation. Stationary and rotating (about a fixed axis) horizontal and vertical cans are considered and the location of the slowest heating zone is of particular interest. One paper focuses on pouches and another on how the long computational times associated with three-dimensional calculations can be reduced by simplification and correlation. These workers used the PHOENICS CFD software for their simulations. Tattiyakul *et al.* (2001, 2002a,b) made use of the FIDAP software to study heat transfer to corn starch solutions in cans experiencing intermittent rotation about a fixed axis, notably the fast axial rotations that occur in reel and spiral cookers.

In the above simulations, various complicating factors, such as temperature and shear rate-dependent apparent viscosity, three-dimensional effects, transient effects and cans rotating about a fixed axis, have been included to add more realism to the simulations. However, off-axis rotation and the influence of the headspace bubble do not appear to have been studied using CFD.

The simulation, via validated CFD, of the flow of a shear-thinning non-Newtonian fluid inside a cylindrical container undergoing end-over-end rotation when there is a headspace bubble present is therefore a relevant and challenging problem with which to demonstrate the impact that CFD can make in the food processing industry. Consequently, a substantial part of the remainder of this chapter describes how validated CFD can be used to determine conditions under which EOE batch processing of liquid foodstuffs can be optimised. The description is based on work published in three journal papers by James *et al.* (2001, 2006), and Hughes *et al.* (2003), although this is the first time that an overview of the combined work of these, and other related results, has been assembled in one place. The results presented in a recent article by Tucker *et al.* (2006), on the implementation in pilot and full-scale plant of conclusions drawn from

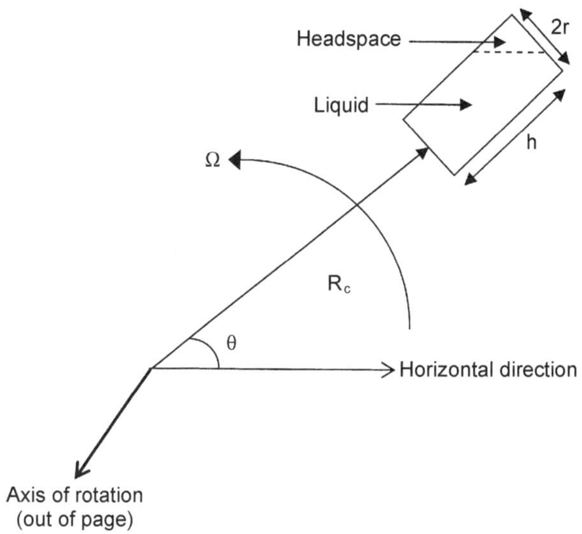

Fig. 12.1 Schematic diagram of a can in end-over-end, off-axis rotation.

computational and associated rheological studies, are then discussed. The chapter ends with a summary and discussion of future trends.

12.2 Methodology

12.2.1 Modelling assumptions and problem formulation

The basic flow problem that has to be simulated numerically is that of the motion of a liquid inside a cylindrical can that rotates in end-over-end fashion about an axis that, in general, does not pass through the centre of the can, i.e. off-axis rotation. This configuration is shown schematically in Fig. 12.1, in which r is the radius of the can, h is the height of the can, R_c is the off-axis radius of rotation and Ω is the angular speed of rotation. The can contains a headspace bubble, which is taken to be 10% by volume in most of the calculations that follow. In practical applications, the fluid physical properties vary with both temperature and shear rate: liquid foodstuffs are often shear-thinning and temperature-thinning. In the work to be described, only shear-thinning behaviour is considered but it would be relatively simple to include the effects of temperature-thinning too if thought necessary. Non-isothermal conditions prevail, of course, in full-scale, industrial processing and this case is considered in Section 12.3.3.

The simulation of this type of problem, assumed to be a laminar, incompressible and unsteady flow of a non-Newtonian fluid in three dimensions, and with a free surface, can now be accomplished with a range of commercially available CFD software. The results to be presented here were obtained using an earlier version of the ANSYS CFX software (CFX4,1995). Off-axis rotation is allowed for by using a coordinate system that rotates with the can and the shear rate dependent fluid properties are specified by choosing an appropriate inelastic apparent viscosity model, as discussed in Section 12.2.3.

The method that CFX uses to deal with a free surface is worth describing in some detail since the software was not specifically designed to deal with such flows. If a flow is known to have a free surface, then a two-phase flow is set up such that each computational cell contains a volume fraction of each phase. Initially the phases are completely separated and so the volume fraction for each cell will be zero or one. In the subsequent motion, the two phases are assumed to have a common velocity, a common density and a common pressure, and in this sense the two-phase mixture is treated as a single homogeneous fluid. However, the two phases retain their separate identities through their volume fractions, which are additional variables to be solved for. In practice, there is only one additional variable since the sum of the volume fractions is one. The resulting description of the flow is known as a homogeneous two-phase flow model (Lo, 1990). As the homogeneous flow evolves, the volume fractions of each cell will take on values other than zero or one. The additional constraint of free surface flow is imposed by interrupting the simulation at frequent, small time intervals and applying a procedure to completely separate the two phases. This involves redistributing mass so that there is a clearly defined interface on either side of which the volume fraction of each phase is zero or one. In this way the interface between the fluids is 'sharpened' and a free surface is obtained. This is the technique used to compute the shape and motion of the headspace bubble and, subsequently, its influence on the remainder of the flow. More recent versions of the ANSYS CFX software not only use a different method of solving the governing equations for mass, momentum and energy transport than that used in CFX4, but also incorporate a more advanced treatment of free surfaces. Nevertheless, to the authors' knowledge, the homogeneous two-phase flow model is still one method used to deal with free surface problems in this particular CFD software product. In Section 12.5, further information and reference material is included, some of which describes alternative methods of dealing with free surfaces.

The governing equations for the homogeneous mixture are the mass conservation equations for each phase,

$$\frac{\partial(\rho_i\alpha_i)}{\partial t}+\frac{\partial}{\partial x}(\rho_i\alpha_i u)\ \frac{\partial}{\partial y}(\rho_i\alpha_i v)+\frac{\partial}{\partial z}(\rho_i\alpha_i w)=0,\quad i=1,2\,, \qquad [12.1]$$

the x, y and z components of the momentum equations for the homogeneous mixture,

$$\frac{\partial}{\partial t}(\rho_m u)+\frac{\partial}{\partial x}(\rho_m u^2-2\eta_m\frac{\partial u}{\partial x})+\frac{\partial}{\partial y}(\rho_m uv-\eta_m(\frac{\partial u}{\partial y}+\frac{\partial v}{\partial x}))+$$
$$\frac{\partial}{\partial z}(\rho_m uw-\eta_m(\frac{\partial u}{\partial z}+\frac{\partial w}{\partial x}))=-\frac{\partial p}{\partial x}+\rho_m g_x\,, \qquad [12.2]$$

$$\frac{\partial}{\partial t}(\rho_m v) + \frac{\partial}{\partial x}(\rho_m uv - \eta_m(\frac{\partial v}{\partial x} + \frac{\partial u}{\partial y})) + \frac{\partial}{\partial y}(\rho_m v^2 - 2\eta_m \frac{\partial v}{\partial y}) +$$

$$\frac{\partial}{\partial z}(\rho_m vw - \eta_m(\frac{\partial v}{\partial z} + \frac{\partial w}{\partial y})) = -\frac{\partial p}{\partial y} + \rho_m g_y , \quad [12.3]$$

and

$$\frac{\partial}{\partial t}(\rho_m w) + \frac{\partial}{\partial x}(\rho_m uw - \eta_m(\frac{\partial w}{\partial x} + \frac{\partial u}{\partial z})) + + \frac{\partial}{\partial y}(\rho_m vm - \eta_m(\frac{\partial v}{\partial z} + \frac{\partial w}{\partial y}))$$

$$+ \frac{\partial}{\partial z}(\rho_m w^2 - 2\eta_m \frac{\partial w}{\partial z}) = -\frac{\partial p}{\partial z} + \rho_m g_z , \quad [12.4]$$

and the energy equation for the homogeneous mixture

$$\frac{\partial}{\partial t}(\rho_m H) + \frac{\partial}{\partial x}(\rho_m uH) + \frac{\partial}{\partial y}(\rho_m vH) + \frac{\partial}{\partial z}(\rho_m wH) =$$

$$\frac{\partial}{\partial x}\left(k_m \frac{\partial T}{\partial x}\right) + \frac{\partial}{\partial y}\left(k_m \frac{\partial T}{\partial y}\right) + \frac{\partial}{\partial z}\left(k_m \frac{\partial T}{\partial z}\right) . \quad [12.5]$$

In these equations t is time, α_i is the volume fraction of the i-th phase, (u, v, w) is the velocity vector, ρ_i is the density of the i-th phase, H is the total enthalpy, defined in terms of the static enthalpy h by

$$H = h + \frac{1}{2}(u^2 + v^2 + w^2) \quad [12.6]$$

$$\rho_m = \alpha_1\rho_1 + \alpha_2\rho_2 \quad [12.7]$$

$$\eta_m = \alpha_1\eta_1 + \alpha_2\eta_2 \quad [12.8]$$

and

$$k_m = \alpha_1 k_1 + \alpha_2 k_2 \quad [12.9]$$

where η_i is the viscosity of the i-th phase and k_i is the thermal conductivity of the i-th phase. The viscosity of the liquid phase, η_1 say, may depend on the shear rate ($\dot{\gamma}$) and temperature (T). The shear rate $\dot{\gamma}$ is defined by

$$\gamma = 2\left(\frac{\partial u}{\partial x}\right)^2 + 2\left(\frac{\partial v}{\partial y}\right)^2 + 2\left(\frac{\partial w}{\partial z}\right)^2 + \left(\frac{\partial u}{\partial y} + \frac{\partial v}{\partial x}\right)^2 + \left(\frac{\partial v}{\partial z} + \frac{\partial w}{\partial y}\right)^2 + \left(\frac{\partial w}{\partial x} + \frac{\partial u}{\partial z}\right)^2 \quad [12.10]$$

The equations have been written in terms of a Cartesian coordinate system, but in practice a polar coordinate system that rotates with the can is employed.

12.2.2 Numerical features

A structured grid comprising five blocks of cells was used to define a mesh in the cross-section of a cylindrical can. There were 1044 cells in a cross-section and 30 cells along the length of the can. Numerical experimentation with simulation of the flow in a cylindrical can in off-axis rotation but with the orientation of the can axis fixed in a horizontal direction indicated that this density of cells over the cross-section leads to results that were sufficiently insensitive to the mesh (James *et al*, 2001). For a Newtonian fluid, there are two time-scales that affect the flow: a diffusion time scale $t_D = r^2\rho_1/(\lambda^2\eta_1)$ and a 'free-surface' time scale $t_s = \sqrt{r/g}$, where ρ_1, η_1, g and λ are, respectively, the fluid density, the fluid dynamic viscosity, the acceleration due to gravity and a numerical constant which we may estimate to be approximately four, from a calculation in Batchelor (1974). The radius of the cans to be considered is 0.038 m and, for water, this leads to $t_D/t_s \approx 10^4$. The surface sharpening algorithm requires small computational time steps, typically of the order of $t_s/6$, and computations need to be continued until either a steady state, in the case of on-axis rotation, or periodic state, in the case of off-axis rotation, is reached. The computations can take several hours when carried out on a typical PC and so numerical tests were carried out to see how well two-dimensional computations could represent a three-dimensional flow. In the case of axial rotation, it was demonstrated in James *et al.* (2001) that, under isothermal conditions, the effect on the flow of the ends of the can was confined to relatively small regions near the ends and that calculations of the degree of fluid mixing obtained with two-dimensional calculations (i.e. assuming infinitely long cans) were similar in many respects to those obtained with three-dimensional calculations. A quantitative analysis, discussed in Section 12.3, in which normalised median shear rates and fluid velocities obtained with two-dimensional and three-dimensional calculations are compared, for Newtonian and non-Newtonian fluids, confirms that two-dimensional calculations in the axial rotation case provide results that are reasonably close to those obtained with a fully three-dimensional calculation.

There is no equivalent simplification for cans rotating end-over-end under isothermal conditions nor for cans undergoing either mode of rotation under non-isothermal conditions. Nevertheless, it will be seen later that some useful information can be obtained by carrying out 'numerical experiments' with two-dimensional calculations in EOE mode.

12.2.3 Liquid characterisation

It is well known that the physical properties of liquid foodstuffs vary with both temperature and shear rate. There is no technical difficulty in including the temperature dependence, if it is known, but the validation of flow simulations is carried out under isothermal conditions and so the influence of variable viscosity is seen only through shear rate dependence. The effect of fluid elasticity is ignored. The most popular model for apparent viscosity in use in the food industry appears to be the power-law model (see, for example, Bird *et al.*, 1987) in which the apparent viscosity $\eta_1(\dot{\gamma})$ depends on shear rate $\dot{\gamma}$ according to the equation

$$\eta_1(\dot{\gamma}) = K(\dot{\gamma})^{n_p - 1}, \qquad [12.11]$$

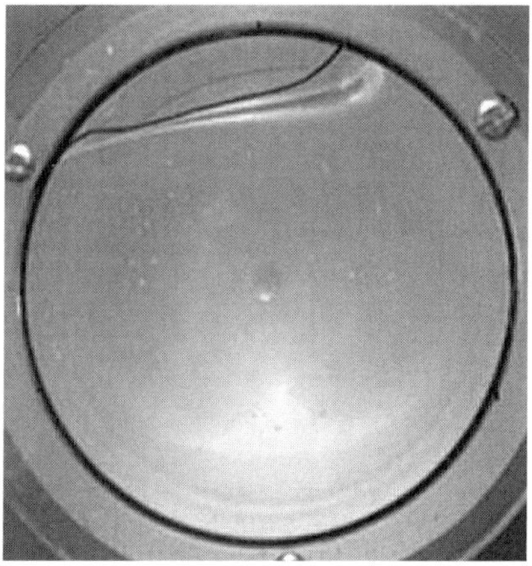

Fig. 12.2 Comparison of experimentally observed and numerically predicted free surface position for a can containing Keltrol rotating at 30 rpm about its fixed axis of symmetry. The dark line shows the predicted headspace bubble surface superimposed on a picture of the bubble position.

where K is a constant, termed the consistency, and n_p is the power law index. A more accurate characterisation of the non-Newtonian fluids used in the validation studies (see next section) is given by a version of the so-called Cross model with a zero viscosity at infinite shear rate:

$$h_1(\gamma) = \frac{\eta_0}{1 + (k\gamma)^{n_c}} \qquad [12.12]$$

Here the symbols η_0, k and n_c are, respectively, the limiting viscosity at small rates of shear, a rate constant and a power law index. Note that, for large rates of shear, the above model approximates a power law model with consistency parameter $K = \eta_0/k^{n_c}$ and power law index $n_p = 1 - n_c$. For the two non-Newtonian fluids used in the validation studies, the material parameters η_0, k and n_c are found using a controlled stress rheometer. Three fluids have been used in experiments to provide data with which to validate isothermal calculations, namely (a) Corena Oil 27, (b) a 2% by weight solution of Keltrol and (c) the so-called A1 fluid. Corena Oil 27 is Newtonian and has a viscosity of 0.092 Pa.s at 20 °C, Keltrol is a food additive and the 2% solution has an apparent viscosity at 20 °C that may be modelled by Equation [12.12] with η_0 = 347 Pa.s, k = 417 s^{-1} and n_c = 0.78. The A1 fluid is a highly elastic liquid that is shear-thinning. Its apparent viscosity at 20 °C may be modelled by Equation [12.12] with η_0 = 9.93 Pa.s, k = 1.18 s^{-1} and n_c = 0.73.

12.2.4 Experimental validation

The main form of experimental validation used is flow visualisation, in particular, observation of the shape and location of the headspace bubble. Much of the validation work was done for cans rotating in axial mode and the simplest case is

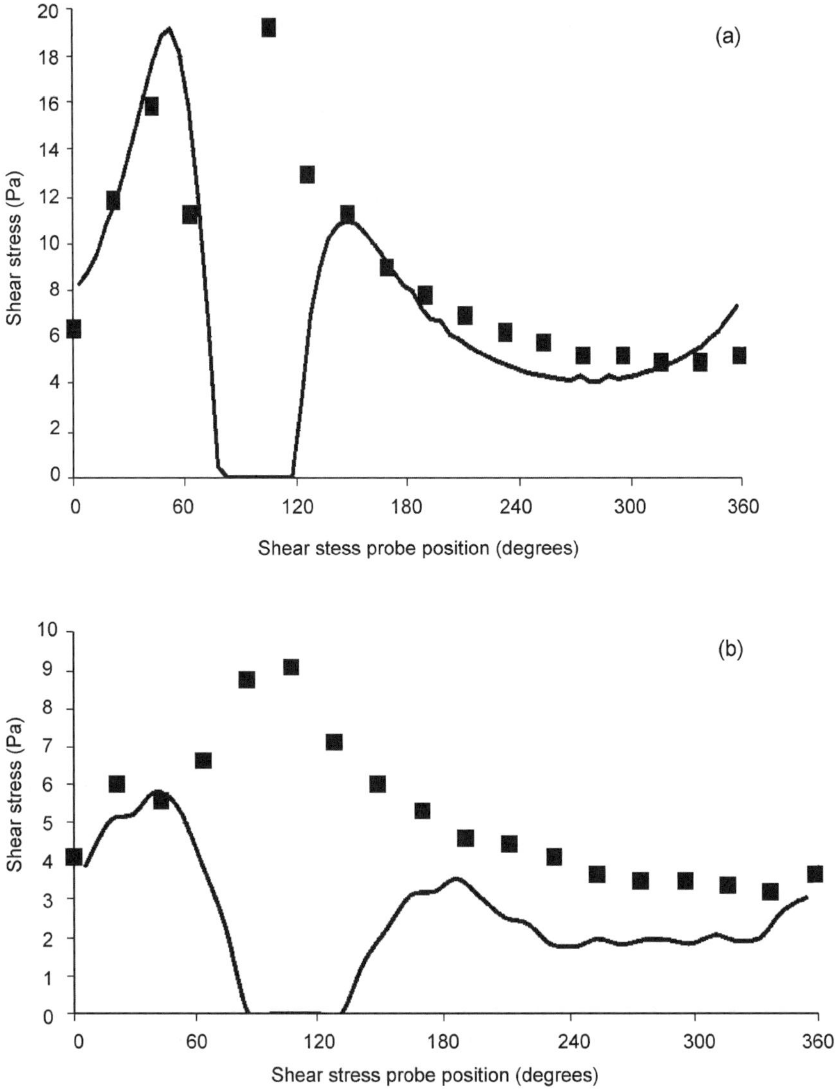

Fig.12. 3 Comparison of measured and predicted (a) curved wall and (b) end wall shear stresses for a can containing A1 fluid rotating at 30 rpm about its fixed axis of symmetry. The angular position of the probe is measured from the horizontal. Experimental (■), numerical (—).

where the axis of rotation is fixed. Figure 12.2 shows a typical comparison between observation and prediction for the Keltrol solution under these conditions. In off-axis axial rotation an additional, quantitative measure of validation was used, namely wall shear stress. The shear stress is measured using hot-film, stick-on probes whose method of operation is described in, for example, Bruun (1995). Figure 12.3 shows a sample of the many experimental and computational results

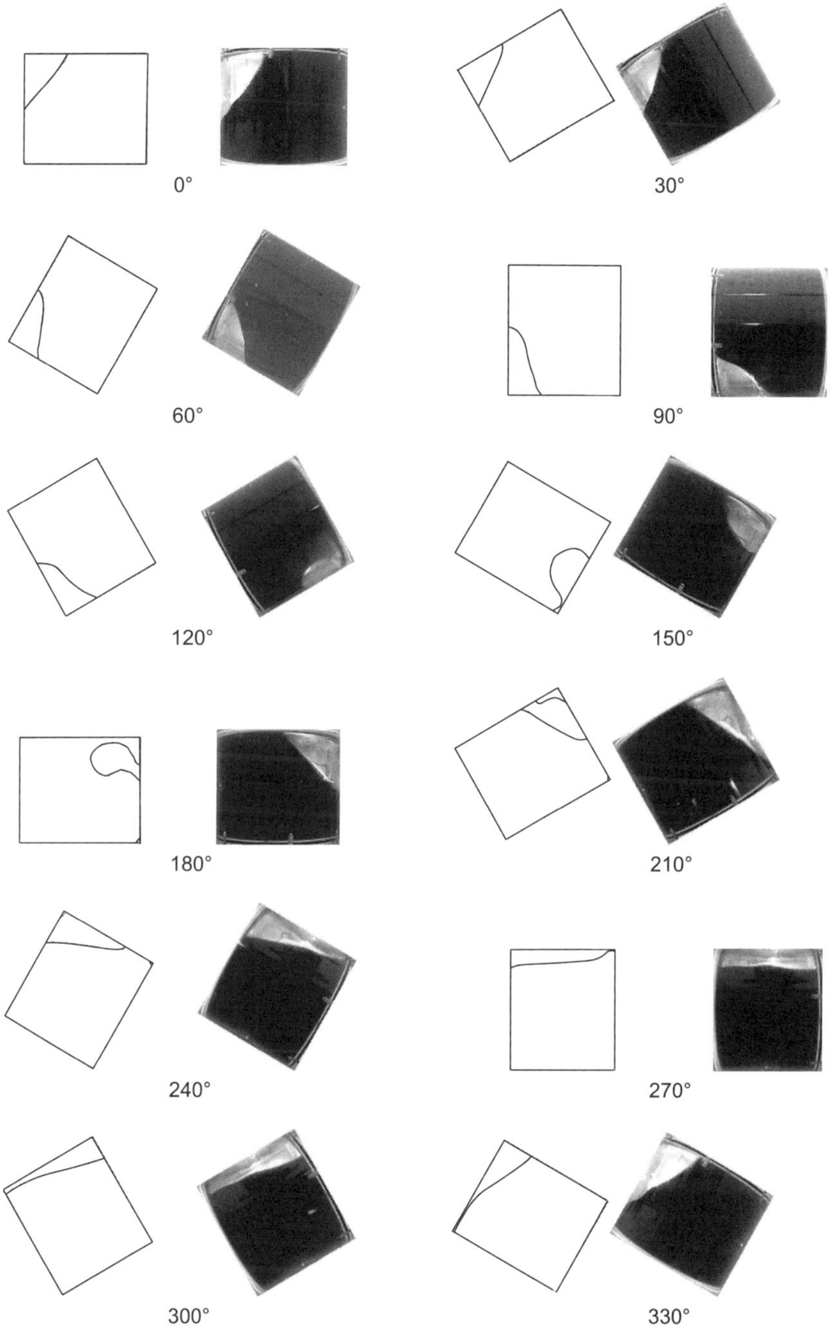

Fig. 12.4 Comparison of computed and observed bubble positions for off-axis rotation of a can containing Keltrol (R_c = 1.0 m, Ω = 30 rpm). The angular position of the can is measured from the horizontal position.

obtained. In general, agreement between measurement and prediction is good for data obtained from the can curved surface but is less good for the end walls, which is to be anticipated because of the difficulty associated with aligning the probes with the flow at these locations. In EOE motion, the shear stress probes were not used and so validation rests solely on comparison of simulated and observed headspace bubble position. The sequence of video images from experiments and computational results in Fig. 12.4 shows that very good agreement can be obtained. Not only are the bubble shape and location well predicted when it sits at the top of the can, but also when it gets trapped at the bottom of the can and subsequently bursts through the liquid core. This feature of the flow plays a key part in the enhancement of mixing, as described in the next section. The overall level of agreement between observation, measurement and numerical predictions accumulated from results in axial and EOE rotational modes lends confidence to the use of numerical simulation in situations where there is no experimental data for comparison, i.e. in numerical experimentation and design.

12.3 Using validated computational fluid dynamics simulations

12.3.1 Characterising mixing

It is of interest to food processors to know how the speed of rotation, Ω, and distance of the centre of the can from the axis of rotation, R_c, affect the sterilisation process in EOE motion. As a first step towards answering this question, the flow can be simulated under isothermal conditions. A great deal of numerical information is easily generated, and to make sense of it a method of characterising the effect of the two key parameters, Ω and R_c, is needed. Heat transfer will depend on the degree of mixing and it is the enhancement of mixing by the motion of the headspace bubble that is of primary interest. The shear rate, $\dot{\gamma}$, is one measure of the degree of mixing and so its value computed at each computational cell is used to generate a distribution of the degree of mixing throughout the can. The computed distribution turns out to be highly skewed and so the median value, $\dot{\gamma}_m$, is chosen as a representative measure of the overall level of $\dot{\gamma}$. This value represents a measure of mixing at a particular angular location in the rotational motion of the can and so will vary with the angular location of the can. An average value over all angular positions, the averaged median shear rate, $\bar{\dot{\gamma}}_m$, is therefore calculated and this gives a single figure representing the degree of mixing for a can rotating at a given angular speed at a given off-axis location. Figure 12.5 shows a typical variation of $\bar{\dot{\gamma}}_m$ with angular position and Fig. 12.6 shows a composite plot of normalised average median shear rate, $\bar{\dot{\gamma}}_m/\Omega$, against $R_c\Omega^2$. Two features of Fig. 12.6 are very important. Firstly, the results for a range of values of R_c and Ω fall, approximately, onto one curve and, secondly, the curve shows a marked peak at a 'critical' value of $R_c\Omega^2$. This indicates that, if operating conditions are such that the value of $R_c\Omega^2$ is near the critical value, then mixing is near optimal. It will be seen in Section

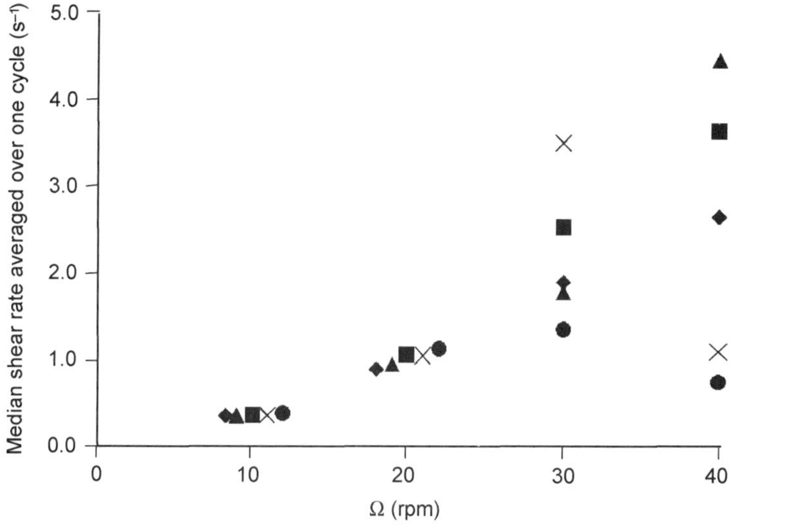

Fig. 12.5 Variation of median shear rate averaged over one cycle of rotation for various off-axis positions and rotational speeds for a can containing Keltrol. R_c = 0.0 m (◆), R_c = 0.25 m (▲), R_c = 0.5 m (■), R_c = 0.75 m (×), R_c = 1.0 m (●). Data points at 10 rpm and 20 rpm are spread horizontally for clarity.

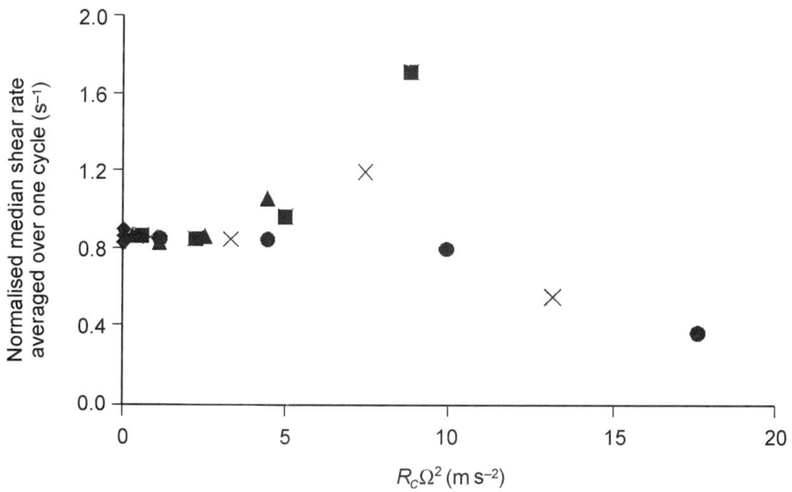

Fig. 12.6 Normalised median shear rate averaged over one cycle for various off-axis positions and rotational speeds (Corena Oil 27). R_c = 0.0 m (◆), R_c = 0.25 m (▲), R_c = 0.5 m (■), R_c = 0.75 m (×), R_c = 1.0 m (●).

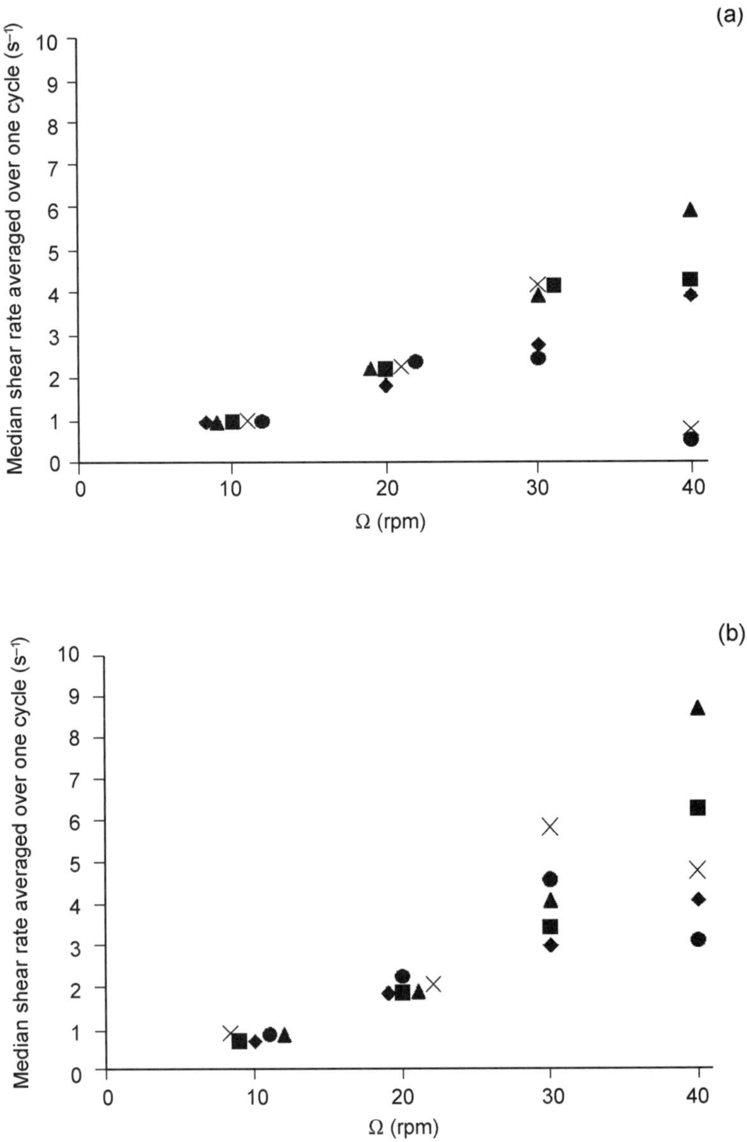

Fig. 12.7 Variation of median shear rate averaged over one cycle of rotation with off axis position and rotational speed for can aspect ratio (a) 0.44 and (b) 1.47. $R_c = 0.0$ m (◆), $R_c = 0.25$ m (▲), $R_c = 0.5$ m (■), $R_c = 0.75$ m (×), $R_c = 1.0$ m (●). Data points at 10 rpm and 20 rpm are spread horizontally for clarity.

12.3.2 that the location of the maximum in mixing does, in fact, correspond well with the computed minimum in heat-up times, and so it really does represent an optimal condition for EOE processing. The results shown in Fig. 12.6 are for a particular Newtonian fluid, but numerical tests for a range of Newtonian and non-

Newtonian fluids show that, while there are detailed changes for different fluids, the important conclusion that there is an optimal value of $R_c\Omega^2$ remains true. There is a physical explanation of why a maximum occurs and why the conditions where the maximum occurs are relatively insensitive to fluid properties, notably shear-thinning. The controlling physical mechanisms revealed by the computations and observations are buoyancy and fluid inertia. At low rotational speeds, buoyancy effects force the bubble to remain at the top of the can and at high speeds inertia forces the fluid away from the centre of the can, and so the bubble is forced to the inside, nearest the axis of rotation. There must therefore be intermediate speeds of rotation, at a given radius of rotation R_c, at which the bubble may be trapped near the inside during part of the motion but bursts through the fluid to regain its position at the top of the can for the rest of the motion. Under these circumstances, the bubble motion provides a very effective disturbance to the flow and so mixing will be enhanced. The numerical computations allow these conditions to be found accurately for a given set of parameters and fluid properties.

12.3.2 Parametric studies

The can aspect ratio (height : diameter) is 0.87 in all of the off-axis work discussed so far and results are now presented to show the effect on mixing of varying the aspect ratio. Figure 12.7 shows the dependence of $\bar{\dot{\gamma}}_m$ on angular speed and radial position, for Corena Oil 27, in cans with a smaller (0.44) and larger (1.44) aspect ratio than the 'standard' can aspect ratio (0.87). However, given the very large changes in aspect ratio, the corresponding change in maximum median shear rate is not very large. It could be concluded that changes in can aspect ratio are unlikely to have a major influence on optimal conditions. This observation suggests that, even when container shapes other than cylindrical cans are used, the optimal conditions for mixing may nevertheless be similar to those for cylindrical cans.

The way a can is placed in the retort could influence optimal conditions. In the work so far presented it has been assumed that when the can is at the highest or lowest point in its rotational path, its axis of symmetry is horizontal. Numerical experiments have been carried out to see what would happen if the can axis of symmetry were vertical at these points. In Fig. 12.8, results are shown for the case when the aspect ratio is 1.47, as in this case any effect due to can orientation would be expected to be greater than for the lower ratios. It is seen, from a comparison of Figures 12.6 and 12.8, that there is an effect but it is not pronounced. Price and Bhowmik (1994) came to a similar conclusion from their experimental work, although the two orientations they considered do not appear to correspond exactly with those above.

12.3.3 Non-isothermal flow

In this section, the thermal response of the fluid inside the can to the application of a constant temperature on the can surface is calculated. The can wall temperature is fixed at 140 °C and the initial temperature of the fluid inside the can is set at

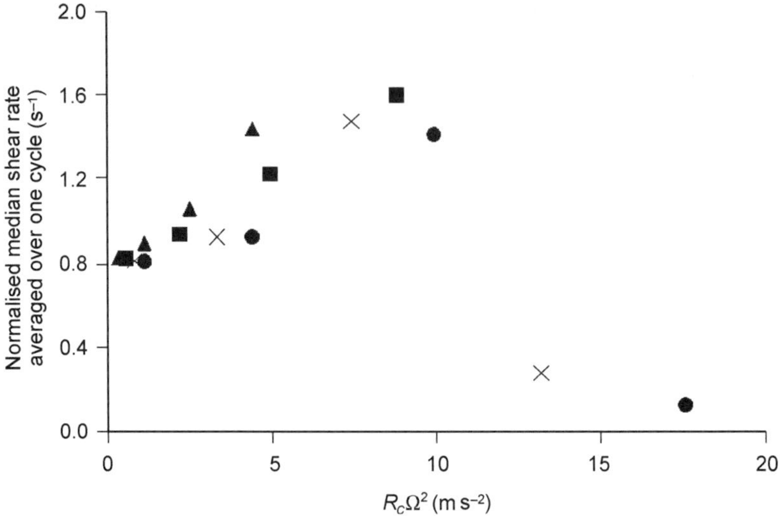

Fig. 12.8 Normalised median shear rate averaged over one cycle for various off-axis positions and rotational speeds when the can axis is vertical, rather than horizontal, at the highest and lowest points in the cycle (results for Corena Oil 27 are shown to facilitate comparison with Fig. 12.6). R_c = 0.25 m (▲), R_c = 0.5 m (■), R_c = 0.75 m (×), R_c = 1.0 m (●).

20 °C. Of primary interest is the time it takes for those fluid particles with the lowest temperature ('cold spots') to reach a given temperature. Numerical simulations reveal several important features. Firstly, from calculations with a Newtonian fluid with thermal properties similar to water, it is confirmed that the inclusion of the headspace bubble in the calculations results in a dramatic reduction in the time it takes for cold spots to reach 90% of the wall temperature. Secondly, it is found that a third timescale, a 'heat up' time t_{HU}, defined as the time it takes the coldest spots to reach 90% of the wall temperature, leads to the need for very long computational run times, especially when three-dimensional EOE calculations are made. For the purposes of undertaking parametric studies, James *et al.* (2006) therefore worked with a fictitious fluid that had an artificially high thermal diffusivity. A comparison of heat-up times using two-dimensional and three-dimensional calculations also indicated that, while there were differences in detail in the results obtained, trends for heat-up times were very similar. Thus, the parametric studies were carried out mainly for a fictitious fluid in a two-dimensional approximation of EOE motion. Numerical simulations reveal that there is an optimum value of $R_c\Omega^2$ at which t_{HU} is minimised. Figure 12.9 shows a typical set of results. Of particular note is the fact that the optimal value of $R_c\Omega^2$ for heating corresponds well with that found for optimal mixing under isothermal conditions. This indicates that optimal conditions for minimising heat up times can be identified from isothermal calculations, which results in a substantial saving in computational time. Of course, if heat-up times for a real liquid foodstuff are required, then the

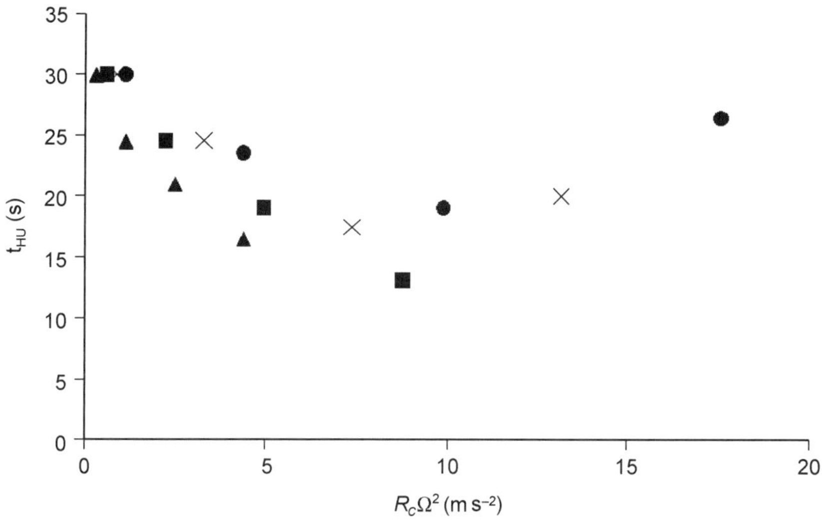

Fig. 12.9 Variation of t_{HU} with $R_c\Omega$, for various off-axis positions and rotational speeds (Newtonian fluid, end-over-end rotation). R_c = 0.25 m (▲), R_c = 0.5 m (■), R_c = 0.75 m (×), R_c = 1.0 m (●).

actual fluid properties need to be used and fully three-dimensional calculations made. The authors have made an attempt to simulate the experimental results of Anantheswaran and Rao (1985), who measured the variation with time of the temperature at the centre of a can of sucrose solution with a 6.25% headspace in EOE motion with constant, hot wall temperature. It was found that, in order to obtain reasonable computational results, very long cpu times are required, at least on a conventional, single processor PC. This problem will, of course, diminish as computer speeds increase and the use of high performance clusters with parallelised codes becomes more widespread.

Figure 12.10 shows the location of cold spots during the heating process. It is seen that, while the cold spots move closer to the centre of the can as heating takes place, it is not necessarily true that the coldest spot is at the geometric centre of the can when a headspace bubble is present. This information could be important when interpreting heating data taken from temperature probes located at the geometrical centre of the can.

12.3.4 Pilot plant and industrial scale implementation

Tucker *et al.* (2006) have built on the results from the numerical simulations by first identifying food simulants whose shear-thinning characteristics at constant temperature mimic the temperature-thinning characteristics of real foodstuffs during heating. They carried out a series of experiments to identify visually the optimum rotational speed for mixing with a range of fluids in isothermal conditions, then transferred the findings to a pilot rig, and finally to an industrial scale rig. The main conclusion was that, by using a programme of variable rotational speeds determined by isothermal experiments, heating times could be considerably

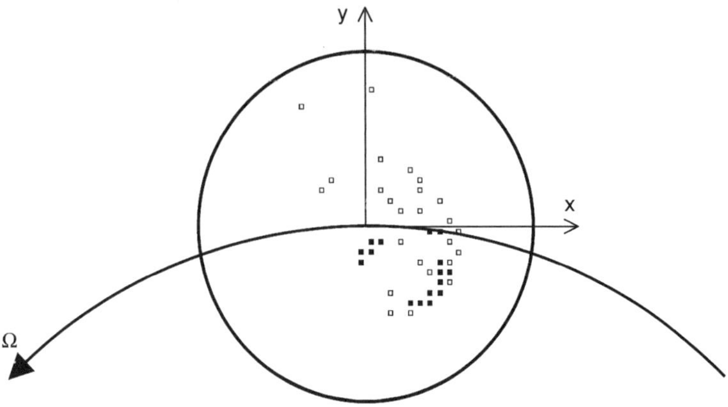

Fig. 12.10 Variation in location of 'cold spots' for flow of a Newtonian fluid (thermal diffusivity = 8.57×10^{-8} m^{2}s^{-1}), off-axis axial rotation (R_c = 1.0 m), 20 rpm. Arrow indicates direction of rotation. □ Locations for first 20% of heating time. ■ Locations for remaining 80% of heating time.

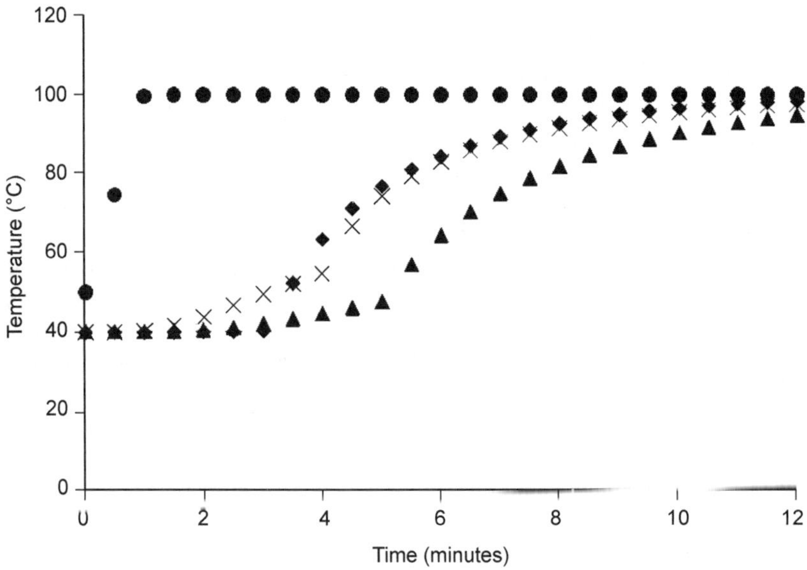

Fig. 12.11 Comparison of temperature time profiles for the standard process (▲) (constant rotational speed of 14 rpm); an optimised process in which the heating programme is interrupted and the rotational speed is adjusted manually (◆); and an optimised process in which the rotational speed is automatically adjusted at timed intervals (×). The retort temperature is indicated by (●).

reduced. In Fig. 12.11, adapted from Tucker *et al.* (2006), the heat-up times measured in a full-scale batch cooker, for a conventional time–temperature profile and constant rotational speed, and one in which the rotational speed is adjusted along the lines indicated by the computational and rheometrical calculations, are shown. The benefits of the modified process, as indicated by the reduced heat-up times, are clearly seen.

12.4 Summary and future trends

The particular food processing application described in the last two sections is just one example showing how validated CFD simulations can be used to provide insight into the detailed flow mechanisms that control one type of rotary thermal processing. The simulations can also be used as a tool with which numerical experiments to assess the effects of parameter changes can be made quickly and cheaply. The reviews by Norton and Sun (2006) and Xia and Sun (2002) give numerous other examples, showing the wide range of applications that have been studied using numerical flow simulation. It may, therefore, confidently be concluded that CFD is now well established as a diagnostic and design tool in the food processing sector and it appears very likely that its usage and range of application will continue to grow. However, there are factors that may slow down the growth of CFD as a design and investigative tool in the food processing industry. Firstly, there are still many physical processes for which mathematical models need to be formulated or refined. These include the need for more accurate turbulence models for those flows that involve fully turbulent flow or transitional flow, under isothermal and non-isothermal conditions, the need for new constitutive equations for complex materials, the modelling of flows in which there are large particulates, and the coupling of internal flows with external temperature fields. The treatment of multi-phase flows and flows with free surfaces or where part of the fluid detaches from the main body of fluid also gives rise to technical difficulties. Numerical methods will also need to continue to be developed, with specialist techniques emerging to deal with complicated specific types of flow. One example here is the emergence of Smoothed Particle Hydrodynamics (see, for example, Liu and Liu, 2003), which is a method that can give insight into free surface flows with significant deformation and separation. As more sophisticated mathematical descriptions of physical processes emerge, the need for more powerful computational resources grows.

Perhaps the single most important factor preventing more widespread take-up of CFD in food processing companies of moderate size is the difficulty of justifying financially the employment of a team of CFD specialists. Even if this barrier were removed, the supply of people with the necessary skills and training to undertake all of the tasks involved with the application of CFD techniques is not plentiful.

12.5 Sources of further information and advice

Computational fluid dynamics

The 'CFD online' web pages (http://www.cfd-online.com/) provide an excellent way to find information about all aspects of CFD, including numerical methods, modelling for CFD, software (commercial and free) and books. There are also links to other sites that describe different aspects of CFD and to conferences and courses.

The web addresses of some of the commercial software products mentioned in this chapter are:

ANSYS CFX	http://www.ansys.com/products/cfx.asp
FLUENT	http://www.fluent.com/
PHOENICS	http://www.cham.co.uk/
FIDAP	http://www.fluent.com/software/fidap/index.htm

Food processing

Holdsworth, S D (1997), *Thermal Processing of Packaged Foods*, London, Blackie Academic and Professional.

Department of Health (2000), *Guidelines for the Production of Heat Preserved Foods,* London: The Stationery Office, Second edition.

Richardson, P (2001), *Thermal Technologies in Food Processing*, Cambridge, Woodhead Publishing Limited.

Richardson, P (2004), *Improving the Thermal Processing of Foods*, Cambridge, Woodhead Publishing Limited.

Rheology

Barnes, H A (2000), *A Handbook of Elementary Rheology*, Institute of Non-Newtonian Fluid Mechanics, University of Wales.

Tanner, R I, (2000), *Engineering Rheology*, Oxford Engineering Science Series, Oxford University Press.

Owens, R G and Phillips, T N (2002), *Computational Rheology*, Imperial College Press, London.

Tanner, R. I. and Walters, K. (1998), *Rheology: An Historical Perspective*, Elsevier.

12.6 References

Abdul Ghani, A G, Farid, M M, Chen, X D and Richards, P (1999), 'Numerical simulation of natural convection heating of canned food by computational fluid dynamics', *Journal of Food Engineering*, **41**, 55–64.

Abdul Ghani, A G, Farid, M M, Chen, X D and Richards, P (2001), 'Thermal sterilization of canned foods in a 3-D pouch using computational fluid dynamics', *Journal of Food Engineering*, **48**, 147–156.

Abdul Ghani, A G, Farid, M M and Chen, X D, (2002), 'Numerical simulation of transient temperature and velocity profiles in a horizontal can during sterilization using computational fluid dynamics', *Journal of Food Engineering*, **51**, 77–83.

Abdul Ghani A G, Farid M M and Zarrouk, S J (2003), 'The effect of can rotation on sterilization of liquid foods using computational fluid dynamics', *Journal of Food Engineering*, **57**, 9–16.

Abdul Ghani, A G and Farid M M (2006), 'Using computational fluid dynamics to analyze the thermal sterilization of solid–liquid food mixtures in cans' *Innovative Food Science and Emerging Technologies*, **7**, 55–61.

Anantheswaran, R C and Rao, M A (1985), 'Heat transfer to model Newtonian liquid foods in cans during end-over-end rotation', *J Food Engng*, **4**, 1–9.

Batchelor, G K (1974), *An Introduction to Fluid Dynamics*, Cambridge University Press.

Berry, M R, Savage, R A and Pflug, I J (1979), 'Heating characteristics of cream style corn processed in a steritort: Effects of headspace, reel speed and consistency', *Journal of Food Science*, **44**, 831–835.

Bird, R B, Armstrong, R C and Hassager, O, (1987), *Dynamics of Polymeric Liquids, Volume 1, Fluid Mechanics*, John Wiley & Sons, USA.

Bruun, H H, (1995), *Hot Wire Anemometry: Principles and Signal Analysis*, Oxford University Press.

CFX4 User guide, (1995), AEA Technology, Harwell, Oxfordshire.

Clifcorn, L E, Peterson, G T, Boyd, J M and O'Neill, J H, (1950), 'A new principle for agitating in processing of canned foods', *Food Technology*, **4**, 450–460.

Farid M M and Abdul Ghani, A G, (2004), 'A new computational technique for the estimation of sterilization time in canned food', *Chemical Engineering and Processing*, **43**, 523–531.

Hughes, J P, Jones, T E R and James, P W, (2003), 'Numerical simulation and experimental visualisation of the isothermal flow of liquid containing a headspace bubble inside a closed cylinder during off-axis rotation', *Trans. IChemE*, **81**, Part C, 119–128.

James, P W, Hughes, J P and Jones, T E R, (2001), 'Numerical and experimental simulation of the flow in rotating containers', *NAFEMS International Journal of CFD Case Studies*, **3**, 31–60.

James, P W, Hughes, J P, Jones, T E R and Tucker, G S, (2006), 'Numerical simulations of non-isothermal flow in off-axis rotation of a can containing a headspace bubble', *Trans. IChemE*, **84**, Part A, 311–318.

Liu, G R and Liu M B, (2003), *Smoothed Particle Hydrodynamics*, World Scientific Publishing Co, Singapore.

Lo, S, (1990), *Modelling of free surface flows using a two fluid model*, UKAEA Report AERE R 13781, Harwell, Oxon, UK.

Naveh, D and Kopelmann, I J (1980), 'Effect of some processing parameters on the heating transfer coefficients in a rotating autoclave', *Journal of Food Processing and Preservation*, **4**, 67–77.

Norton, T and Sun, Da-Wen (2006), 'Computational fluid dynamics (CFD) – An effective and efficient design and analysis tool for the food industry: A review', *Trends in Food Science and Technology*, **17**, 600–620.

Owens, R G and Phillips, T N (2002), *Computational Rheology*, Imperial College Press, London.

Parchomchuk, P (1977), 'A simplified method for agitation processing of canned foods', *Journal of Food Science*, **42**, 265–268.

Price, R B and Bhowmik, S R (1994), 'Heat transfer in canned foods undergoing agitation', *Journal of Food Engineering*, **23** (4), 621–629

Quarini, G L (1995), 'Applications of computational fluid dynamics in food and beverage production', *Food Sci. Technol. Today*, **4**, 234–236.

Scott, S and Richardson, P (1999), 'The application of computational fluid dynamics in the food industry', *Trends in Food Science and Technology*, **8**, 119–124.

Sujatha, K S, Webster, M F, Binding, D M and Couch, M A (2003), 'Modelling and experimental studies of rotating flows in part-filled vessels: Wetting and peeling', *Journal of Food Engineering*, **57**, 67–79.

Tattiyakul, J, Rao, M A and Datta, A K (2001), 'Simulation of heat transfer to a canned corn starch dispersion subjected to axial rotation,' *Chemical Engineering and Processing*, **40**, 391–399

Tattiyakul, J, Rao, M A and Datta, A K (2002a), 'Heat ransfer to a canned corn starch dispersion under intermittent agitation,' *Journal of Food Engineering*, **54**, 321–329.

Tattiyakul, J, Rao, M A and Datta, A K (2002b), 'Heat transfer to three canned fluids of different thermo-rheological behaviour under intermittent agitation', *Food and Bioproducts Processing*, **80**, 20–27.

Tucker, G S, Emond, S P, Hughes, J P, Jones, T E R and James, P W (2006), 'Maximising the headspace bubble action via rotary thermal processing of packaged foods', *Food Manufacturing Efficiency*, **1**, 25–33.

Tucker, G S (2004), 'Improving rotary thermal processing', in Richardson, P (ed.), *Improving the Thermal Processing of Foods*, Woodhead Publishing Ltd, Cambridge.

Vandenberghe, M (2001), 'Agitation in axially rotated cans'. Industrial Food Processing: Experiments and Numerical Simulation,, *Joint British Society of Rheology/University of Wales Institute of Non-Newtonian Fluid Mechanics Conference*, Plymouth.

Wilcox, D C (2000), *Turbulence Modeling for CFD*, (2nd Edition), DCW Industries, California, USA.

Xia, B and Sun, Da-Wen (2002), 'Applications of computational fluid dynamics (CFD) in the food industry: A review', *Computers and Electronics in Agriculture*, **34**, 5–24.

Yang,W H and Rao, M A (1998), 'Transient natural convection heat transfer to starch dispersion in a cylindrical container: Numerical solution and experiment', *Journal of Food Engineering*, **36,** 395–415.

Part III

Safety of in-pack processed foods

13

Emerging pathogens of concern in in-pack heat-processed foods

P. McClure, Unilever, UK

13.1 Introduction

The range of in-pack heat-processed foods in the market today is wide and includes packs that are minimally processed, receiving a relatively low pasteurisation process such as 70 °C for 2 min, up to packs that receive a full commercial sterilisation process, exceeding 3 min at 121 °C. These packs will also have a wide variety of shelf-lives ranging from very short periods, e.g. <10 days, to long times, such as those typically required for ambient stability, that are commonly greater than 12 months. When considering pathogens that are relevant to this wide range of products, they will include organisms that are relatively sensitive to heat and those that are more resistant. The purpose of this chapter is to focus on emerging pathogens of concern to these products and to describe some of their characteristics and methods for control.

The term 'emerging pathogen' can have a number of different meanings and various interpretations and definitions can be found in the literature. These include:

- Pathogens that have newly appeared in a population or have existed previously but are rapidly increasing in incidence or geographic range (adapted from Potter *et al.*, 1997);
- One that is linked to a 'new' disease, i.e. a disease that is perceived to be a novel, immediate and serious threat to health (Smith and Fratamico, 1995);
- Disease of infectious origin whose incidence in humans has increased within the past two decades or threatens to increase in the near future (IOM, 1992);

- An infectious disease whose incidence is increasing following its first introduction into a new host population or whose incidence is increasing in an existing host population as a result of long-term changes in its underlying epidemiology (Woolhouse and Dye, 2001)

In a World Health Organisation consultation on emerging zoonotic diseases, emerging pathogens were defined by the working group (http://whqlibdoc.who.int/hq/2004/WHO_CDS_CPE_ZFK_2004.9.pdf) as:

- A zoonosis that is newly recognised or newly evolved or that has occurred previously but shows an increase in incidence or expansion in geographical, host or vector range.

There are other definitions of emerging disease that parallel those for emerging pathogens, that are based on emergence in the past two decades or on diseases that have newly appeared in a population. These are all imprecise definitions and are intended to cover pathogens that may be emerging for a population, a subpopulation, a single host or group of hosts. The severity of illness, morbidity and mortality rates are often used to compare pathogens and also to focus resources in order to minimise the risk to public health. The term 'emerging pathogen' is often based on subjective judgement of criteria that reflect an increase in awareness, improved identification and/or diagnosis, or identification of new vehicles of transmission. That said, considering the reports (Slutsker *et al.*, 1998; Doyle, 1994; Altekruse *et al.*, 1997; Tauxe, 2002; Meng and Doyle, 1998; Smith and Fratamico, 1995) that have been published on this subject in the past 20 years or so, the list of foodborne pathogens that have been characterised as 'emerging' has not changed dramatically. The organisms listed include bacteria, viruses and protozoa. The major agents that have emerged in recent years are shown in Table 13.1.

13.2 Changing patterns in foodborne disease

In an analysis of foodborne disease statistics in the US, Mead *et al.* (1999) listed 27 major known pathogens associated with foodborne disease in the 1990s and calculated that these pathogens accounted for only 19% of the total estimated number of cases attributable to foodborne infections and 36% of the deaths. It has been suggested that the gap between identified and unidentified agents could indicate a large contribution by unrecognised pathogens. Considering the list of 27 pathogens, a significant number have been identified as foodborne pathogens in only the last 30 years, and include bacteria, viruses and protozoan parasites. Of these, the 'emergence' of the protozoal parasite *Cyclospora cayetanensis* and unusual serotypes of salmonellae (often not considered as 'emerging') in the US has been attributed to increasing world trade in ready-to-eat foodstuffs. *Campylobacter jejuni* and *Listeria monocytogenes* are also relatively recent, and are sometimes referred to as 'emerging', but the increasing reports of their incidence in different environments is undoubtedly related to improved recovery/

Table 13.1 Emerging and re-emerging foodborne pathogens, reasons for their emergence, and foods commonly associated with transmission

Pathogen group	Pathogen	Possible reasons for emergence	Foods commonly associated with disease
Bacteria	*Campylobacter jejuni*	Increased awareness as a foodborne pathogen due to improved methods of isolation/detection/identification; economic development and land use; changes in human demographics	Poultry, beef, unpasteurised milk
	Escherichia coli O157:H7	Changes in human behaviour, farming practices	Ground beef, fresh produce, fermented meat
	Listeria monocytogenes	Changes in human demographics and human behaviour	Ready-to-eat foods, e.g. soft cheeses, paté
	Salmonella spp.	Changes in farming/industry practices, microbial adaptation, economic development and land use, changes in human behaviour	Poultry, eggs, meat, fresh produce
	Vibrio spp.	International travel, human demographics, changes in farming/harvesting practices, international distribution, lack of water chlorination	Seafood (raw, undercooked or cross-contaminated)
	Yersinia enterocolitica	Changes in human behaviour, farming/harvesting practices	Milk, water, pork
Viruses	Noroviruses	Changes in human behaviour, farming practices, economic development and land use	Fresh/frozen produce, seafood
	Hepatitis A and E	International distribution and commerce, farming/harvesting practices	Fresh fruit
	Rotavirus	Improved methods of detection, food handling practices	Raw seafood, water
Protozoa	*Cyclospora cayetanensis*	International distribution, farming/harvesting practices	Fresh fruit
	Cryptosporidium parvum	Land use, farming/harvesting practices	Water, pasteurised milk, apple juice
	Giardia lamblia	Land use, farming/harvesting practices, changes in human behaviour/preferences	Water, fresh produce

detection methods. In some regions, reports of campylobacteriosis continue to grow but in the case of *Listeria monocytogenes*, numbers appear to have peaked in many regions. In the late 1980s and early 1990s, some scientists even referred to *Clostridium botulinum* as 'emerging' following foodborne outbreaks associated with baked potatoes, grilled onions and garlic-in-oil. In these cases, it is the food vehicle that is the surprise and not the pathogen itself.

The nature of outbreaks has also changed over the past 20 years. Prior to this, outbreaks were relatively localised, following local and common exposure. More recently, many outbreaks are identified as highly dispersed, involving different regions and/or countries, spread over many hundreds of miles. This is due to centralised manufacture/processing, followed by distribution over large distances. Such outbreaks would not be detected without the aid of sophisticated characterisation tools that allow 'fingerprinting' of the particular strains responsible. In many cases, such as those involving *E. coli* O157:H7 and fresh produce or raw meat, there is low level contamination of very large volumes of product, making traceability very difficult. In addition, multiple stock keeping units (SKUs) are often produced in the same factory and may be affected by a common problem in processing, resulting in many different product types having the same contamination.

13.3 Reasons for emergence

Many of the recently emerging foodborne pathogens are associated with meat from poultry, cattle and other animals. The appearance of these pathogens is, generally speaking, a global trend and is not restricted to particular geographic locations. The reasons for their appearance and spread are poorly understood and it is suspected that the shift to a global economy, international trade and changes in the livestock industry may have contributed to these recent developments. No doubt, some of this is also due to improved surveillance, reporting and methods of detection.

Foodborne pathogens that have emerged in recent years share a number of characteristics. Nearly all of these have an animal reservoir from which they spread to man, i.e. they are foodborne zoonoses, but unlike established zoonoses, they do not often cause illness in the animal host. Another worrying trend is that these pathogens are able to spread globally in a short period of time. Many of the emerging pathogens are becoming increasingly resistant to antibiotics and this has been attributed, partly, to the use of antibiotics in animals. The practice of using antibiotics in animal production is coming under increasing pressure and there have been recent legislative changes that address this issue in particular parts of the world. Unfortunately, it is likely that some of these practices will continue in those areas that are not properly regulated or policed.

13.3.1 Changes in microorganisms

There are some pathogens that are regarded to have recently emerged as a function of evolution. Some bacteria are capable of evolving relatively quickly because of their ability to transfer genes and mutate, affecting 'contingency' genes that are highly mutable compared to the more stable 'housekeeping' genes. Horizontal gene transfer between separate lineages of bacterial pathogens may occur at a higher frequency in more highly mutable organisms, such as some members of the Enterobacteriaceae. A review of the genome plasticity in Enterobacteriaceae (Brunder and Karch, 2000) refers to a move away from the view of bacterial genomes as static, monolithic structures, towards relatively variable, fluid structures. This is thought to be a major factor in the emergence of completely new pathotypes of *Escherichia coli*, such as verocytotoxigenic *E. coli* O157 and other serovars, and causes confusion with classical taxonomic approaches and the concept of a 'species'. Wain *et al.* (2001) provided a useful summary of genomic studies that have contributed to the understanding of how pathogenic *E. coli* and *S. enterica* have evolved.

There are common features that some recently emerged pathogens share. Many have an animal reservoir but, as stated, do not necessarily cause disease in those animals; some can infect multiple hosts; many are rapidly and widely spread; some have low infectious dose, carry multiple antibiotic resistance genes and can cause serious illness. In addition, there are host factors that may lead to increased susceptibility, such as age, chronic or other debilitating diseases, immunosuppression and pregnancy.

13.3.2 New vehicles of transmission

New food vehicles have been identified in recent years. These new vehicles include foods that were once thought to be 'safe', such as eggs, apple juice, fresh fruit, fresh vegetables and fermented meats. With consumer preferences for fresher, less heavily processed foods likely to continue, it is probable that new food vehicles for foodborne disease will continue to emerge. Globalisation of the food supply introduces hazards from these regions into other areas and disseminates pathogens over wide geographical areas. In addition to their increasing popularity in consumption patterns, fresh fruits and vegetables have also become increasingly important vehicles in foodborne disease statistics. In the US, these products were responsible for only 1% of foodborne disease cases in the 1970s, but in the 1990s this had increased to 12% (Sivapalasingam *et al.*, 2004). Between 1990 and 2003, the US Centre for Science in the Public Interest showed that fresh fruit and vegetables caused 554 foodborne disease outbreaks in the US, outnumbering those linked to poultry for the same period. In a recent outbreak of botulism in North America (US and Canada), the vehicle identified was a chilled carrot juice product contaminated with spores of proteolytic *C. botulinum*. It is thought that the products were temperature abused during distribution and/or by the consumer.

Alternative processes, if incorrectly assessed, may also provide an additional source of infection or intoxication. Continued consolidation within the food industry is likely to lead to increasingly large markets and wider distribution from centralised manufacturing operations. With increasing demand from increasing populations, we are likely to see more re-use and recycling of water and waste, and this may have an impact on the microbiological hazards that must be controlled.

13.3.3 Improved epidemiology and surveillance

Fortunately, improved epidemiological capability, provided through better detection methods and better cooperation/coordination between different surveillance networks, is likely to allow quicker detection of geographically widespread outbreaks of foodborne disease. Cases of illness that may have been identified as 'sporadic' in the past can now be linked with other cases, identifying outbreaks associated with common sources. Molecular methods are transforming taxonomy; also our understanding of the genomes of particular pathogens and groups of pathogens, such as the Enterobacteriaceae. This has already let us gain some insight into evolutionary processes and should allow us to better anticipate the potential of microorganisms to incorporate new genetic material and develop new virulence characteristics. An improved understanding of pathogenesis of foodborne disease and colonisation of animals may also allow development of new intervention strategies.

13.3.4 Changes in the human host

With anticipated increases in the average life expectancy, through improved medical treatment of chronic disease and other advances, there is likely to be an increase in the proportion of persons with age-related susceptibility to foodborne disease. Also, there is likely to be a continuing increase in the number of immunosuppressed individuals, due to infection with HIV and other chronic illnesses. Human host factors that are important in foodborne disease are described in more detail by Green-Johnson (2006), Gahan (2006) and Kobayashi (2006).

13.3.5 Changes in industry, trade and consumer lifestyle

Changes in the food industry, including global sourcing of foodstuffs, intensified animal production and changes in the eating habits of consumers may account for the wider range of micro-organisms and foods being implicated in foodborne disease. Regional differences in growing, harvesting and processing practices can lead to familiar pathogens being found in a wider range of foods or contamination of specific foods with emerging or uncharacterised pathogens. Lifestyle changes – fast food, eating out-of-home, preference for 'fresh' foods (often eaten raw) and reduced awareness of food safety may also increase exposure. Risk assessment can provide a tool for examining the impact of all these changes on the emergence of new hazards and changes in levels of risk. As assessments are only as good as the information input, and cover a changing area, they are likely to need frequent review as the occurrence, distribution or transmission rates of an agent may change with changes in the supply chain.

13.4 Emerging pathogens

There are some key attributes of emerging pathogens that should be considered in assessing the risk they may pose to in-pack heat processed foods. These include their:

- ecology – where they are found, e.g. in raw materials or primary produce, in food manufacturing environments and on food handlers, and the numbers likely to be present;
- growth and survival characteristics, including heat resistance;
- pathogenic potential, including severity of disease, population(s) affected and infective dose.

The recently emerged pathogens listed in Table 13.1 are discussed in more detail below and their relevance to in-pack heated foods outlined.

13.4.1 *Campylobacter* spp. and *Arcobacter* spp.

Campylobacter spp. are microaerophilic, Gram-negative, spiral-shaped, non-spore forming, motile bacteria. *Campylobacter* spp. were not recognised as foodborne pathogens until the late 1970s and are one of the leading causes of foodborne disease in developed and developing countries. They do not survive in food as well as other vegetative pathogens and are capable of growth only in a relatively limited temperature range and in microaerophilic conditions. As with other infectious bacteria, survival of *Campylobacter* spp. is better at lower temperatures (Curtis *et al.*, 1995), although they are sensitive to drying and freezing (McClure, 2000). *Campylobacter jejuni* and other *Campylobacter* spp. are relatively sensitive to heat, with typical D_{55} values ranging from 0.6 to 2.3 minutes. The heating menstruum appears to have only a small influence on heat sensitivity but cells heated in 0.1 M phosphate buffer (pH 7.0) exhibit significantly faster loss of viability compared with cells heated in peptone solutions or in foods. Maximum heat resistance occurs at near pH 7.0 and decreases as pH moves away from neutrality.

Food-associated illness usually results from eating foods that are re-contaminated after cooking, or eating foods of animal origin that are raw or inadequately cooked. *Campylobacter* spp. are part of the normal intestinal flora of a wide variety of wild and domestic animals and have a high level of association with poultry (Shane, 1992).

The organisms appear to be particularly well adapted to the avian gut (Wassenaar *et al.*, 1998), reaching levels of 10^{10} cfu/g caecal contents without animals having symptoms of disease. Once one bird acquires *Campylobacter*, the rest of the flock becomes colonised very quickly and currently there are no effective methods of control to ensure that flocks will not become colonised. A factor in emergence is increased recognition through improved methods of isolation and identification. Another important feature in terms of clinical disease in humans is the emergence of quinolone-resistant strains attributed to use of fluoroquinolones in the poultry industry (Teuber, 1999). Consumption of poultry and poultry products have been frequently quoted as risk factors for *Campylobacter* infection (Schorr *et al.*, 1994; Solomon and Hoover, 1999; Rautelin and Hanninen, 2000). Various surveys of raw chicken indicate contamination levels up to 86% (Jorgensen *et al.*, 2002; Kramer *et al.*, 2000; Wilson, 2002). Levels of *C. jejuni* on retail raw poultry of between 10^2 and 10^4 per gram have been reported (McClure, 2000). From these data, it should be expected that all raw chicken could be contaminated with *Campylobacter* spp., with a worst case level of *Campylobacter* spp. on chicken at about 2.4×10^4 cfu/g. Other meats (e.g. beef) will also harbour *Campylobacter* spp., usually at lower levels.

There are four main species of *Campylobacter* that are associated with human illness. *C. jejuni* is the cause of most human cases of campylobacteriosis. *C. coli* is the cause of around 10% of cases. The infective dose to cause foodborne illness is thought to be relatively low. An infectious dose of 500 organisms is most frequently quoted for *C. jejuni* (Corry, 1999). Therefore, *Campylobacter* spp. should be absent from ready-to-eat foods. In practice, this means absent in a 25 g sample. The level of *Campylobacter* spp. at the start of the cooking step in a food to be heated in-pack will be the same or lower than levels found on the raw material (usually raw meats). There is usually no opportunity for growth on these materials during storage, and some conditions (e.g. freezing) may reduce levels present on raw produce. A 6 log reduction in numbers will easily control this pathogen in foods that may be contaminated and since heat resistance is lower than for many other infectious pathogens (such as *Salmonella* spp. and *Listeria monocytogenes*) that must also be controlled in foods, heat processes used to control these pathogens (e.g. 70 °C for 2 min) will also control *Campylobacter* spp.

Prior to 1991, *Arcobacter butzleri* and *A. cryaerophilus* were known as aerotolerant *Campylobacter*. These organisms have been associated with abortions and enteritis in animals and enteritis in man. Although both species are known to cause disease in man, most human isolates come from the species *A. butzleri*. There is very little known about the epidemiology, pathogenesis and real clinical significance of arcobacters, but it is thought that consumption of contaminated food may play a role in transmission of this group of organisms to man. Although arcobacters have never been associated with outbreaks of foodborne illness, they have been isolated from domestic animals, poultry, ground pork and water. There are relatively few studies describing thermal inactivation. D'Sa and Harrison (2005) reported D_{60} values between 0.07 and 0.12 min in phosphate-buffered saline (pH 7.3) and 2.18 min at 55 °C in ground pork. Hilton *et al.* (2001) reported *D*-values at 50 and 55 °C of 2.1 and 1.1 min, respectively, for exponential cells in phosphate buffer (more resistant than stationary-phase cells). Both studies indicate that *Arcobacter butzleri* strains tested are more thermotolerant than *Campylobacter* spp. but still far less tolerant than many other vegetative infectious foodborne pathogens that are controlled by the standard '70 °C for 2 min'.

13.4.2 *Escherichia coli* O157:H7 and other enterohaemorrhagic *E. coli*

Escherichia coli O157:H7 is the most common serovar of enterohaemorrhagic *E. coli* (EHEC) associated with foods. Like other *E. coli*, it is a motile, Gram-negative, rod-shaped, non-sporing bacterium. Cases of illness linked to *E. coli* O157:H7 were first recognised in the early 1980s and probably stretch back to the 1960s; reasons for its emergence are not clear but appear to be related to a number of factors (Armstrong *et al.*, 1996). Considering its growth and survival characteristics, *E. coli* O157:H7 and other haemorrhagic *E. coli* behave similarly to non-pathogenic types, with evidence of some strains linked to foodborne disease being able to survive low pH conditions. Some strains of non-pathogenic *E. coli* also demonstrate this ability. One of the most important control measures for *E. coli* O157, other VTEC and other infectious pathogens, is heating. Numerous studies (reviewed by Stringer *et al.*, 2000) have shown that *E. coli* O157:H7 is not unusually heat-tolerant, compared to other *E. coli*, and other members of the Enterobacteriaceae. The D_{60} values reported in this review range between 0.8 and 1.9 min, with z values ranging between 4.6 and 5.5 C°. A more recent review by O'Bryan *et al.* (2006) reports D_{60} values between 0.38 and 4.4 min in meat and poultry products.

Surveys of raw meats for sale have revealed *E. coli* O157:H7 in 2–4% of ground beef, 1.5% of pork and poultry, and 2–5.9% of lamb and lamb products (Doyle and Schoeni, 1987; Sekla *et al.*, 1990; Chapman *et al.*, 1996). Todd *et al.* (1988) reported that levels of 10–6200 *E. coli* O157:H7 per gram have been detected in beef samples associated with illness. The other food type that has been systematically tested for *Escherichia coli* O157:H7 is milk. More recently, fresh produce has been linked to outbreaks of foodborne disease, and this is likely to be associated with produce coming into contact with contaminated water and/or faeces. Therefore levels are likely to be lower than those found in faeces and on carcasses, providing that temperatures are properly controlled and are sufficiently low to prevent growth.

To control *E. coli* O157:H7 and other pathogenic *E. coli* in in-pack heated foods, the temperature regimes (e.g. 70 °C for 2 min) used to control other infectious pathogens are more than adequate for eliminating large numbers (e.g. 10^6 cells/g).

13.4.3 *Listeria monocytogenes*

Listeriosis is an atypical foodborne disease that has attracted a great deal of attention since the early 1980s, mainly because of the severity, high mortality rate and non-enteric nature of disease. Listeriosis is caused by *L. monocytogenes*, which is found in many environments and is frequently carried in the intestinal tract of many animals, including man. *Listeria monocytogenes* is a Gram-positive non-sporing bacterium that is ubiquitous. It is more heat resistant than many other vegetative infectious pathogens and heat processes for in-pack pasteurised foods are often set to control this particular agent. Several reviews of published

D and *z*-values are available (Doyle *et al.*, 2001; Mackey and Bratchell, 1989; O'Bryan *et al.*, 2006), reporting *D* values between 0.31 and 16.7 min at 60 °C and between 0.27 and 0.41 min at 68 °C, with *z*-values generally between 5 and 7.4 C°.

The organism is often found in healthy animals and humans, with a carrier rate of 10–50% in cattle, poultry and swine. The organism has been isolated from a variety of foods, at levels of 13% in raw meat, 3–4% in raw milk and 3–4% in dairy products (Farber and Peterkin, 1991). In some raw materials, such as raw poultry, up to 100% of samples have tested positive for presence of *L. monocytogenes* (Antunes *et al.*, 2002) and other studies report prevalence rates of around 50% (Lewis and Corry, 1991; Hudson and Mead, 1989), with numbers generally less than 100 cfu/g but some reaching 10^3–10^4 cfu/g (Rorvik *et al.*, 2003). Some of the major outbreaks in man have been attributed to meat products such as pork tongue and meat pâté. Foods associated with outbreaks have largely been refrigerated, processed and are ready-to-eat, and tend to be foods that have been recontaminated after processing. *Listeria monocytogenes* is able to grow at chill temperatures and this must be taken into account when considering target log reductions required for intervention techniques such as heat processing.

The disease in man is commonly associated with meningitis, septicaemia and abortion. Recent outbreaks, however, have been associated with a milder form of disease characterised by gastroenteritis and flu-like symptoms. Host factors are likely to play an important role in the susceptibility to listeriosis, together with presence of virulence factors in the organism. Many individuals frequently ingest *L. monocytogenes* without any apparent ill effects and harmful levels are generally regarded to be well above 100 cfu/g. Consequently, foods containing this level (or below) are considered safe for the general population. For in-pack heated foods, it is important that they will not allow multiplication to levels above 100 cfu/g; therefore, complete elimination is usually sought so that other factors do not have to be relied on to establish safety. To ensure complete elimination, taking into account the highest levels expected on raw materials and appropriate storage conditions preventing growth, a 6-log reduction is commonly used in the food industry in the UK, and in the US a target log reduction of ≥ 5 logs is required (Lado and Yousef, 2007). Although listeriosis is a severe disease, the number of cases, compared to some of the other foodborne diseases, is relatively low. Interestingly, the isolates associated with processed meats more often originate from the processing environment than from the animal itself.

13.4.4 *Salmonella* spp.

Although *Salmonella* spp. have been recognised as significant foodborne pathogens for many years, they are sometimes included in lists of emerging pathogens because of new vehicles implicated in disease or because they are a cause of increasing numbers of cases of illness, or because new subtypes have emerged, such as multiple-antibiotic-resistant *S. typhimurium* DT (definitive type) 104 or a combination of these, e.g. *S. enteritidis* PT (phage type) 4 in eggs. One factor thought to play a role in the emergence of particular phage types/strains is increased invasiveness in host animals, and some of these subtypes also cause more severe illness in vulnerable populations, such as the young and the elderly.

Salmonellae are primarily intestinal parasites of humans and many other animals, including rodents, wild birds and domestic animals. Recently, the nomenclature of salmonellae has been revised since modern taxonomic methods suggested that all serotypes of *Salmonella* probably belonged to one DNA-hybridisation group. *S. enterica* was originally subdivided into seven sub-groups, *Salmonella enterica* subspp. *enterica*, *salamae*, *arizonae*, *diarizonae*, *houtenae*, *bongori* and *indica*. *S. enterica* subsp. *bongori* has since been elevated to species level. Only serotypes of subsp. *enterica* are still named (e.g. *S. enterica* subsp. *enterica* serotype *typhimurium* or *S. typhimurium* or simply *typhimurium*) indicating that the named serotype is a member of subsp. *enterica*.

The general physiological characteristics of *Salmonella* spp. (such as heat resistance and tolerance to other stress conditions) tend to remain the same or similar, although there are some traits, such as antibiotic resistance, that differ and are more of a concern in recently emerged strains and subtypes. Therefore, the principal controls that have been used in food processing, including heat treatment, remain the same. A comparison of reported *D*-values for *Salmonella* spp. is included in the study by O'Bryan *et al.* (2006), reporting D_{60} and D_{70} values between 3.8 and 8.5 min, and 0.07 and 0.32 min, respectively, for meat and poultry products. The highest levels found in contaminated raw materials tend to occur in poultry where numbers in the gut contents of chickens are reported as 100–1000 cfu/g (WHO/FAO, 2002). The target log reduction for these pathogens in foods where they are a significant hazard is 6 logs in the UK and some other European countries and 6.5 for beef and 7.0 logs for poultry, in the US (O'Bryan *et al.*, 2006).

13.4.5 *Vibrio* spp.

Vibrio spp. are Gram-negative non-spore-forming bacteria that have become important food and water-borne pathogens in particular regions of the world, causing disease in particular sub-populations. *Vibrio cholerae* strains belonging to the O1 and O139 serogroups have been responsible for epidemics of cholera, sustained by poor hygiene/sanitation, lack of chlorination, poor immunity and consumption of raw fish and other seafood. Outbreaks in South America (e.g. Peru) first appeared in the early 1990s and were linked to decisions to terminate the chlorination of water supplies because of concerns over cancer risks. The epidemic strains were thought to be introduced through discharge of ballast tank contaminated water from cargo ships. Vibrios are commonly found in sea water and are found contaminating seafood in temperate waters. Transmission of cholera is primarily through the faecal/oral route and the role of food as a vehicle is less important compared to other pathogenic vibrios.

Vibrio vulnificus has emerged as a foodborne pathogen, first reported in 1979 in the US, causing serious disease in individuals with liver disease and other clinical conditions that make them more susceptible to illness through immunosuppression. Cases of *V. vulnificus*-associated illness have the highest mortality rate in the US (Mead *et al.*, 1999) and are increasing at an alarming rate in other regions, such as Taiwan, due to incidence of hepatitis B and C, and the popularity of eating raw or undercooked seafood. The cause of the increase in cases associated with consumption of oysters harvested from the Gulf of Mexico (Motes *et al.*, 1998) has been attributed to a change in harvesting practices in the region, to the summer months in order to compete better with nearby oyster beds that were not productive because of parasitism, pollution and other factors (Slutsker *et al.*, 1998). There has also been an increase in consumption of seafood and the aquaculture industry is now very large in many regions. In Japan, cases of illness associated with *V. parahaemolyticus* overtook *Salmonella* as the leading cause of disease in the late 1990s, and this species is also a significant cause of pandemic gastroenteritis (Nair *et al.*, 2006).

Vibrios are not particularly resistant to factors, such as heating, that are traditionally used to control infectious pathogens in foods. Thermal inactivation studies show that these organisms will not survive thorough heating of shellfish to an internal temperature of at least 60 °C for several minutes, even if present in high numbers (West, 1989). *D*-values for 52 strains of *V. vulnificus* averaged 78 s at 47 °C (Cook and Ruple, 1992). The foods contaminated with vibrios are mainly shellfish, crustaceans and fish. The levels of organisms can be as high as 10^3–10^4 cfu/g in produce from warmer waters and have been recorded at 10^5 cfu/g (Jackson *et al.*, 1997) and 10^6 cfu/g (Tamplin, 1994) for oysters during the summer months, whereas produce harvested during colder periods tend to have much less, e.g. < 10 cfu/g. Fortunately, the infectious dose for vibrios is generally high and this is borne out by disease statistics that do not record infections during winter months when many raw oysters are consumed. For seafoods heat processed in-pack, it could well be that processes should be targeted towards vibrios that may be present in high numbers if harvested from warm waters likely to harbour the organisms.

13.4.6 *Yersinia enterocolitica*

Although *Yersinia enterocolitica* has been associated with foodborne disease only for the past 30 years or so, it is likely that this organism has been the cause of foodborne disease for a much longer period of time. The organism is Gram-negative, non-sporing and rod-shaped, and is a member of the Enterobacteriaceae family. The organism is frequently associated with pigs, and some serovars causing human disease tend to reflect those found in pigs (Nesbakken, 1992). Compared with some of the other organisms featured in this chapter, *Y. enterocolitica* is not a major cause of foodborne illness but it is acknowledged that there has been a genuine increase in the number of cases reported (WHO, 1983, 1995) and improvements in reporting systems and methods for detection are likely to have had an impact here.

Studies show that pigs are a reservoir for *Y. enterocolitica* with frequencies of contamination as high as 63% in pig carcasses (Nesbakken, 1988) and, in some cases, higher incidence (up to 85%) has been reported in healthy pigs at levels up to 1.72×10^3 per cm^2 (Nesbakken, 1988). Heat resistance of yersinias is similar to other members of the Enterobacteriaceae, with a pasteurisation temperature/time of 70 °C for 2 min being more than enough to effect inactivation of the high numbers that may be present in contaminated foods. The organism is also able to grow at low temperatures, which is an important consideration if contaminated foods are held at chill temperatures prior to heat processing. Mathematical models can be used to predict likely increases in numbers and the maximum numbers likely to be present on foods, so *Y. enterocolitica* may feature as a target pathogen in some in-pack heat-processed foods. Heat treatment of meat at 60 °C for 1–3 min completely destroys *Y. enterocolitica* (Lee *et al.*, 1981) and a D_{60} value of 27 s has been reported in scalding water (Sörqvist and Danielsson-Tham, 1990).

13.4.7 Foodborne viruses

The major foodborne viruses in terms of numbers of recognised cases are noroviruses (formerly Norwalk-like viruses, also known as NoV) and hepatitis A virus (HAV). Rotavirus is also associated with foodborne illness, with some large outbreaks caused by rotavirus groups B and C, and hepatitis E is associated with waterborne disease. The burden of foodborne disease caused by viral agents is relatively poorly understood because of the limitations of reporting/surveillance systems and detection methods. In terms of their ecology and routes/vehicles for transmission, viruses are much less well understood than to most bacterial foodborne pathogens. Most documented outbreaks of viral-associated foodborne disease are linked to food that has been manually handled by infected workers somewhere between primary production (e.g. on the farm) and final preparation of the food. Industrially-processed foods are not normally associated with outbreaks of disease. There have been increases in reports of norovirus activity in Europe in October/November period in 2006 compared with 2004 and 2005 (Kroneman *et al.*, 2006), reported by 9 out of 11 country authorities: outbreaks of norovirus infections in Denmark from frozen raspberries imported from Poland (Falkenhorst *et al.*, 2005) affected more than 1000 people. The increase in norovirus activity may be due to improved surveillance/awareness and methodology.

Considering in-pack heated foods, the primary concern is handling of produce prior to processing, and the key controls to prevent contamination of foods include good agricultural practice (GAP), good hygienic practice (GHP) and good manufacturing practice (GMP). The main reservoirs of foodborne enteric viruses are thought to be humans, although there are some similar viruses found in other animals, and zoonotic transmission might occur if the right conditions arise (Koopmans and Duiver, 2004). Bivalve molluscs are known to be contaminated from time to time, because of exposure to human faecal material. Most foodborne viruses are more resistant to heat than most vegetative bacteria. It is recommended that inactivation profiles of the most stable RNA viruses (such as HAV) are used as indicators of those viruses that are not easily studied, such as noroviruses, because of the absence of culture methods. The information available suggests that < 2 log inactivation of HAV would be expected with a pasteurisation of 70 °C for 2 min (Millard *et al.*, 1987), and data from contamination studies shows that more than 1000 virus particles may be transferred from faecally contaminated fingers to foods, so inactivation of at least 3 logs would be required to inactivate these agents (Koopmans and Duizer, 2004). This emphasises the importance of hygienic handling prior to processing.

13.4.8 Protozoan parasites

The protozoan parasites listed in Table 13.1 as having emerged in recent years include *Cyclospora cayetanensis*, *Cryptosporidium parvum*, and *Giardia lamblia*. These agents produce resistant forms, termed cysts or oocysts, that enable the organisms to persist in the environment (e.g. surface waters) until they are ingested by hosts that enable them to continue their life-cycle, producing disease and multiplying so that further cysts or oocysts are released to infect other hosts. These organisms are similar to viruses in that they do not multiply in foods and are present in foods due to faecal contamination. The natural reservoirs of *Cryptposoridium parvum* are mammals such as cattle, goats, other farm animals and humans. *Cyclospora cayetanensis* is known to infect only humans. *Giardia lamblia* (also called *G. intestinalis*) is a sub-species of *G. duodenalis*, infecting mammals.

Foods known to be affected by these agents include vegetables, tripe, raspberries, lettuce, apple juice, and salads, and contamination from infected food handlers may occur on fresh produce when it is grown or harvested in the field or during processing/handling. Therefore GAP, GHP and GMP are all important to prevent contamination occurring. Fortunately, these agents are relatively heat sensitive so they will not survive in in-pack heated foods that receive a pasteurisation of 70 °C for 2 min. Fayer (1994) reported that bovine *C. parvum* oocysts are not infectious after 2 min heating at 64.2 °C.

13.5 Effect of reducing severity of heat treatments in heat-processed foods

The standard heat processes used in the food industry range from relatively mild processes, such as low-temperature pasteurisation (e.g. 70 °C for 2 min), that will inactivate vegetative infectious pathogens, to higher temperature pasteurisations such as the non-proteolytic *C. botulinum* cook (typically 90 °C for 10 min for 6*D* inactivation) commonly used in medium shelf-life chilled foods, up to full sterilisation processes, effecting 12*D* inactivation of proteolytic *C. botulinum*. These are prescribed in legislation or ingrained in industry practice and have generally been set to accommodate the 'worst case' situation in terms of contamination levels, heat resistance characteristics of target pathogens and other elements that are considered in food safety management. Consequently, it is considered that many processes are overly conservative and could be reduced if there were tools available that would allow an objective quantitative assessment of hazards and their control, and include consideration of any uncertainty and variability in elements used to assess risk. Importantly, the survival characteristics of the emerging pathogens listed are comparable with those of existing pathogens, which means that there is no requirement to increase the thermal processes used for in-pack heated foods. This is not necessarily the case for foods using different controls, such as fermented foods, where the survival characteristics of some emerging pathogens have required a change to current processes.

Recent developments in food safety techniques and approaches have allowed food scientists to apply new tools in safety assessment. The progress made in risk assessment and modelling the fate of microorganisms in foods has enabled advisory bodies (such as Codex Alimentarius) and regulatory agencies (such as the European Food Safety Authority) to establish a framework for food safety linked to public health targets, with the ultimate aim of reducing the burden of foodborne disease. These recent developments (see CCFH, 2005; CCGP, 2005) have led to the introduction of concepts of Food Safety Objectives (FSOs), Performance Objectives (POs) and Performance Criteria (PC) that are intended to help in the process by translating risk management decisions on public health targets to measures that the food industry needs to implement on a daily basis. The application of these relatively new concepts is still being discussed at an international level and requires further development before they are adopted, but there are recent studies that have considered these concepts with the objective of setting new standards for heat-processed foods.

Recent studies by Membré *et al.* (2007) and Tuominen *et al.* (2007) have described how these new concepts may be applied. In the first of these studies, the authors outline an approach that considers the highest levels of infectious pathogens of concern to heat-processed poultry products and conclude that a safe heat process can be derived that is significantly lower than the standard process used currently, without compromising consumer safety. Our laboratory has also been applying modelling approaches to optimise thermal processes typically used for in-pack heated foods that are stored at chill (Membré *et al.*, 2006). Such approaches take account of levels of contamination, heat resistance characteristics of target microorganisms and in-pack variability of thermal kinetics during heat processing. These are key considerations for food manufacturers wishing to introduce changes to standard processes that have an established safety record. Such approaches will allow food producers to reduce costs in processing, linked to reduced energy consumption and shorter processing times, increasing yield and minimising other losses, for example, reduction in levels of natural components such as vitamins. A general tool for deriving FSOs has been published recently (Rieu *et al.*, 2007), considering prevalence of contamination of the product, average concentration of the hazard per serving and dispersion of the concentration among those servings.

The main food safety management tool that is used in the food industry is HACCP and this is likely to remain the case for many years to come. Two of the key steps in HACCP are hazard identification and continuous review of HACCP plans. It is vital that food producers keep a continuous lookout for emerging pathogens that may be relevant to their raw materials and processes. This includes pathogens that may be found on raw produce and materials, those that are transmitted by infected food handlers and those that are transmitted by other materials used in food manufacture, including water. Water is an important raw material and food processing aid that is used in cleaning, and cooling of in-pack heated foods. Many of the emerging pathogens described above are relevant to in-pack heated foods in one way or another and it is essential that food producers consider these in their HACCP plans and also in their product and process designs.

13.6 Future trends

From the description of emerging pathogens and the factors that have (or may have) led to their development, it is clear that there is often more than one factor involved. At the present time we are seeing a decrease in the number of cases of some common foodborne pathogens, such as salmonellae, in developed countries such as the US, UK, and other parts of Europe. This is encouraging and suggests that some disease prevention strategies may be beginning to take effect. Despite this, the incidence of foodborne illnesses and deaths caused by unsafe food are increasing. The genetic plasticity of microorganisms poses a serious threat for the future, and will undoubtedly lead to the emergence of novel infectious diseases. At the genetic and molecular level, the virulence traits of pathogens clearly show us that pathogenicity does not arise by slow adaptive evolution but rather by step changes.

Global data indicate that the epidemiology of foodborne diseases is changing and that an increased range of microorganisms and foods is causing foodborne illness. A better understanding of the distribution, epidemiology and threat posed by emerging and uncharacterised pathogens is needed because they can have rapid, and poorly controlled, global spread through the food chain and hence the population. The detection, reporting and characterisation of food- and waterborne illnesses play an important role in identifying the origins and incidence of disease when links can be made to the causative agents and the foods involved. This information needs to feed back to producers and regulators to ensure that product and process designs cover microbiological hazards that are realistically expected to be present.

Although it is generally the case that the list of emerging pathogens has changed little over the past 20 years, there have been some interesting studies pointing towards some other organisms that have the potential to become recognised as enteropathogens. Janda and Abbott (2006) used clinical, epidemiological, pathogenicity, veterinary, laboratory and molecular data to rank a number of Gram-negative organisms from five different families for evidence of enteropathogenicity. The organisms in decreasing order of evidence (from strong to weak) were *Providencia alcalifaciens*, *Vibrio mimicus*, *Campylobacter upsaliensis*, *Aeromonas hydrophila/caviae/veronii*, *Bacteroides fragilis*, *Citrobacter freundii*, *Grimontia* (*Vibrio*) *hollisae*, *Klebsiella pneumoniae*, *Vibrio fluvialis/furnissii*, *Campylobacter lari/hyointestinalis*, *Hafnia alvei/Arcobacter/ Laribacter* and *Aeromonas jandei/media/trota*. Such studies offer the potential to use data/information from different sources collectively, that will also make use of recently developed techniques such as DNA microarray chip technology. They also emphasise the importance of infectious intestinal disease studies and the ability to distinguish between different potential causative agents of gastrointestinal disease.

It is clear that some pathogens have emerged due, in part, to changes in practice in the food industry, including primary production and subsequent processing. The increasing demand for fresher, healthier foods will continue to drive food producers to develop and apply milder processes that can meet the changing needs of consumers and of society, where there will be further pressures in the areas of sustainability and reduced energy costs to minimise the impact on the environment. It is difficult to anticipate the potential effects of changes made to the ways that foods are grown, harvested, processed, handled and transported but there are some key learnings that can be made from the pathogens that have emerged in the past 30 years. Established pathogens will also continue to appear in novel food vehicles. The recent outbreak of botulism is the US linked to chilled carrot juice is a good example of this. The cause of illness was identified as proteolytic *C. botulinum*, which is not capable of growth at refrigeration temperatures, so temperature abuse of the product appears to be the root cause of this outbreak. It is interesting to note that this outbreak has led the Food and Drug Administration to modify its guidance for fresh vegetable juice manufacturers, advising that they introduce a secondary factor in products, that will control growth of proteolytic *C. botulinum* in case these chilled products are temperature abused (FDA, 2007).

With zoonotic pathogens, trying to control and reduce their levels and incidence in farm animals will inevitably lead to opportunities for other organisms to adapt and develop and therefore it is essential that there is continued close monitoring of the flora associated with these animals, to allow us to be forewarned of impending challenges. Unfortunately, there will always be a smaller number of foodborne agents and diseases that we know little or nothing about, as is the case with transmissible spongiform encephalopathy. These pose more difficult challenges and mean that we will be confronted with situations where we need new methods of isolation, detection and identification, improved epidemiology, surveillance and clinical diagnosis and perhaps most importantly, unequivocal evidence of cause and effect.

13.7 Sources of further information and advice

There are a number of excellent sources of information relevant to emerging foodborne pathogens. The pathogens referred to here are described in a number of texts and more detail on their ecology, epidemiology, methods of detection and control can be found in Motarjemi and Adams (2006) and other sources such as Doyle *et al.* (2001) and Blackburn and McClure (2000). Factors that may have led or contributed to emergence of pathogens, including more susceptible sub-populations, globalization of food supply, changes in practice in primary production and food processing, changes in consumer practice and habits, and changes in microorganisms and agents of foodborne disease are covered in more detail by Potter (2006). Current information on foodborne disease outbreaks can be found at national and international websites, such as those of *Morbidity and Mortality Weekly Reports* (*MMWR*, US), Health Protection Agency (HPA, UK), Basic Surveillance Network (BSN, Sweden), WHO (WHO Network-of-Networks on Foodborne Diseases), Eurosurveillance (Europe), Enter-net (Europe), Foodborne Viruses in Europe network (FBVE, Europe). Recent developments in food safety management concepts can be found on the Codex Alimentarius website, European Food Safety Authority website and also in ICMSF Book 7 (ICMSF, 2002).

13.8 References

Altekruse, S.F., Cohen, M.L., Swerdlow, D.L. (1997) Emerging foodborne diseases. *Emerging Infectious Diseases*, **3,** 285–293.

Armstrong, G.L., Hollingsworth, J., Morris, J.G. (1996) Emerging foodborne pathogens: *E. coli* O157:H7 as a model of entry of a new pathogen into the food supply of the developed world. *Epidemiologic Reviews*, **18**, 29–51.

Antunes, P., Reu, C., Sousa, J.C., Pestana, N. and Peixe, L. (2002) Incidence and susceptibility to antimicrobial agents of *Listeria* spp. and *Listeria monocytogenes* isolated from poultry carcasses in Porto, Portugal. *Journal of Food Protection* **65**, 1888–1893.

Blackburn, C. de W., McClure, P.J. (2000) *Foodborne Pathogens*, Woodhead Publishing Ltd, Cambridge, UK.

Brunder, W. and Karch, H. (2000) Genome plasticity in Enterobacteriaceae. *International Journal of Medical Microbiology*, **290**, 153–165.

Codex Committee on Food Hygiene (CCFH) (2005) Proposed draft principles and guidelines for the conduct of microbiological risk management. In *Report of the 37th Session of the Codex Committee on Food Hygiene, Buenos Aires*, Argentina, 14 to 19 March 2005. ALINORM 05/28/13, Appendix III.

Codex Committee on General Principles (CCGP) (2005) Proposed draft working principles for risk analysis for food safety. In *Report of the 21st Session of the Codex Committee on General Principles*, Paris, France, 11 to 15 April 2005. ALI NORM 05/28/33A.

Chapman, P.A., Siddons, C.A., Cerdan Malo, A.T. and Harkin, M.A. (1996) Lamb products as a potential source of *E. coli* O157. *The Veterinary Record*, October 26th, 427–428.

Cook, D.W., Ruple, A.D. (1992) Cold storage and mild heat treatment as processing aids to reduce numbers of *Vibrio vulnificus* in raw oysters. *Journal of Food Protection*, **55**, 985–989.

Corry, J.E.L. (1999) Detection by Cultural and Modern Techniques. In: *Encyclopedia of Food Microbiology*, eds. Robinson, R.K., Batt, C.A., Patel, P.D. 1, 341–347.

Curtis, L.M., Patrick, M., Blackburn, C.D. (1995) Survival of *Campylobacter jejuni* in foods and comparison with a predictive model. *Letters in Applied Microbiology,* **21** (3), 194–197, 1995.

Doyle, M. P. and Schoeni, J. L. (1987) Isolation of *Escherichia coli* O157:H7 from retail fresh meats and poultry. *Applied and Environmental Microbiology,* **53,** 2394–2396.

Doyle, M.E., Mazzotta, A.S., Wang, T, Wiseman, D.W., Scott, V.N. (2001) Heat resistance of *Listeria monocytogenes. Journal of Food Protection*, **64**, 410–429.

Doyle, M.P. (1994) The emergence of new agents of foodborne disease in the 1980s. *Food Research International*, 27, 219–226.

Doyle, M.P. (2000) *Food Microbiology: Fundamentals and Frontiers*, 2nd ed., ASM Press, Washington, D.C.

D'Sa, E., Harrison, M.A. (2005) Effect of pH, NaCl content, and temperature on growth and survival of *Arcobacter* spp. *Journal of Food Protection*, **68**, 18–25.

Falkenhorst, G., Krusell, L., Lisby, M., Madsen, S.B., Böttiger, B., Mølbak, K. (2005) Imported frozen raspberries cause a series of norovirus outbreaks in Denmark. *Eurosurveillance*, **10** (9), 2–4, http://www.eurosurveillance.org/ew/2005/050922.asp.

Farber, J. M., Peterkin, P.I. (1991) *Listeria monocytogenes*, a foodborne pathogen. *Microbiology Reviews*, **55,** 476–511.

Fayer, R. (1994) Effect of high temperature on infectivity of *Cryptosporidium parvum* oocysts in water. *Applied and Environmental Microbiology*, **60**, 2732–2735.

Food and Drug Administration (FDA) (2007) *Guidance for Industry on Refrigerated Carrot Juice and other Refrigerated Low-acid Juices*. http://www.cfsan.fda.gov/~dms/juicgu15.html.

Gahan, C.G.M. (2006) Specific immune mechanisms of defence against foodborne pathogens. In F*ood Consumption and Disease Risk*, ed. Potter, M., Woodhead Publishing Ltd, Cambridge, UK, pp. 214–232.

Green-Johnson, J.M. (2006) Nonspecific host defences against foodborne pathogens. In *Food Consumption and Disease Risk*, ed. Potter, M., Woodhead Publishing Ltd, Cambridge, UK, pp. 183–213.

Hilton, C.L., Mackey, B.M., Hragreaves, A.J., Forsythe, S.J. (2001) The recovery of *Arcobacter butzleri* NCTC 12481 from various temperature treatments. *Journal of Applied Microbiology*, **91**, 929–932.

Hudson, W.R. and Mead, G.C. (1989*) Listeria* contamination at a poultry processing plant. *Letters in Applied Microbiology*, **9**, 211–214.

Institute of Medicine (IOM) (1992) *Emerging infections: Microbial threats to health in the United States*. Washington, DC: National Academy Press..

International Commission on Microbiological Specifications for Foods (ICMSF) (2002) *Microorganisms in Foods 7, Microbiological Testing in Food Safety Management.* Kluwer Academic/Plenum, London, UK.

Jackson, J.K., Murphree, R.L., Tamplin, M.L. (1997) Evidence that mortality from *Vibrio vulnificus* infection results from single strains among heterogeneous populations in shellfish. *Journal of Clinical Microbiology*, **35**, 2098–2101.

Janda, J.M., Abbott, S.L. (2006) New Gram-negative enteropathogens: Fact or fancy? *Reviews in Medical Microbiology*, **17**, 27–37.

Jorgensen, F., Bailey, R., Williams, S., Henderson, P., Wareing, D.R.A., Bolton, F.J., Frost, J.A., Ward, L., Humphrey, T.J. (2002) Prevalence and numbers of *Salmonella* and *Campylobacter* spp. on raw, whole chickens in relation to sampling methods. *International Journal of Food Microbiology*, **76**, 151–164.

Kobayashi, J. (2006) Enhanced susceptibility to foodborne infections and disease due to underlying illnesses and pregnancy. In *Food Consumption and Disease Risk*, ed. Potter, M., Woodhead Publishing Ltd, Cambridge, UK, pp. 233–250..

Koopmans, M., Duiver, E. (2004) Foodborne viruses: An emerging problem. *International Journal of Food Microbiology*, **90**, 23–41.

Kramer, J.M., Frost, J.A, Bolton, F.J., Wareing, D.R.A. (2000) *Campylobacter* contamination of raw meat and poultry at retail sale: Identification of multiple types and comparison with isolates from human infection. *Journal of Food Protection*, **63**, 1654–1659.

Kroneman, A., Vennema, H., Harris, J., Reuter, G., von Bonsdorff, C.-H., Hedlund, K.-O.,

Vainio, K., Jackson, V., Pothier, P., Koch, J., Schreier, E., Böttiger, B., Koopmans, M. (2006) Increase in norovirus activity reported in Europe. *Eurosurveillance*, **11** (12), 1–8, http://www.eurosurveillance.org/ew/2006/061214.asp.

Lado, B.H. and Yousef, A.E. (2007) Characteristics of *Listeria monocytogenes* important to food processors. In *Listeria, Listeriosis and Food Safety*, eds. Ryser, E.T. and Marth, E.M., 3rd edition, CRC Press, London.

Lee, W.H., Vanderzant, C., Stern, N. (1981) The occurrence of *Yersinia enterocolitica* in foods. In: *Yersinia enterocolitica*, ed. Bottone, E.J., CRC Press, Boca Raton, FL, 161–171.

Lewis, S.J. and Corry, J.E.L. (1991) Survey of the incidence of *Listeria monocytogenes* and other *Listeria* spp. in experimentally irradiated and in matched unirradiated raw chickens. *International Journal of Food Microbiology* **12**, 257–262.

Mackey, B.M. and Bratchell, N. (1989) The heat resistance of *Listeria monocytogenes. Letters in Applied Microbiology*, **9**(3), 89–94.

McClure, P.J. (2000) Microbiological hazard identification in the meat industry. In: Brown, M (ed.) *HACCP in the Meat Industry*, Woodhead Publishing, Cambridge, UK, 157–176.

Mead, P.S., Slutsker, L., Dietz, V., McCaig, L.F., Bresee, J.S., Shapiro, C., Griffin, P.M., Tauxe, R.V. (1999) Food-related illness and death in the United States. *Emerging Infectious Disease*, **5** (5) 607–625.

Membré, J.-M., Amezquita, A., Bassett, J., Giavedoni, P., Blackburn, C. De-W., Gorris, L.G.M. (2006) A probabilistic modeling approach in thermal inactivation: Estimation of postprocess *Bacillus cereus* spore prevalence and concentration. *Journal of Food Protection*, **69,** 118–129.

Membré, J.-M., Bassett, J., Gorris, L. (2007) Applying the food safety objective and related standards to thermal inactivation of *Salmonella* in poultry meat. *Journal of Food Protection*, **70**, 163–171.

Meng, J., Doyle, M.P. (1998) Emerging and evolving microbial foodborne pathogens. *Bulletin Institute Pasteur*, **96**, 151–164.

Millard, J., Appleton, H., Parry, J.V. (1987) Studies on heat inactivation of hepatitis A virus with special reference to shellfish. *Epidemiology and Infections*, **98**, 397–414.

Motarjemi, Y., Adams, M. (2006) *Emerging Foodborne Pathogens*, Woodhead Publishing, Cambridge, UK.

Motes, M.L., Depaola, A., Cook, D.W., Veazey, J.E., Hunsucher, J.C., Garthright, W.E., Blodgett, R.J., Chirtel, S.J. (1998) Influence of water temperature and salinity on *Vibrio vulnificus* in northern Gulf and Atlantic coast oysters (*Crasostrea virginica*). *Applied and Environmental Microbiology*, **64**, 1459–1465.

Nair, G.B., Faruque, S.M., Sack, D.A. (2006) Vibrios. In *Emerging Foodborne Pathogens*, eds Motarjemi, Y., Adams, M., Woodhead Publishing, Cambridge, UK, pp. 332–372.

Nesbakken, T. (1988) Enumeration of *Yersinia enterocolitica* O:3 from the porcine oral cavity, and its occurrence on cut surfaces of pig carcasses and the environment in a slaughterhouse. *International Journal of Food Microbiology*, **8**, 287–293.

Nesbakken, T. (1992) *Epidemiological and food hygienic aspects of* Yersinia enterocolitica *with special reference to the pig as a suspected source of infection*. Thesis, Norwegian College of Veterinary Medicine, Oslo, Norway.

O'Bryan, C.A., Crandall, P.G., Martin, E.M., Griffis, C.L. and Johnson, M.G, (2006) Heat resistance of *Salmonella* spp., *Listeria monocytogenes*, *Escherichia coli* O157:H7, and *Listeria innocua* M1, a potential surrogate for *Listeria monocytogenes*, in meat and poultry: A review. *Journal of Food Science*, 71 (3), R23–30.

Potter, M.E., Motarjemi, Y., Käferstein, F.K. (1997). Emerging foodborne diseases. *World Health*, January–February:16–17.

Potter, M.E. (2006) *Food Consumption and Disease Risk*, Woodhead Publishing Ltd, Cambridge, UK.

Rautelin, H. and Hanninen, M.L. (2000) *Campylobacters*: The most common bacterial enteropathogens in the Nordic countries. *Annals of Medicine*, 32, 440–445.

Rieu, E., Duhem, K., Vindel, E., Sanaa, M. (2007) Food safety objectives should integrate the variability of the concentration of pathogen. *Risk Analysis*, **27**, 373–386.

Rorvik, L.M., Aase, B., Alvestad, T. and Caugant, D.A. (2003) Molecular epidemiological survey of *Listeria monocytogenes* in broilers and poultry products. *Journal of Applied Microbiology*, **94**, 633–640.

Schorr, D., Schmid, H., I., Baumgartner, A., Vorkauf, H., Burnens, A. (1994) Risk factors for *Campylobacter* enteritidis in Switzerland. *Zentralblatt fur Hygiene und Umweltmedizin*, **196**, 327–337.

Sekla, L., Milley, D., Stackiw, W., Sisler, J., Drew, J. and Sargent, D. (1990) Verotoxin-producing *Escherichia coli* in ground beef – Manitoba. *Canadian Diseases Weekly Report* **16,** 103–105.

Shane, S.M. (1992) The significance of *Campylobacter jejuni* infection in poultry: A review. *Avian Pathology*, **21**, 189–213.

Sivapalasingam, S., Friedman, C.R., Cohen, L., Tauxe, R.V. (2004) Fresh produce: A growing cause of outbreaks of foodborne illness in the United States, 1973 through 1997. *Journal of Food Protection*, **67**, 2342–2353.

Slutsker, L., Altekruse, S.F., Swerdlow, D.L. (1998) Foodborne diseases: Emerging pathogens and trends. *Emerging Infectious Diseases*, **12**, 199–216.

Smith, J.L., Fratamico, P.M. (1995) Factors involved in the emergence and persistence of food-borne diseases. *Journal of Food Protection*, **58**, 696–708.

Solomon, E.B. and Hoover, D.G. (1999) *Campylobacter jejuni*: A bacterial paradox. *Journal of Food Safety*, **19**, 121–136.

Sörqvist, S. and Danielsson-Tham, M.L.(1990) Survival of *Campylobacter*, *Salmonella* and *Yersinia* spp. in scalding water used at pig slaughter. *Fleischwirtschaft*, **70**, 1460–1466.

Stringer, S.C., George, S.M. and Peck, M.W. (2000) Thermal inactivation of *Escherichia coli* O157:H7. *Journal of Applied Microbiology* **88** 79S–89S, Suppl. S.

Tamplin, M.L. (1994) The ecology of *Vibrio vulnificus*. In *Proceedings of the 1994* Vibrio vulnificus *Workshop*, eds Watkins, W., McCarthy, S., Office of Seafood, Washington, D.C.

Tauxe, R.V. (2002) Emerging foodborne pathogens. *International Journal of Food Microbiology*, **78**, 31–41.

Teuber, M. (1999) Spread of antibiotic resistance with food-borne pathogens. *Cellular and Molecular Life Sciences*, **56**, 755–763.

Todd, E.C.D., Szabo, R.A., Peterkin, P., Sharpe, A.N., Parrington, L., Bundle, D., Gidney, M.A.J. and Perry, M.B. (1988) Rapid hydrophobic grid membrane filter-enzyme-labelled antibody procedure for identification and enumeration of *Escherichia coli* O157 in foods. *Applied and Environmental Microbiology* **54**, 2536–2540.

Tuominen, P., Ranta, J., Maijala, R. (2007) Studying the effects of POs and MCs on the *Salmonella* ALOP with a quantitative assessment model for beef production. *International Journal of Food Microbiology*, **118**, 35–51.

Wain, J., House, D., Pickard, D., Dougan, G., Frankel, G. Acquisition of virulence-associated factors by the enteric pathogens *Escherichia coli* and *Salmonella enterica*. *Philosophical Transactions of the Royal Society London* B, **356**, 1027–1034.

Wassenaar, T.M., Geilhausen, B. and Newell, D.G. (1998) Evidence of genome instability in *Campylobacter jejuni* isolated from poultry. *Applied and Environmental Microbiology*, **64**, 1816–1821.

West, P.A. (1989) The human pathogenic vibrios – A public health update with environmental perspectives. *Epidemiology and Infection*, **103**, 1–34.

Wilson, I.G. (2002) *Salmonella* and *Campylobacter* contamination of raw retail chickens from different producers: A six-year survey. *Epidemiology and Infection*, **129** (3): 635–645.

World Health Organisation (WHO) (1983) *Yersiniosis: Report on a WHO meeting*, Paris, 1981, WHO Regional Office for Europe, Euro Reports and Studies 60, Copenhagen, pp. 31.

World Health Organisation (WHO) (1995) *Report of the WHO Consultation on Emerging Foodborne Diseases*, Berlin.

World Health Organisation/Food and Agricultural Organisation of the United Nations (WHO/FAO) (2002) *Risk assessment of* Salmonella *in eggs and broiler chickens*. http://www.who.int/foodsafety/publications/micro/en/salmonella.pdf

Woolhouse, M.E.J, Dye, C. (2001) Population biology of emerging and re-emerging pathogens. *Philosophical Transactions of the Royal Society London* B, **356**, 981–982.

14

Foodborne clostridia and the safety of in-pack preserved foods

S. C. Stringer and M. W. Peck, Institute of Food Research, UK

14.1 Introduction

14.1.1 Clostridia of public health significance

The clostridia are a group of obligatory anaerobic spore forming, large rod-shaped bacteria that have a strictly fermentative metabolism. Most are saprophytes but a small number are pathogenic for humans. *Clostridium tetani* grows in wounds and is responsible for tetanus, *Clostridium difficile* is contracted by person-to-person spread and causes symptoms that can vary from mild diarrhoea to colitis and peritonitis, while *C. botulinum* and *C. perfringens* are the two clostridia most commonly associated with foodborne illness. Foodborne botulism is a severe but rare disease while *C. perfringens* is responsible for a generally milder disease but is the second most common cause of foodborne illness and associated death in the UK (Adak *et al.*, 2002). One major feature of clostridia is their ability to form endospores. This dormant resting form of the cell has greatly increased resistance to environmental stresses such as heat, chemicals and dehydration, which makes them an important target in food processing.

Spores of proteolytic *C. botulinum* are the most heat resistant produced by foodborne pathogens. Thermal processes for in-pack foods stored at ambient temperature are therefore scrutinised with respect to the effect they would have on spores of this organism. Non-proteolytic *C. botulinum* is the *Clostridium* of greatest concern in products intended for chilled storage as neither proteolytic *C. botulinum* nor *C. perfringens* grow below 10 °C. *C. perfringens* tends to cause poisoning when foods have been cooled too slowly or held warm for prolonged periods. With these comments in mind, this chapter will focus primarily on the hazards presented by proteolytic *C. botulinum* and by non-proteolytic *C. botulinum* to the safety of in-pack processed foods.

14.1.2 In-pack processing

The ability of *C. botulinum* to produce toxin in foods can be controlled by destroying any spores present or by controlling environmental conditions so they are not conducive to growth. Successful processing relies on understanding the lethal effect of any processes applied and its interactions with the conditions within the product to support subsequent growth. Where processes are designed to destroy spores in food that could support growth, it is also important to prevent post-process contamination. This should be achieved through the implementation, verification and validation of a hazard analysis and critical control points (HACCP) system.

14.2 Characteristics of *Clostridium botulinum* and foodborne botulism

14.2.1 *Clostridium botulinum* and other neurotoxin-producing clostridia

Clostridium botulinum is a heterogeneous species consisting of four physiologically and genetically distinct groups of bacteria that share the ability to produce botulinum neurotoxin (BoNT). Occasional strains of two other species, *Clostridium baratii* and *Clostridium butyricum*, are also capable of forming BoNT, and they have been associated with foodborne botulism (Anniballi *et al.*, 2002; Harvey *et al.*, 2002). However, Group I (proteolytic) *C. botulinum* and Group II (non-proteolytic) *C. botulinum* are responsible for nearly all cases of foodborne botulism (Lund and Peck, 2000). These two organisms differ physiology (Table 14.1), and present different hazards in foods. Proteolytic *C. botulinum* produces spores of high heat resistance, but it is not able to grow at refrigeration temperatures. It is therefore mainly of concern in heat-treated, ambient stable, low acid foods. Canning processes for low-acid foods are designed to ensure that spores of this organism are inactivated. Spores of non-proteolytic *C. botulinum* are of moderate heat resistance, but this organism can multiply and form neurotoxin at temperatures as low as 3.0 °C. It is therefore of most concern in pasteurised, refrigerated, low acid food.

14.2.2 Neurotoxin and characteristics of botulism

Proteolytic *C. botulinum* and non-proteolytic *C. botulinum* are a severe hazard in food as they have the ability to produce the most poisonous substances known, botulinum neurotoxin (BoNT), the causative agent of botulism. Botulism is a rare,

Table 14.1 Characteristics of the four clostridia associated with foodborne botulism (compiled from: Morton *et al.*, 1990; Anniballi *et al.*, 2002; Peck and Stringer, 2005 and Stringer and Peck, unpublished results)

Organism	Spore heat resistance ($D_{100\,°C}$)	Minimum growth temperature	Minimum growth pH
Proteolytic *Clostridium botulinum*	>15 min	10–12 °C	4.6
Non-proteolytic *Clostridium botulinum*	<0.1 min	2.5–3.0 °C	5.0
Neurotoxigenic *Clostridium baratii*		10–15 °C	
Neurotoxigenic *Clostridium butyricum**	<0.1 min	12 °C	4.8

*Non-neurotoxigenic strains form spores of greater heat resistance ($D_{100\,°C}$ = 4.7 min) and are capable of growth at pH 4.4.

but severe disease, with an estimated oral lethal dose of 30 ng of neurotoxin (Peck and Stringer, 2005). Foodborne botulism is an intoxication resulting from consumption of food in which neurotoxin-producing clostridia have grown. The consumption of as little as 0.1g of contaminated food can result in botulism (Lund and Peck, 2000). There are seven botulinum neurotoxins (A to G), originally designated according to their antigenic response. More recently, the amino acid sequence and mode of action of all the neurotoxins have been established (Dodds and Austin, 1997; Lund and Peck, 2000). Each toxin cleaves at a specific site on proteins involved in the release of the neurotransmitter, acetylcholine. This prevents the passage of nerve impulses to the associated muscle, thus leading to flaccid paralysis. Symptoms usually occur within 12–36 h of toxin consumption and may include blurred vision, dysphagia (difficulty swallowing), generalised weakness, nausea/vomiting, dysphonia (difficulty speaking), dizziness/vertigo, and muscle weakness. Flaccid paralysis of the respiratory muscles can result in death if respiration is not maintained artificially.

14.2.3 Incidence of spores of *C. botulinum* in foods and the environment

Spores of both proteolytic and non-proteolytic *C. botulinum* are found in many environments including soils, fresh-water and marine sediments and the gastrointestinal tract of animals. They have been isolated from a vast range of foods including prepared fish, shellfish, meat, poultry, honey, dairy products, fruit and vegetables (Dodds, 1992; Dodds and Austin, 1997; Lund and Peck, 2000; Fach *et al.*, 2002; Carlin *et al.*, 2004). The food survey data, although not extensive, indicates a pattern of a low background level of contamination, with the occasional more heavily contaminated sample. This widespread distribution means food raw materials cannot be guaranteed to be free of spores. Foods which are, or can become, anaerobic may allow growth of *C. botulinum* and must therefore be subjected to treatments that destroy spores, or stored under conditions that prevent growth and toxin formation.

14.2.4 Outbreaks of foodborne botulism associated with in-pack processed foods

In Europe, more than 2500 cases of foodborne botulism were reported in 1999/2000 (Peck, 2006). The frequency of the disease varies greatly from country to country, depending to a large extent on the traditions of food preparation. For example, Poland historically has reported very high rates of botulism; between 1984 and 1987 there were 1301 outbreaks (Hauschild, 1992). This high number of incidences is thought to relate to high levels of home bottling of foods, practised in the difficult economic conditions prevailing at that time. The incidence of botulism in Poland has fallen considerably in recent years, with 53 cases being registered in 2004 (Czerwinski *et al.*, 2006). Many other countries have lower but still significant rates of foodborne botulism. For example, over the past 20 years, approximately 20–40 cases have been reported annually in Italy, Germany, France and USA (Peck, 2006). Again, many of these are associated with home preservation techniques. The proportion of cases related to commercial rather than home-preserved products depends largely on prevalence of botulism from home prepared foods. In the UK, which has a low level of botulism, the majority of cases have been associated with commercial food products. The proportion of outbreaks associated with commercial products also varies with the food type. The system for reporting botulism in Poland is more efficient than many countries, and data have been gathered on the number of cases of botulism acquired from home-bottled or industrially canned products. Data for 1993–2004 are shown in Table 14.2. There were 351 cases of botulism associated with red meat thermally processed in cans or glass jars, and commercial products accounted for 25% of these cases. Home bottling of fish is less prevalent than bottling of meat. In 1993–2004 there were 124

Table 14.2 Cases of botulism associated with thermally processed red meats and fish in Poland, 1993–2004, and the percentage of cases associated with commercially prepared foods

Year	Red meat		Fish		Reference
	Number of cases	% cases from commercial foods	Number of cases	% cases from commercial foods	
1993	45	31	12	75	(Przybylska, 1995)
1994	32	34	23	61	(Przybylska, 1996)
1995	38	18	16	?	(Przybylska, 1997)
1996	35	20	9	?	(Przybylska, 1998)
1997	32	50	5	?	(Przybylska, 1999)
1998	26	15	9	100	(Przybylska, 2000)
1999	33	15	10	100	(Przybylska, 2001)
2000	15	13	9	100	(Przybylska, 2002)
2001	21	10	7	100	(Przybylska, 2003)
2002	30	23	9	22	(Przybylska, 2004)
2003	32	19	10	30	(Czerwinski *et al.*, 2005)
2004	12	58	5	20	(Czerwinski *et al.*, 2006)
Total	351	25	124	68	

cases of botulism associated with thermally preserved fish and 68% of the cases, where specified, were associated with commercial products (Table 14.2). Cases of botulism have great medical and economic impact. It is therefore important that new food processes continue to be developed safely, and that the foodborne botulism hazard is minimised.

Commercial shelf-stable foods

Some well-documented examples of botulism outbreaks associated with commercial, shelf-stable, in-pack preserved foods are shown in Table 14.3. Cases of botulism associated with shelf-stable foods are usually caused either by failure to implement known control measures, poor process control or recontamination following damage to the container. These causes are illustrated by the outbreaks in Denmark 2003, Italy 1998 and South Africa 2002, respectively. In the case of botulism associated with eating garlic in chilli oil that occurred in Denmark in 2003 (Krusell and Lohse, 2003), jars containing cloves of garlic under oil, pH 4.7, were heated to a temperature of 83–85 °C for an unspecified period of time, and assigned a two-year shelf-life at ambient temperature. It was also reported that 134 jars from the implicated lot were rejected at the time of production but no further action was taken. In Italy in 1998, a case of botulism arose following consumption of a jar of a local organic vegetable soup (Aureli *et al.*, 1998). The soup had been produced by a small factory where jars were sterilised for 35 minutes in an autoclave at 121 °C whatever their size, and the functioning of the autoclave was neither verified nor documented. It is thought that the fatal cases of botulism that occurred in South Africa in 2002 resulted from corrosion damage, allowing entry of environmental bacteria into a can of pilchards in tomato sauce (Frean *et al.*, 2004).

Commercial foods intended for refrigerated storage

Foods that are intended to be stored chilled (rather than at ambient), having low acid, high water activity and which have not been subjected to a sterilisation heat treatment have also been responsible for outbreaks of botulism. In these cases temperature abuse is nearly always a contributing factor. Examples of outbreaks of botulism associated with refrigerated in-pack preserved foods intended for refrigerated storage are shown in Table 14.4. In many cases, botulism has been as the result of toxin produced by proteolytic *C. botulinum*. Proteolytic *C. botulinum* does not grow at 10 °C or below, so this indicates that gross temperature abuse has occurred. In other cases, foods have been consumed after being stored for longer than recommended. In 2004 a German patient contracted botulism after eating vacuum-packed smoked salmon that was three days after the use-by date (Dressler, 2005). It is also important to ensure the quality of the ingredients used in processed foods. The largest recorded outbreak of botulism in the UK occurred in 1989 and was associated with the consumption of hazelnut yoghurt (O'Mahony *et al.*, 1990). This product was intended for refrigerated storage and labelled to be consumed within two days of purchase. However, the product was toxic from manufacture as it was made using toxic hazelnut conserve that had not been processed adequately before ambient storage.

Table 14.3 Examples of outbreaks of botulism associated with industrially produced in-pack preserved foods intended for ambient stable storage

Outbreak	Product	*C. botulinum* toxin type	Contributory factors	Reference
1973 USA	Canned peppers in oil	Type B*		(Barker *et al.*, 1977)
1978 UK	Canned salmon	Non proteolytic E	Post processing recontamination	(Ball *et al.*, 1979)
1986 Taiwan	Peanuts in water	Not stated	Under processing. Unlicensed factory.	(Chou *et al.*, 1988)
1989 UK	Hazelnut conserve	Proteolytic B	Unsafe process	(O'Mahony *et al.*, 1990)
1993 USA	Aseptically canned cheese sauce	Proteolytic A	Possibly contaminated after opening	(Townes *et al.*, 1996)
1993 Italy	Bottled aubergine in oil	Proteolytic B	Unsafe process	(D'Argnio *et al.*, 1995)
1997 Italy	Roast mushrooms in oil	Not stated	Not stated	(Therre, 1999)
1997 Italy	Canned truffle cream	Type B*	Not stated	(Therre, 1999)
1997 Spain	Canned asparagus	Not stated	Not stated	(Therre, 1999)
1998 Italy	Organic vegetable soup	Proteolytic A	Under processing. No quality control	(Aureli *et al.*, 1998)
1998 Japan	Bottled green olives in brine	Not stated	Not stated	(Matsuki and Goto, 1998)
2002 South Africa	Tinned pilchards in tomato sauce	Proteolytic A	Post processing recontamination	(Frean *et al.*, 2004)
2003 South Korea	Canned sausage	Not stated	Not stated	(Anon, 2003)
2003 Denmark	Garlic in chilli oil	Type B	Unsafe process	(Krusell and Lohse, 2003)
2007 Spain	Bottled artichokes	Not stated	Not stated	(Anon, 2007a)
2007 USA	Hot dog chilli sauce	Not stated	Under processing	(Anon, 2007b)

*not stated whether proteolytic or non-proteolytic strain.

Table 14.4 Examples of outbreaks of botulism associated with commercially produced in-pack preserved foods intended for refrigerated storage

Outbreak	Product	*C. botulinum* toxin type	Contributory factors	Reference
1970 Germany	Vacuum-packed smoked trout	Non proteolytic E	?	(Baumgart, 1970)
1982Madagascar	Commercial pork sausage	Non proteolytic E	Inadequate preservation	(Viscens *et al.*, 1985)
1985 Canada	Bottled garlic-in-oil	Proteolytic B	Temperature abuse, no secondary protection	(St Louis *et al.*, 1988)
1989 USA	Garlic in oil	Proteolytic A	Temperature abuse	(Morse *et al.*, 1990)
1994 USA	Commercial clam chowder	Proteolytic A	Temperature abuse, no secondary protection	(Sobel *et al.*, 2004)
1994 USA	Commercial black bean dip	Proteolytic A	Temperature abuse, no secondary protection	(Sobel *et al.*, 2004)
1996 Italy	Mascarpone cheese	Proteolytic A	Presumed temperature abuse	(Aureli *et al.*, 2000)
1997 Germany	Hot-smoked vacuum-packed fish	Non proteolytic E	Suspected temperature abuse	(Korkeala *et al.*, 1998)
1998 Argentina	Matambre (Meat roll)	Proteolytic A	Temperature abuse	(Villar *et al.*, 1999)
1998 Germany	Vacuum-packed smoked trout	Non proteolytic E	?	(Therre, 1999)
1999 Japan	Boil-in-the bag curry	Proteolytic A(B)	Temperature abuse	(Kobayashi *et al.*, 1999)
1999 France	Commercial chilled fish soup	Proteolytic A	Temperature abuse.	(Carlier *et al.*, 2001)
2004 UK	Hummus	Not stated	Temperature abuse.	(McLauchlin *et al.*, 2006)
2004 Germany	Vacuum-packed smoked salmon	Non proteolytic E	Consumed after use-by date	(Dressler, 2005)
2006 Canada/USA	Carrot juice	Proteolytic A	Presumed temperature abuse	(Shuler *et al.*, 2006)
2006 Finland	Vacuum-packed smoked fish	Non proteolytic E	Temperature abuse suggested	(Lindstrom *et al.*, 2006)

14.3 Control of foodborne botulism hazard presented by proteolytic *Clostridium botulinum*

As foodborne botulism is caused by the consumption of preformed toxin in food, it can be prevented by either subjecting foods to sufficient heat treatment to destroy spores or by ensuring that the environment in the food and storage conditions are not conducive to growth of *C. botulinum*.

14.3.1 Thermal inactivation of spores

The ability of *C. botulinum* to form heat-resistant spores means it is an important target for thermal inactivation in low-acid canned foods. Studies on the heat resistance of spores of proteolytic *C. botulinum* began in the early part of the twentieth century in response to outbreaks of botulism associated with the consumption of inadequately heat-treated foods. Esty and Meyer (1922) studied the heat resistance of 109 strains of *C. botulinum* from a variety of sources in a range of media. They showed that the measured heat resistance of spores depended on many factors, such as the strain tested and the composition, pH and water activity of the heading menstruum and the recovery media. These early studies formed the basis of standards used in the canning industry today. It was decided that the heat treatments for low-acid, ambient stable foods, which do not inhibit growth should reduce the population of *C. botulinum* by a factor of 10^{12} (a 12D – 12 decima process) (Stumbo *et al.*, 1975). The most heat resistant spores had a $D_{121\,°C}$ value of 0.21 min, so a 12D process would be achieved in 2.52 min at 121 °C. This is rounded up to three minutes for commercial practice. Commercial sterility is the process required to prevent the growth of pathogens and spoilage organisms in a product under normal conditions of storage or distribution. It is usually expressed as the sterilisation process equivalent time in min at 121 °C (F_0 value). Many foods are processed to more than the F_0=3 required for a 12D reduction of spores of the pathogen proteolytic *C. botulinum*, as the processes are designed to destroy the spores of mesophilic and thermophilic spoilage organisms that can be more heat resistant. When calculating equivalent lethality, the *z*-value of proteolytic *C. botulinum* spores is usually taken to be 10 °C (Lund and Peck, 2000).

14.3.2 Conditions preventing growth and toxin formation

The limits for growth of proteolytic *C. botulinum* have been established for many individual factors under otherwise optimum conditions. Refrigeration is a very effective way of preventing growth of proteolytic *C. botulinum*; the minimum temperature at which growth and toxin production occurs is within the range of 10° to 12 °C. Growth and toxin formation have been described at 12 °C or 12.5 °C in three to four weeks, while attempts to establish growth at 10 °C or below have been unsuccessful (e.g. Ohye and Scott, 1953; Smelt and Haas, 1978; Peck and Stringer, 2005). Although Tanner and Oglesby (1936) reported growth from vegetative inocula at 10 °C, this observation was marred by a temporary rise in the incubator temperature to 20 °C. While refrigerated storage will prevent growth, there is concern that proteolytic *C. botulinum* may grow and form toxin in products that have been manufactured using a sub-lethal heat process if the product is temperature abused. Indeed, temperature abuse has caused several outbreaks of foodborne botulism associated with proteolytic *C. botulinum* in foods intended for refrigerated storage (Table 14.4).

The dividing line between acid and low-acid foods is at pH 4.5–4.6, as it is generally accepted that growth is prevented at or below this pH. Outbreaks of botulism have occurred with acid products, such as home-canned pears pH 3.86 (Meyer and Gunnison, 1929) and tomato juice pH 4.2 (Odlaug and Pflug, 1978), but these cases have been associated with the growth of spoilage organisms in the product that are thought to have raised the pH in localised microenvironments. It is important that processes for low-acid canned foods should be sufficient to prevent spoilage activity that could change the conditions in the container.

Concentrations of NaCl above 10% will prevent growth, and the minimum water activity permitting growth is 0.93 and 0.94 with glycerol and NaCl, respectively, as the humectants (Hauschild, 1989). *C. botulinum* is a strict anaerobe and both the absence of oxygen and a low redox potential are necessary for growth to initiate from low inocula. This does not mean that *C. botulinum* cannot grow in food under a headspace containing oxygen, as conditions suitable for growth (i.e. absence of oxygen and a low redox potential) can occur inside foods (Snyder, 1996). The redox potential in a food will be influenced by the chemical composition of that food and growth by spoilage organisms as well as the gaseous atmosphere (Lund, 1993).

14.3.3 Using combinations of preservative factors that prevent growth

Some foods may be unacceptable organoleptically if produced using a single preservative factor, such as thermal sterilisation, to ensure their safety. Instead, preservation often relies on combinations of preservative factors, each at sub-inhibitory concentrations, that together reduce the probability of growth by an appropriate factor yet maintain product quality. Examples of where sub F_0 thermal processes are combined with additional preservative factors to create safe products include the acidification of canned vegetables and fruit and the use of nitrite and salt in canned cured meats.

The additional protection against proteolytic *C. botulinum* provided to refrigerated products by sub-lethal cooking processes may be marginal. For example, spores of a mixture of strains of proteolytic *C. botulinum* in meat slurry heated at 95 °C for 23 min produced toxin within 28 days when subsequently incubated at 12 °C, compared to 21 days for unheated spores. The times to toxin were the same for both heated and unheated spores incubated at 16 or 25 °C (Peck and Stringer, 2005).

Non-thermal factors can also be combined to prevent growth of proteolytic *C. botulinum*. The effect of combinations of pH and NaCl concentration or water activity on time to growth has been determined in laboratory medium (Table 14.5),

Table 14.5 The combined effect of NaCl concentration and pH on the growth/no growth boundary of proteolytic *C. botulinum* at 20 °C and 30 °C in laboratory medium (from data in: Ohye and Christian, 1967; Baird-Parker and Freame, 1967; Richardson and Peck, unpublished results)

NaCl concentration (%) [Water activity]	Temperature	pH 5	pH 6	pH 7
0 [0.997]	30 °C	Growth	Growth	Growth
	20 °C	Growth	Growth	Growth
1.2 [0.990]	30 °C	Growth	Growth	Growth
	20 °C	No growth	Growth	Growth
3.0 [0.980]	30 °C	Growth	Growth	Growth
	20 °C	No growth	Growth	Growth
4.8 [0.970]	30 °C	No growth	Growth	Growth
	20 °C	No growth	Growth	Growth
6.5 [0.960]	30 °C	No growth	Growth	Growth
	20 °C	No growth	No growth	No growth
8.2 [0.950]	30 °C	No growth	No growth	Growth
	20 °C	No growth	No growth	No growth
9.8 [0.940]	30 °C	No growth	No growth	No growth
	20 °C	No growth	No growth	No growth

vacuum-packed potatoes (Dodds, 1989) and in lumpfish caviar (Hauschild and Hilsheimer, 1979).

It is impossible to measure all the different combinations of factors that could be used to prevent growth. Instead, it is feasible to measure the behaviour associated with selected combinations of conditions and then create mathematical descriptions of the results that allow interpolation to all combinations of conditions. Several predictive models have been developed for proteolytic *C. botulinum* (Baker and Genigeorgis, 1992; Lund, 1993; Dodds, 1993; Lund and Peck, 2000).

14.3.4 Control of infant botulism hazard presented by proteolytic *Clostridium botulinum* in in-pack processed foods

Unlike foodborne botulism, infant botulism is an infection. Spores are ingested by the infant and their immature intestinal flora are unable to prevent colonisation by proteolytic *C. botulinum* (and also neurotoxigenic strains of *C. baratii* and *C. butyricum*). Toxin is produced by cells growing in the gut. A disease similar to infant botulism also very rarely affects adults, and occurs when competing bacteria in the normal intestinal flora have been suppressed (e.g. by antibiotic treatment).

While some canned baby foods will receive the $F_0=3$ heat treatment, other foods are likely to receive a lower heat treatment in order to preserve the quality and nutritional content of the product. The lower heat treatments used in the production of acidic foods such as fruit purees, and pasteurisation processes used for in-pack chilled products, will not guarantee destruction of all proteolytic *C. botulinum* spores. While low pH and refrigerated storage control the foodborne hazard by preventing growth during product storage, viable spores may remain in the food and present a hazard. Concern has been voiced about the possibility of viable spores remaining in in-pack processed food targeted at infants of less than 12 months of age (ACMSF, 2006b). To date, there have been no reported cases of infant botulism associated with processed in-pack foods. The only sources that have been positively identified are honey and general environmental contamination (e.g. soil, dust), although it should be noted that there was a possible link between a case of infant botulism and consumption of powdered infant milk (Brett *et al.*, 2005). It is estimated that between 10 and 100 spores are sufficient to bring about infection.

14.4 Control of foodborne botulism hazard presented by non-proteolytic *Clostridium botulinum*

14.4.1 Thermal inactivation of spores

Spores of non-proteolytic *C. botulinum* are of only moderate heat resistance, but survive pasteurisation treatments and may be a hazard in food heated to less than 100 °C. Their heat resistance has been measured in phosphate buffer and a range of foods (Table 14.6). Estimates of the highest *D*-values in phosphate buffer have generally been reported as within the range of 0.5 to 2.4 min at 82.2 °C. Higher heat resistances have been reported in various foods (Table 14.6). Based on measurements of a type B strain in cod, the Advisory Committee on the Microbial Safety of Food (ACMSF) in the UK suggested that a thermal process of 10 min at 90 °C was sufficient to achieve the desired 6D reduction required for refrigerated products (ACMSF, 1992). These criteria have also been adopted by other organisations (ECFF, 2006; CFA, 2006). A 6D reduction is considered an acceptable reduction with respect to non-proteolytic *C. botulinum* (ACMSF, 1992). A thermal process of 10 min at 90 °C will have little effect on spores of proteolytic *C. botulinum*; by extrapolation the *D*-value of proteolytic *C. botulinum* spores at 90 °C would be 251 min.

The measured heat resistance of spores will depend on many factors such as the strain tested and the composition, pH and water activity of both the heating menstrum and the recovery media. The effect of intrinsic properties of foods and other environmental factors on the thermal inactivation of spores of non-proteolytic *C. botulinum* has been studied, and predictive models of thermal inactivation have been produced (Juneja and Eblen, 1995; Juneja *et al.*, 1995a, b; Stringer and Peck, 1997; Stringer *et al.*, 1997; Lindstrom *et al.*, 2003). The *z*-value for spores of non-proteolytic *C. botulinum* is less than the value of 10 °C used for proteolytic *C. botulinum*. Values of between 6 °C and 9 °C have been proposed.

One factor that has been shown to have a large effect on measured heat resistance is the presence of lysozyme following heat treatment. Mild heat treatments can sub-lethally damage spores of non-proteolytic *C. botulinum* by inactivating the germination system. Spores damaged in such a way cannot germinate and outgrow on some nutrient media, but if lysozyme is present, it can diffuse through the coat

Table 14.6 Heat resistance of spores of non-proteolytic *C. botulinum* in phosphate buffer and foods

Strain[a] (toxin type)	Heating menstruum	Temperature range (°C)	$D_{82.2\ °C}$ (min)	*z*-value (°C)	Recovery method[b]	Reference
Saratoga (E)	0.02 M phosphate buffer (pH 7)	70–80	0.48	5.6	TDT	(Bohrer *et al.*, 1973)
Saratoga (E)	0.07 M phosphate buffer (pH 7)	70–80	0.69	7.2	TDT	(Bohrer *et al.*, 1973)
610 (F)	0.07 M phosphate buffer (pH 7)	71–85	0.84	6.3	TDT	(Lynt *et al.*, 1979)
2129 (B)	0.07 M phosphate buffer (pH 7)	92–102	32	9.7	TDT	(Scott and Bernard, 1982)
Nanaimo (E)	0.07 M phosphate buffer (pH 7)	70–80	1.9	9.6	counts	(Ohye and Scott, 1957)
Minneapolis (E)	0.07 M phosphate buffer (pH 7)	74–86	1.2	8.3	counts	(Schmidt, 1964)
1304 (E)	0.07 M phosphate buffer (pH 7)	80	0.52	–	counts	(Alderton *et al.*, 1974)
Elkund 17B (B)	0.07 M phosphate buffer (pH 7)	85	0.36	–	counts	(Peck *et al.*, 1992)
Alaska (E)	0.1 M phosphate buffer (pH 7)	70–90	2.4	10.0	count	(Juneja *et al.*, 1995a)
Eklund 17B (B)	0.07 M phosphate buffer (pH 7)	77.5–90	54	12.4	counts (+LYS)[c]	(Smelt, 1980)
Eklund 17B (B)	0.07 M phosphate buffer (pH 7)	75–95	231	7.6	counts (+LYS)[c]	(Peck *et al.*, 1993)
Alaska (E)	Whitefish chubs	74–85	2.2	7.6	TDT	(Crisley *et al.*, 1968)
Saratoga (E)	Tuna in oil	72.2–82.2	6.6	6.1	TDT	(Bohrer *et al.*, 1973)
Saratoga (E)	Sardines in tomato sauce	72.2–82.2	2.9	6.3	TDT	(Bohrer *et al.*, 1973)
Beluga (E)	Blue crabmeat	74–85	0.74	8.3	TDT	(Lynt *et al.*, 1977)
202 (F)	Crabmeat	76.6–85	1.2	6.4	TDT	(Lynt *et al.*, 1979)
Beluga (E)	Oyster	50–80	0.82	7.6	TDT	(Bucknavage *et al.*, 1990)
Minnesota (E)	Oyster	73.9–82.2	0.43	6.6	TDT	(Chai and Liang, 1992)
Mixed strains	Beef	70–90	18	6.8	TDT	(Fernandez and Peck, 1997)
Mixed strains	Crabmeat	88.9–94.4	80	8.5	TDT	(Peterson *et al.*, 1997)
Mixed strains	Menhaden surimi	73.9–82.2	1.2	9.8	TDT	(Rhodehamel *et al.*, 1991)
Eklund 17B (B)	Cod	75–92	8.7	8.6	counts	(Gaze and Brown, 1990)
Mixed strains	Beef	70–90	139	9.8	counts (+LYS)[c]	(Fernandez and Peck, 1999)
Beluga (E)	Rainbow trout	75	4.6	–	counts	(Lindstrom *et al.*, 2003)
Beluga (E)	Whitefish	81–90	1.6	10.1	counts	(Lindstrom *et al.*, 2003)
Beluga (E)	Rainbow trout	75–93	76	10.4	counts (+LYS)[c]	(Lindstrom *et al.*, 2003)
Beluga (E)	Whitefish	81–90	44	10.1	counts (+LYS)[c]	(Lindstrom *et al.*, 2003)

[a] Where multiple strains were tested, data are reported for the most heat resistant strain.

of a proportion of spores (for most strains 0.1–1%) and induce germination by hydrolysing peptidoglycan in the spore cortex (Gould, 1989). The measured $D_{85\,°C}$ value can differ by two orders of magnitude between lysozyme sensitive and insensitive spores (Peck, 2006). Lysozyme activity can be found in many foods, including hen egg white and other birds' eggs, microbiological culture media containing egg yolk emulsion or horse blood (Peck *et al.*, 1992), fruit and vegetable extracts (Stringer *et al.*, 1999), red meat and crab (Lund and Peck, 1994). It has also been proposed that hen egg white lysozyme can be added to foods as a preservative (Nielsen, 1991; Fuglsang *et al.*, 1995). The importance of the presence of enzymes with lytic activity in foods pasteurised in-pack depends on their ability to withstand the heat treatments causing damage to the spores. Hen egg white lysozyme is relatively heat stable in buffer (Proctor and Cunningham, 1988), but its ability to refold after heat treatment depends on the food it is in, making prediction of post-heat treatment lytic activity difficult. Some post-heat treatment activity has been shown to remain for hen egg white lysozyme added to meat slurry. Fernandez and Peck (1999) showed that substantially more spores survived, and growth occurred more rapidly and over a wider range of conditions, following heat treatment at 85° or 95 °C when lysozyme had been added to a meat slurry pre-heat treatment. A similar effect may also have been noted with endogenous lysozyme in crabmeat: heat treatments of 90.6 °C for 65 min, 92.2 °C for 35 min, or 94.4 °C for 15 min were required to prevent growth and toxin formation from 10^6 spores of non-proteolytic *C. botulinum* when incubated at 27 °C for 150 days (Peterson *et al.*, 1997).

These experiments have shown that the heat treatments advocated by the ACMSF (ACMSF, 1992) and ECFF (ECFF, 2006) do not, in themselves, deliver a 6*D* process in all circumstances. However, where such processes are combined with storage at refrigeration temperatures (required to prevent growth from the heat-resistant proteolytic *C. botulinum*), growth from non-proteolytic *C. botulinum* can be prevented or greatly delayed.

14.4.2 Conditions preventing growth and toxin formation

Non-proteolytic *C. botulinum* can grow and produce toxin at refrigeration temperatures, with a growth minimum around 3 °C. Graham *et al.* (1997) showed growth from 10^5 spores after seven weeks at 3.0 °C, six weeks at 3.1 °C, and five weeks at 3.2 °C. Earlier studies had demonstrated growth and toxin production at 3.3 °C within five weeks (Schmidt *et al.*, 1961; Eklund *et al.*, 1967a,b). Growth and toxin production have not been detected within 13 weeks at 2.1–2.5 °C (Ohye and Scott, 1957; Schmidt *et al.*, 1961; Eklund *et al.*, 1967a,b; Graham *et al.*, 1997).

It is generally recognised that growth and toxin production do not occur below pH 5.0 or at a NaCl concentration above 5%, and that the minimum water activity permitting growth is 0.94 and 0.97 with glycerol and NaCl, respectively, as the humectants (Baird-Parker and Freame, 1967; Hauschild, 1989; Graham *et al.*, 1996; Lund and Peck, 2000). The effect of other preservative factors on growth of non-proteolytic *C. botulinum* has been reviewed elsewhere (e.g. Hauschild, 1989;

Lund and Peck, 2000). Again, although redox potential and oxygen concentration affects the ability of *C. botulinum* to grow and form toxin (Lund, 1993), the presence of oxygen in a product's headspace cannot be relied on to prevent growth and toxin formation. Studies with meat slurry found that an initial aerobic atmosphere (20% oxygen) did not restrict growth compared to an atmosphere of N_2 (Peck, unpublished results). In a study with flounder at 12 °C, the fish became toxic in 11 days when packed in air, in 10 days when packed in CO_2, 14 days in N_2 and 15 days when vacuum packed (Post *et al.*, 1985). The use of oxygen as a preservative factor is therefore cautioned since, despite the presence of oxygen, the food itself may be sufficiently reduced and support growth and toxin production by non-proteolytic *C. botulinum* (Snyder, 1996).

14.4.3 Using combinations of preservative factors that prevent growth

Whilst maintaining processed in-pack products at a temperature below 3.0 °C would prevent growth and toxin production by non-proteolytic *C. botulinum*, there is substantial doubt that this temperature can be guaranteed to be maintained throughout the distribution chain, particularly in products intended for domestic use. Permitted refrigerated storage temperatures in retail are greater than 3 °C in many countries, and temperatures above 3 °C are the norm in domestic refrigerators. Long-life retail products often contain additional hurdles to prevent botulism. Safety with respect to non-proteolytic *C. botulinum* often depends upon more than one preservative factor and it is important to define combinations of preservative factors (e.g. heat treatment, pH, a_w and subsequent storage time at refrigeration temperatures) that provide an appropriate degree of protection. An important development in describing the effect of combinations of preservative factors has been that of predictive models. Examples of models that have been described include the effect of combinations of pH and NaCl concentration on time-to-growth at chilled temperatures (Graham *et al.*, 1996; Whiting and Oriente, 1997), the effect of heat treatment and storage temperature on time-to-growth (Fernandez and Peck, 1997) and the combined effects of heat treatment, pH, NaCl and storage temperature (Peck and Stringer, 2005).

14.5 Recommendations and guidelines to ensure the safe production of in-pack processed foods with respect to *Clostridium botulinum*

14.5.1 Shelf-stable foods

The term low-acid foods is applied to foods with a pH > 4.5 in the UK and to foods with a pH > 4.6 in the USA (Lund and Peck, 2000). These are considered to be foods with a pH that could allow growth of *C. botulinum*. Canned products without additional controlling factors, generally receive a heat treatment equivalent to 121 °C for three minutes (an $F_0 3$ process). This is known as the 'botulinum cook'

Table 14.7 Recommended procedures to ensure the safety of minimally heated refrigerated foods with respect to non-proteolytic *C. botulinum* (ACMSF 1992, 2006)

- storage at <3.0 °C
- storage at ≤8 °C and a shelf-life of ≤10 days
- storage at chill temperature[a] combined with heat treatment of 90 °C for 10 min or equivalent lethality (e.g. 80 °C for 129 min, 85 °C for 36 min)[b]
- storage at chill temperature combined with ≤pH 5.0 throughout the food
- storage at chill temperature combined with a salt concentration ≥3.5% throughout the food
- storage at chill temperature combined with ≤a_w 0.97 throughout the food
- storage at chill temperature combined with a combinations of heat treatment and other preservative factors which can be shown consistently to prevent growth and neurotoxin production by *C. botulinum.*

Notes:
[a] chill temperature is specified as 8 °C in England, Wales and Northern Ireland.
[b] Alternative heat treatments of 80 °C for 270 min, 85 °C for 52 min are recommended by the European Chilled Food Federation and UK Chilled Food Association (ECFF, 2006; CFA, 2006).

for low-acid foods, and is intended to reduce the number of viable spores of proteolytic *C. botulinum* by a factor of 10^{12} (a 12D process). The use of other factors to control *C. botulinum* in shelf-stable foods has been reviewed (Lund and Peck, 2000).

14.5.2 Foods intended for refrigerated storage

The UK Advisory Committee on the Microbiological Safety of Food have made recommendations on the safe production of vacuum and modified atmosphere-packed refrigerated foods with respect to *C. botulinum* and the associated foodborne botulism hazard (Table 14.7) (ACMSF, 1992; ACMSF, 2006a), and best practice guidelines for the production of chilled food have been published by the European Chilled Food Federation (ECFF, 2006)) and the UK Chilled Food Association (CFA, 2006). These recommend that heat treatments, or combination processes, deliver a safety factor of 10^6 (a 6D process) with regard to spores of non-proteolytic *C. botulinum.*

14.6 Improving prediction of the behaviour of *Clostridium botulinum* in food environments

14.6.1 Thermal death and growth models

An important development in food microbiology has been the creation of predictive models. These models allow interpolation of interactions between two or more factors, and can be used to reduce the amount of challenge testing required to ensure product safety. Predictive microbiology is based on the hypothesis that growth is an intrinsic characteristic of the organism and will occur reproducibly in the same environment. Thus, by measuring growth once, you can tell how a future population will develop under the same conditions. Measured trends in bacterial behaviour are described with a mathematical function. A secondary model can be used to describe the relationship between the parameters of the primary model and environmental factors such as temperature, pH and water activity. Some predictive models for *C. botulinum* have been developed in food. These tend to predict well for the foods on which they were based but may be of limited use in other types of food. Other models have been developed in laboratory media and are more generic in application. Models have been developed, using single or multiple factors, based on growth kinetics, probability of growth or time to toxin (Baker and Genigeorgis, 1990, 1992; Baker, 1993; Dodds, 1993; Meng and Genigeorgis, 1993; Lund, 1993; McClure *et al.*, 1994; Graham *et al.*, 1996; Whiting and Oriente, 1997; Lund and Peck, 2000; Fernandez *et al.*, 2001). Where tested, predictions from these models generally compare well with observed growth and toxin production in independent datasets, giving the user confidence that these models can be used to target, more effectively, challenge testing. Some of these models are freely available on the web, including ComBase Predictor (www.combase.cc). This site also allows access to published (and in some cases also unpublished) original growth and death curves.

14.6.2 Stochastic modelling based on single-spore lag times

While predictive modelling has successfully reduced the numbers of growth studies and challenge tests required, refinements in modelling techniques are ongoing. One area where improvements are being sought is in lag-phase modelling. Current models usually describe population kinetics, but this approach can be problematic when the initial population is very small. Spores of *C. botulinum* are widespread in the environment but are usually present at low concentrations; thus, if growth occurs in food packs, it is likely to initiate from just a few spores. This will have consequences when trying to predict growth, as uncertainty around the estimation of growth times of *C. botulinum* increases as the initial spore concentration decreases (Whiting and Oriente, 1997). Such variability cannot be determined from observations made at the population level; instead, it requires data for individual spores and the use of stochastic modelling (Baranyi, 2002). In order to better understand and predict lag-time variability, studies have been undertaken to quantify the heterogeneity of the duration of individual stages in lag from single spores of *C. botulinum* (Stringer *et al.*, 2005; Webb *et al.*, 2007). These studies showed that times to growth from individual spores are highly variable. Considerable variability was observed in all stages of lag, and the duration of the stages was not necessarily correlated with each other. Additional uncertainty arises as, unlike growth rate, the duration and variability of lag time also depended on the historic treatment of the spores. It is hoped that new modelling techniques can use this type of data to improve predictions of time-to-growth and create estimates of the level of uncertainty associated with a prediction.

14.6.3 Process risk modelling

To date, the development of foods safe from toxin production by *C. botulinum* has depended on the use of default performance criteria that are fail-safe and intended to control the foodborne botulism hazard under 'worst-case' situations. Default criteria (guidelines) include the use of a 6D (6-decimal) process for the safe production of minimally heat-processed refrigerated foods or a 12D (12-decimal) process for ambient-stable products. The limitation of using such default criteria is that they fail to take account of many contributory factors, such as the quality of raw materials. More recently, quantitative process risk models have been created using a probabilistic modelling based on Markov Chain Monte Carlo methods or Bayesian Belief Networks. Such approaches use probability distributions to represent variables rather than a single-point value, and allow diverse, variable and uncertain information to be combined into a single consistent calculation to give a clear end-point and an associated representation of uncertainty. The outputs from probabilistic risk models are fully inspectable in both forward and backward directions and so can be used for structured enquiries, sensitivity analyses and 'what if' enquiries. They can therefore be used to identify potential control areas, highlight areas where improvements should be targeted to effect greatest reduction in risk, determine areas where research should be targeted to have the greatest impact on safety and show how new information impacts on established beliefs.

A process risk model describing the safety of gnocchi with respect to non-proteolytic *C. botulinum* has been published (Barker *et al.*, 2005). Here, initial contamination levels were combined with production processes (such as mixing, partitioning, heating, cooling and packaging) and consumer treatment of the packs to give a single output, the probability of toxicity in an individual retail pack at the time of use. This analysis showed that correctly stored gnocchi was safe with respect to non-proteolytic *C. botulinum*, despite none of the controlling factors being sufficient to prevent growth in isolation.

14.7 Recent advances in understanding of the functional genomics and physiology of foodborne clostridia

Recent advances in molecular biology have the potential to increase greatly our understanding of the ability of foodborne clostridia to grow in food or cause illness. Information from sequenced genomes offers an excellent opportunity for comparative genomic analysis of the clostridia, provides a framework for functional genomics and is expected to provide valuable insights into the metabolic diversity, lifestyle, mechanisms of pathogenicity and evolution of the clostridia.

14.7.1 Clostridial genome sequences

Sequencing the genomes of prokaryotic and eukaryotic organisms is currently a very active area of research. Genomes have been published for four species of clostridia that are human pathogens; one strain of proteolytic *C. botulinum*

(Sebaihia *et al.*, 2007), three strains of *C. perfringens* (Shimizu *et al.*, 2002; Myers *et al.*, 2006), one strain of *C. tetani* and one strain of *C. difficile* (Bruggemann *et al.*, 2003; Sebaihia *et al.*, 2006). The genome of one strain of the non-pathogenic solvent-producing species, *C. acetobutylicum*, has also been sequenced (Nolling *et al.*, 2001). This sequencing has confirmed the heterogeneity of the clostridia. It was found that just 568 (16%) of the proteolytic *C. botulinum* coding sequences (CDSs) were shared with all the other sequenced clostridia, 1511 (41%) of the CDSs had orthologues in at least one other sequenced clostridia and 1571 (43%) of the CDSs that are present in the proteolytic *C. botulinum* genome were absent from all other sequenced clostridial genomes (Sebaihia *et al.*, 2007). Genome sequencing of 36 other clostridial strains is currently in progress. This includes five further strains of proteolytic *C. botulinum*, one strain of non-proteolytic *C. botulinum*, and six strains of *C. perfringens* (http://www.genomesonline.org/gold.cgi, accessed July 13, 2007). Clearly, knowledge of clostridial genomes is currently very limited, compared to what it will be when all these additional sequences become available.

14.7.2 Comparative genomic indexing of clostridia

The use of typing tools such as 16S rRNA sequencing, ribotyping, pulsed-field gel electrophoresis (PFGE) and amplified fragment length polymorphism (AFLP) have provided important information on the phylogeny of *C. botulinum* and other clostridia (Nevas *et al.*, 2005; Lindstrom and Korkeala, 2006). However, these methods provide no information on the genetic basis for the differences between strains and species. The first example of comparative genomic indexing of *C. botulinum* was recently published (Sebaihia *et al.*, 2007). Here, a DNA microarray was constructed from PCR products based on the predicted CDSs of the sequenced strain (proteolytic *C. botulinum* strain Hall A (ATCC 3502)), with a 100–500 bp probe for each CDS. Comparative genomic hybridisation experiments were then performed comprising a competitive hybridisation between Hall A DNA (as a reference), DNA of the test strain, and the immobilised DNA probes of the microarray. For each test strain, it was established which of the CDSs were either shared with the reference strain, or were absent/highly divergent. A higher degree of relatedness was identified between strains of proteolytic *C. botulinum* than was reported previously for *C. difficile* strains (Sebaihia *et al.*, 2007). Approximately 87%–96% of the Hall A CDSs were possessed by nine other strains of proteolytic *C. botulinum*, and 84%–87% of the CDSs were shared with two strains of *C. sporogenes*. Two prophages present in the Hall A strain were absent from all 11 test strains (Sebaihia *et al.*, 2007). A further, more detailed, comparative genomic indexing study of *C. botulinum* is currently in progress in the authors' laboratory.

14.7.3 Clostridial transcriptomics

A DNA microarray can also be used for transcriptomic studies; that is, to measure transcriptional activity of all CDSs simultaneously. In this procedure, RNA is extracted from a culture, converted to DNA (cDNA), and then applied to the DNA microarray. This enables transcriptional snapshots of all the CDSs to be made simultaneously, which is useful for assessing the effect of stress treatments (such as heat, acid or oxidative shock) and for systems biology studies. The effect of mutations in specific genes on global gene expression can also be studied using this functional genomic approach. Transcriptomic studies of *C. botulinum* are currently in progress, although results have not yet been published. In the longer term, this new information may form the basis for new knowledge-led interventions to improve the effectiveness of food manufacturing processes and to ensure food safety.

14.8 Future trends

Patterns of food production and consumption are continuously changing, requiring risk assessments to be continuously updated. Current trends include the use of new technologies (for example, high-pressure processing), increased consumption of chilled foods, a reduction in the use of preservatives and a move towards longer shelf-life. Many of these developments favour the growth of clostridia. As botulism is a very severe disease, it is important to understand where *C. botulinum* poses a hazard and how to prevent growth and toxin production in any new food or where processes, formulation or storage conditions have been changed.

To ensure the continued safe development of foods, it will be necessary for quantitative microbiological food safety to evolve. The ability of *C. botulinum* to grow and produce toxin in a food will depend on many properties of that food and the processing. Not all effects and their interactions can be quantified in current models. Failure to take into account factors such as the pattern and level of contamination makes it very difficult to assess the true safety margin associated with a food product. Hopefully, the development and use of probalistic-based risk modelling will aid assessment of the true magnitude of any risk and help identify, and minimise, potential new hazards. It is also likely that studies on the fundamental physiology and genetics of clostridia will improve our knowledge of important processes, such as the ability of spores to germinate, and the conditions under which toxin is produced. Improved understanding of growth and toxin production might even lead to novel preservation methods.

14.9 Sources of further information and advice

The following articles provide further information on *Clostridium botulinum* in foods:

Hauschild A H W and Dodds K L (1993) *Clostridium botulinum: Ecology and control in foods*. New York, Marcel Deker.

Lund B M and Peck M W (2000) '*Clostridium botulinum*'. in Lund B M , Baird-Parker T C and Gould G W, *The microbiological safety and quality of food*, Gaithersburg, Aspen Publishers, 1057–1109.

14.10 References

ACMSF (Advisory Committee on the Microbiological Safety of Food) (1992), *Report on Vacuum Packaging and Associated Processes*. London, Her Majesty's Stationery Office.

ACMSF (Advisory Committee on the Microbiological Safety of Food) (2006a), Minutes of meeting held on 8th June 2006. (www.food.gov.uk/science/ouradvisors/microbiogsafety/acmsfmeets/acmsf2006/acmsfmeet080606/acmsfmin080606; accessed 13/07/2007).

ACMSF (Advisory Committee on the Microbiological Safety of Food) (2006b), *Report on minimally processed infant weaning foods and the risk of infant botulism*. London, Food Standards Agency.

Adak G K, Long S M and O'Brien S J (2002), 'Trends in indigenous foodborne disease and deaths, England and Wales: 1992 to 2000', *Gut*, **51**, 832–841.

Alderton G, Chen J K and Ato K A (1974), 'Effect of lysozyme on the recovery of heated *Clostridium botulinum* spores', *Applied Microbiology*, **27**, 613–615.

Anniballi F, Fenicia L, Franciosa G and Aureli P (2002), 'Influence of pH and temperature on the growth of and toxin production by neurotoxigenic strains of *Clostridium butyricum* type E', *Journal of Food Protection*, **65**, 1267–1270.

Anon (2003), 'Botulism, canned sausage suspected – South Korea', www.promedmail.org (29 Jun 2003).

Anon (2007a), 'Botulism, canned artichokes – Spain (Guipuzcoa)', www.promedmail.org (7 Jan 2007)

Anon (2007b), 'Botulism, canned food, human, pet – USA: Alert, recall', www.promedmail.org (23 Jul 2007).

Aureli P, Di Cunto M, Maffei A, De Chiara G, Franciosa G, Accorinti L, Gambardella A M and Greco D (2000), 'An outbreak in Italy of botulism associated with a dessert made with mascarpone cream cheese', *European Journal of Epidemiology*, **16**, 913–918.

Aureli P, Franciosa G and Salmaso S (1998), 'Organic vegetable soup associated with a case of botulism – Northern Italy', *Eurosurveillance Weekly*, **2**, 10 September 1998.

Baird-Parker A C and Freame B (1967), 'Combined effect of water activity, pH and temperature on the growth of *Clostridium botulinum* from spore and vegetative inocula', *Journal of Applied Bacteriology*, **30**, 420–429.

Baker D A (1993), 'Probability-models to assess the safety of foods with respect to *Clostridium botulinum*', *Journal of Industrial Microbiology*, **12**, 156–161.

Baker D A and Genigeorgis C (1990), 'Predicting the safe storage of fresh fish under modified atmospheres with respect to *Clostridium botulinum* toxigenicity by modeling length of the lag phase of growth', *Journal of Food Protection*, **53**, 131–140, 153.

Baker D A and Genigeorgis C (1992), 'Predictive modelling', in Hauschild A H W and Dodds K L, *Clostridium botulinum. Ecology and control in foods,* New York, Marcel Dekker, 343–406.

Ball A P, Hopkinson R B and Farrell I D (1979), 'Human botulism caused by *Clostridium botulinum* type E: The Birmingham outbreak', *Quarterly Journal of Medicine,* **48**, 473–491.

Baranyi J (2002), 'Stochastic modelling of bacterial lag phase', *International Journal of Food Microbiology*, **73**, 203–206.

Barker G C, Malakar P K, Del Torre M, Steechini M L and Peck M W (2005), 'Probabilistic representation of the exposure of consumers to *Clostridium botulinum* neurotoxin in a minimally processed potato product', *International Journal of Food Microbiology,* **100**, 345–357.

Barker W H, Weissman J B, Dowell V R, Gutmann L and Kautter D A (1977), 'Type-B Botulism Outbreak Caused by A Commercial Food Product – West Virginia and Pennsylvania, 1973', *Journal of the American Medical Association*, **237**, 456–459.

Baumgart J (1970), 'Nachweis von *Clostridium botulinum* typ E bei handelsfertigen Forellen', *Die Fleischwirtschaft*, **11**, 1545–1546.

Bohrer C W, Denny C B and Yao M G (1973), '*Thermal destruction of type E* Clostridium botulinum. *Final report on RF 4603*', Washington DC, National Canners Association Research Foundation.

Brett M M, McLauchlin J, Harris A, O'Brien S, Black N, Forsyth R J, Roberts D and Bolton F J (2005), 'A case of infant botulism with a possible link to infant formula milk powder: Evidence for the presence of more than one strain of *Clostridium botulinum* in clinical specimens and food', *Journal of Medical Microbiology*, **54,** 769–776.

Bruggemann H Baumer S, Fricke W F, Wiezer A, Liesegang H, Decker I, Herzberg C, Martinez-Arias R, Merkl R, Henne A and Gottschalk G (2003), 'The genome sequence of *Clostridium tetani*, the causative agent of tetanus disease', *Proceedings of the National Academy of Sciences of the United States of America*, **100**, 1316–1321.

Bucknavage M W, Pierson M D, Hackney C R and Bishop J R (1990), 'Thermal inactivation of *Clostridium botulinum* type E spores in oyster homogenates at minimal processing temperatures', *Journal of Food Science*, **55**, 372–373.

Carlier J P, Henry C, Lorin V and Popoff M R (2001), 'Le botulisme en France a la fin du deuxieme millenaire (1998–2000)', *Bulletin Epidemiologique Hebdomadaire*, **2001**, 37–39.

Carlin F, Broussolle V, Perelle S, Litman S and Fach P (2004), 'Prevalence of *Clostridium botulinum* in food raw materials used in REPFEDs manufactured in France', *International Journal of Food Microbiology*, **91**, 141–145.

CFA (Chilled Food Association) (2006), *Best Practice Guidelines for the Production of Chilled Foods. 4th edition*. Norwich. The Stationery Office.

Chai T J and Liang K T (1992), 'Thermal-resistance of spores from 5 type E *Clostridium botulinum* strains in Eastern oyster homogenates', *Journal of Food Protection*, **55**, 18–22.

Chou J H, Hwang P H and Malison M D (1988), 'An outbreak of type A foodborne botulism in Taiwan due to commercially preserved peanuts', *International Journal of Epidemiology*, **17**, 899–902.

Crisley F D, Peeler J T, Angelotti R and Hall H E (1968), 'Thermal resistance of five strains of *Clostridium botulinum* type E in ground whitefish chubs', *Journal of Food Science*, **33**, 411–416.

Czerwinski M, Czarkowski M P and Kondej B (2005), 'Zatrucia jadem kielbasianym w 2003 roku' [Botulism in Poland in 2003], *Przeglad Epidemiologiczny*, **59**, 281–287.

Czerwinski M, Czarkowski M P and Kondej B (2006), 'Zatrucia jadem kielbasianym w 2004 roku' [Botulism in Poland in 2004], *Przeglad Epidemiologiczny*, **60**, 456–470.

D'Argnio P, Palumbo F, Ortolani R, Pizzuti R, Russo M, Carducci R, Soscia M, Aureli P, Fenicia L, Franciosa G, Parella A and Scala V (1995), 'Type B botulism associated with roasted eggplant in oil – Italy, 1993', *Morbidity and Mortality Weekly Report,* **44**, 33–36.

Dodds K L (1989), 'Combined effect of water activity and pH on inhibition of toxin production by *Clostridium botulinum* in cooked, vacuum-packed potatoes', *Applied and Environmental Microbiology*, **55**, 656–660.

Dodds K L (1993), 'An introduction to predictive microbiology and the development and use of probability models with *Clostridium botulinum*', *Journal of Industrial Microbiology*, **12**, 139–143.

Dodds K L (1992), '*Clostridium botulinum* in the environment', in Hauschild A H W and Dodds K L, *Clostridium botulinum: Ecology and Control in Foods*, New York, Marcel Dekker, 21–52.

Dodds K L and Austin J W. (1997), '*Clostridium botulinum*', in Doyle M P, Beuchat L R and Montville T J. *Food Microbiology, Fundamentals and Frontiers*, Washington DC, ASM, 288–304.

Dressler D (2005), 'Botulismus durch Räucherlachsverzehr', *Der Nervenarzt*, 7**6,** 763–766.

ECFF (European Chilled Food Federation) (2006), *Recommendations for the production of pre-packaged chilled foods*, European Chilled Food Federation. (www.ecff.net).

Eklund M W, Poysky F T and Wieler D I (1967a), 'Characteristics *of Clostridium botulinum* type F isolated from the Pacific coast of the United States', *Applied Microbiology*, **15**, 1316–1323.

Eklund M W, Wieler D I and Poysky F T (1967b), 'Outgrowth and toxin production of non-proteolytic type B *Clostridium botulinum* at 3.3 °C to 5.6 °C', *Journal of Bacteriology*, **93**, 1461–1462.

Esty J R and Meyer K F (1922), 'The heat resistance of the spores of *B. botulinus* and allied anaerobes. XI', *Journal of Infectious Diseases*, **31**, 650–663.

Fach P, Perelle S, Dilasser F, Grout J, Dargaignaratz C, Botella L, Gourreau J M, Carlin F, Popoff M R and Broussolle V (2002), 'Detection by PCR-enzyme-linked immunosorbent assay of *Clostridium botulinum* in fish and environmental samples from a coastal area in northern France', *Applied and Environmental Microbiology*, **68**, 5870–5876.

Fernandez P S, Baranyi J and Peck M W (2001), 'A predictive model of growth from spores of non-proteolytic *Clostridium botulinum* in the presence of different CO_2 concentrations as influenced by chill temperature, pH and NaCl', *Food Microbiology*, **18**, 453–461.

Fernandez P S and Peck M W (1997), 'Predictive model describing the effect of prolonged heating at 70 to 80 °C and incubation at refrigeration temperatures on growth and toxigenesis by nonproteolytic *Clostridium botulinum*', *Journal of Food Protection*, **60,** 1064–1071.

Fernandez P S and Peck M W (1999), 'A predictive model that describes the effect of prolonged heating at 70 to 90 °C and subsequent incubation at refrigeration temperatures on growth from spores and toxigenesis by nonproteolytic *Clostridium botulinum* in the presence of lysozyme', *Applied and Environmental Microbiology*, **65**, 3449–3457.

Frean J, Arntzen L, van den Heever J and Perovic O (2004), 'Fatal type A botulism in South Africa, 2002', *Transactions of the Royal Society of Tropical Medicine and Hygiene*, **98**, 290–295.

Fuglsang C C, Johansen C, Christgau S and AdlerNissen J (1995), 'Antimicrobial enzymes: Applications and future potential in the food industry', *Trends in Food Science and Technology*, **6**, 390–396.

Gaze J E and Brown G D (1990), *Determination of the heat resistance of a strain of non-proteolytic Clostridium botulinum type B and a strain of type E, heated in cod and carrot homogenate over the temperature range 70 to 92 °C.*, Chipping Campden, Campden Food and Drink Research Association.

Gould G W (1989), 'Heat-induced injury and inactivation', In Gould G W, *Mechanisms of action of food preservation procedures*, London, Elsevier, 11–42.

Graham A F, Mason D R, Maxwell F J and Peck M W (1997), 'Effect of pH and NaCl on growth from spores of non-proteolytic *Clostridium botulinum* at chill temperature', *Letters in Applied Microbiology*, **24**, 95–100.

Graham A F, Mason D R and Peck M W (1996), 'Predictive model of the effect of temperature, pH and sodium chloride on growth from spores of non-proteolytic *Clostridium botulinum*', *International Journal of Food Microbiology*, **31**, 69–85.

Harvey S M, Sturgeon J and Dassey D E (2002), 'Botulism due to *Clostridium baratii* type F toxin', *Journal of Clinical Microbiology*, **40**, 2260–2262.

Hauschild A H W (1989), '*Clostridium botulinum*', in Doyle M P, *Foodborne bacterial pathogens*, New York, Marcel Dekker, 112–189.

Hauschild A H W (1992), 'Epidemiology of human foodborne botulism', in Hauschild A H W and Dodds K L, *Clostridium botulinum. Ecology and control in foods*, New York, Marcel Dekker, 69–104.

Hauschild A H W and Hilsheimer R (1979), 'Effect of salt content and pH on toxigenesis by *Clostridium botulinum* in caviar', *Journal of Food Protection*, **42**, 245–248.

Juneja V K and Eblen B S (1995), 'Influence of sodium chloride on thermal inactivation and recovery of nonproteolytic *Clostridium botulinum* type B strain Kap B5 spores – a research note', *Journal of Food Protection*, **58**, 813–816.

Juneja V K, Eblen B S, Marmer B S, Williams A C, Palumbo S A and Miller A J (1995a), 'Thermal-resistance of nonproteolytic type B and type E *Clostridium botulinum* spores in phosphate buffer and turkey slurry', *Journal of Food Protection*, **58**, 758–763.

Juneja V K, Marmer B S, Phillips J G and Miller A J (1995b), 'Influence of the intrinsic properties of food on thermal inactivation of spores of nonproteolytic *Clostridium botulinum*: Development of a predictive model', *Journal of Food Safety*, **15**, 349–364.

Kobayashi H, Kurokawa N, Kobayashi M, Saito Y, Kobayashi N, Tsuda T, Fujisawa K, Kubo M and Eto Y (1999), 'A 12-year-old case of botulism with acute flaccid paralysis, August 1999 – Chiba', *Infectious Agents Surveillance Report*, **20**, 272–273, 301–302.

Korkeala H, Stengel G, Hyytia E, Vogelsang B, Bohl A, Wihlman H, Pakkala P and Hielm S (1998), 'Type E botulism associated with vacuum-packaged hot-smoked whitefish', *International Journal of Food Microbiology*, **43**, 1–5.

Krusell L and Lohse N (2003), 'A case of human botulism in Denmark after consumption of garlic in chilli oil dressing produced in Germany', *Eurosurveillance Weekly*, **7**, 13 February 2003.

Lindstrom M and Korkeala H (2006), 'Laboratory diagnostics of botulism', *Clinical Microbiology Reviews*, **19**, 298–314.

Lindstrom M, Nevas M, Hielm S, Lahteenmaki L, Peck M W and Korkeala H (2003), 'Thermal inactivation of nonproteolytic *Clostridium botulinum* type E spores in model fish media and in vacuum-packaged hot-smoked fish products', *Applied and Environmental Microbiology*, **69**, 4029–4036.

Lindstrom M, Vuorela M, Hinderink K, Korkeala H, Dahlsten E, Raahenmaa M and Kuusi M (2006), 'Botulism associated with vacuum-packed smoked whitefish in Finland', *Eurosurveillance Weekly*, **11**, 060720.

Lund B M (1993), 'Quantification of factors affecting the probability of development of pathogenic bacteria, in particular *Clostridium botulinum*, in foods', *Journal of Industrial Microbiology*, **12**, 144–155.

Lund B M and Peck M W (2000), '*Clostridium botulinum*', in Lund B M, Baird-Parker T C and Gould G W, *The Microbiological safety and quality of food,* Gaithersburg, Aspen Publishers, 1057–1109.

Lund B M and Peck M W (1994), 'Heat-resistance and recovery of spores of nonproteolytic *Clostridium botulinum* in relation to refrigerated, processed foods with an extended shelf-life', *Journal of Applied Bacteriology*, **76**, S115–S128.

Lynt R K, Kautter D A and Solomon H M (1979), 'Heat resistance of nonproteolytic *Clostridium botulinum* type F in phosphate buffer and crabmeat', *Journal of Food Science*, **44**, 108–111.

Lynt R K, Solomon H M, Lilly T J and Kautter D A (1977), 'Thermal death time of *Clostridium botulinum* type E in meat of the blue crab', *Journal of Food Science*, **42**, 1022–1025, 1037.

Matsuki K and Goto A (1998), 'Botulism, human – Japan (Tokyo)'. www.promedmail.org (16 Aug 1998).

McLauchlin J, Grant K A and Little C L (2006), 'Food-borne botulism in the United Kingdom', *Journal of Public Health*, **28**, 337–342.

McClure P J, Blackburn C W, Cole M B, Curtis, P S, Jones J E, Legan J D, Ogden I D, Peck M W, Roberts T A, Sutherland J P and Walker S J (1994), Modeling the growth, survival and death of microorganisms in foods – The UK Food Micromodel approach. *International Journal of Food Microbiology*, **23(3–4)**, 265–275.

Meng J H and Genigeorgis C A (1993), 'Modeling lag phase of nonproteolytic *Clostridium botulinum* toxigenesis in cooked turkey and chicken breast as affected by temperature, sodium lactate, sodium chloride and spore inoculum', *International Journal of Food Microbiology*, **19**, 109–122.

Meyer K F and Gunnison J B (1929), 'Botulism due to home canned Bartlett pears. XXXIX', *Journal of Infectious Diseases*, **45**, 135–147.

Morse D L, Pickard L K, Guzewich J J, Devine B D and Shayegani M (1990), 'Garlic-in-oil associated botulism: Episode leads to product modification', *American Journal of Public Health*, **80**, 1372–1373.

Morton R D, Scott V N, Bernard D T and Wiley R C (1990), 'Effect of heat and pH on toxigenic *Clostridium butyricum*', *Journal of Food Science*, **55**, 1725–1727.

Myers G S A, Rasko D A, Cheung J K, Ravel J, Seshadri R, Deboy R T, Ren Q H, Varga J, Awad M M, Brinkac L M, Daugherty S C, Haft D H, Dodson R J, Madupu R, Nelson W C, Rosovitz M J, Sullivan S A, Khouri H, Dimitrov G I, Watkins K L, Mulligan S, Benton J, Radune D, Fisher D J, Atkins H S, Hiscox T, Jost B H, Billington S J, SongerJ G, McClane B A, Titball R W, Rood J I, Melville S B and Paulsen I T (2006), 'Skewed genomic variability in strains of the toxigenic bacterial pathogen, *Clostridium perfringens*', *Genome Research*, **16**, 1031–1040.

Nevas M, Lindstrom M, Hielm S, Bjorkroth K J, Peck M W and Korkeala H (2005), 'Diversity of proteolytic *Clostridium botulinum* strains, determined by a pulsed-field gel electrophoresis approach', *Applied and Environmental Microbiology*, **71**, 1311–1317.

Nielsen H K (1991), 'Novel bacteriolytic enzymes and cyclodextrin glycosyl transferase for the food industry', *Food Technology*, **45**, 102–104.

Nolling J, Breton G, Omelchenko M V, Makarova K S, Zeng Q D, Gibson R, Lee H M, Dubois J, Qiu D Y, Hitti J, Wolf Y I, Tatusov R L, Sabathe F, Doucette-Stamm L, Soucaille P, Daly M J, Bennett G N, Koonin E V and Smith D R (2001), 'Genome sequence and comparative analysis of the solvent-producing bacterium *Clostridium acetobutylicum*', *Journal of Bacteriology*, **183**, 4823–4838.

Odlaug T E and Pflug I J (1978), '*Clostridium botulinum* and acids foods', *Journal of Food Protection*, **41**, 566–573.

Ohye D F and Christian J H B (1967), 'Combined effects of temperature, pH and water activity on growth and toxin production by *Cl. botulinum* types A, B and E', in Ingram M and Roberts T A, *Botulism 1966*, London, Chapman and Hall, 217–231.

Ohye D F and Scott W J (1953), 'The temperature relations of *Clostridium botulinum* types A and B', *Australian Journal of Biological Science*, **6**, 178–189.

Ohye D F and Scott W J (1957), 'Studies in the physiology of *Clostridium botulinum* type E', *Australian Journal of Biological Science*, **10**, 85–94.

O'Mahony M, Mitchell E, Gilbert R J, Hutchinson D N, Begg N T, Rodhouse J C and Morris J E (1990), 'An outbreak of foodborne botulism associated with contaminated hazelnut yoghurt', *Epidemiology and Infection*, **104**, 389–395.

Peck M W (2006), '*Clostridium botulinum* and the safety of minimally heated, chilled foods: An emerging issue?', *Journal of Applied Microbiology*, **101**, 556–570.

Peck M W, Fairbairn D A and Lund B M (1992), 'The effect of recovery medium on the estimated heat-inactivation of spores of nonproteolytic *Clostridium botulinum*', *Letters in Applied Microbiology*, **15**, 146–151.

Peck M W, Fairbairn D A and Lund B M (1993), 'Heat-resistance of spores of nonproteolytic *Clostridium botulinum* estimated on medium containing lysozyme', *Letters in Applied Microbiology*, **16**, 126–131.

Peck M W and Stringer S C (2005), 'The safety of pasteurised in-pack chilled meat products with respect to the foodborne botulism hazard', *Meat Science*, **70**, 461–475.

Peterson M E, Pelroy G A, Poysky F T, Paranjpye R N, Dong F M, Pigott G M and Eklund M W (1997), 'Heat-pasteurization process for inactivation of nonproteolytic types of *Clostridium botulinum* in picked Dungeness crabmeat', *Journal of Food Protection*, **60**, 928–934.

Post L S, Lee D A, Solberg M, Furgang D, Specchio J and Graham C (1985), 'Development of botulinal toxin and sensory deterioration during storage of vacuum and modified atmosphere packaged fish fillets', *Journal of Food Science*, **50**, 990–996.

Proctor V A and Cunningham F E (1988), 'The chemistry of lysozyme and its use as a food preservative and a pharmaceutical', *CRC Critical Reviews in Food Science and Nutrition*, **26**, 359–395.

Przybylska A (1995), 'Zatrucia jadem kielbasianym w 1993 roku' [Botulism in Poland 1993], *Przeglad Epidemiologiczny*, **XLIX**, 139–145.

Przybylska A (1996), 'Zatrucia jadem kielbasianym w 1994 roku' [Botulism in Poland 1994], *Przeglad Epidemiologiczny*, **50**, 143–149.

Przybylska A (1997), 'Zatrucia jadem kielbasianym w 1995 roku' [Botulism in Poland 1995], *Przeglad Epidemiologiczny*, **51**, 111–118.

Przybylska A (1998), 'Zatrucia jadem kielbasianym w 1996 roku' [Botulism in Poland 1996], *Przeglad Epidemiologiczny*, **52**, 87–92.
Przybylska A (1999), 'Zatrucia jadem kielbasianym w 1997 roku' [Botulism in Poland 1997], *Przeglad Epidemiologiczny*, **53**, 115–120.
Przybylska,A. (2000), Zatrucia jadem kielbasianym w 1998 roku. [Botulism in 1998], *Przeglad Epidemiologiczny* **54**, 115–121.
Przybylska A (2001), 'Zatrucia jadem kielbasianym w 1999 roku' [Botulism in Poland in 1999], *Przeglad Epidemiologiczny*, **55**, 103–109.
Przybylska A (2002), 'Zatrucia jadem kielbasianym w 2000 roku' [Botulism in Poland in 2000], *Przeglad Epidemiologiczny*, **56**, 305–310.
Przybylska A (2003), 'Zatrucia jadem kielbasianym w 2001 roku' [Botulism in Poland in 2001], *Przeglad Epidemiologiczny*, **57**, 99–105.
Przybylska A (2004), 'Zatrucia jadem kielbasianym w 2002 roku' [Botulism in Poland in 2002], *Przeglad Epidemiologiczny*, **58**, 103–110.
Rhodehamel E J, Solomon H M, Lilly T, Kautter D A and Peeler J T (1991), 'Incidence and heat-resistance of *Clostridium botulinum* type E spores in menhaden surimi', *Journal of Food Science*, **56**, 1562–1563, 1592.
Schmidt C F (1964), 'Spores of *C. botulinum*: Formation, resistance, germination', in Lewis K H and Cassel K, *Botulism 1964*, Cincinatti, US Public Health Service, 69–82.
Schmidt C F, Lechowich R V and Folinazzo J F (1961), 'Growth and toxin production by type E *Clostridium botulinum* below 40°F', *Journal of Food Science*, **26**, 626–630.
Scott V N and Bernard D T (1982), 'Heat-resistance of spores of non-proteolytic type-B *Clostridium botulinum*', *Journal of Food Protection*, **45**, 909–912.
Sebaihia M, Peck M W, Minton N P, Thomson N R, Holden M T G, Mitchell W J, Carter A T,Bentley S D, Mason D R, Crossman L, Paul C J, Ivens A, Wells-Bennik M H J, Davis I J, Cerdeno-Tarraga A M, Churcher C, Guail M A, Chillingsworth T, Feltwell T, Fraser A, Goodhead I, Hance Z, Jagels K, Larke N, Maddison M, Moule S, Mungall K, Norbertczak H, Rabbinowitsch E, Sanders M, Simmonds M, White B, Whitehead S and Parkhill J (2007), 'Genome sequence of a proteolytic (Group I) *Clostridium botulinum* strain Hall A and comparative analysis of the clostridial genomes', *Genome Research*, **17**, 1082–1092.
Sebaihia M, Wren B W, Mullany P, Fairweather N F, Minton N, Stabler R, Thomson N R, Roberts A P, Cerdeno-Tarrraga A M, Wang H W, Holden M T G, Wright A, Churcher C, Quail M A, Baker S, Bason N, Brooks K, Chillingworth T, Cronin A, Davis P, Dowd L, Fraser A, Feltwell T, Hance Z, Holroyd S, Jagels K, Moule S, Mungall K, Price C, Rabbinowitsch E, Sharp S, Simmonds M, Stevens K, Unwin L, Whithead S, Dupuy B, Dougan G, Barrell B and Parkhill J (2006), 'The multidrug-resistant human pathogen *Clostridium difficile* has a highly mobile, mosaic genome', *Nature Genetics*, **38**, 779–786.
Shimizu T, Ohtani K, Hirakawa H, Ohshima K, Yamashita A, Shiba T, Ogasawara N, Hattori M, Kuhara S and Hayashi H (2002), 'Complete genome sequence of *Clostridium perfringens*, an anaerobic flesh-eater', *Proceedings of the National Academy of Sciences of the United States of America*, **99**, 996–1001.
Shuler C, Drenzek C, Lance S, Gonzalez G, Miller J, Tobin-D'Angelo M, Gabel J, Burnett C, Atrubin D, Sobel J, Juliao P, Maslanka S, Wiersma P and Sheth A (2006), 'Botulism associated with commercial carrot juice – Georgia and Florida, September 2006', *Morbidity and Mortality Weekly* **55**, 1198–1190.
Smelt J P P M (1980), '*Heat resistance of* Clostridium botulinum *in acid ingredients and its significance for the safety of chilled foods*', Doctoral Thesis, University of Utrecht, The Netherlands.
Smelt J P P M and Haas H (1978), 'Behaviour of proteolytic *Clostridium botulinum* types A and B near the lower temperature limits of growth', *European Journal of Applied Microbiology and Biotechnology*, **5**, 143–154.
Snyder O P (1996), 'Redox potential in deli foods: Botulism risk?', *Dairy, Food and Environmental Sanitation*, **16**, 546–548.

Sobel J, Tucker N, Sulka A, McLaughlin J and Maslanka S (2004), 'Foodborne botulism in the United States, 1990–2000', *Emerging Infectious Diseases*, **10**, 1606–1611.

St Louis M E, Peck S H S, Bowering D, Morgan G B, Blatherwick J, Banerjee S, Kettyls G D M, Black W A, Milling M E, Hauschild A H W, Tauxe R V and Blake P A (1988), 'Botulism from Chopped Garlic – Delayed Recognition of A Major Outbreak', *Annals of Internal Medicine*, **108**, 363–368.

Stringer S C, Fairbairn D A and Peck M W (1997), 'Combining heat treatment and subsequent incubation temperature to prevent growth from spores of non-proteolytic *Clostridium botulinum*', *Journal of Applied Microbiology*, **82**, 128–136.

Stringer S C, Haque N and Peck M W (1999), 'Growth from spores of nonproteolytic *Clostridium botulinum* in heat-treated vegetable juice', *Applied and Environmental Microbiology*, **65**, 2136–2142.

Stringer S C and Peck M W. (1997), 'Combinations of heat treatment and sodium chloride that prevent growth from spores of nonproteolytic *Clostridium botulinum*', *Journal of Food Protection*, **60**, 1553–1559.

Stringer S C, Webb M D, George S M, Pin C and Peck M W (2005), 'Heterogeneity of times required for germination and outgrowth from single spores of nonproteolytic *Clostridium botulinum*', *Applied and Environmental Microbiology*, **71**, 4998–5003.

Stumbo C R, Purohit K S and Ramakrishna T V (1975), 'Thermal process lethality guide for low-acid foods in metal containers', *Journal of Food Science*, **40**, 1316–1323.

Tanner F W and Oglesby E W (1936), 'Influence of temperature on growth and toxin production by *Clostridium botulinum*', *Food Research*, **1**, 481–494.

Therre H (1999), 'Botulism in the European Union' *Eurosurveillance*, **4(1)**, 2–7.

Townes J M, Cieslak P R, Hatheway C L, McCroskey L M, Griffin P M, Solomon H M, Holloway J T, Keller C F and Baker M P (1996), 'An outbreak of type A botulism associated with a commercial cheese sauce', *Annals of Internal Medicine*, **125**, 558–563.

Villar R G, Shapiro R L, Busto S, Riva-Posse C, Verdejo G, Farace M I, Rosetti F, San Juan J A, Julia C M, Becher J, Maslanka S E and Swerdlow D L (1999), 'Outbreak of type A botulism and development of a botulism surveillance and antitoxin release system in Argentina', *Journal of the American Medical Association*, **281**, 1334–1338, 1340.

Viscens R, Rasolofonirina N and Coulanges P (1985), 'Premiers cas humains de botulisme alimentaire a Madagascar', *Archives of the Institute Pasteur, Madagascar*, **52**, 11–22.

Webb M D, Pin C, Peck M W and Stringer S C (2007), 'Historical and contemporary NaCl concentrations affect the duration and distribution of lag times from individual spores of nonproteolytic *Clostridium botulinum*', *Applied and Environmental Microbiology*, **73**, 2118–2127.

Whiting R C and Oriente J C (1997), 'Time-to-turbidity model for non-proteolytic type B *Clostridium botulinum*', *International Journal of Food Microbiology*, **36**, 49–60.

15

Hazardous compounds in processed foods

C. Perez-Locas and V. A. Yaylayan, McGill University, Canada

15.1 Introduction

For several decades, synthetic chemicals have been the focus of attention, from consumer groups to government regulatory agencies, as potential risk hazards in food. Concerns pertaining to their safety in foods have long been studied in relation to their immediate toxicity as well as to their long-term mutagenicity and carcinogenicity. However, the focus began shifting with the discovery of a series of mutagens and carcinogens (Figures 15.1 and 15.2) found to be naturally produced in foods as a result of thermal treatment, particularly at high temperature frying, baking, barbecuing and roasting (Sugimura, 1986). Moreover, in 2002, a world-wide concern was raised with the discovery of acrylamide, a suspected carcinogen, in a variety of foods (Tareke *et al.*, 2002). Despite the lack of conclusive evidence supporting the degree of toxicity of several thermally generated food hazards, the potential threat of possible carcinogenicity resulting from the presence of these compounds in foods provides enough incentive for further research. Numerous studies have been conducted on thermally generated food toxicants, ranging from identification, formation, toxicology and analysis (Wenzl *et al.*, 2007). This chapter will summarize these findings.
A list of abbreviations used in this chapter will be found in Section 15.8.

15.2 Polycyclic aromatic hydrocarbons

15.2.1 Sources

Polycyclic Aromatic Hydrocarbons, also referred to as Polynuclear Aromatic

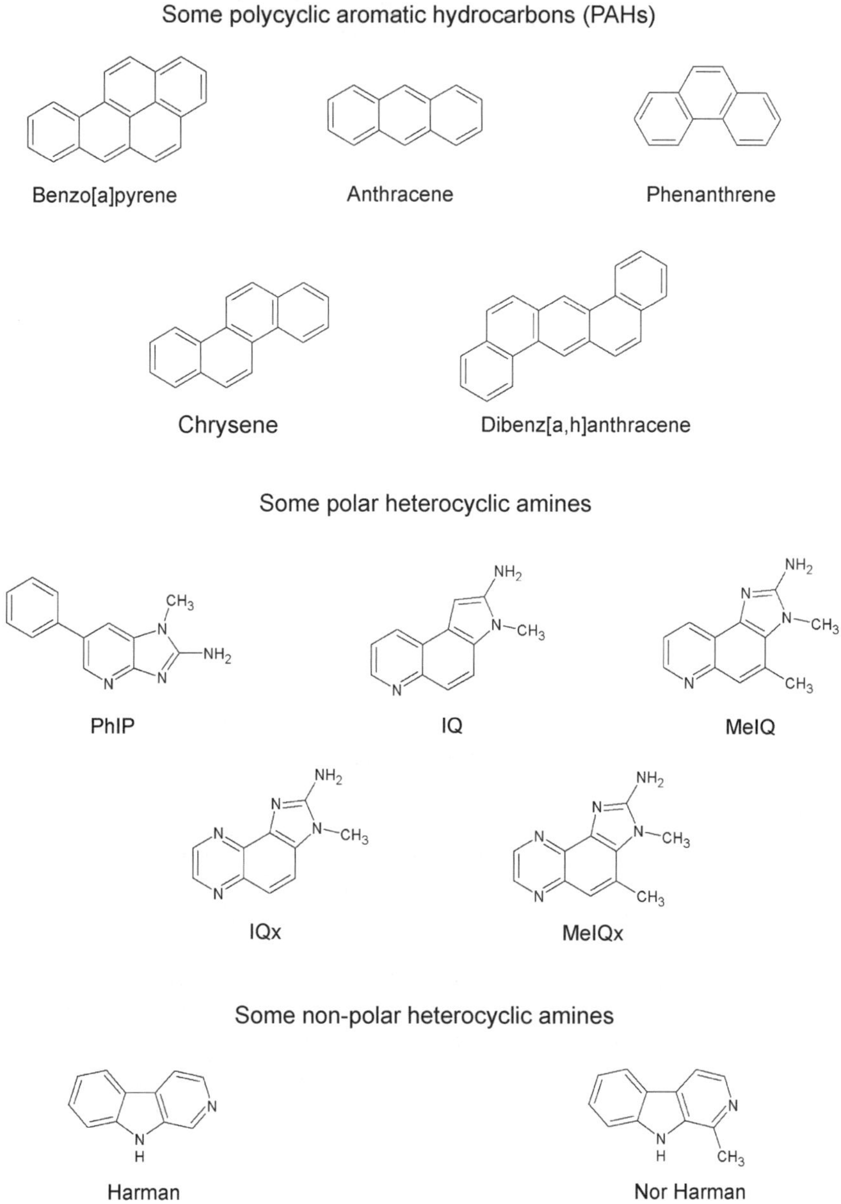

Fig. 15.1 Chemical structure of various polycyclic aromatic hydrocarbons and heterocyclic amines.

Hydrocarbons (PAH) are a group of compounds primarily associated with environmental contamination (Fig. 15.1). As indicated by their classification, their structures are composed of two or more aromatic rings, typically fused together. They appear to originate from pyrolysis of almost any organic material and are

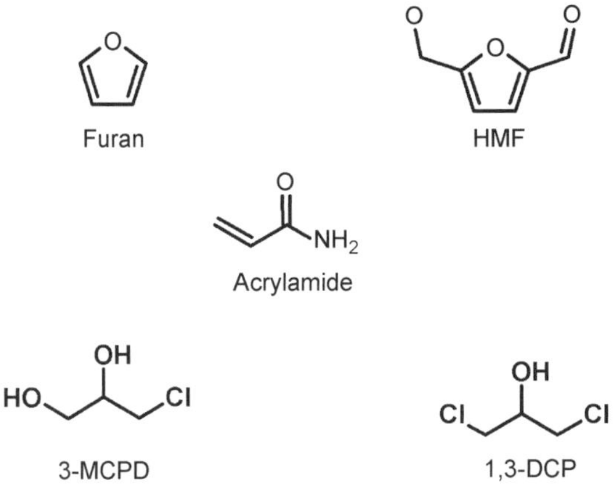

Fig. 15.2 Chemical structure of selected heat-induced toxicants.

hence practically unavoidable (Higman *et al.*, 1970). Most types of smoke, such as tobacco, oil, wood and volcano are known to harbor a variety of these (IARC, 1983; Jacob, 1996). Environmental PAHs are also found in soils and in airborne particles originating from natural or anthropogenic processes. Cross-contamination into the food source is therefore readily known to occur. However, endogenous formation of PAHs in foods as a result of thermal processing is also an important contributor to these daily ingested heterocycles (Phillips, 1999; Janoszka *et al.*, 2004).

Exposure of fats and oils to high temperatures, as well pyrolysis of proteins, certain amino acids and carbohydrates, can lead to endogenous dietary PAH formation (Higman *et al.*, 1970; Sharma *et al.*, 2003; Britt *et al.*, 2004). Clearly, charring of food from high temperature barbecuing, roasting or broiling, especially on an open flame, can certainly contribute to foodborne PAH levels arising from smoke surface contamination as well as from resulting degradation of proteins, sugars and oil (Phillips, 1999; Janoszka *et al.*, 2004). Smoking and drying processes also contribute to PAH levels (Panalaks, 1976; Gomaa *et al.*, 1993).

15.2.2 Diversity of PAH

Over 10 000 PAH related compounds are present in the environment, from which some have been identified and studied. In foods, it is rather difficult to treat PAHs individually as each member of this classification is generally present at very low ppb or ppt levels (Janoszka *et al.*, 2004; Wenzl *et al.*, 2006b). However, although their individual contribution might appear insignificant, their cumulative impact is important. Benzo[a]pyrene (B[a]P), one of the PAHs exhibiting the highest carcinogenic potency, is often used as a marker for PAH contamination. Estimation of overall PAH content can also be achieved by monitoring a variety of PAHs, as established by the IPCS recommendations (IPCS, 1998). In addition to parent PAHs, many PAH derivatives have also been identified, such as alkylated PAHs and nitrogen-containing PAHs, that tend to have a greater carcinogenic potential than their unsubstituted counterparts. Although their specifics are not discussed here, a number of reviews and government documents on polycyclic aromatic compounds (PACs) have been published, addressing their occurrence and health effects (Barnes *et al.*, 1983; Phillips, 1999; EC, 2002).

15.2.3 Toxicology

Although trends explaining the higher toxicity of some PAHs relative to others are not well defined, the stability of structures containing more than three rings (ideally 3 to 6) tend to make them more harmful (Wenzl *et al.*, 2006b). Moreover, among the many theories proposed, toxicity can be partially explained by the presence of many structural regions associated with an increased reactivity of the molecule. For example, the occurrence of a so-called biochemically active 'bay region' is seen as one of the main determining structural features associated with carcinogenicity. Less carcinogenic PAHs typically lack this bay region (Jacob, 1996) and, although exceptions do arise, this theory is among the most favored. The ease of epoxidation and resistance to enzymatic detoxification of this area is the hypothesized explanation.

The body absorbs PAHs from virtually all routes of exposure, including orally, via the respiratory tract and through the skin. In response to their non-polar nature, ingested PAHs are stored in adipose tissue and are slowly released to form various metabolites, most of which are excreted (IPCS, 1998). Not all PAHs have been classified; nonetheless, several members of this grouping are recognized as being mutagenic, carcinogenic and teratogenic, as is the case with dibenz[a,h]anthracene, benzo[b]fluoranthene, indeno[1,2,3-cd]pyrene, benzo[a]pyrene and dibenzo[a,l] pyrene, among others (Phillips, 1999). A number of animal studies have associated diets high in PAH with an increased number of tumors, such as in the foregut, bladder and lungs. Moreover, assessment of occupational exposure has shown positive correlation between exposure to polynuclear hydrocarbons and an increased risk of lung cancer in workers (Lloyd, 1971; Gibbs and Horowitz, 1979). A more detailed overview of PAH toxicology was presented by the Europeean Commission's Scientific Committee on Foods (EC, 2002).

15.2.4 Formation

The formation pathways of PAHs have not yet been established. In their quest for a better mechanistic understanding, Britt's group (Britt *et al.*, 2004) undertook the pyrolytic degradation of proline Amadori compound, providing evidence of a possible Maillard reaction implication. Formation of PAHs from amino acids indicates the occurrence of decarboxylation and deamination reactions. Badger's group (Badger *et al.*, 1960) had long ago suggested dimerization of three-carbon units and Hurd and Simon (1962) hypothesized on possible trimerization of two-carbon units from proline degradation products. Moreover, PAHs have been proposed to be formed via a Diels–Alder cycloaddition reaction between benzene and lipid degradation products such as butadiene. Despite the passing of time, only limited information is available as to the details of PAH formation particularly related to food systems, and complete mechanisms have not yet been presented.

Endogenous formation of PAHs in food can occur as a result of high temperature cooking such as grilling and roasting, where levels of up to 164ppb and even 320ppb can be found (Panalaks, 1976; Chen and Lin, 1997). Uncooked meats only contain trace amounts. A variety of PAH yielding-components are presented in Table 15.1.

15.2.5 Analysis and quantitation

It is difficult to analyze food for each PAH present. However, a strategy must be devised on obtaining insight on food's total PAH load. Therefore, depending on the objective of the analysis, the analysts may look into extraction of B[a]P, considered by many as a marker for the overall carcinogenic impact of PAH in the food. As previously mentioned, B[a]P is seen as one of the most toxic PAHs and its levels are important to determine. Alternatively, a few common PAHs can also be isolated and quantified for added insight (Phillips, 1999). Analytical precautions are therefore taken according to the information required.

From a general perspective, analysis of PAHs in foods is considered to be a complex process. Low ppb concentrations and interference from compounds of similar properties, as well as separation of the many different PAHs, are challenging issues. Moreover, their susceptibility to oxidation and photodecomposition requires special handling procedures and the tendency of low molecular weight PAHs to sublimate is yet another analytical consideration. Overcoming these issues requires adequate purification such as, to avoid interferences, the use of internal ^{13}C-labeled standards, the use of UV protected lamps, the addition of antioxidants, as well as minimizing solvent drying operations (Janoszka *et al.*, 2004; Tamakawa, 2004).

Extraction and adequate purification are essential components for PAH quantification in food. Over the years, a number of analytical approaches have been studied and compared in the hope of obtaining maximum recovery of low-level PAHs. Methodologies used largely depend on the type of matrix. Low-fat foods, such as fruits and vegetables, can undergo a mild liquid–liquid extraction, soxhlet or sonication treatment, whereas foods with a higher lipid content or proteinaceous material, such as meats, require a harsher alkaline saponification prior to extraction (soxhlet, sonication, SPE)(Bartle, 1991; Janoszka *et al.*, 2004). As explained in Tamakawa's review (2004), methanolic KOH saponification with solvent extraction is one of the most common procedures as it will eliminate a number of interfering substances including fats and pigments, although addition of Na_2S antioxidant can be necessary to counteract the potentially destructive impact that basicity might have on recovery of the more labile PAHs. Solid–phase extraction is also a popular means of removing matrix and has been known to be used without saponification as in the case of oils (Moret and Conte, 2002; Moret *et al.*, 2005). More recent methodologies have also been evaluated, including Microwave

Table 15.1 Formation of polyaromatic hydrocarbons from carbohydrates, amino acids and others

Precursors	Pyrolysis temperature	Identified PAHs	Reference
Glucose	840 °C	Indene; Nap thalene; Acenap thalene; Fluorene; Phenanthrene; Anthracene; Benzo(a)pyrene	Higman *et al.*, 1970
Fructose	840 °C	Indene; Nap thalene; Acenap thalene; Fluorene; Dibenzofuran; Phenanthrene; Anthracene; Fluroanthene; Pyrene; Benzo(a)pyrene	Higman *et al.*, 1970
Cellulose	840 °C	Indene; Nap thalene; Acenap thalene; Fluorene; Phenanthrene; Anthracene; Benzo(a)pyrene; Pyrene	Higman *et al.*, 1970
Casein	840 °C	Indene; Nap thalene; Fluorene	Higman *et al.*, 1970
Collagen	840 °C	Indene; Nap thalene;	Higman *et al.*, 1970
Valine	650 °C/850 °C	Benzo[a]pyrene	Patterson *et al.*, 1978
Asparagine	920 °C	Acenaphthylene; Fluorene; Phenanthrene; Anthracene; Fluoroanthene; Pyrene	Sharma *et al.*, 2003
Proline	850 °C/30s	Nap thalene ; Phenanthracene; Anthracene; Pyrene	Smith *et al.*, 1975
Proline	840 °C/10s	Acenapthylene; Acenap thene; Fluorene; Fluoroanthene; Pyrene;	Britt *et al.*, 2004
Glucose	850 °C/10 or 30s	Phenanthracene; Anthracene; Pyrene	Smith *et al.*, 1975 and Britt *et al.*, 2004
1-[(2'-carboxy) pyrrolidinyl] -1-deoxy-D-fructose (Proline Amadori)	840 °C/10s and 800 °C/1s	Benzofuran; Nap thalene; Acenap thylene; Acenap thene; Fluorene; Phenanthrene; Fluoroanthene; Pyrene; Benz(a)anthracene; Benzo-fluoroanthenes; Benzo(a)pyrene; Anthracene; Pyrene; Chrysene	Britt *et al.*, 2004
Proline–Glucose	840 °C/10s	Phenanthracene; Anthracene; Pyrene	Britt *et al.* 2004

Assisted Extraction (MAE) (Filipkowska *et al.*, 2005; Pakou *et al.*, 2007), Super Critical Fluid Extraction (SFE) (Rivas, 2006), Accelerated Solvent Extraction (ASE) (Wang *et al.*, 1999; Ziegenhals *et al.*, 2007) and Pressurized Liquid Extraction (PLE) (Houessou *et al.*, 2006). Most of these newer analytical procedures, however, are mainly focused on environmental samples from soil, air and marine.

Prior to analytical separation, further purification steps are often required to achieve maximum recovery and to minimize interfering substances. Liquid–liquid partition, as well as number of chromatographic methodologies, have been frequently used following extraction for additional refinement. Moreover, the convenience of commercial SPE cartridges have been exploited for dietary PAH analysis (Tamakawa *et al.*, 1986; Janoszka *et al.*, 2004; Moret *et al.*, 2005)).

Quantitative separation and identification of dietary PAHs from purified samples are typically achieved by Gas Chromatography (GC) coupled with FID or MS. GC/MS on a non-polar columns with SIM mode can be a particularly useful tool for identification and quantitation. High Performance Liquid Chromatography (HPLC) with UV, MS or fluorescence are other strategies used for PAH analysis and are particularly useful for high molecular weight, heat-labile PAHs (EC, 2002). Moreover, the selectivity and sensitivity that fluorescence provides have their advantages, particularly in minimizing interferences. Thin layer chromatography (TLC) coupled with fluorescence has also been used (Tamakawa, 2004), although it is not as popular as other chromatographic techniques.

15.2.6 *De minimis* principle

Although aiming for complete elimination of PAHs in food would be difficult to achieve, strategies have been proposed to keep thermal generation of PAH in food to a minimum. Favoring lower temperature cooking, avoiding charring of foods and contact with direct flames, as well as cooking leaner products (low fat meats) can certainly reduce the amounts of PAHs formed (Phillips, 1999). In response to the toxicological data obtained, a number of European countries imposed or recommended limits for various PAHs. Wenzyl and co-workers have presented a summary of the set limits (2006b).

15.3 Heterocyclic aromatic amines

15.3.1 Sources

Heterocyclic aromatic amines (HAs) represent a large group of compounds, many of which exhibit carcinogenic, mutagenic and other toxic properties. They are typically produced as a result of high-temperature thermal processing and can be found in a wide variety of protein-rich food products (Bartoszek, 1997). HAs can be subdivided into two groups: polar and non-polar. Polar HAs are characteristically found in meats and fish, as they require the presence of creatine for their formation. Non-polar HAs are also found in cooked meat and fish, as well as in alcoholic drinks, coffee brews and cigarettes.

Table 15.2 Classification, properties and formation of common heterocyclic amines (adapted from Skog (1993) and Jägerstad *et al.* (1998)).

Sub-classification	Chemical name (abbreviation	Molecular weight	Structure	Model systems, conditions and yield			References
Polar heterocyclic amines							
Imidazoquinoline-type	2-Amino-3-methylimidazo [4,5-*f*]quinoline (IQ)	198	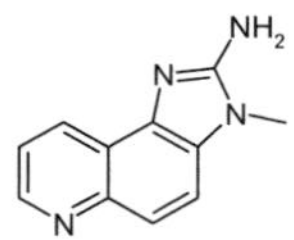	Pro	Dry	0.4	(Yoshida *et al.*, 1984; Grivas *et al.*, 1986; Felton and Knize, 1990, 1991a,b; Knize *et al.*, 1988)
				Gly; Fru	DEG-H_2O	1.0	
				Phe	Dry	3.0	
				Phe; Glu	Dry	13.5	
				Ser	Dry	3.7	
Imidazoquinoline-type	2-Amino-3,4-dimethyl-imidazo[4,5-*f*] quinoline (MeIQ)	212	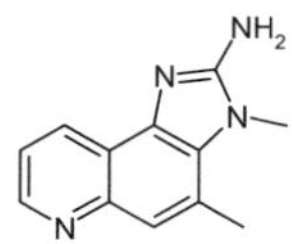	Ala; Fru	DEG-H_2O	nd	Grivas *et al.*, 1985
Imidazoquinoxaline-type	2-Amino-3-methylimidazo [4,5-*f*]quinoxaline (IQx)	199	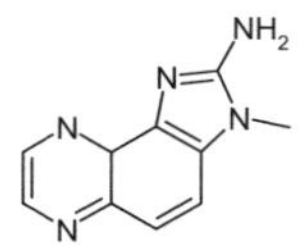	Ser	Dry	2.7	Knize *et al.*, 1988; Skog and Jagerstad 1993.
				Thr; Glu	H_2O	nd	
Imidazoquinoxaline-type	2-Amino-3,8-dimethylimidazo [4,5-*f*] quinoxaline (MeIQx)	213	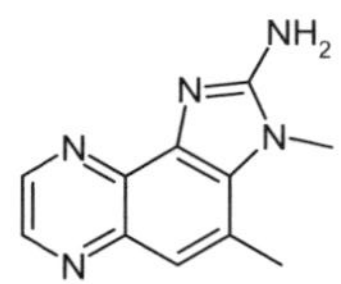	Gly; Glu	DEG-H_2O	4.4	Jagerstad *et al.*, 1984; Muramatsu and Matsushima 1985; Negishi *et al.*, 1985; Skog and Jägerstad, 1990, 1991 and 1993; Grivas *et al.*; 1986; Overik *et al.* 1989; Skog *et al.*, 1992; Johansson *et al.*, 1993;
				Gly; Glu	H_2O	7–17.9	
				Gly; Glu	DEG-H_2O*	4.0	
				Gly; Fru	DEG-H_2O	6–7	
				Ala	Dry	nd	
				Ala; Glu	DEG-H_2O	0.9	
				Ala; Rib	DEG-H_2O	1.8	
				Ala; Thr; Rib	DEG-H_2O*	10	
				Lys; Rib	DEG-H_2O	4.2	
				Phe; Glu	DEG-H_2O*	nd	

				Ser	Dry	nd	
				Thr; Glu	DEG-H_2O	nd	
				Thr; Glu	H_2O	9	
				Tyr	Dry	nd	
Imidazoquinoxaline-type	2-Amino-3,4,8-trimethyl-imidazo [4,5-*f*] quinoxaline (4,8-MeIQx)	227	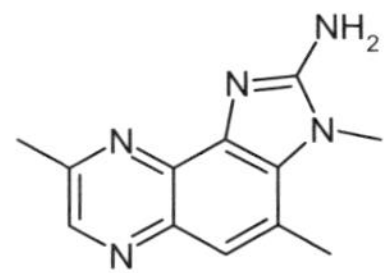	Ala; Fru	DEG-H_2O	1.9–2.6	Negishi *et al.*, 1985; Grivas *et al.*, 1985; Muramatsu and Matsushima, 1985; Skog and Jägerstad, 1990, 1991 and Gly; 1993; Skog *et al.*, 1992; Johansson *et al.*, 1993;
				Ala; Rib	DEG-H_2O	1.5	
				Ala; Glu	DEG-H_2O	4.2	
				Ala; Thr; Rib	DEG-H_2O	36	
				Gly; Glu	DEG-H_2O*	nd	
				Glu	H_2O	nd	
				Lys; Rib	DEG-H_2O	26.1	
				Phe; Glu	DEG-H_2O*	nd	
				Thr; Glu	DEG-H_2O	nd	
				Thr; Glu	H_2O	30	
Imidazoquinoxaline-type	2-Amino-3,7,8-trimethyl-imidazo[4,5-*f*] quinoxaline (7,8-MeIQx)	227	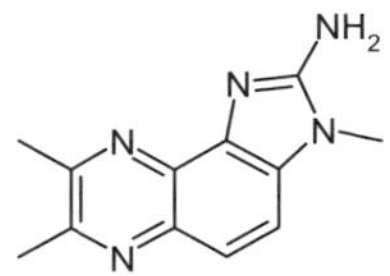	Gly; Glu	DEG-H_2O	1.1	Negishi *et al.*, 1985; Skog and Jägerstad, 1990
				Gly; Glu	DEG-H_2O*	nd	
Imidazoquinoxaline-type	2-Amino-3,4,7,8-trimethyl-imidazo[4,5-*f*] quinoxaline (4,7,8-TriMeIQx)	241	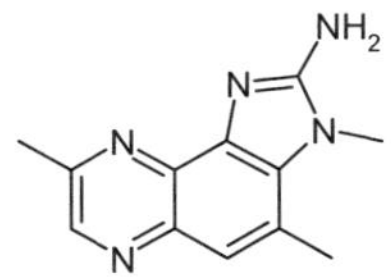	Ala; Thr; Glu	DEG-H_2O*	6	Skog *et al.*, 1992

Table 15.2 *Cont'd*

Sub-classification	Chemical name (abbreviation)	Molecular weight	Structure	Model systems, conditions and yield	References
Imidazopyridine-type	2-Amino-1-methyl-6-phenylimidazo[4,5-*b*] pyridine (PhIP)	224	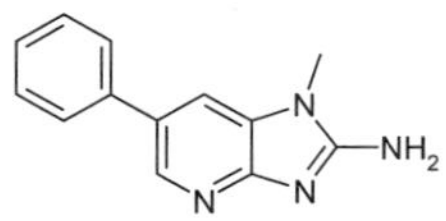	Leu Dry nd Phe Dry 735 Phe DEG-H_2O* 6.4 Phe; Glu Dry 560 Glu DEG-H_2O 3.6 Phe; Glu DEG-H_2O* 20.9	Shioya *et al.*, 1987; Felton and Knize, 1990; Overik *et al.*, 1989; Skog and Phe; Jägerstad, 1991
Imidazopyridine-type	2-Amino-1-methyl-6-(4-hydroxyphenyl)imidazo [4,5-*b*]pyridine (4'OH-PhIP)	240	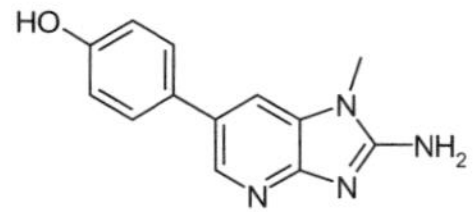	Tyr; Glu	Kurosaka *et al.*, 1992.
Imidazopyridine-type	Dimethylimidazopyridine (DMIP)	162	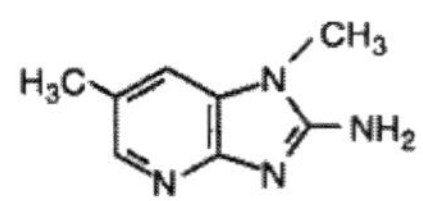	None so far	Becher *et al.*, 1988, 1989; Felton *et al.*,1984
Imidazopyridine-type	Dimethylimidazopyridine (TMIP)	176	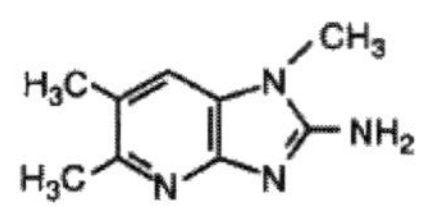	None so far	Becher *et al.*, 1988, 1989; Felton *et al.*,1984
Non-polar heterocyclic amines – amino carbolines					
α-Carbolines	2-amino-9*H*-pyrido[2,3-*b*] indole (AaC)	183	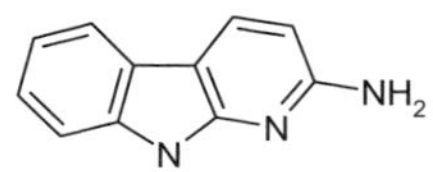	Soybean Globulin, Tryptophan or Albumin pyrolysis	Jägerstad *et al.*, 1998
α-Carbolines	2-amino-3-methyl-9*H*-pyrido [2,3-*b*]indole (MeAaC)	197	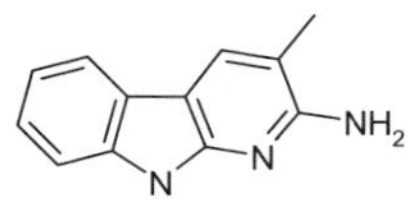		

β-Carbolines	9*H*-pyrido[4,3-*b*] indole (Norharman)	168	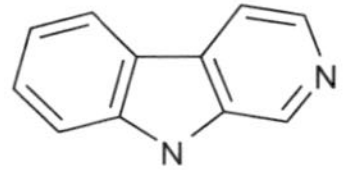	Tryptophan pyrolysis and various polar HA forming model systems containing create(ni)ne, amino acids and sugars.	Jägerstad *et al.*, 1998
β-Carbolines	1-methyl-9*H*-pyrido [4,3-*b*] indole (harman)	182	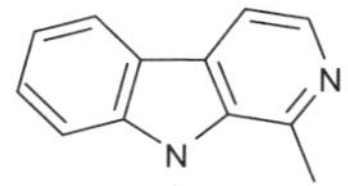		
γ-Carbolines	3-Amino-1,4-dimethyl-5*H*-pyrido[4,3-*b*]indole (Trp-P-1)	211	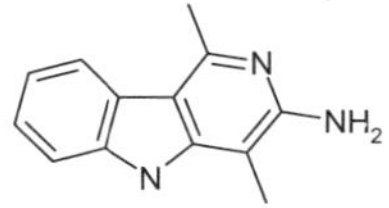	Tryptophan or various protein pyrolysis	Jägerstad *et al.*, 1998
γ-Carbolines	3-Amino-1-methyl-5*H*-pyrido[4,3-*b*]indole (Trp-P-2)	197	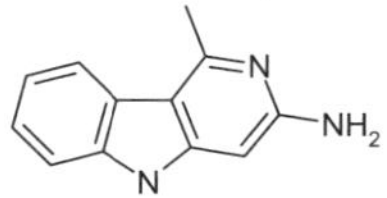		
δ-Carbolines	2-Amino-6-methyl-dipyrido[1,2-*a*:3',2'-*d*] imidazole (Glu-P-1)	198	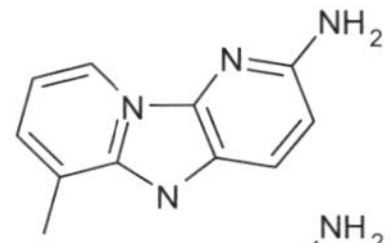	Glutamate, glutamic acid or casein Pyrolysis	Jägerstad *et al.*, 1998
δ-Carbolines	2-Amino-dipyrido [1,2-*a*:3',2'-*d*]imidazole (Glu-P-2)	184	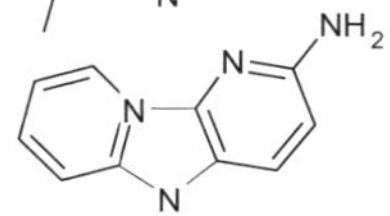		

Model systems: Amino Acids – Al : Alanine; Gly : Glycine ; Pro : Proline; Ser: Serine; Leu: Leucine; Phe: Phenylalanine; *Sugars* – Glu: Glucose; Rib: Ribose; Fru: Fructose. All Polar HA were heated in the presence of Creatin(in)e.
Conditions – Dry: Dry heating at 180 or 200 °C for 1h; DEG- H_2O: Refluxed in Diethylene Glycol-Water (5:1) 128 °C/2h; *DEG- H_2O: Heated in Diethylene Glycol-Water (5:1) 180 °C/10min ; H_2O: Heated in water in closed tubes at 180 °C for 10 or 30 minutes.
Yield – Given in nmol/mmol of Creatin(in)e.

15.3.2 Structure of heterocyclic aromatic amines

Several HAs of concern have been classified based on their polarity and are presented in Table 15.2. Polar HAs possess a common 2-aminoimidazole moiety in their structure, most likely originating from creatine. Also referred to as 'thermic mutagens', 'Amino imidazoazaarenes (AIAs)' or 'IQ-type HAs', they are classified into three subgroups based on their structure; thus, the imidazoquinolines, imidazoquinoxalines and imidazopyridines (Felton *et al.*, 1986; Chen and Chiu, 1998; Jagerstad *et al.*, 1998).

Non-polar HAs, commonly referred to as carboline-type HAs or pyrolytic mutagens, are typically formed at much higher temperatures, often exceeding 250 °C, resulting from amino acid or protein pyrolysis (Wakabayashi *et al.*, 1997; Chen and Chiu, 1998). They contain a common 2-aminopyridine moiety and are further subdivided into the pyridoindole and the dipyridoimidazoles, based on their structures as presented in Table 15.2.

15.3.3 Toxicology

To date, most imidazole-type HAs, except for diMeIQx, have been studied and found to exhibit long-term carcinogenicity and genotoxicity in rodents (Yoshimi *et al.*, 1988; Adamson *et al.*, 1990; Ohgaki *et al.*, 1991; Wakabayashi *et al.*, 1992; Schut and Snyderwine, 1999). PhIP is the most abundant AIA, but also the least carcinogenic. MeIQ is present in foods in smaller quantities, but results in the strongest carcinogenic consequences in rodents. Moreover, IQ has been tested and has proved to be carcinogenic as well as a promoter of cardiovascular disease in primates (Adamson *et al.*, 1990; Adamson and Thorgeirsson, 1995; Schut and Snyderwine, 1999). *In vitro*, various HAs have been shown to genetically damage mammalian cells (Barnes *et al.*, 1983). They act by modifying DNA, leading to the formation of neoplastic cells from normal cells (Weisburger and Jones, 1989). Thus, scientific consensus shows that heterocyclic amines are genotoxic, interacting with DNA and forming adducts, affecting its biological behaviour, inducing genetic damage and playing a role in carcinogenesis (Barnes *et al.*, 1983; Weisburger and Jones, 1989; Wakabayashi *et al.*, 1993). Schut and Snyderwine (1999) were able to relate HCA–DNA adducts to their possible role in cancer development and elucidate the pathways by which AIAs are activated in the body prior to the formation of DNA adducts. As a result of the toxicological information obtained over a number of years, IQ is classified as a type-2A carcinogen by the IARC, whereas MeIQ, MeIQx and PhIP are classified as type 2B possible carcinogens by the IARC, and 'Reasonably anticipated to be Human Carcinogens' by the National Toxicology Program (NTP) (IARC, 1993; NTP, 2002). The common 2-aminoimidazo group originating from creatine is the main contributor to the carcinogenic and genotoxic properties of the molecules (Grivas and Jägerstad, 1984).

Among the non-polar HAs, particular attention has been attributed to the β-carbolines such as harman (1-methyl-9H-pyrido[3,4-b]-indole) and norharman (9H-pyrido[3,4-b]-indole). The importance of β-carbolines stems from their

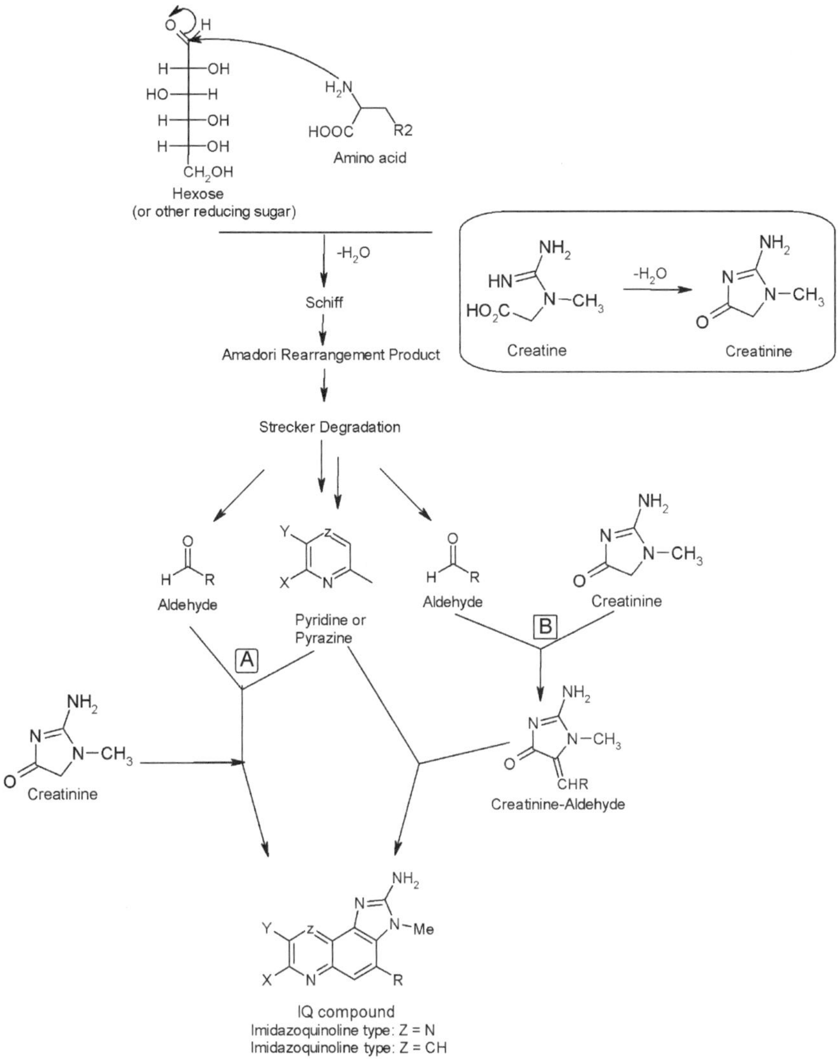

Fig. 15.3 General mechanism of formation of IQ and IQx-type as proposed by (a) Jägerstad *et al.*, 1983 and (b) Nyhammer, 1984.

implication in various diseases such as Parkinson's, tremors, addiction, as well as cancer. Although individually non-mutagenic to the Ames test, they are indirect contributors to carcinogenesis, acting as co-mutagens in the presence of aromatic amines such as aniline and toluidine, 3-aminopyridine, 2-amino-3-methylpyridine and N,N-diphenylamine (Nagao *et al.*, 1978; Sugimura *et al.*, 1982), many of which are found in cigarette smoke, vegetables, environment, detected in breast milk and urine, overall making the interaction rather likely. Several mutagens have been isolated, from reacting β-carbolines with aniline and toluidine. The mutagenic compounds arising from the coupling are grouped into aminophenyl β-carbolines derivatives and include APNH, APH, AMPH, 2'AMPNH and 3'AMPNH, APNH being the most potent with an activity matching that of MeIQx and Trp-P-2 (Totsuka *et al.*, 2004).

15.3.4 Formation

Polar HAs require the presence of amino acids and creatine/creatinine, and are therefore found mainly in cooked meat and fish, although traces have also been isolated from cigarette smoke. The presence of reducing sugars has been shown to enhance as well as inhibit HA formation, the effect varying according to the type of HA, the amount and type of sugars added, and the heating conditions (Skog and Jägerstad, 1990). Although several mechanistic pathways of formation of imidazoquino(xa)lines have been proposed, they remain mainly speculative. Besides identification of the precursors, little has been shown to prove their route of formation. The two main mechanisms are presented in Fig. 15.3 as proposed by Jägerstad and co-workers (1983) and Nyhammer (1986). Both are based on Maillard formation of aldehydes, pyridines and pyrazines via Strecker degradation reaction. The first is initiated by an aldol condensation between an aldehyde and a pyridine (if quinoline) or a pyrazine (if quinoxaline), yielding the corresponding vinylpyridine/vinylpyrazine which can undergo conjugated addition to creatinine, thus resulting in formation of the IQ or IQx compound after dehydration and aromatization. Alternatively, Nyhammer (1986) also suggested aldol condensation of creatinine with an aldehyde followed by a conjugated addition of the pyridine/pyrazine onto the vinyl moiety. Jones and Weisberger (1989) also supported this alternative pathway. Indirect evidence for the involvement of the proposed intermediates was provided by Jägerstad and co-authors (1983), showing that addition of vinylpyridines or vinylpyrazines resulted in a two-fold increase in mutagenicity in model systems. However, Jones and Weisberger (1989) did not observe this effect, thus the information still remains questionable. More efforts have been directed towards investigating the mechanism of formation of the aminoimidazopyridine PhIP. PhIP is formed in greater amounts in model systems as opposed to the other polar HAs, and labeling studies are therefore more easily undertaken. Felton and Knize (1991a) observed the intact incorporation of a phenyl ring from phenylalanine and provided evidence of creatine forming the imidazole moiety of PhIP by observing the isotope label incorporation pattern. However, later studies (Murkovic *et al.*, 1999; Zochling and Murkovic, 2002) actually proposed a mechanism of formation from phenylalanine, by reacting intermediates and ^{13}C or ^{15}N-labeled precursors. The proposed mechanism is presented in Fig. 15.4. According to this mechanism, phenylalanine is deaminated to phenylacetaldehyde, a known product of phenylalanine. Phenylacetaldehyde appears to be a more direct precursor than its amino acid as it leads to significantly higher levels of PhIP. Phenylacetaldehyde is then expected to undergo aldol addition with creatinine, yielding an unstable molecule, highly prone to dehydration into its aldol condensation product. The authors claim successful isolation of

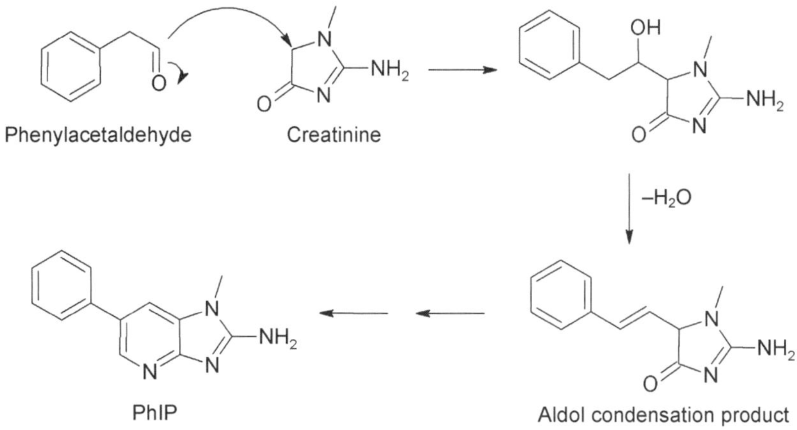

Fig.15. 4 Proposed mechanism of PhIP formation (adapted from Zochling Murkovic, 2002)

the intermediate in both model and meat systems. The final steps in PhIP formation remain highly speculative as many reactions are proposed to explain ring closure and aromatization of the pyridine. Possible reactions with creatinine, phenylacetaldehyde or free radicals are suggested, although little evidence is available to support the hypotheses.

Non-polar HAs were among the first food mutagens identified, originally isolated mainly between 1975 and 1980 (Sugimura, 1997) from protein and amino acid pyrolysis and later found in the charred parts of cooked meats and other proteinaceous foods. They are generated in cooked foods at ppb levels, as well as in tobacco smoke, and are found in various plants. They are also formed in biological systems (Totsuka *et al.*, 1999). Although mainly produced at high temperatures, the β-carbolines can also be formed at temperatures as low as 40 °C, as opposed to other pyrolytic mutagens. However, flame-broiled meats generate up to 800 μg/g of norharman and 170 μg/g harman, a few hundred folds greater than amounts found under milder cooking conditions such as frying or roasting (Pfau and Skog, 2004). Although little is known about their mechanisms of formation, they are tentatively proposed to form via a series of fragmentation and free radical reactions. Initial reaction is suggested to occur via high temperature-induced free radical reactions fragmenting the amino acid or protein precursors and to be finalized by condensation of the various fragments formed into 2-aminopyridine-containing heterocyclic structures (Matsushima, 1982; Yoshida *et al.*, 1986). A mechanism of formation via the amino acid tryptophane, a precursor to harman and norharman β-carbolines has also been presented by Yaylayan and co-workers (1990) as depicted in Fig. 15.5.

15.3.5 Analysis and quantitation

Due to the low ppb/ppt levels found in the complex matrix and poor chromato-

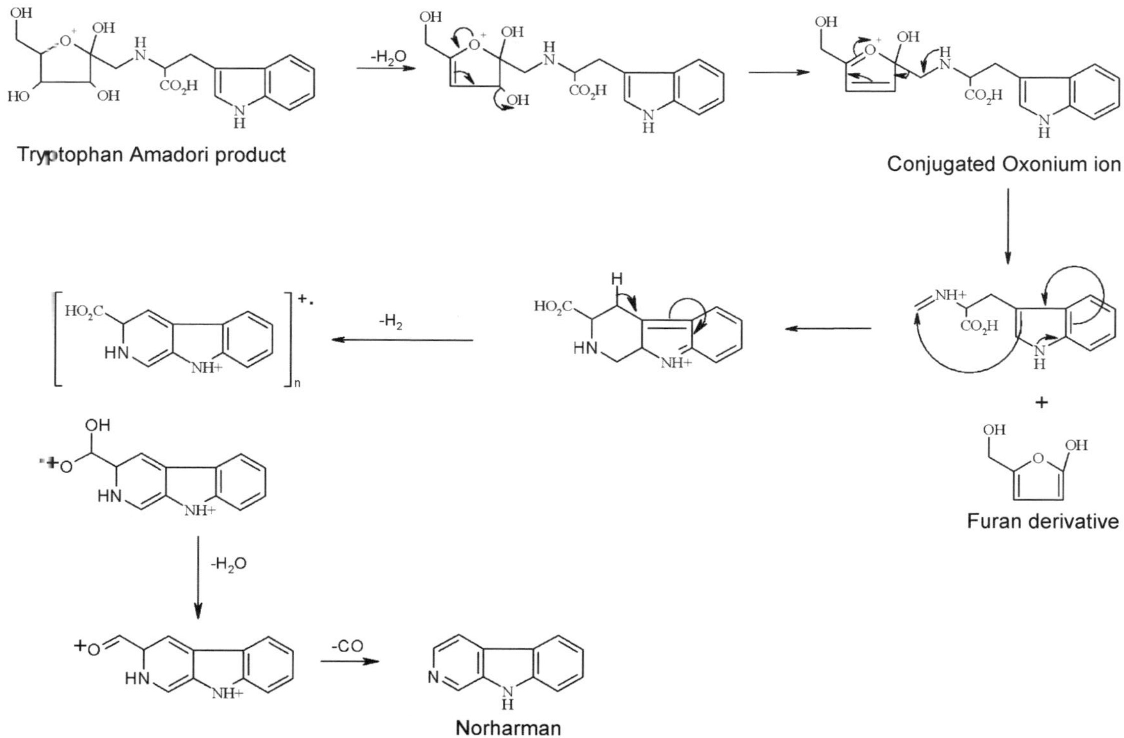

Fig. 15.5 Proposed mechanism of formation of norharman (adapted from Yaylayan *et al.*, 1990).

graphic properties, the isolation and detection of HAs remains challenging and requires several extraction and purification steps. To date, the internationally accepted method for the extraction and analysis of both polar and non-polar HAs remains that of Gross and Gruter (1992). This procedure includes liquid–liquid alkaline extraction from the food matrix on a diatomaceous earth cartridge, from which the HAs are eluted into a cation exchange PRS cartridge that will concentrate basic compounds. The apolar HAs are eluted with acidified methanol and the polar HAs are subsequently desorbed with ammonium acetate buffer. Both eluants are passed through a C18 silica cartridge. Further concentration can be achieved by passing the eluant through an additional cation exchange TSK cartridge (Gross *et al.*, 1992). Recently, however, simplified purification methods have been proposed using blue cotton, blue chitin or blue rayon, which selectively bind the HA. A review by Skog (Skog, 2004) presents the numerous advantages of this methodology as it allows a much simpler approach with results comparable to those obtained from Gross's the more labor-intensive method.

Separation is mainly performed with HPLC, as it does not require an additional derivatization step prior to analysis. Detection is typically done by UV or mass spectrometry as well as fluorescence in some instances (PHIP and β-carbolines). In the case of GC/MS analysis, low volatility and tailing attributed to the highly polar imidazole ring is a major issue. Casal and co-authors (2004) suggested a *tert*-butyl silylation procedure using N-methyl-N-(*tert*-butyldimethylsilyl)trifluoroacetamide (MTBSTFA) reagent prior to injection, successfully yielding mono and disilyl derivatives of each polar HA.

15.3.6 *De minimis* principle

Besides reducing cooking temperature and avoiding ingestion of particularly charred foods, HA levels can be reduced by a number of additives. Various natural extracts such as grape seed and oleoresins have proven to be effective, as well as addition of organosulfur-containing compounds (such as garlic) prior to cooking (Shin *et al.*, 2002; Ahn and Grun, 2005). Moreover, wine marinades can help reduce levels of some carcinogenic HAs, resulting, however, in an increase in others (Busquets *et al.*, 2006). Other marinades were also evaluated and showed significant inhibition of HA formation (Shin and Lee, 2004, 2005).

15.4 Acrylamide

15.4.1 Sources

Acrylamide is a water-soluble chemical known for its presence in tap water, cigarette smoke, laboratories and several industries (Schumacher *et al.*, 1977; IARC, 1994). Despite having been classified in 1994 by the IARC as a type IIA probable human carcinogen, scientists were unaware of its presence in food. For decades, industrial synthesis of large quantities of this chemical served multiple functions, primarily revolving around its conversion to polyacrylamide, a multipurpose, non-toxic impermeable solid matrix. This harmless polymeric form of acrylamide is widely used in cosmetics, as a grouting agent in the building and repairing of water tunnels, sewers and fountains, and as a water flocculating agent for waste water treatment, as well as for solid support in protein or nucleic acid separation (PAGE) (IARC, 1994; Friedman, 2003). In 2002, much controversy arose from the discovery of acrylamide in a variety of thermally processed foods (Tareke *et al.*, 2002). High temperature cooking of staple foods, particularly potato-based products, has received most attention due to their high acrylamide content (150–4000 μg/kg) combined with significant consumption. Other carbohydrate-rich foods as well as other foods such as coffee are also considered important sources of acrylamide dietary intake.

Despite a variety of acrylamide-generating precursors, general consensus among researchers targets the amino acid asparagine as the main precursor, requiring the presence of a carbonyl to yield significant levels of the toxicant in foods. Other amino acids have also been shown to produce acrylamide. Such amino acids include methionine, cysteine, serine, β-alanine and carnosine (Stadler *et al.*, 2002; Yaylayan *et al.*, 2003, 2004, 2005). Moreover, the fatty acid oxidation product, acrolein commonly found in oils, produces acrylamide when reacted with ammonia. However, it has not been shown to be of major importance in the frying of model food systems (Mestdagh *et al.*, 2004).

15.4.2 Structure

Acrylamide's reactivity stems from the presence of two reactive groups. Its double bond conjugated to a carbonyl (Fig.15.2) is an ideal site for the formation of Michael adduct by nucleophile-containing molecules such as amines, thiols or alcohols. Hence acrylamide has the potential for biological alkylation of peptides, proteins, glutathione, nucleic acids and DNA. Moreover, the amide group provides acrylamide with additional reactive characteristics (Friedman, 2003).

15.4.3 Toxicology

Acrylamide is toxic to humans on various levels, via inhalation, dermal absorption and ingestion. Rodent studies have proven high level exposure to be neurotoxic and even deadly (Schaumburg *et al.*, 1974; Friedman *et al.*, 1995; Friedman, 2003). However, the low levels of food-derived acrylamide are clearly not expected to produce such an acute response. Daily exposure to acrylamide hovers around 0.5 μg/kg wgt/day (WHO, 2002), far below the threshold for acute toxicity. Therefore, concern lies mainly upon the possible mutagenic or carcinogenic effects of low dose, long-term exposure potentially leading to mutagenesis and carcinogenesis. Despite numerous studies, results and conclusions remain contradictory and lack of information with respect to human cancer risk from population studies leaves much room for debate, especially considering the exaggerated extrapolations made from animal studies (Fuhr *et al.*, 2006). Therefore, it remains highly questionable whether or not the chemical taken in from food actually poses a risk to human health (Granath and Törnqvist, 2003). Many data have been published pertaining to cancer risk assessment of acrylamide. Evidence to date, including a few large-scale comprehensive human case-control studies, has failed to show any association between acrylamide intake and increased incidence of various types of cancers (Pelucchi *et al.*, 2006). However, regardless of the debate, maintaining carcinogen formation as low as possible is still considered of great importance. Consequently, in response to high demand, analytical companies are providing kits for the monitoring of acrylamide levels in foods (Walz and Trinh, 2005). A more detailed review on acrylamide toxicology and its implications is presented by Shipp and his co-workers (Shipp *et al.*, 2006).

Levels of acrylamide encountered in food range anywhere between 0 and 3.5mg per kilogram. Due to the various possible routes of exposure, including smoke, food and water, hemoglobin (Hb) adduct of acrylamide and its reactive epoxide metabolite glycidamide is often used to assess human exposure to acrylamide. Acrylamide and glycidamide are electrophilic chemicals due to their α,β-unsaturated amide group. They therefore have the capacity of reacting *in vivo* and *in vitro* with nucleophilic compounds, forming adducts with the sulfydryl or α-amino group of proteins such as Hb. Hb adducts (formed with acrylamide or glycidamide attaching to the α-NH_2 group of the N-terminal of valine of Hb) are used as biomarkers as they are distributed in the blood and can be measured by gas chromatography/mass spectrometry (GC/MS) and correlated to the levels of exposure. One advantage of using Hb adducts is that they provide a time-averaged estimate of exposure (Perez *et al.*, 1999; Törnqvist and Ehrenberg, 2001).

15.4.4 Formation

From the initial studies, the Maillard reaction was suspected as the main pathway for the generation of acrylamide in food (Mottram *et al.*, 2002). However, these studies clearly indicated that the mechanism of formation of acrylamide did not conform to the known Maillard reaction sequence. The low yield resulting from the pyrolysis of the Amadori compound as opposed to the high yield generated by its Schiff base precursor (or N-glycosylasparagine) strongly suggested that the acrylamide formation pathway occurs prior to an Amadori step (Stadler *et al.*, 2002). Moreover, pyrolysis of N-glycosylasparagine yielded significantly higher levels of acrylamide than asparagine–glucose models, clearly indicating involvement of the Maillard reaction. Subsequently, use of dicarbonyl moieties as opposed to α-hydroxy carbonyls also leads to acrylamide formation (Stadler *et al.*, 2002). The lack of α-hydroxy moiety blocks the Maillard reaction at the Schiff base stage. Taking these premises into consideration, general pathways were proposed (Yaylayan *et al.*, 2003, Zyzak *et al.* 2003) describing the formation of acrylamide from asparagines (Fig. 15.6). The main feature of these proposals is the decarboxylation of the Schiff base formed between the asparagine and sugars. Initially, Yaylayan *et al.* (2003) proposed a sugar-assisted decarboxylation step through the formation of 5-oxazolidinone intermediate, leading to acrylamide after an Amadori

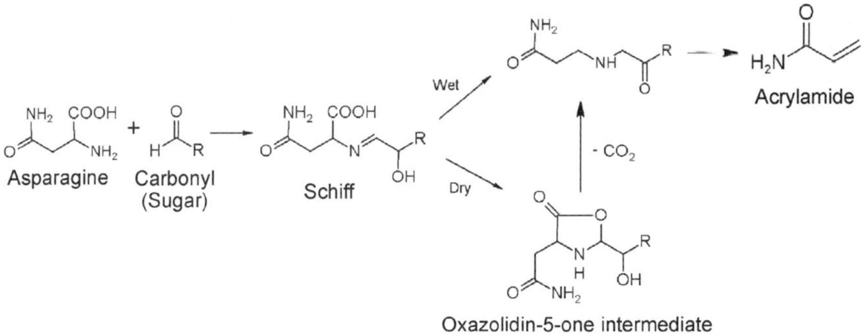

Fig. 15.6 Proposed mechanisms of acrylamide formation.

rearrangement and an independent decarboxylation mechanism of the Schiff base as proposed by Zyzak and co-workers (2003). Studies published by Stadler *et al.* (2004) provided further evidence for the Amadori product of the decarboxylated asparagine as having the potential for being the main precursor of acrylamide. However, despite some evidence, details surrounding acrylamide formation remain primarily speculative due to the lack of conclusive information regarding the immediate precursors involved. It is likely that both decarboxylated Amadori and Schiff products, as well as 3-aminopropionamide, contribute to the formation of acrylamide in food, depending on the matrix, temperature, moisture content and pH.

Several other minor precursors have also been studied and shown to generate acrylamide. Such precursors include various amino acids such as aspartic acid, β-alanine, carnosine, serine and cysteine (Yaylayan *et al.*, 2003, 2004), as well as fat and oil degradation products. Although mechanisms of acrylamide formation from other amino acids or from fatty acid oxidation derived from acrolein reaction with ammonia (Yasuhara *et al.*, 2003) have been proposed, they are considered as being relatively insignificant. This fact was demonstrated when asparaginase enzyme was added to model potato systems and processed, resulting in a 99% drop in acrylamide content (Zyzak *et al.*, 2003).

15.4.5 Analysis and quantitation

The hydrophilic character of acrylamide allows for a simple water or methanol extraction from the food (Rosen and Hellenas, 2002; Tateo and Bononi, 2003). A saline solution can also be used to avoid emulsification during the sample pretreatment, water–acetone solution or a pressurized liquid extraction. Depending on the food matrix, a defatting step with an organic solvent, or a deproteinating step, could be necessary to remove interfering components (Delatour *et al.*, 2004). An internal standard is also added to allow for a more reliable quantification. This standard is typically an isotopically labeled acrylamide. Further clean-up, using combinations of solid phase cartridges or mixed mode cation exchange (MCX) or hydrophylic lipophilic balance (HLB) are widely used. Numerous analytical procedures have been published (Wenzl *et al.*, 2007) over the years focusing on maximizing extraction efficiency with removal of interferants. These methodologies also vary as a function of the type of food matrix analysed. A thorough review has been presented by Zhang and co-workers (2005)

Analysis of acrylamide is generally achieved by one of two main methodologies: GC/MS or LC/MS-MS. In the case of gas chromatography, derivatization of the analyte is required in order to enhance volatility, selectivity, sensitivity and retention time. Bromination has been the derivatization methodology of choice, allowing for good sensitivity, although silylation has also proven itself valuable when combined with headspace SPME analysis. It is important to note, however, that derivatization is not necessarily essential and some groups have chosen to eliminate this lengthy procedure using a more polar GC phase, although it does not afford as low a limit of detection as the derivatization techniques (Wenzl *et al.*, 2006a).

LC-MS/MS is analytically advantageous as it does not require any derivatization step prior to analysis, yet retains selectivity. The use of a modified dC18 reverse phase minimizes retention problems, whereas ion exchange chromatography reduces interferences. Both approaches have been reported as being adequate. Tandem mass spectrometry is the most common detection mode due to its high selectivity. Mass spectrometry can be used instead; however, derivatization with 2-mercaptobenzoic acid is necessary for optimal confirmation (Jezussek and Schieberle, 2003).

15.4.6 *De minimis* principle

Since the discovery of acrylamide's impact in our food supply, a number of investigations have been conducted targeting practical means of reduction. A key factor that has been repeatedly shown to influence acrylamide levels is temperature. Microwaving and boiling generates minimal amounts, whereas harsher treatments such as frying and baking lead to high levels (Matthaus *et al.*, 2004). Particularly in the case of potatoes, soaking them in water prior to baking or frying, significantly reduces acrylamide levels as it leaches away its precursors on the surface. Additional impact has also been observed by addition of salts (Goekmen and Senyuva, 2007), citric acid or amino acids (glycine, lysine, cysteine) to the soaking solution. In some instances, however, a balance must be created to avoid a corresponding inhibition of flavor and color development (Pedreschi *et al.*, 2004; Ishihara *et al.*, 2005, 2006; Kim *et al.*, 2005; Kolek *et al.*, 2006; Low *et al.*, 2006). Addition of amino acids such as glycine to dough in the making of snacks also has a negative impact on acrylamide formation (Kim *et al.*, 2005). Addition of asparaginase enzyme is an alternative way of reducing acrylamide levels in foods, although its cost makes it commercially not viable. The most systematic approach to the mitigation problem, designed to help food manufacturers, is the CIAA 'Toolbox' which integrates the results of the food industry's cooperation in understanding acrylamide formation and mitigation steps. Its aim is to provide information regarding the mitigation steps evaluated by different food manufacturers and other partners in the food chain (CIAA, 2005). This approach allows the individual manufacturer to determine which of the mitigation procedures can be applied to their specific production parameters.

15.5 Furans

15.5.1 Sources

The parent furan and hydroxymethylfurfural (5-HMF) are the two most important members in this group (Fig. 15.2). Hydroxymethylfurfural is found in many foods produced from either sugar degradation alone or in combination with Maillard reaction. Common food sources include coffee, honey, breads and cookies, caramels, heat-treated milk, dried fruits as well as fruit juices and it is formed as a result of thermal processing, drying or storage. In some instances it has been found in levels exceeding 1g/kg of food and up to 6.2g/kg in coffee (Schultheiss *et al.*, 1999; Janzowski *et al.*, 2000; Murkovic and Pichler, 2006). Discussion will focus on thermally generated HMF, although the principle remains the same in other instances. On the other hand, formation of the parent furan, a known toxicant, has been known to occur in foods since the 1960s. Maga (1979) had previously reported its formation from various carbohydrates. However, recent awareness of the widespread levels of furan in canned and jarred products raised questions as to its overall toxicity, precursors and formation in foods (FDA, 2003). Furan is a volatile aromatic heterocyclic organic compound, exhibiting a boiling point of 31.4 °C. Consequently, once generated as a result of thermal processing, it tends to immediately be released into the atmosphere, unless processing occurs in an enclosed environment. Understandably, in the case of in-can (or jar) processing, the furan formed remains in the vicinity and is subsequently consumed with the product. In addition to food, sources of furan also include cigarette smoke, coal and wood smoke (NTP, 1993). In addition to carbohydrates, Health Canada (2004) was able to produce very high furan levels via thermal treatment of ascorbic acid, dehydroascorbic acid and thiamine. Amino acids alone were also found to produce trace amounts, although sugar–amino acid mixtures proved to be far more efficient. Also, furan was generated from thermal oxidation of polyunsaturated fatty acids.

15.5.2 Toxicology

The potential toxicity of HMF remains a questionable issue. Janowski's group (Janzowski *et al.*, 2000) provided evidence of weak mutagenicity and genotoxicity arising from HMF levels comparable to those found in certain foods. However, experiments were conducted *in vitro* and the relatively low consumption of HMF-rich foods (dried fruits, for example) are to be considered. An old estimate of HMF intake showed up to 2.5mg/kg body weight daily dietary HMF intake per person (Ulbricht *et al.*, 1984). However, in the light of more recent studies, Janzowski proposes that a more realistic assessment is an average intake of about 30–60 mg per person (0.5–1 g/kg wgt/day) and, overall, concludes that HMF poses an insignificant health risk. However, the results are controversial because some animal studies show dose-related tumor development (Schoental *et al.*, 1971; Zhang *et al.*, 1993; Surh *et al.*, 1994), whereas others do not (Miyakawa *et al.*, 1991). The response also varies according to the region studied. No human data are available and overall toxicology studies are scarce. In the case of furan, listed by the IARC as a type II B possible human carcinogen (IARC, 1995) based on a series of animal studies indicating formation of malignant tumors, there is also still much room for debate as to its contribution to long-term cancer development in humans. Nevertheless, it is also part of the Department of Health and Human Services (DHHS) list of carcinogens. Among the many effects, chromosomal aberration, DNA damage and hepatotoxicity have been observed in mice as a result of ingested furan (NTP, 1993). Metabolic activation of furan into one of its reactive metabolites, *cis*-2-butene-1,4-dial, is suspected to occur via cytochrome P450 and is probably responsible for some of furan's cancer-inducing properties (Chen *et al.*, 1995). The European Food Safety Authority (EFSA) have estimated that the intake from baby foods is <0.2 μg/kg bw per day for a 6-month old baby weighing 7.5 kg (EFSA, 2004). For adults, the daily intake from canned or jarred vegetables was calculated to be 1.1–23 μg/person and from beer the value was 1.3–50 μg/person and from coffee, the major dietary source of furan, was 2.4–116 μg/person (EFSA, 2004). Lack of human epidemiological studies on the risk invoked by furan exposure prevents any further speculation, although in this case no data indicates a difference in the pathways to tumor development between rodent and humans. Several toxicological studies are currently under way in the hope of realistically assessing the risk posed by furan.

15.5.3 Formation

Both caramelization and Maillard reaction pathways contribute to HMF formation. Acid-catalyzed sugar dehydration and cyclization lead to formation of the furan derivatives, including HMF. Heat-catalyzed Maillard reaction between amino acids and sugars also produce HMF, via dehydration and enolization reactions. In addition, HMF is often used as an indicator of the extent of the thermal treatment in many foods. Excessive accumulation of HMF in certain foods (milk, bread, juice) can be correlated to excessive heat treatment of the product (Rada-Mendoza *et al.*, 2004). HMF generation is highly temperature and precursor (type of sugar) dependent. The pH will have an impact on the relative rate of the Maillard reaction (favored at low pH 5–6) and caramelization (favored at higher pH), consequently impacting on the pathway and the extent of HMF formation. In the case of furan, Perez-Locas and Yaylayan (2004) were the first to provide detailed mechanistic information regarding furan formation from a variety of precursors, using an isotope labeling technique. These studies have indicated that furan can be formed from amino acids, carbohydrates and ascorbic acid degradations (Yaylayan, 2006). A summary of these pathways is presented in Fig. 15.7. In addition, an oxidative pathway from polyunsaturated fatty acids (PUFA) was proposed by

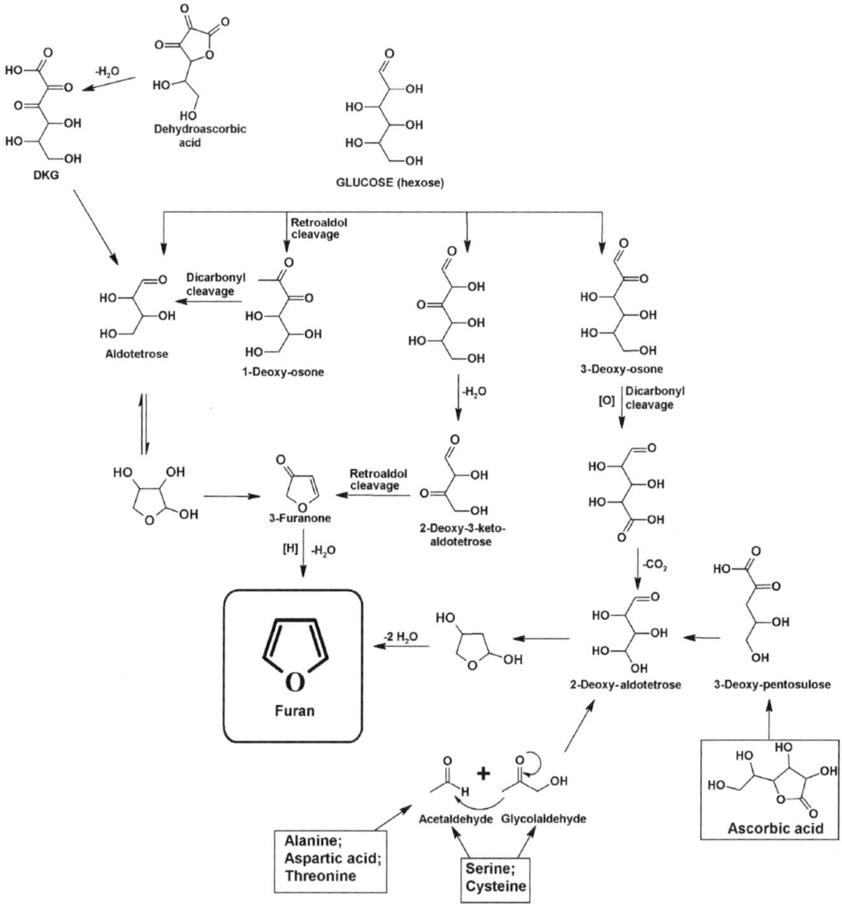

Fig. 15.7 Furan formation from sugars, ascorbic acid and amino acids (adapted from Perez-Locas and Yaylayan, 2005).

Becalski and Seaman, (2005). Recently, using ^{13}C-labelled ascorbic acid, the pathways proposed by Perez-Locas and Yaylayan (2004) were verified by Limacher *et al.* (2007).

15.5.4 Analysis and quantitation

Analytical procedures for HMF determination vary slightly as a function of the food matrix and the separation tools. The HPLC methods typically involve water or methanol extraction, precipitation with clarifying agents such as Carrez solutions or TCA if necessary, filtration and direct injection into HPLC with UV or diode array detection. Wavelengths for detection appear to vary among authors between 280 and 285 nm (Ramirez-Jimenez *et al.*, 2003; Ameur *et al.*, 2006, 2007; Murkovic and Pichler, 2006. Some more complex matrices, such as coffee, can require an SPE purification prior to injection to remove interfering compounds (Murkovic and Pichler, 2006). A GC-MS methodology was also proposed and validated by Teixido and co-workers (2006). Briefly, the sample is extracted with acidified water and passed through an SPE cartridge, followed by concentration and derivatization. The authors tested a variety of cartridges, evaluating ENV+ as giving the best response and recovery. Among the derivatizing agents tested, N,O-bis (trimethylsilyl) trifluoroacetamide (BSTFA) was found to provide a good response within a short time-frame. The procedure is slightly more labor intensive, however, and provides a significantly lower level of detection but it was shown to work on many types of sample, including solids, liquids and semi-liquids (Teixido *et al.*, 2006).

The volatility of furan makes it a good compound to analyze as it does not require complex extraction and purification steps. Liquids can usually be analyzed 'as is' and solids require addition of water. Common procedure consists of sealing the homogenized food sample with Furan-d_4 internal standard in a headspace vial with an SPME fiber (Caboxen-PDMS). The headspace is then simply inserted into the GC-MS injection port where the adsorbed furan can be desorbed into a DB-5 or a non-polar PLOT column. Mass spectrometry is then used for sensitive and selective identification (Becalski *et al.*, 2005; Goldmann *et al.*, 2005). Several reviews on furans have recently been published (Yaylayan, 2006; Crews and Castle, 2007). In addition, a single issue of the journal *Food Additives and Contaminants* was dedicated to furan and acrylamide in food (2007, Supplement 1, Volume 24).

15.5.5 *De minimis* principle

Little effort has been made to reduce the levels of furan and HMF in food besides changing processing conditions (lower temperatures), which could create a bigger issue affecting the microbial safety of canned and pasteurized foods. No recommendations or limits have been set from a carcinogen perspective as they are not treated as potential health hazards. HMF content in several fluids are set in order to fulfill its role as an indicator in monitoring the overall quality parameters, representative of the degree of processing of the food as opposed to its potential health effects. Some attempts have been made to reduce furan levels by irradiation, and although successful in liquids, were not very effective in minimizing furan in actual food matrices (Fan and Mastovska, 2006; Fan and Sommers, 2006).

15.6 Chloropropanols

15.6.1 Sources

Chloropropanols are a group of foodborne contaminants mainly present as a result of processing. The most abundant and the most studied member of this group is 3-monochloropropanediol (3-MCPD) followed by 1,3-dichloropropanol (1,3-DCP). The former was isolated from various Asian products containing acid-hydrolyzed vegetable proteins (HVP), soy and oyster sauces, as well as different packaged soups. After further investigation, 3-MCPD was also found to occur in a variety a food, ranging from bakery and cereal products (0.01–0.134 mg/kg) to dairy (0.016–0.031 mg/kg) and meat (0.01–0.081). Thermal processing has a significant impact on staple foods as well as meats. The greater the thermal treatment (toasting, roasting, grilling), the more 3-MCPD is produced (Crews *et al.*, 2002). Although 3-MCPD levels are far greater in products containing acid-HVP, consumption of bread and other grain products is high and regular (Breitling-Utzmann *et al.*, 2003) in North American and European cultures, thus making its overall contribution to dietary 3-MCPD significant. Roasting of coffee beans will also generate important amounts of this toxicant, despite most of it being present in its mono or diester forms. However, enzyme-catalyzed hydrolysis may release some free 3-MCPD *in vivo*; therefore, caution should be exercised in estimating their amounts. In its free form, Dolezal and co-workers (Dolezal *et al.*, 2005) found coffee beans to contain 10.1 and 18.5 μg/kg, the amounts varying according to the type of bean and temperature used.

15.6.2 Toxicology

Although 3-MCPD is no longer considered to be genotoxic *in vivo* (COM, 2000), it is still believed to exhibit carcinogenic characteristics (COC, 2000). Animal studies have provided much evidence of the tumor formation potential of MCPD intake. A review of its carcinogenicity has been presented by Lynch and co-workers (Lynch *et al.*, 1998). Overall, it is encouraged to keep levels as low as possible and a maximum tolerable daily intake has been set at 2 μg per kg body weight (Joint FAO/WHO Expert Committee, 2001). Moreover, a number of countries have also set their own limits as to the maximal tolerable levels of 3-MCPD in certain soy products (Xing and Cao, 2007). Although thermally generated chloropropanols are not regulated, however, the EU has set a limit of 0.02 mg kg^{-1} of 3-MCPD in acid-HVP and soy sauce.

15.6.3 Formation

Formation of 3-MCPD resulting from acid hydrolysis of proteinaceous material has been investigated and its mechanism understood. Essentially, the added acid (HCl) would act on residual lipids of the protein matrix by hydrolyzing them into chloropropanediol diesters, which can be hydrolyzed further into MCPD (Collier *et al.*, 1991). However, much information is lacking regarding thermal MCPD formation in non-acidified foods (Hamlet and Sadd, 2005).

An investigation on MCPD formation resulting from domestic cooking was undertaken by Crews *et al.* (2001). Comparisons of MCPD levels in a variety of raw versus cooked food products provided evidence of the catalytic power of thermal treatment on its generation. Foods such as bread, meat and cheeses showed increasing levels of MCPD as a result of cooking. Not much explanation is provided, although Hamlet and co-workers' review (2002) suggested some migration from meat casings in the case of sausages, as well as heat-induced transformation of the corresponding mono and diesters in the case of cheese, which are known components in the raw matrix. Some cereal products (such as breads) are known to produce glycerol from yeast fermentation, which can then react with chloride ions from the salt during the baking (Hamlet and Sadd, 2005). Collier and co-workers (1991) also observed MCPD formation in modified starches due to heating, particularly in the presence of mineral acids, and propose a reaction between residual lipids and chloride ions from added salt. Various malt products are also known to contain MCPD following thermal treatment (Hamlet *et al.*, 2002). Besides speculation, lack of solid mechanistic information targeting the details of thermal 3-MCPD formation clearly remains an issue.

15.6.4 Analysis and quantitation

Similar to many thermally generated foodborne toxicants, the relatively low levels of MCPD formed make it difficult to analyze from complex food matrices. Moreover, several analytical barriers need to be overcome as a result of the low molecular weight and high boiling point. Absence of a fluorescence and UV/VIS absorbing chromophore makes it a poor candidate for HPLC analysis. The official method, validated by Brereton and co-workers (2001) and applicable to most food products, consists in extracting MCPD from the matrix using a salt solution, and passing the extract through an Extrelut diatomaceous earth commercial cartridge. The compound is eluted using a mixture of hexane–ether, concentrated, derivatized by heptafluorobutyrylimidazole (HFBI) and analyzed by GC/MS. Derivatization is key as it allows for an increase in molecular weight, a drop in polarity and improved sensitivity and selectivity, allowing for better analytical response. The use of an internal standard (MCPD-d_5) is also crucial as it accounts for losses and allows for a more reliable quantification. A number of authors use this approach or a slightly modified version of it (Crews *et al.*, 2002, 2003; Breitling-Utzmann *et al.*, 2003, 2005; Abu-El-Haj *et al.*, 2007). Another competing analytical methodolgy uses phenylboronic acid (PBA) with GC/MS/MS. Although it is simpler and less time consuming, PBA reacts only with *cis* diols, allowing for MCPD analysis but impeding on simultaneous determination of other chloropropanols of possible interest such as 1,3-dichloropropanol (Breitling-Utzmann *et al.*, 2005; Dolezal *et al.*, 2005). A recent review by Wenzl and co-workers (2007) summarizes different analytical approaches.

15.6.5 *De minimis* principle

A number of methods designed to reduce MCPD levels have been patented (Hamlet *et al.*, 2002). These patents, however, tend to target MCPD generated from the HVP processing. Minimizing thermally generated MCPDs can be achieved by decreasing time and temperature of cooking. Moreover, it has been shown that sugars have the potential of enhancing MCPD formation (Breitling-Utzmann *et al.*, 2005). Hamlet and Sadd (2005) speculate that amino acids and ammonia, known to prevent MCPD generation by reacting with the glycerol precursor, will instead undergo Maillard reaction with sugars present, thus indirectly allowing greater production. Minimizing the use of baking agents, which also show a marked increase in MCPD levels, is also a suggested approach. No mechanism has been proposed to explain this phenomenona; the main ingredient, sucrose, is believed partly to be the cause (Breitling-Utzmann *et al.*, 2005), possibly via the same Maillard mechanism proposed earlier. Furthermore, reducing the use of organic acids (such as citric acid) that have shown to increase MCPD formation, can also minimize its presence.

15.7 Conclusion

In the case of thermally-generated toxicants, consensus clearly indicates cooking at lower temperature for longer time leads to lower levels. Burning or charring of any food can significantly increase the load of various potential carcinogens. Numerous food additives, antioxidants, marinades and other components have been tested over the years and some have been shown to reduce considerably the formation of certain toxicants, although their practicality at times is questionable. The main impact, however, comes from the method chosen for processing. Very few carcinogens are produced when food is boiled, steamed or microwaved. Harsher treatments, such as frying, roasting, grilling and barbecuing, harbor greater concern, particularly when very high and uncontrolled temperatures are reached (such as cooking on an open flame). Despite numerous animal studies, only few human case studies have been conducted. Various attempts to correlate different toxicants from dietary sources with increased cancer risk have led to no specific conclusions. Some guidelines and recommendations have been presented over the years, based on research. However, not much has been imposed by governments as regulatory measures, possibly due to the lack of epidemiological evidence.

15.8 Abbreviations used

PAH:	Polycyclic Aromatic Hydrocarbon
B[a]P:	Benzo[a]pyrene
HA:	Heterocyclic amine
IQ:	2-Amino-3-methylimidazo[4,5-*f*]quinoline
IQx:	2-Amino-3-methylimidazo [4,5-*f*]quinoxaline
MeIQ:	2-Amino-3,4-dimethylimidazo[4,5-*f*] quinoline
MeIQx:	2-Amino-3,8-dimethylimidazo[4,5-*f*] quinoxaline
diMeIQx:	2-Amino-3,4,8-dimethylimidazo[4,5-*f*]quinoxaline
Trp-P-2:	3-Amino-1-methyl-5*H*-pyrido[4,3-*b*]indole
PhIP:	2-Amino-1-methyl-6-phenylimidazo[4,5-*b*]pyridine
HMF:	5-Hydroxymethylfurfural
3-MCPD:	3-Monochloropropane-1,2-diol
1,3-DCP:	1,3-Dichloropropan-2-ol
Py-GC/MS:	Pyrolysis Gas Chromatography/Mass Spectrometry
GC/MS:	Gas Chromatography/Mass Spectrometry
LC/MS:	Liquid Chromatography/Mass Spectrometry

15.9 References

Abu-El-Haj, S., Bogusz, M. J., Ibrahim, Z., Hassan, H. and Al Tufail, M. (2007). Rapid and simple determination of chloropropanols (3-MCPD and 1,3-DCP) in food products using isotope dilution GC-MS. *Food Control* **18**(1), 81–90.

Adamson, R. H. and Thorgeirsson, U. P. (1995). Carcinogens in Foods: Heterocyclic Amines and Cancer and Heart Disease. *Nutrition and Biotechnology in Heart Disease and Cancer* **369**, 211–220.

Adamson, R. H., Thorgeirsson, V. P., Snyderwine, E. G., Thorgeirsson, S. S., Reeves, J., Dalgard, D. W., Takayama, S. and Sugimura, T. (1990). Carcinogenicity of 2-amino-3-methyl-imidazo[4,5-f]quinoline in Nonhuman Primates: Induction of Tumors in Three Macaques. *Japanese Journal of Cancer Research* **81**, 10–14.

Ahn, J. and Grun, I. U. (2005). Heterocyclic amines: 2. Inhibitory effects of natural extracts on the formation of polar and nonpolar heterocyclic amines in cooked beef. *Journal of Food Science* **70**(4), C263–C268.

Ameur, L. A., Mathieu, O., Lalanne, V., Trystram, G. and Birlouez-Aragon, I. (2007). Comparison of the effects of sucrose and hexose on furfural formation and browning in cookies baked at different temperatures. *Food Chemistry* **101**(4), 1407–1416.

Ameur, L. A., Trystram, G. and Birlouez-Aragon, I. (2006). Accumulation of 5-hydroxymethyl-2-furfural in cookies during the baking process: Validation of an extraction method. *Food Chemistry* **98**(4), 790–796.

Badger, G. M., Kimber, R. W. L. and Spotswood, T. M. (1960). Mode of Formation of 3,4-Benzopyrene in Human Environment. *Nature* **187**(4738), 663–665.

Barnes, W., Spingarn, N. E., Garvie-Gould, C., Vuolo, L. L., Wang, Y. Y. and Weisburger, J. H. (1983). Mutagens in Cooked Foods: Possible Consequences of the Maillard Reaction. *The Maillard Reaction in Foods and Nutrition*. G. R. Waller and M. S. Feather. Washington D.C, ACS Symposium Series. **215**, 485–506.

Bartle, K. D. (1991). Analysis and Occurrence of Polycyclic Aromatic Hydrocarbons in

Food. *Food Contaminants: Sources and Surveillance*. R.P.C.S. Creaser. Cambridge, The Royal Society of Chemistry: 41–60.

Bartoszek, A. (1997). Mutagenic and Carcinogenic Components in Foods. *Chemical and Functional Properties of Food Components*. Z. E. Sikorski. Pennsylvania, Technomic Publications: 263–284

Becalski, A., Forsyth, D., Casey, V., Lau, B. P. Y., Pepper, K. and Seaman, S. (2005). Development and validation of a headspace method for determination of furan in food. *Food Additives and Contaminants* **22**(6), 535–540.

Becalski, A. and Seaman, S. (2005). Furan precursors in food: A model study and development of a simple headspace method for determination of furan. *Journal of AOAC International* **88**(1), 102–106.

Becher G., Knize M. G. and Felton J. S. (1989) Identification and synthesis of new mutagens from a fried Norwegian product. *Var Foda* **42**, Suppl. 2, 85–90.

Becher, G., Knize, M. G., Nes, I. F. and Felton, J. S. (1988). Isolation and Identification of Mutagens from a Fried Norwegian Meat Product. *Carcinogenesis* **9**(2), 247–253.

Breitling-Utzmann, C. M., Hrenn, H., Haase, N. U. and Unbehend, G. M. (2005). Influence of dough ingredients on 3-chloropropane-1,2-diol (3-MCPD) formation in toast. *Food Additives and Contaminants* **22**(2), 97–103.

Breitling-Utzmann, C. M., Kobler, H., Herbolzheimer, D. and Maier, A. (2003). 3-MCPD – Occurrence in bread crust and various food groups as well as formation in toast. *Deutsche Lebensmittel-Rundschau* **99**(7), 280

Brereton, P., Kelly, J., Crews, C., Honour, S., Wood, R. and Davies, A. (2001). Determination of 3-chloro-1,2-propanediol in foods and food ingredients by gas chromatography with mass spectrometric detection: Collaborative study. *Journal of AOAC International* **84**(2), 455–465.

Britt, P. F., Buchanan, A. C., Owens, C. V. and Skeen, J. T. (2004). Does glucose enhance the formation of nitrogen containing polycyclic aromatic compounds and polycyclic aromatic hydrocarbons in the pyrolysis of proline? *Fuel* **83**(11–12), 1417–1432.

Busquets, R., Puignou, L., Galceran, M. T. and Skog, K. (2006). Effect of red wine marinades on the formation of heterocyclic amines in fried chicken breast. *Journal of Agricultural and Food Chemistry* **54**(21), 8376–8384.

Casal, S., Mendes, E., Fernandes, J. O., Oliveira, M. and Ferreira, M. A. (2004). Analysis of heterocyclic aromatic amines in foods by gas chromatography-mass spectrometry as their tert.-butyldimethylsilyl derivatives. *Journal of Chromatography A* **1040**(1), 105–114.

Chen, B. H. and Chiu, C. P. (1998). Analysis, Formation and Inhibition of Heterocyclic Amines in Foods: An Overview. *Journal of Food and Drug Analysis* **6**, 625–636.

Chen, B. H. and Lin, Y. S. (1997). Formation of polycyclic aromatic hydrocarbons during processing of duck meat. *Journal of Agricultural and Food Chemistry* **45**(4), 1394–1403.

Chen, L. J., Hecht, S. S. and Peterson, L. A. (1995). Identification of *cis*-2-butene-1,4-dial as a Microsomal Metabolite of Furan. *Chemical Research in Toxicology* **8**(7), 903–906.

CIAA (2005). Technical Report '*The CIAA Acrylamide Toolbox*': *A summary of the efforts and progress achieved to date by the European Food and Drink Industry (CIAA) in lowering levels of acrylamide in food*. Brussels, September (http://www.ciaa.be)

COC, Committee on Carcinogenicity of Chemicals in Food, Consumer Products and the Environment. (2000). *Carcinogenicity of 3-monochloropropane-1,2-diol (3-MCPD)*. Statement, COC/00/S5 (Update of COC/99/S5), December

Collier, P. D., Cromie, D. D. O. and Davies, A. P. (1991). Mechanism of Formation of Chloropropanols Present in Protein Hydrolysates. *Journal of the American Oil Chemists Society* **68**(10), 785–790.

COM, Committee on Mutagenicity of Chemicals in Food, Consumer Products and the Environment. (2000). *Mutagenicity of 3-monochloropropane-1,2-diol (3-MCPD)*. Statement, COM/00/S4, October.

Crews, C., Brereton, P. and Davies, A. (2001). The effects of domestic cooking on the levels of 3-monochloropropanediol in foods. *Food Additives and Contaminants* **18**(4), 271–280.

Crews, C., Hasnip, S., Chapman, S., Hough, P., Potter, N., Todd, J., Brereton, P. and Matthews, W. (2003). Survey of chloropropanols in soy sauces and related products purchased in the UK in 2000 and 2002. *Food Additives and Contaminants* **20**(10), 916–922.

Crews, C., Hough, P., Brereton, P., Harvey, D., Macarthur, R. and Matthews, W. (2002). Survey of 3-monochloropropane-1,2-diol (3-MCPD) in selected food groups, 1999-2000. *Food Additives and Contaminants* **19**(1), 22–27.

Crews, C and L. Castle. (2007). A review of the occurrence, formation and analysis of furan in heat-processed foods. *Trends in Food Science and Technology*. **18**, 365–372

Delatour, T., Perisset, A., Goldmann, T., Riediker, S. and Stadler, R. H. (2004). Improved sample preparation to determine acrylamide in difficult matrixes such as chocolate powder, cocoa, and coffee by liquid chromatography tandem mass spectroscopy. *Journal of Agricultural and Food Chemistry* **52**(15), 4625–4631.

Dolezal, M., Chaloupska, M., Divinova, V., Svejkovska, B. and Velisek, J. (2005). Occurrence of 3-chloropropane-1,2-diol and its esters in coffee. *European Food Research and Technology* **221**(3–4), 221–225.

EC (2002). European Commission, Opinion of the Scientific Committee on Food, (http://europa.eu.int/comm/food/fs/sc/scf/out153_en.pdf).

EFSA (2004). Report of the Scientific Panel on Contamination in the Food Chain on provisional findings on furan in food to the European Safety Authority. *EFSA Journal*, **137**, 1–2 (available at : http://www.efsa.eu.int/science/contam_documents/760_en.html).

Fan, X. T. and Mastovska, K. (2006). Effectiveness of ionizing radiation in reducing furan and acrylamide levels in foods. *Journal of Agricultural and Food Chemistry* **54**(21), 8266–8270.

Fan, X. T. and Sommers, C. H. (2006). Effect of gamma radiation on furan formation in ready-to-eat products and their ingredients. *Journal of Food Science* **71**(7), C407–C412.

FDA (2003). http://www.fda.gov.

Felton, J. S. and Knize, M. G. (1990) Heterocyclic-amine mutagens/carcinogens in foods. In *Handbook of Experimental Pharmacology*, 94/1. Edited by C. S. Cooper and P. L. Graver. pp. 471–502. Springer-Verlag, Berlin.

Felton, J. S. and Knize, M. G. (1991a). Mutagen Formation in Muscle Meats and Model Heating Systems. *Mutagens in Food Detection and Prevention*. H. Hayatsu. Boston, CRC Boca Raton: 57–66.

Felton J. S. and Knize M. G. (1991b) Occurrence, identification, and bacterial mutagenicity of heterocyclic amines in cooked foods. *Mutation Research* **259**, 205–217.

Felton, J. S., Knize, M. G., Shen, N. H., Andresen, B. D., Bjeldanes, L. F. and Hatch, F. T. (1986). Identification of the Mutagens in Cooked Beef. *Environmental Health Perspectives* **67**, 17–24.

Felton, J. S., Knize, M. G., Wood, C., Wuebbles, B. J., Healy, S. K., Stuermer, D. H., Bjeldanes, L. F., Kimble, B. J. and Hatch, F. T. (1984). Isolation and Characterization of New Mutagens from Fried Ground-beef. *Carcinogenesis* **5**(1), 95–102.

Filipkowska, A., Lubecki, L. and Kowalewska, G. (2005). Polycyclic aromatic hydrocarbon analysis in different matrices of the marine environment. *Analytica Chimica Acta* **547**(2), 243–254.

Friedman, M. (2003). Chemistry, biochemistry, and safety of acrylamide. A review. *Journal of Agricultural and Food Chemistry* **51**(16), 4504–4526.

Friedman, M. A., Dulak, L. H. and Stedham, M. A. (1995). A Lifetime Oncogenicity Study in Rats with Acrylamide. *Fundamental and Applied Toxicology* **27**(1), 95–105.

Fuhr, U., Boettcher, M. I., Kinzig-Schippers, M., Weyer, A., Jetter, A., Lazar, A., Taubert, D., Tomalik-Scharte, D., Pournara, P., Jakob, V., Harlfinger, S., Klaassen, T., Berkessel, A., Angerer, J., Sorgel, F. and Schomig, E. (2006). Toxicokinetics of acrylamide in humans after ingestion of a defined dose in a test meal to improve risk assessment for acrylamide carcinogenicity. *Cancer Epidemiology Biomarkers and Prevention* **15**(2), 266–271.

Gibbs, G. W. and Horowitz, I. (1979). Lung Cancer Mortality in Reduction Plant Workers. *Journal of Occupational Medicine* **21**, 347–353.

Goekmen, V., Senyuva, H. Z. (2007). Effects of some cations on the formation of acrylamide and furfurals in glucose–asparagine model system. *European Food Research and Technology* **225**(5–6),815–820.

Goldmann, T., Perisset, A., Scanlan, F. and Stadler, R. H. (2005). Rapid determination of furan in heated foodstuffs by isotope dilution solid phase micro-extraction–gas chromatography–mass spectrometry (SPME–GC–MS). *Analyst* **130**(6), 878–883.

Gomaa, E. A., Gray, J. I., Rabie, S., Lopezbote, C. and Booren, A. M. (1993). Polycyclic Aromatic Hydrocarbons in Smoked Food-products and Commercial Liquid Smoke Flavorings. *Food Additives and Contaminants* **10**(5), 503–521.

Granath, F. and Törnqvist, M. (2003). Who Knows Whether Acrylamide in Food Is Hazardous to Humans? *Journal of the National Cancer Institute* **95**, 842–843.

Grivas, S., Nyhammar, T., Olsson, K. and Jägerstad, M. (1986). Isolation and Identification of the Food Mutagens IQ and MeIQx from a Heated Model System of Creatinine, Glycine and Fructose. *Food Chemistry* **20**(2), 127–136.

Grivas, S. and Jägerstad, M. (1984). Mutagenicity of Some Synthetic Quinolines and Quinoxalines Related to Iq, Meiq or Meiqx in Ames Test. *Mutation Research* **137**(1), 29–32.

Grivas, S., Nyhammar, T., Olsson, K. and Jägerstad, M. (1985). Formation of a new mutagenic DiMelQx compound in a model system by heating creatinine, alanine and fructose. *Mutation Research* **151**, 177–183.

Gross, G. A. and Gruter, A. (1992). Quantitation of Mutagenic Carcinogenic Heterocyclic Aromatic Amines in Food-products. *Journal of Chromatography* **592**(1–2), 271–278.

Gross, G. A., Gruter, A. and Heyland, S. (1992). Optimization of the Sensitivity of High-performance Liquid Chromatography in the Detection of Heterocyclic Aromatic Amine Mutagens. *Food and Chemical Toxicology* **30**(6), 491–498.

Hamlet, C. G., Jayaratne, S. M. and Matthews, W. (2002). 3-Monochloropropane-1,2-diol (3-MCPD) in food ingredients from UK food producers and ingredient suppliers. *Food Additives and Contaminants* **19**(1), 15–21.

Hamlet, C. G. and Sadd, P. A. (2005). Effects of yeast stress and pH on 3-monochloropropanediol (3-MCPD)-producing reactions in model dough systems. *Food Additives and Contaminants* **22**(7), 616–623.

Hamlet, C. G., Sadd, P. A., Crews, C., Velisek, J. and Baxter, D. E. (2002). Occurrence of 3-chloro-propane-1,2-diol (3-MCPD) and related compounds in foods: A review. *Food Additives and Contaminants* **19**(7), 619–631.

Health Canada. (2004). http://www.hc-sc.gc.ca.

Higman, E. B., Schmeltz, I. and Schlotzhaver, W. S. (1970). Products from Thermal Degradation of Some Naturally Occurring Materials. *Journal of Agricultural and Food Chemistry* **18**(4), 636

Houessou, J. K., Delteil, C. and Camel, V. (2006). Investigation of sample treatment steps for the analysis of polycyclic aromatic hydrocarbons in ground coffee. *Journal of Agricultural and Food Chemistry* **54**(20), 7413–7421.

Hurd, C. D. and Simon, J. I. (1962). Pyrolytic Formation of Arenes. 3. Pyrolysis of Pyridine, Picolines and Methylpyrazine. *Journal of the American Chemical Society* **84**(23), 4519.

IARC (1983). Polynuclear Aromatic Compounds. Part 1, Chemical, Environmental and Experimental Data. . *IARC Monographs on the Evaluation of the Carcinogenic Risk of Chemicals to Humans*. International Agency for Research on Cancer. Lyon, **32**.

IARC (1993). Some Naturally Occuring Substances: Food Items and Constituents,

Heterocyclic Aromatic Amines and Mycotoxins. *IARC Monographs on the Evaluation of the Carcinogenic Risk of Chemicals to Humans.* International Agency for Research on Cancer. Lyon, **56,** 165–228.

IARC (1994). Some Industrial Chemicals. *IARC Monographs on the Evaluation of Carcinogenic Risk of Chemicals to Humans.* International Agency for Research on Cancer. Lyon, **60,** 560.

IARC (1995). Dry Cleaning, Some Chlorinated Solvents and Other Industrial Chemicals. *IARC Monographs on the Evaluation of Carcinogenic Risk of Chemicals to Humans.* International Agency for Research on Cancer. Lyon, **63,** 558.

IPCS (1998). *Selected Non-heterocyclic Polycyclic Aromatic Hydrocarbons.* International Programme on Chemical Safety, Geneva, World Health Organization. (Environmental Health Criteria, No. 49).

Ishihara, K., Matsunaga, A., Miyoshi, T., Nakamura, K., Nakayama, T., Ito, S. and Koga, H. (2005). Formation of acrylamide in a processed food model system, and examination of inhibitory conditions. *Journal of the Food Hygienic Society of Japan* **46**(2), 33–39.

Ishihara, K., Matsunaga, A., Nakamura, K., Sakuma, K. and Koga, H. (2006). Examination of conditions inhibiting the formation of acrylamide in the model system of fried potato. *Bioscience, Biotechnology and Biochemistry* **70**(7), 1616–1621.

Jacob, J. (1996). The significance of polycyclic aromatic hydrocarbons as environmental carcinogens. *Pure and Applied Chemistry* **68**(2), 301–308.

Jägerstad, M., Laser Reutersward, A., Olsson, R., Grivas, S., Nyhammar, T., Olsson, K. and Dahlqvist, A. (1983). Creatin(in)e and Maillard Reaction Products as Precursors of Mutagenic Compounds: Effects of Various Amino Acids. *Food Chemistry* **12**, 255–264.

Jägerstad, M., Skog, K., Arvidsson, P. and Solyakov, A. (1998). Chemistry, formation and occurrence of genotoxic heterocyclic amines identified in model systems and cooked foods. *Zeitschrift Fur Lebensmittel-Untersuchung Und Forschung A – Food Research and Technology* **207**(6), 419–427.

Jägerstad, M., Olsson, K., Grivas, S., Negishi, C., Wakabayashi, K., Tsuda, M., Sato, S., and Sugimura, T. (1984). Formation of 2-Amino-3,8-Dimethylimidazol[4,5-f] quinoxaline in a Model System by heating Creatinine, Glycine and Glucose. *Mutation Research* **126**(3), 239–244.

Janoszka, B., Warzecha, L., Blaszczyk, U. and Bodzek, D. (2004). Organic compounds formed in thermally treated high-protein food Part I: Polycyclic aromatic hydrocarbons. *Acta Chromatographica* **14**, 115–128.

Janzowski, C., Glaab, V., Samimi, E., Schlatter, J. and Eisenbrand, G. (2000). 5-hydroxymethylfurfural: Assessment of mutagenicity, DNA-damaging potential and reactivity towards cellular glutathione. *Food and Chemical Toxicology* **38**(9), 801–809.

Jezussek, M. and Schieberle, P. (2003). A new LC/MS-method for the quantitation of acrylamide based on a stable isotope dilution assay and derivatization with 2-mercaptobenzoic acid. Comparison with two GC/MS methods. *Journal of Agricultural and Food Chemistry* **51**(27), 7866–7871.

Johansson, M., Skog, K. and Jägerstad, M. (1993). Effects of Edible Oils and Fatty Acids on the Formation of Mutagenic Heterocyclic Amines in a Model System. *Carcinogenesis* **14**(1), 89–94.

Joint FAO/WHO Expert Committee on Food Additives (2001). *Summary of the fifty-seventh meeting of the joint FAO/WHO expert committee on food additives* (JECFA), Rome, 5–14 June (pp. 20–24).

Jones, R. C. and Weisburger, J. H. (1989). Characterization of aminoalkylimidazol-4-one mutagens from liquid–relux models. *Mutation Research* **222**, 43–51.

Kim, C. T., Hwang, E. S. and Lee, H. J. (2005). Reducing acrylamide in fried snack products by adding amino acids. *Journal of Food Science* **70**(5), C354–C358.

Knize M. G., Shen N. H. and Felton J. S. (1988) A comparison of mutagen production in fried ground chicken and beef: Effect of supplemental creatine. *Mutagenesis* **3**, 503–508.

Kolek, E., Simko, P. and Simon, P. (2006). Inhibition of acrylamide formation in asparagine/D-glucose model system by NaCl addition. *European Food Research and Technology* **224**(2), 283–284.

Kurosaka, R., Wakabayashi, K., Ushiyama, H., Nukaya, H., Arakawa, N., Sugimura, T. and Nagao, M. (1992). Detection of 2-Amino-1-Methyl-6-(4-Hydroxyphenyl)Imidazo [4,5-b]Pyridine in Broiled Beef. *Japanese Journal of Cancer Research* **83**(9), 919–922.

Limacher, A., J. Kerler, *et al.* (2007). Formation of furan and methylfuran from ascorbic acid in model systems and food. *Food Additives and Contaminants*, Supplement 1, **24**(S1),122–135

Lloyd, J. W. (1971). Long-term morbidity study of steelworkers V. Respiratory Cancer in Coke Plant Workers. *Journal of Occupational Medicine* **13**, 53–68.

Low, M. Y., Koutsidis, G., Parker, J. K., Elmore, J. S., Dodson, A. T. and Mottram, D. S. (2006). Effect of citric acid and glycine addition on acrylamide and flavor in a potato model system. *Journal of Agricultural and Food Chemistry* **54**(16), 5976–5983.

Lynch, B. S., Bryant, D. W., Hook, G. J., Nestmann, E. R. and Munro, I. C. (1998). Carcinogenicity of Monochloro-1,2-Propanediol *International Journal of Toxicology* **17**(1), 47–76.

Maga, J. A. (1979). Furans in Foods. *CRC Critical Reviews in Food Science and Nutrition* **11**(4), 355–400.

Matsushima, T. (1982). Mechanisms of conversion of food components to mutagens and carcinogens. *Molecular Interrelations of Nutrition and Cancer*. Arnott, M. S., Van Eys, J. and Wang, Y-M. (editors). New York, Raven Press**,** 35–42.

Matthaus, B., Haase, N. U. and Vosmann, K. (2004). Factors affecting the concentration of acrylamide during deep-fat frying of potatoes. *European Journal of Lipid Science and Technology* **106**(11), 793–801.

Mestdagh, F., de Meulenaer, B. van Peteghem, C., Cromphout, C. and Thas, O. (2004). Towards a Better Understanding in Acrylamide Formation, Degradation and Reduction in Model Systems (and Foodstuffs). *Czech Journal of Food Sciences* **Vol. 22** (Special Issue), 11–14.

Miyakawa, Y., Nishi, Y., Kato, K., Sato, H., Takahashi, M. and Hayashi, Y. (1991). Initiating Activity of 8 Pyrolysates of Carbohydrates in a 2-Stage Mouse Skin Tumorigenesis Model. *Carcinogenesis* **12**(7), 1169–1173.

Moret, S. and Conte, L. S. (2002). A rapid method for polycyclic aromatic hydrocarbon determination in vegetable oils. *Journal of Separation Science* **25**(1–2), 96–100.

Moret, S., Purcaro, G. and Conte, L. S. (2005). Polycyclic aromatic hydrocarbons in vegetable oils from canned foods. *European Journal of Lipid Science and Technology* **107**(7–8), 488–496.

Mottram, D. S., Wedzicha, B. L. and Dodson, A. T. (2002). Acrylamide is formed in the Maillard reaction. *Nature* **419**(6906), 448–449.

Muramatsu M. and Matsushima T. (1985) Formation of MelQx and 4,8-DiMeIQx by heating mixtures of creatinine, amino acids and monosaccharides. *Mutation Research* **147**(5), 266–267.

Murkovic, M. and Pichler, N. (2006). Analysis of 5-hydroxymethylfurfual in coffee, dried fruits and urine. *Molecular Nutrition & Food Research* **50**(9), 842–846.

Murkovic, M., Weberb, H.-J., S., G., Frohlich, K. and W., P. (1999). Formation of the Food Associated Carcinogen 2-Amino-1-Methyl-6-Phenylimidazo[4,5-b]pyridine (PhIP) in Model Systems. *Food Chemistry* **65**, 233–237.

Nagao, M., Yahagi, T. and Sugimura, T. (1978). Differences in effects of norharman with various classes of chemical mutagens and amounts of S-9. *Biochemical and Biophysical Research Communications* **83**, 373–378

Negishi, C., Wakabayashi, K., Yamaizumi, J., Saito, H., Sato, S., Sugimura, T. and

Jägerstad, M. (1985). Identification of 4,8-DiMeiQx, a New Mutagen. *Mutation Research*, **147**(5), 267–268.

NTP (1993). *Toxicology and Carcinogenesis Studies of Furan (CAS No. 110-00-9) in F344 Rats and B6C3F1 Mice (Gavage Studies)*. Technical Report Series No 402. NIH Publication No. 93-2857. Research Triangle Park, NC: National Toxicology Program.: 286.

NTP (2002). *Report on Carcinogens. Background Document for Heterocyclic Amines: PhIP, MeIQ and MeIQx*. National Toxicology Program. *http://ntp-server.niehs.nih.gov/newhomeroc/roc11/HCAsPub.pdf*.

Nyhammer, T. (1986). *Studies on the Maillard Reaction and its Role in the Formation of Food Mutagens*, Swedish University of Agricultural Sciences. Doctoral Thesis.

Ohgaki, H., Takayama, S. and Sugimura, T. (1991). Carcinogenicities of Heterocyclic Amines in Cooked Food. *Mutation Research* **259**, 399–412.

Overvik E., Kleman M., Berg I. and Gustavsson J. A. (1989) Influence of creatine, amino acids and water on the formation of the mutagenic heterocyclic amines found in cooked meat. *Carcinogenesis* **10**, 2293–2301.

Pakou, C., Fountoulakis, M., Drillia, P., Kampioti, A., Stamatelatou, K. and Lyberatos, G. (2007). Occurrence and fate of polycyclic aromatic hydrocarbons in sewage treatment plants using microwave-assisted extraction followed by liquid chromatography coupled with fluorescence detector. *Fresenius Environmental Bulletin* **16**(2), 127–132.

Panalaks, T. (1976). Determination and Identification of Polycyclic Aromatic Hydrocarbons in Smoked and Charcoal-broiled Food Products by High-Pressure Liquid Chromatography and Gas Chromatography. *Journal of Environmental Science and Health, Part B – Pesticides, Food Contaminants and Agricultural Wastes* **11**(4), 299–315.

Patterson, J. M., Haidar, N. F., Smith, W. T., Benner, J. F., Burton, H. R. and Burdick, D. (1978). Benzo[a]Pyrene Formation in Pyrolysis of Selected Amino Acids, Amines, and Maleic Hydrazide. *Journal of Agricultural and Food Chemistry* **26**(1), 268–270.

Pedreschi, F., Kaack, K. and Granby, K. (2004). Reduction of acrylamide formation in potato slices during frying. *Lebensmittel-Wissenschaft Und-Technologie-Food Science and Technology* **37**(6), 679–685.

Pelucchi, C., Galeone, C., Levi, F., Negri, E., Franceschi, S., Talamini, R., Bosetti, C. N., Giacosa, A. and La Vecchia, C. (2006). Dietary acrylamide and human cancer. *International Journal of Cancer* **118**(2), 467–471.

Perez-Locas, C. and Yaylayan, V. A. (2004). Origin and mechanistic pathways of formation of the parent furan – A food toxicant. *Journal of Agricultural and Food Chemistry* **52**(22), 6830–6836.

Perez, H. L., Cheong, H. K., Yang, J. S. and Osterman-Golkar, S. (1999). Simultaneous analysis of hemoglobin adducts of acrylamide and glycidamide by gas chromatography–mass spectrometry. *Anal. Biochem.* **274**, 59–68.

Pfau, W. and Skog, K. (2004). Exposure to beta-carbolines norharman and harman. *Journal of Chromatography B-Analytical Technologies in the Biomedical and Life Sciences* **802**(1), 115–126.

Phillips, D. H. (1999). Polycyclic aromatic hydrocarbons in the diet. *Mutation Research – Genetic Toxicology and Environmental Mutagenesis* **443**(1–2), 139–147.

Rada-Mendoza, M., Sanz, M. L., Olano, A. and Villamiel, M. (2004). Formation of hydroxymethylfurfural and furosine during the storage of jams and fruit-based infant foods. *Food Chemistry* **85**, 605–609.

Ramirez-Jimenez, A., Guerra-Hernandez, E. and Garcia-Villanova, B. (2003). Evolution of non-enzymatic browning during storage of infant rice cereal. *Food Chemistry* **83**(2), 219–225.

Rivas, F. J. (2006). Polycyclic aromatic hydrocarbons sorbed on soils: A short review of chemical oxidation based treatments. *Journal of Hazardous Materials* **138**(2), 234–251.

Rosen, J. and Hellenas, K. E. (2002). Analysis of acrylamide in cooked foods by liquid chromatography tandem mass spectrometry. *Analyst* **127**(7), 880–882.

Schaumburg, H. H., Wisniewski, H. M. and Spencer, P. S. (1974). Ultrastructural Studies of Dying-back Process. 1. Peripheral Nerve Terminal and Axon Degeneration in Systemic Acrylamide Intoxication. *Journal of Neuropathology and Experimental Neurology* **33**(2), 260–284.

Schoental, R., Hard, G. C. and Gibbard, S. (1971). Histopathology of Renal Lipomatous Tumors in Rats Treated with Natural Products Pyrrolizidine Alkaloids and Alpha, Beta-Unsaturated Aldehydes. *Journal of the National Cancer Institute* **47**(5), 1037–1044.

Schultheiss, J., Jensen, D. and Galensa, R. (1999). Hydroxymethylfurfural und Furfural in Kaffeeproben: HPLC-Biosensor-Kopplung mit Supressionstechnik. *Lebensmittelchemie* **53**, 159.

Schumacher, J. N., Green, C. R., Best, F. W. and Newell, M. P. (1977). Smoke Composition: An Extensive Investigation of the Water Soluble Portion of Cigarette Smoke. *Journal of Agricultural and Food Chemistry* **25**, 310–320.

Schut, H. A. J. and Snyderwine, E. G. (1999). DNA Adducts of Heterocyclic Amine Food Mutagens: Implications for Mutagenesis and Carcinogenesis. *Carcinogenesis* **20**, 353–368.

Sharma, R. K., Chan, W. G., Seeman, J. I. and Hajaligol, M. R. (2003). Formation of low molecular weight heterocycles and polycyclic aromatic compounds (PACs) in the pyrolysis of alpha-amino acids. *Journal of Analytical and Applied Pyrolysis* **66**(1–2), 97–121.

Shin, H. S. and Lee, S. (2004). Influence of honey-containing marinades on heterocyclic aromatic amine formation and overall mutagenicity in fried pork chops. *Food Science and Biotechnology* **13**(5), 651–656.

Shin, H. S. and Lee, Y. (2005). Influence of commercial marinades on heterocyclic aromatic amine formation and overall mutagenicity in fried beef steak. *Food Science and Biotechnology* **14**(3), 323–327.

Shin, H. S., Rodgers, W. J., Strasburg, G. M. and Gray, J. I. (2002). Reduction of heterocyclic aromatic amine formation and overall mutagenicity in fried ground beef patties by organosulfur compounds. *Journal of Food Science* **67**(9), 3304–3308

Shioya, M., Wakabayashi, K., Sato, S., Nagao, M. and Sugimura, T. (1987). Formation of a Mutagen, 2-Amino-1-Methyl-6-Phenylimidazo[4,5-B]Pyridine (Phip) in Cooked Beef, by Heating a Mixture Containing Creatinine, Phenylalanine and Glucose. *Mutation Research* **191**(3–4), 133–138.

Shipp, A., Lawrence, G., Gentry, R., McDonald, T., Bartow, H., Bounds, J., Macdonald, N., Clewell, H., Allen, B. and Van Landingham, C. (2006). Acrylamide: Review of toxicity data and dose-response analyses for cancer and noncancer effects. *Critical Reviews in Toxicology* **36**(6–7), 481–608.

Skog, K. (1993). Cooking Procedures and Food Mutagens – A Literature Review. *Food and Chemical Toxicology* **31**(9), 655–675.

Skog, K. (2004). Blue cotton, blue rayon and blue chitin in the analysis of heterocyclic aromatic amines – A review. *Journal of Chromatography, B–Analytical Technologies in the Biomedical and Life Sciences* **802**(1), 39–44.

Skog, K. and Jägerstad, M. (1990). Effects of Monosaccharides and Disaccharides on the Formation on Food Mutagens in Model Systems. *Mutation Research* **251**, 59–69.

Skog K. and Jägerstad M. (1991) Effect of glucose on the formation of PhlP in a model system. *Carcinogenesis* **12** (12), 2297–2300.

Skog K. and Jägerstad M. (1993) Incorporation of carbon atoms from glucose into the food mutagens MelQx and 4,8-DiMelQx using 14C-labelled glucose in a model system. *Carcinogenesis* **14** (10), 2027–2031

Skog, K., Knize, M. G., Felton, J. S. and Jägerstad, M. (1992). Formation of New

Heterocyclic Amine Mutagens by Heating Creatinine, Alanine, Threonine and Glucose. *Mutation Research* **268**(2), 191–197.

Smith, W. T., Haidar, N. F., Patterson, J. M. (1975). Pyrolysis of valine, α-aminobutyric acid and proline. *Tobacco Science* **19,** 142.

Stadler, R. H., Blank, I., Varga, N., Robert, F., Hau, J., Guy, P. A., Robert, M. C. and Riediker, S. (2002). Acrylamide from Maillard reaction products. *Nature* **419**(6906), 449–450.

Stadler, R. H., Robert, F., Riediker, S., Varga, N., Davidek, T., Devaud, S., Goldmann, T., Hau, J. and Blank, I. (2004). In-depth mechanistic study on the formation of acrylamide and other vinylogous compounds by the Maillard reaction. *Journal of Agricultural and Food Chemistry* **52**(17), 5550–5558.

Sugimura, T. (1986). Past, Present, and Future of Mutagens in Cooked Foods. *Environmental Health Perspectives* **67**, 5–10.

Sugimura, T. (1997). Overview of carcinogenic heterocyclic amines. *Mutation Research* **376**, 211–219

Sugimura, T., Nagao, M. and Wakabayashi, K. (1982). Metabolic aspects of comutagenic action of norharman. *Biological Reactive Intermediates – II, Chemical Mechanisms and Bio-logical Effects (Part B).* R. Snyder, D. J. Jollow, D. V. Parkeet *et al.* NY, Plenum Press**:** 1011.

Surh, Y. J., Liem, A., Miller, J. A. and Tannenbaum, S. R. (1994). 5-Sulfo-oxymethylfurfural as a Possible Ultimate Mutagenic and Carcinogenic Metabolite of the Maillard Reaction Product, 5-Hydroxymethylfurfural. *Carcinogenesis* **15**(10), 2375–2377.

Tamakawa, K. (2004). Polycyclic aromatic hydrocarbons (PAHs). *Pesticide, Veterinary and Other Residues in Food.* D. Watson. Cambridge (UK), Woodhead Publishing Ltd**:** 473–518.

Tamakawa, K., Ohgane, Y., Katoh, Y., Mishima, Y., Seki, T. and Tsunoda, A. (1986). Simplified Analysis of Polycyclic Aromatic Hydrocarbons in Daily Diet with Sep-Pak Silica Cartridge. *Eisei Kagaku–Japanese Journal of Toxicology and Environmental Health* **32**(6), 475–477.

Tareke, E., Rydberg, P., Karlsson, P., Eriksson, S. and Tornqvist, M. (2002). Analysis of acrylamide, a carcinogen formed in heated foodstuffs. *Journal of Agricultural and Food Chemistry* **50**(17), 4998–5006.

Tateo, F. and Bononi, M. (2003). A GC/MS method for the routine determination of acrylamide in food. *Italian Journal of Food Science* **15**(1), 149–151.

Teixido, E., Santos, F. J., Puignou, L. and Galceran, M. T. (2006). Analysis of 5-hydroxymethylfurfural in foods by gas chromatography–mass spectrometry. *Journal of Chromatography A* **1135**(1), 85–90.

T´ örnqvist, M. and Ehrenberg, L. (2001). Estimation of cancer risk caused by environmental chemicals based on *in vivo* dose measurement. *Journal of Environmental Patholology Toxicology and Oncology* **20**, 263–271.

Totsuka, Y., Takamura-Enya, T., Nishigaki, R., Sugimura, T. and Wakabayashi, K. (2004). Mutagens formed from beta-carbolines with aromatic amines. *Journal of Chromatography B* **802**, 135–141.

Totsuka, Y., Ushiyama, H., Ishihara, J., Sinha, R., Goto, S., Sugimura, T. and Wakabayashi, K. (1999). Quantification of the co-mutagenic β-carbolines, norharman and harman, in cigarette smoke condensates and cooked foods. *Cancer Letters* **143**, 139.

Ulbricht, R. J., Northup, S. J. and Thomas, J. A. (1984). A Review of 5-Hydroxymethylfurfural (Hmf) in Parenteral Solutions. *Fundamental and Applied Toxicology* **4**(5), 843–853.

Wakabayashi, K., Nagao, M., Esumi, H. and Sugimura, T. (1992). Food-derived Mutagens and Carcinogens. *Cancer Research* **52**(7), S2092–S2098.

Wakabayashi, K., Totsuka, Y., Fukutome, K., Oguri, A., Ushiyama, H. and Sugimura, T. (1997). Human Exposure to Mutagenic/Carcinogenic Heterocyclic Amines and Comutagenic β-Carbolines. *Mutation Research* **376**, 253–259.

Wakabayashi, K., Ushiyama, H., Takahashi, M., Nukaya, H., Kim, S.-B., Hirose, M., Ochiai, M., Sugimura, T. and Nagao, M. (1993). Exposure to Heterocyclic Amines. *Environmental Health Perspectives* **99**, 129–133.

Walz, R. and Trinh, A. (2005). Acrylamide in food: Kits, columns and tubes for sensistive GC–MS and LC–MS analyses and solid phase extraction. *Advanced Analytical Chemistry* **2**, 4–6.

Wang, G. D., Lee, A. S., Lewis, M., Kamath, B. and Archer, R. K. (1999). Accelerated solvent extraction and gas chromatography mass spectrometry for determination of polycyclic aromatic hydrocarbons in smoked food samples. *Journal of Agricultural and Food Chemistry* **47**(3), 1062–1066.

Weisburger, J. H. and Jones, R. C. (1989). Nutritional Toxicology: On the Mechanism of Inhibition of Formation of Potent Carcinogens during Cooking. *The Maillard Reaction in Ageing, Diabetes, and Nutrition*, 377–390.

Wenzl, T., Karasek, L., Rosen, J., Hellenaes, K. E., Crews, C., Castle, L. and Anklam, E. (2006a). Collaborative trial validation study of two methods, one based on high performance liquid chromatography–tandem mass spectrometry and on gas chromatography–mass spectrometry for the determination of acrylamide in bakery and potato products. *Journal of Chromatography A* **1132**(1–2), 211–218.

Wenzl, T., Lachenmeier, D. W. and Gokmen, V. (2007). Analysis of heat-induced contaminants (acrylamide, chloropropanols and furan) in carbohydrate-rich food. *Analytical and Bioanalytical Chemistry* **389**(1), 119–137.

Wenzl, T., Simon, R., Kleiner, J. and Anklam, E. (2006b). Analytical methods for polycyclic aromatic hydrocarbons (PAHs) in food and the environment needed for new food legislation in the European Union. *Trac-Trends in Analytical Chemistry* **25**(7), 716–725.

WHO. (2002). 'Health Implications of Acrylamide in Food. Report of a Joint FAO/WHO Consultation.' from http://www.who.int/fsf/acrylamide.

Xing, X. P. and Cao, Y. H. (2007). Determination of 3-chloro-1,2-propanediol in soy sauces by capillary electrophoresis with electrochemical detection. *Food Control* **18**(2), 167–172.

Yasuhara, A., Tanaka, Y., Hengel, M. and Shibamoto, T. (2003). Gas chromatographic investigation of acrylamide formation in browning model systems. *Journal of Agricultural and Food Chemistry* **51**(14), 3999–4003.

Yaylayan, V., Pare, J. R. J., Laing, R. and Sporns, P. (1990). Formation of Beta-Carbolines from 1-[(1'-Carboxy-2'-Indol-3'-Yl-Ethyl)Amino]-1-Deoxy-D-Fructose under Electron-impact Conditions. *Organic Mass Spectrometry* **25**(3), 141–145.

Yaylayan, V. A. (2006). Precursors, Formation and Determination of Furan in Food. *Journal für Verbraucherschutz und Lebensmittelsicherheit* **1**, 5–9.

Yaylayan, V. A., Perez-Locas, C., Wnorowski, A. and O'Brien, J. (2004). The role of creatine in the generation of N-methylacrylamide: A new toxicant in cooked meat. *Journal of Agricultural and Food Chemistry* **52**(17), 5559–5565.

Yaylayan, V. A., Perez-Locas, C., Wnorowski, A. and O'Brien, J. (2005). Mechanistic pathways of formation of acrylamide from different amino acids. *Chemistry and Safety of Acrylamide in Food*. **561,** 191–203.

Yaylayan, V. A., Wnorowski, A. and Perez-Locas, C., (2003). Why asparagine needs carbohydrates to generate acrylamide. *Journal of Agricultural and Food Chemistry* **51**(6), 1753–1757.

Yoshida, D., Matsumoto, T., Okamoto, H., Mizusaki, S., Kushi., A. and Fukuhara, Y. (1986). Formation of Mutagens by Heating Foods and Model Systems. *Environmental Health Perspectives* **67**, 55–58.

Yoshida D., Sato Y. and Mizusaki S. (1984). Isolation of 2-amino-3-methyl-imidazo[4,5-f]-quinoline as a mutagen from the heated product of a mixture of creatine and proline. *Agricultural and Biological Chemistry* **48**, 241–243.

Yoshimi, N., Sugie, S., Iwata, H., Mori, H. and Williams, G. M. (1988). Species and sex differences in genotoxicity of heterocyclic amine pyrolysis and cooking products in the hepatocyte primary culture/DNA repair test using rat, mouse, and hamster hepatocytes. *Environmental and Molecular Mutagenesis* **12**, 53–64.

Zhang, X. M., Chan, C. C., Stamp, D., Minkin, S., Archer, M. C. and Bruce, W. R. (1993). Initiation and Promotion of Colonic Aberrant Crypt Foci in Rats by 5-Hydroxymethyl-2-Furaldehyde in Thermolyzed Sucrose. *Carcinogenesis* **14**(4), 773–775.

Zhang, Y., Zhang, G. Y. and Zhang, Y. (2005). Occurrence and analytical methods of acrylamide in heat-treated foods – Review and recent developments. *Journal of Chromatography A* **1075**(1–2), 1–21.

Ziegenhals, K., Jira, W. and Speer, K. (2007). Analysis of polycyclic aromatic hydrocarbons (PAH) in meat products and spices. *Fleischwirtschaft* **87**(6), 98–103.

Zochling, S. and Murkovic, M. (2002). Formation of the heterocyclic aromatic amine PhIP: Identification of precursors and intermediates. *Food Chemistry* **79**(1), 125–134.

Zyzak, D. V., Sanders, R. A., Stojanovic, M., Tallmadge, D. H., Eberhart, B. L., Ewald, D. K., Gruber, D. C., Morsch, T. R., Strothers, M. A., Rizzi, G. P. and Villagran, M. D. (2003). Acrylamide formation mechanism in heated foods. *Journal of Agricultural and Food Chemistry* **51**(16), 4782–4787.

Part IV

Improving the quality of particular in-pack processed products

16

Use of the natural food preservatives, nisin and natamycin, to reduce detrimental thermal impact on product quality

J. Delves-Broughton, Danisco, UK

16.1 Introduction

Thermal treatment of food products to render them free of spoilage and pathogenic micro-organisms has been practised for more than five thousand years, although the science and mechanism of the process in killing micro-organisms has been understood only for approximately the last 100 years. Only a few food preservation techniques rely on killing the relevant infecting micro-organisms and thermal inactivation is the most widely used process in this category. Other procedures rely mainly on slowing the growth of the food spoilage micro-organisms. Ramesh (1999) described both the advantages and disadvantages of using heat for food preservation. Advantages are that heat is economical, it is safe and chemical-free foods can be produced, the product becomes tender with the desired cooked flavour and texture, the majority of the spoilage and pathogenic organisms present in foods are killed and, when packed in sterile containers, the foods have a very long shelf-life. The disadvantages of using heat are that overcooking may lead to textural disintegration and undesired cooked flavour, nutritional deterioration can result from high temperature processing, and in some cases heat-resistant micro-organisms in the form of vegetative cells or more likely as spores can survive the heat process and, on subsequent storage, cause the foods to spoil or result in the growth of pathogenic or toxin-producing organisms.

The use of preservatives as an adjunct to heat processes or the preservation of foods by using a combination of several factors is often described as 'hurdle technology' (Leistner and Gorris,1995). Successful implementation of hurdle technology or combination-preservation can allow improvements in the safety and quality of foods, as well as possible economic savings in food production. This latter point may become even more relevant in the future as the world's energy sources diminish and energy becomes more expensive. The use of food preservatives as an adjunct to heat processes, allowing a reduction in the heat process, can work in one of two ways (or in some cases as a combination of both). These are (i) the application of the preservative will decrease the heat resistance of the micro-organism or (ii) the residual levels of the preservative in the food following the heat process will prevent the growth of surviving micro-organisms.

16.2 Heat processes

Basically, heat processes can be conveniently sub-divided into pasteurisation and sterilisation.

Pasteurisation is a process of heat treatment used to inactivate enzymes and kill relatively heat-sensitive micro-organisms that cause spoilage. Pasteurisation results in minimal changes in food properties. It can be achieved by a combination of time and temperature and can be broadly divided into either (i) heating foods to a relatively low temperature and maintaining for a longer time or (ii) heating foods to a higher temperature and maintaining for a short time only. It is used to extend the shelf-life of food at low temperatures (usually 2–6° C) for several days. Pasteurisation can be performed by two basic methods: (i) by first filling sterile containers with the product and then pasteurising or (ii) by pasteurizing the product first and then filling sterile containers. Bacterial spores and more heat-resistant vegetative bacteria can survive pasteurisation processes. For more details on the type of pasteurisation processes available, the reader is referred to Ramesh (1999).

Sterilisation is a far more severe heat process than pasteurisation and produces foods that can have a prolonged shelf-life at ambient temperature. Like pasteurisation, sterilisation can be achieved by one of two methods. In the canning process, for example, the filled cans are exposed to a temperature of 110–125 °C for a prolonged period or in Ultra-High Temperature (UHT) processes for a higher temperature of 132–145 °C for a short period of 3–7 seconds. As with pasteurisation, foods can be sterilised inside the container or first sterilised and then filled into aseptic containers. Again, the reader is referred to Ramesh (1999) for more details of sterilisation processes and equipment.

The degree of a sterilisation heat process is expressed as the F_0 value. An F_0 value of 1 is equivalent to a heat treatment of 121.1 °C for 1 minute. When determining F_0 for canned foods, it is important that the temperature is monitored inside the container in a position where the heat penetration is least, and similarly in the retort where the heat penetration is least. An F_0 value of 3 is considered as the minimum level of heat to ensure a 'botulinum kill' in low acid (pH above 4.5)

Table 16.1 Effect of heat processing on major nutritional components (Ramesh, 1999)

Nutrient	Effect
Dry matter	Loss of total solids into canning liquor Dilution Dehydration
Protein	Enzymic inactivation Loss of certain essential amino acids Loss of digestibility
Carbohydrate	Starch gelatinisation and increased digestibility No apparent change in content of carbohydrate Sugars browning due to Maillard reaction
Dietary fibre	Generally no loss of physiological value
Lipids	Conversion of *cis* fatty acids to *trans* by oxidation Loss of essential fatty acid activity
Water-soluble vitamins	Large losses of vitamins C and B_1 due to leaching and heat inactivation Increased bioavailability of biotin and niacin due to enzyme inactivation
Fat-soluble vitamins	Mainly heat stable Losses due to oxidation of lipids
Minerals	Losses due to leaching Possible increase in sodium and calcium levels by uptake from canning liquor

canned foods. As *Clostridium botulinum* cannot grow below pH 4.5, heat processing of high acid (pH below 4.5) foods need not necessarily require the 'botulinum kill' process. Thus a heat process of F_0 of 3 will ensure a botulinum kill and hence the safety of heat processed canned foods. Such foods are considered 'commercially sterile' and can have a long shelf-life under ambient temperatures of storage. However, as will be explained in more detail later, such heat-processed foods can contain low numbers of non-pathogenic/food poisoning heat-resistant spores belonging to mainly thermophilic species that can cause spoilage of foods when stored at high ambient temperatures.

16.3 Effect of heat processing on product quality

Heat processing can adversely affect the quality of foods and this will vary from food to food. A significant problem that can occur with the over heat processing of foods is the Maillard reaction, which can affect the colour and flavour as well as the nutritional status, and product disintegration, which can be a particular concern in fish and shellfish. Egg products subjected to too much heat can coagulate. These and other effects of heat processing on the nutritional components of foods are summarised in Table 16.1.

16.4 Effect of heat against micro-organisms

The susceptibility of micro-organisms in the form of both vegetative cells and spores is dependent on a number of factors. These include the pH of the product, levels and types of preservatives, water activity, previous conditions of the micro-organisms of concern, product composition, and competitive micro-organisms. Thermal destruction of micro-organisms tends to follow first-order rate reaction kinetics and has been traditionally described by the time at a specific temperature required to reduce a population of organisms by 90%, or 1 logarithmic value. This value is referred to as the '*D*' value or decimal reduction value. The change in *D* value with temperature will also follow a first-order relationship. The temperature increase required to reduce a micro-organism's *D* value by a factor of 10 is referred to as the '*z*' value. Thus for thermal processes, understanding a micro-organism's *D* and *z* value under specific conditions of pH, water activity, and food composition will allow a processor to measure the amount of microbial destruction delivered by the process. In the event that temperature can be neither modelled nor physically measured, microbial destruction by a process can be physically measured by inoculating a pack of the product with a known number of indicator organisms and then measuring the number of micro-organisms that remain in the food after the process. Thus reference to the *D* value of micro-organisms and spores can give an indication of their relative heat resistance, which varies significantly between various types of micro-organisms. The *D* values of a range of micro-organisms are shown in Table 16.2. It can be seen that vegetative non-spore-forming bacteria such as Gram-negative *Escherichia coli, Pseudomonas* and *Salmonella* and Gram-positive *L. monocytogenes* are relatively sensitive to heat. Of the non-spore-forming bacteria, the Gram-positive bacterium, *Enterococcus faecalis,* shows the highest degree of heat resistance. The presence of this organism in heat-processed foods can be indicative of a marginal heat process. Bacterial spores are far more heat resistant. It should be noted that spores of proteolytic *Cl. botulinum* are far more resistant to heat than non-proteolytic types (see Chapter 14 for further information). Of the bacterial spores, the ones that show the greatest heat resistance are those produced by thermophilic spore-forming bacteria, namely *Geobacillus stearothermophilus* (previously named *Bacillus stearothermophilus*), the cause of flat sour spoilage, and *Thermoanaerobacterium thermosaccharolyticum* (previously known as *Clostridium thermosaccharolyticum*) that causes can swells. An unusual bacterial spore-former, in that, for a mesophile, its spores show very high heat resistance, is *Bacillus sporothermodurans* (Hammer *et al.*, 1995; Scheldeman *et al.*, 2006). Another interesting spore-former, in that it can grow at low pH and cause off-flavours due to guaicol production in fruit juice, is *Alicyclobacillus acidoterrestris* (Splittstoesser *et al.,* 1994; Yamazaki *et al.*,2000; Jensen, 1999).

Yeasts and moulds are relatively sensitive to heat as long as they do not produce ascospores. Ascospores are produced by fungi of the genera *Byssochlamys, Neosartorya,* and *Talaromyces.* Ascospores are not as resistant to heat as bacterial spores and will not survive sterilisation processes, but they can survive pasteurisation (Pitt and Hocking, 1999).

Table 16.2 Approximate *D* values (minutes) at $A_w > 0.95$ and ph *ca.*7 of bacteria, moulds and yeasts of importance in foods. (Modified from Mossel *et al*. (1995) with permission from John Wiley and Sons.)

Microbial species	55 °C	60 °C	65 °C	70 °C	80 °C	90 °C	100 °C	110 °C	120 °C
Bacterial vegetative cells									
E .coli	4–6	*ca.* 2							
P. fluorescens	5–6								
Salmonella		0.1–2.5	0.07	0.03					
S. seftenberg		6–10							
Ent. faecalis		5–20	1.6–2.3	0.3					
L. monocytogenes		3–8	0.8	0.05					
Bacterial spores									
Psychrophile					17.5				
B. cereus					4.5	0.1–0.5			
Mesophile									
B. polymyxa									
B. cereus					1.0				
Cl. botulinum A and B – proteolytic						5–35	0.3 –18	2.3–5.2	
Thermophile									
G. stearothermophilus							800–1600		1.0–5.8
T. thermosaccharolyticum							400		3–4
Yeasts and moulds									
Vegetative cells/Asexual spores									
Penicillium veniculatum		4							
Saccharomyces									
Ascospores of moulds and yeasts									
Hansenula anomala		4							
S. cerevisiae		> 4							
Byssochlamys fulva					5–8				
Taralomyces spp.					8–200				

16.5 Use of the bacteriocin, nisin, as an adjunct to heat processes, enabling improvement in product quality

Nisin is an antibacterial polypeptide or bacteriocin produced by certain strains of *Lactococcus lactis* subsp. *lactis*. Nisin shows antibacterial activity against Gram-positive bacteria but shows no or little activity against Gram-negative bacteria, yeasts and moulds. For reviews on nisin, the reader is referred to Delves-Broughton and Gasson (1984); Delves-Broughton *et al.* (1996); Thomas *et al.* (2000); Abee and Delves-Broughton (2003); and Thomas and Delves-Broughton (2005).

Nisin was discovered as a result of problems encountered in cheese-making, due to the growth of 'inhibitory streptococci' in milk stored overnight. This bacteriocin is now used as a food preservative world-wide, available commercial preparations, the first of which was Nisaplin®, developed between 1962 and 1965. In 1969, a joint FAO/WHO expert committee concluded nisin was safe to be used as a food additive (FAO/WHO, 1969), and in 1988 it was declared GRAS (FDA, 1988). There are many reasons for nisin's acceptance as a natural antimicrobial for food. Nisin is produced by fermentation using a food grade starter culture, *Lactococcus lactis* subsp. *lactis*. Since its activity spectrum is limited to Gram-positive bacteria, it is not suitable for therapeutic use and cannot be used to cover up poor manufacturing practice (which usually results in Gram-negative bacterial contamination). Another reason that it is not suitable as a therapeutic drug is due to its rapid digestion and breakdown in the body. Nisin differs from most therapeutic antibiotics since it is a primary metabolite produced by a process involving ribosomal transcription and translation. It has also been shown that nisin does not contribute to antibiotic drug resistance.

16.5.1 Mode of action

Vegetative cells

For a bacteriocin, nisin has an unusually broad-spectrum of activity against Gram-positive bacteria, both spoilage and pathogenic. Nisin acts on vegetative cells by absorbing to the cell wall precursor Lipid II, inserting into the cytoplasmic membrane and forming pores. If the cell is actively growing and the membrane is energised, then nisin can cause cell death as a result of leakage of low molecular weight compounds and dissipation of the proton motive force. Cells are often in a starved or resting physiological state in food, in which case nisin causes growth stasis. Nisin also inhibits peptidoglycan synthesis, a component of bacterial cell walls and acts against bacterial spores. Bacterial spores are far more sensitive to nisin than their vegetative forms.

Spores

Although the mode of action against vegetative cells is well understood, its action against bacterial spores has been not so intensively studied and it is still uncertain as to its precise mode of action, and even whether it is sporostatic or sporicidal. The consensus of opinion is that it is predominantly sporostatic in action, in that it usually does not kill spores but exerts an inhibitory action preventing their outgrowth. This was demonstrated by Thorpe (1960), who showed that, when nisin is bound onto spores of *Geobacillus stearothermophilus,* the reduction of heat-resistance observed was apparent rather than real and was due to the adsorption of nisin on the spores. When the nisin was removed by application of trypsin, the heat viability was restored. Hitchins *et al.* (1963) and Gould (1964) showed that, when nisin acts sporostatically, it does not prevent germination but prevents post-germination swelling and subsequent outgrowth of the spore. Gould and Hurst (1962) classified *Bacillus* spores into two groups. The small-spored species (e.g. *B. subtilis*), which appeared to rupture their envelopes by mechanical pressure, were inhibited by about 0.125 μg/ml nisin. The large-spored species (e.g. *B. cereus*) outgrew by a lytic mechanism. The spore envelope disappeared during outgrowth and these spores were much more nisin resistant, requiring more than 2.5 μg/ml nisin for inhibition.

Sporicidal effects have been reported as being dependent on the target strain and the degree of heat treatment (de Vuyst and Vandamme, 1994). It is possible that some workers have been misled that nisin is sporocidal by apparent loss of viability that is in fact reversible. It is an area that requires further research. Nisin action against spores at the molecular level has been been reported only by one group of researchers (Morris *et al.,* 1984). They showed that nisin action against spores was caused by binding of the nisin with sulphydryl groups on protein residues on the spore surface. What is clear is that nisin is far more effective against bacterial spores that have been heat damaged. For example, spores of *Clostridium* Putrefactive Anaerobe PA3679, which have survived a heat treatment equivalent to F_0 3, are 10 times more heat sensitive to nisin than those that are not heat damaged (Fowler and Gasson, 1991).

16.5.2 Assay methods

Nisin can be quantified either in terms of the molecule or its biological activity, which is expressed as International Units and is defined as 0.001 mg of a standard nisin preparation. The horizontal agar diffusion test using *Micrococcus luteus* as the indicator organism is widely used to measure nisin levels in food. To measure high-potency commercial extracts, HPLC or a liquid activity assay using the dye resazurin may be utilised. Many other assays have been described in the literature, e.g. microtitre plate bioassays, ELISA, green fluorescent protein and bioluminescence.

16.5.3 Applications of nisin to heat-processed foods

The above provides a background to how nisin can be used as a food preservative to allow a reduction in the heat process, thus reducing the thermal impact on product quality. A number of examples on such use of nisin will be given. Space does not permit all examples of nisin in combination with heat processes to be included.

UHT milk

UHT milk is typically processed at 142 °C for 2 s. When packaged aseptically, it results in a product that is commercially sterile. UHT milk has a distinct cooked flavour that is not liked by people used to drinking pasteurised milk. It is widely acknowledged that consumer acceptance of milk is influenced more by its flavour than any other attribute (Lewis,1994; Nursten, 1997). The intensity of this cooked flavour could be reduced by decreasing the processing temperature. Although this would improve the flavour, there is then the possibility of survival of heat-resistant spoilage and pathogenic spore formers. Wirjantoro *et al.* (2001) investigated a reduction of the heat treatment of milk from 142 °C for 2 s down to 117 °C for 2 s, with and without the addition of nisin to the reduced heat-processed milk (RHT). The RHT milk was preferred by a trained taste panel to the UHT milk. Nisin at 1.875 and 3.75 μg/ml increased the shelf-life at 30 °C of the RHT milk from less than 4 days (control) to more than 25 days.

B. sporothermodurans produces highly heat-resistant spores which can survive UHT processing (Scheldeman *et al.*, 2006). Massive contaminations of UHT and of sterilised milk due to this mesophilic aerobic spore former were reported in Italy, Austria, and Germany in 1985 and 1990. Unpublished studies by Danisco have shown that the organism is sensitive to low levels of nisin, with 0.25 μg/ml controlling 10^3 spores/ml in both microbiological growth media and milk. Efficacy of nisin has also been demonstrated against *B. sporothermodurans* spores in soya milk (Thomas and Delves-Broughton, 2001b).

Dairy and other desserts

The quality of pasteurised dairy desserts such as crème caramel and chocolate dairy desserts can be improved by reducing heat processing and using nisin to maintain the shelf-life. Inclusion of rice, tapioca, sago, and semolina, common ingredients of canned puddings, increases the difficulty of preservation by heating alone. These puddings are made by adding dry cereal and sugar to empty cans, then filling with hot milk. This may be made from powdered milk and may vary in fat content. The cans are sealed and heat treated, thickening the contents and reducing heat transfer. As with other canned products (see later), a heat treatment severe enough to kill all bacterial spores (especially thermophiles) will sometimes impair flavour and texture. Nisin addition at 1–3 μg/g avoids the necessity of such severe heat treatment while still maintaining protection against spoilage. Nisin is approved for use in canned semolina, tapioca and similar puddings at a maximum level of 3 μg/g in the EU (Turtell and Delves-Broughton, 1998).

Processed cheese products.

The degree of heat processing that can be applied to processed cheese and processed cheese products during its manufacture is limited due to the onset of browning (Maillard reaction), and flavour deterioration (Meyer, 1973; Thomas,1977). The traditional heat processes in the melt process of 85 °C–105 °C, combined with a hot-fill temperature of above 75 °C into jars, tubes, or foil packs, are sufficient to kill vegetative bacteria yeasts and moulds but are insufficient to ensure the complete destruction of bacterial spores belonging to both *Clostridium* and *Bacillus* spp. The anaerobic environment, high moisture and high pH range usually found in processed cheese, favours the growth of *Clostridium* spp., in particular *C. butyricum, C. cochlearium, C. tyrobutyricum,* and *C. sporogones* (Meyer,1973; Delves-Broughton,1998; Lycken and Borch, 2006). Facultative anaerobic *Bacillus* may also cause problems. Spoilage is observed as blowing, off-odours and liquefaction. Incidence of *C. botulinum* is rare but is possible, particularly when temperature abuse occurs (Kosikowski,1977; Briozzo *et al.*,1983; Collins-Thompson and Woods, 1992; Glass and Johnson, 2004)

The use of nisin as an effective preservative in processed cheese was first demonstrated in 1952 by McClintock *et al.* (1952). Efficacy in the use of nisin at 2.5 and 5 µg/g in 40%-moisture processed cheese and 60%-moisture processed spreads (cheddar, cheddar with ham, and emmental) inoculated with a spore cocktail of *C. butyricum, C. tyrobutyricum,* and *C. sporogones* spores and incubated at 37 °C was reported by Delves-Broughton and Gasson (1994). Efficacy was also reported by Plocková *et al.* (1996), who found that similar levels of nisin inhibited *Clostridium* and *Bacillus* during a 3-month storage period. Somers and Taylor (1987) investigated the inhibition of *Cl. botulinum* types A and B, and found that higher nisin levels were required at lower pH, low levels of sodium chloride and phosphate, and high moisture content. The work of Somers and Taylor (1987) provided the basis of approval for nisin use in the USA and its GRAS approval (FDA,1988) Use of nisin in processed cheese products is approved in approximately 60 countries (Turtell and Delves-Broughton, 1998) and this application is the largest commercial use of the preservative.

Canned vegetables

Nisin has been extensively used as a preservative in both low acid (pH above 4.5) and high acid (pH below 4.5) canned foods. In the majority of cases, low acid canned foods should receive a heat treatment to ensure the destruction of *C. botulinum* (heat treatment $F_0 = 3$), but even at processes above $F_0 = 3$, heat-resistant spores of spore-forming thermophilic bacteria such as *Geobacillus steraothermophilus* (previously named *Bacillus stearothermophilus*), the cause of flat sour spoilage, and *Thermoanaerobacterium thermosaccharolyticum* (previously named *Clostridium thermosaccharolyticum)*, the cause of can swelling, can survive and cause spoilage, especially when the cans are stored at warm temperatures (Eyles and Richardson,1988). For some reason, possibly due to the modification of the phospholipid content of their cytoplasmic membranes allowing them to grow at high temperatures themophilic bacteria are remarkably sensitive to nisin, with residual levels of nisin less than 0.05 µg/g of can contents providing protection for prolonged storage periods in warm climates. It is fortunate that such a low residual nisin is effective as up to 98% can be lost during retort processing. An addition level of 2.5 µg/g, resulting in a residual level of 0.05 µg/g after autoclaving, is usually more than sufficient to control thermophilic spoilage. This is because thermophilic spores are remarkably sensitive to nisin, and because of the increased sensitivity to nisin of bacterial spores that have been damaged by heat. The loss of nisin during heat processing can be prevented, to an extent, by acidification of the brine, using citric acid. Optimum heat stability of nisin is at pH 3.5 (Davies *et al.,* 1998). Thus, in low acid canned vegetables, nisin can be used to decrease the heat process to $F_0 = 3$ or slightly higher, allowing a reduction in energy consumption and the possible improvement of nutritional value, appearance, and texture of the food. Such use of nisin in allowing reduced heat processing without increasing the incidence of thermophilic spoilage has been reported for asparagus (Hernandez *et al.,*1964), mushrooms (Funan *et al.,*1990), okra (El-Samehy and Elias, 1977), peas (Gillespy,1953; O'Brien *et al.,* 1956; Nekhotenova,1961; Vas *et al.,*1967), peppers (Duran *et al.,* 1964) and potatoes (Maslennikova and Loshina, 1968).

Cl. botulinum cannot grow in high-acid food (<pH 4.5), which therefore can be safely processed at F_0 levels below 3. Spoilage of these products can occur as a result of growth of acid-tolerant spore formers such as *Cl. pasteurianum* and *B. macerans.* That the growth of such bacteria can be controlled by nisin has been reported in high-acid canned foods such as canned tomatoes (Vas,1963; Maslennikova *et al.,* 1968) and tomato juice (Poretta *et al.,*1968).

Pasteurised liquid egg

Severe heat processing of liquid egg and liquid egg products (scrambled egg, omelet, pancake mixes, etc.) is not possible as the egg will coagulate and lose its functional properties. Regulations exist to ensure that the heat process applied is sufficient to ensure the complete destruction of *Salmonella.* In the UK, for example, pasteurised egg must be processed for at least 2.5 min at a temperature of 64.4 °C. Although such a heat treatment will kill most of the bacteria present, bacterial spores will survive. *B. cereus,* for example, is commonly found in liquid egg (Payne *et al.,*1979; Wood and Waites,1988). Some surviving spore-formers may be capable of growth at refrigerated temperatures. A nisin level of 5 μg/g was reported to significantly increase shelf-life from 6–11days to 17–20 days and prevented the growth of *B. cereus* (Delves-Broughton *et al.,*1992). More recent unpublished studies using nisin levels of 2.5 and 5 μg/g increased shelf-life from less than 20 days to more than 80 days at 8° C.

The temperature and duration of liquid egg pasteurisation can also be reduced with incorporation of nisin (Boziaris *et al.,*1998; Ponce *et al.,*1998). Nisin has also been reported to increase the effectiveness of heat treatment in killing *S. enteriditis* PT4 (Boziaris *et al.,*1998) and *L. monocytogenes* Scott A (Knight *et al.,*1999) in liquid egg. The latter researchers found that nisin was far more effective in liquid egg in increasing the heat sensitivity of *L. monocytogenes* when added to the liquid egg 2h prior to heat treatment.

Use of nisin in pasteurised liquid egg products is approved in the USA, Australia, and New Zealand. It is mainly used in convenience packs of liquid egg products destined for use in the hotel – institute – restaurant (HIR) sector.

Shellfish

The food pathogen *L. monocytogenes* can be a problem in shellfish. Shucked lobster meat produced on the eastern coastline of Canada is sold in cans, packed in brine, and frozen. Problems occurred in the Canadian lobster industry in 1989 when the USFDA. set a zero tolerance for the presence of *L. monocytogenes* in ready-to-eat sea food products (Crawford,1989; Klima and Montville,1995). Subsequently, *L. monocytogenes* was detected in frozen cans of Canadian frozen shucked lobster imported into the USA, which resulted in a costly product recall. Heat processing of the canned lobster was introduced in an attempt to overcome the problem. It was found that the heat processes that ensured complete destruction (log 4–5 kill) were too severe in that they resulted in product shrinkage, thermal breakdown of the lobster meat, and a loss in drained weight which affected the profitability of the product. The maximum heat treatment with which no product deterioration occurred was 60 °C for 5 min. This achieved only a 2 log kill of *L. monocytogenes*, which was considered insufficient. Addition of nisin (25 μg/ g) to the brine, in combination with a heat treatment of 60 °C for 5 min, resulted in a log 4–5 kill. Nisin alone resulted in a log 2 kill (Budu-Amoako *et al.,*1999). Such increased heat sensitivity in combination with nisin for *L. monocytogenes* was confirmed by Modi *et al.* (2000).

Other studies carried out with nisin to reduce *L. monocytogenes* contamination of crustacea include that of Degnan *et al.* (1994); who reported the use of nisin and other bacteriocins as a wash for inoculated steam-sterilized crab meat. Walker and Ferrandini (1974) described a process of preserving cooked shellfish meat by dipping in chlorine solution and submerging in an aqueous solution of inorganic chloride, nisin, and strong organic acid. This treatment was effective in preventing toxin formation and growth of *C. botulinum.*

Fresh pasteurised chilled soups

Fresh, minimally heat processed, pasteurised soups with refrigerated shelf-life are becoming increasingly popular. The soups have the taste and appearance of a home-cooked product. Nisin has proved to be an effective preservative, which has led to its use in the USA, where it has been allowed for certain manufacturers by the process of self-affirmation. Nisin at a level of 2.5 to 6.25 μg/g controls the growth of spoilage *Bacillus* and extends the shelf-life.

Pasteurized fruit juice

The acid-tolerant spore forming Gram-positive bacterium *A. acidoterrestris* can cause flat sour spoilage of fruit juices, due to production of guaicol (Cerny *et al.,*1984; Pettipher *et al.,*1997; Splittstoesser *et al.,*1994). *A. alicycidoterresris* grows in a pH range of 2.5–6.0 at temperatures of 25–60 °C. Both the vegetative cells and spores of this heat-resistant bacterium are extremely sensitive to nisin (Komitopoulou *et al.,* 1999; Yamazaki *et al.,* 2000). The presence of nisin decreases the *D*-value by approximately 40% (Table 16.3).

16.5.4 Future developments on the use of nisin in heat-processed foods

Refrigerated minimal heat processed foods have been developed in response to

Table 16.3 The effect of nisin (1.25 μg/g) on heat resistance of *Alicyclobacillus acidoterrestris* spores in apple juice at pH 3.51

Temperature	Without nisin *D*	*z*	With nisin *D*	*z*
80 °C	41.15 ±0.24 °C	12.2 °C	23.75 ±0.12 °C	13.8 °C
90 °C	7.38 ±0.85 °C		4.56 ±0.09 °C	
95 °C	2.30 ±0.03 °C		1.95 ±0.02 °C	

From Komitopolou *et al.* (1999). Reproduced in modified form with permission from Blackwell Publishing.

consumer demand for lightly processed foods that are safe and have a good shelf-life but at the same time are not over heat processed (Peck, 2006). This presents a challenge for both food technologists and microbiologists to ensure that such foods are of acceptable quality, have sufficient shelf-life, and are safe.

Nisin, in combination with other hurdles, can play a part in the preservation of such foods. Thomas and Isak (2006) describe a strong synergy between nisin and natural extracts of the herb *Rosamarinus officinalis*, which enhanced both the cidal and static antibacterial effects of nisin against Gram-positive bacteria, Synergies against *L. monocytogenes* inoculated into a Bolognese pasta sauce (Fig.16.1) and *B. cereus* inoculated into a carbonara pasta sauce (Fig. 16.2) were reported. Numerous other synergies with nisin have been reported, including sucrose fatty acid esters (Thomas *et al.*, 1998), carvacrol (Pohl and Smid, 1999) and garlic (Singh *et al.*, 2001). Such combinations can increase safety and shelf-life of refrigerated minimal heat-processed foods.

16.6 Use of natamycin as an adjunct to heat processes, enabling improvement in product quality

The polyene macrolide antifungal compound, natamycin (formerly known as pimaricin), like nisin, can be considered natural because it is produced by fermentation of the bacterium *Streptomyces natalensis.* The name derives from the discovery of this strain in 1955 in a soil sample from the Natal province in South Africa (Struyk *et al.*, 1959). Natamycin is now produced as a commercial food preservative. An example is a product produced by Danisco, marketed under the brand name Natamax™. Natamycin is used as an approved food preservative worldwide, mainly for surface treatment of cheese and dried sausages (the only authorisations in the EU) (Stark,1999; Thomas and Delves-Broughton, 2001a; Delves-Broughton *et al.*, 2005). In the USA, natamycin has GRAS status and is allowed in cheese. More recently, its use has been extended to yoghurt, cottage cheese, sour cream, cream cheese, salad dressing and soft tortillas (FDA, 2003a and b). Use is much broader in South Africa, where it was first discovered. In 2002, the World Health Organisation and the Food Agriculture Organisation JECFA committee re-evaluated natamycin with a specific request to assess the issue of

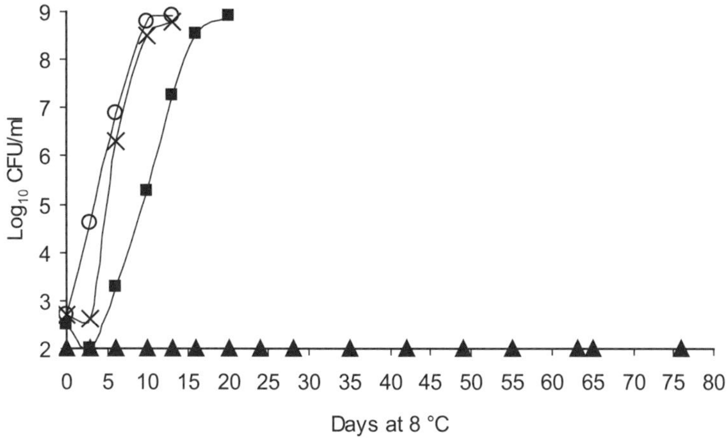

Fig. 16.1 Growth of a cocktail of *L. monocytogenes strains* in Bolognese pasta sauce, pH 5.13, in the presence of (○) no additions; (×) 16.8 ppm phenolic diterpenes; (■) nisin at 100 IU/ml; (▲) combination of nisin at 100 IU/ml and 8.6 ppm phenolic diterpenes. Minimum detection level 100 CFU/ml.

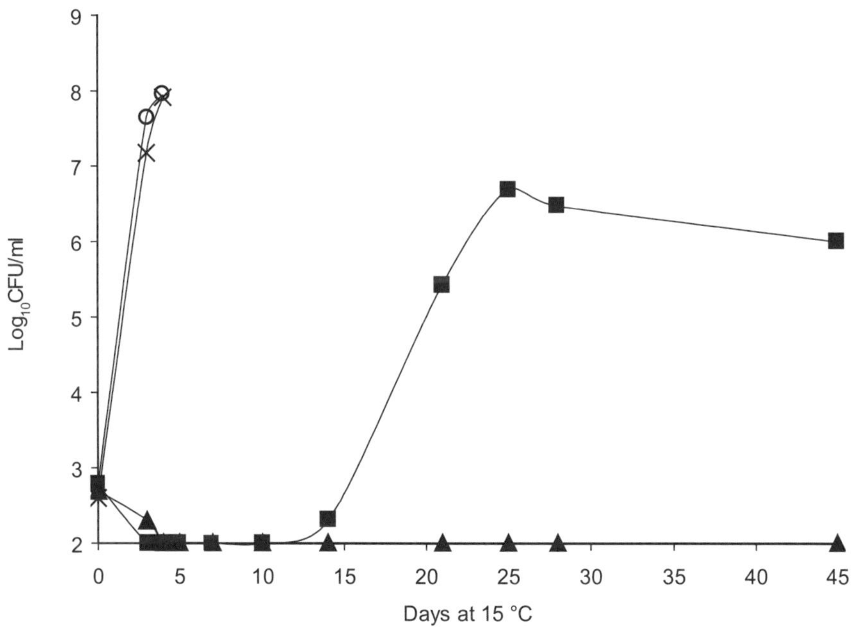

Fig. 16.2 Growth of a cocktail of spores of *B. cereus* strains in a carbonara pasta sauce, pH 5.86, in the presence of (○) no additions; (×) 21 ppm phenolic diterpenes; (■) nisin at 100 IU/ml; (▲) combination of nisin (100 IU/ml) and rosemary extract (8.4 ppm phenolic diterpenes). Minimum detection level 100 CFU/ml. (Thomas and Isak, 2006, reproduced with permission from *Acta Hort.*)

antibiotic resistance (WHO, 2002). Their monograph reviewed all aspects of the toxicity and safety of natamycin and its degradation products, and reached very positive conclusions.

Natamycin is active against nearly all yeasts and moulds but has no effect on bacteria, protozoa or viruses. Its lack of activity against bacteria is of particular benefit in bacterial fermented or ripened food. The antifungal is active at very low concentrations: most moulds are inhibited at 0.5 to 6 μg/ml, although some species require higher concentrations. Yeasts are generally inhibited at 1.0 to 5.0 μg/ml. It is believed that natamycin acts by binding to ergosterol in fungal cell membranes, which leads to increased cell permeability and usually results in cell death. Ergosterol is not present in the cell membranes of viruses, bacteria and protozoa (Hamilton-Miller, 1974).

The natamycin content of food can be determined by several procedures. A bioassay method has been developed, using *Saccharomyces cerevisiae* as the indicator strain (Shirk *et al.*, 1962). A method has been developed for detection of natamycin in cheese and cheese rind by extraction with methanol and UV spectrum determination at 311 nm (minima), 317 nm (maxima) and 329 nm, or by liquid chromatographic separation followed by detection at 303 nm (maxima) (De Ruig *et al.*, 1987). HPLC methods are now widely used, and immunochemical methods have been reported.

16.6.1 Pasteurised fruit juice

Yeasts predominate in spoilage of acid fruit products because they have high acid tolerance and also because many can grow anaerobically. Some yeasts also produce ascospores of varying degrees of heat resistance, that may survive pasteurisation (Pitt and Hocking,1999). Moulds belonging to the genus *Byssochlamys* are almost uniquely associated with spoilage of heat-processed acid foods (Pitt and Hocking,1999). This genus was first recognised as a spoilage agent of canned fruit in the UK (Olliver and Rendle,1934), but has now been found throughout the world, and has been associated with spoilage of strawberries in cans or bottles, blended juices, and fruit gel baby foods (Pitt and Hocking,1999). Other heat-resistant fungi that can be associated with fruit and fruit products include *Talaromyces*, *Neosartorya* and *Eupenicillium*. Ascospores of these organisms can survive pasteurisation treatments (e.g. 90 °C for 3 min) routinely applied to most fruits and fruit products. The natural habitat for these organisms is the soil, and fruit such as grapes, passion fruit, pineapple and mango juices and pulps, strawberries and other berries can easily come into contact with soil directly or as a result of rain splash. Once formed, ascospores may remain dormant for years in decaying fruit debris or soil, and can be isolated from fruits harvested from near the ground and from containers and equipment used to transport and process fruit (Beuchat, 1986). Some strains of *Byssochlamys* can produce patulin, a potent mycotoxin (Rice,1980; Roland and Beuchat, 1984). Unusually for a mould, *Byssochlamys* is tolerant of reduced oxygen conditions and/or elevated carbon dioxide, a characteristic that gives it a selective advantage in products such as canned, bottled or carton fruits and juices. In the presence of very low oxygen levels, these species appear to grow anaerobically and produce carbon dioxide, causing swelling and spoilage of the product. Such spoilage has been observed in pasteurised pineapple juice manufactured in Saudi Arabia. This was effectively controlled by addition of 25 ppm natamycin (as 50 mg/l Natamax™).

16.7 Future trends

Expansion in the use of the natural preservatives nisin and natamycin is dependent upon their acceptance by regulatory authorities as ingredients that are safe. If, as by some lobby groups, such natural preservatives are mistakenly considered to be antibiotics, it is unlikely that their use as food preservatives will expand significantly from the current amopunt. Their utilisation in combination with other natural preservation techniques (hurdle technology) and the use of predictive modelling will further improve the preservation and safety of our food supply and enable reduction in heat processes, thereby reducing thermal damage to our foods and also allowing energy savings to be made. It is also possible that more efficient use of both nisin and natamycin can be achieved by protection of the preservatives from heat by use of encapsulation and controlled release (Pegg and Shahidi, 1999). This is a topic of current research.

16.8 References

Abee, T., and Delves-Broughton, J. (2003). Bacteriocins – Nisin. In *Food Preservatives*, edited by N.J. Russell and G.W.Gould. Kluwer Academic, New York. pp146–178.

Beuchat. L.R. (1986). Extraordinary heat resistance of *Talaromyces flavus* and *Neosartorya fischeri* ascospores in fruit products. *J. Food. Sci* 51, 1506–1510.

Boziaris,I.S., Humpheson, L. and Adams, M.R. (1998). Effect of nisin and heat on injury and inactivation of *Salmonella enteridis* PT4. *Int. J. of Food Microb.* 43, 7–13.

Briozzo, J., da Lagarde, E.A., Chirfe, J. and Prada, J.L. (1983). *Clostridium botulinum* growth and toxin production in media and processed cheese spread. *Appl.Envion. Microbiol.* 45, 1150–1152.

Budu-Amoako, E., Ablett, R. F., Harris, J. and Delves-Broughton, J. (1999). Combined effect of nisin and moderate heat on the destruction of *Listeria monocytogenes* in cold-pack lobster meat. J. Food Prot. 62, 46–50.

Cerny, G., Hennlich, W. and Poralla, K. (1984). Spoilage of fruit juice by bacilli: Isolation and characterization of the spoiling micro-organism. *Zeitschrift für Lebensmittel Untersuchung und Forschung.* 179, 224–227.

Collins-Thompson, D. L. and Woods, D. S. (1992). Control in dairy products. In *Clostridium botulinum, Ecology and Control in Foods*, edited by A. H. W. Hauschild and K. L. Dodds, Marcel Dekker, USA, pp 261–277.

Crawford, L.M. (1989). *Revised policy for controlling Listeria monocytogenes.* Fed. Register 54, 22345–22346.

Davies, E. A., Bevis, H., Potter, R., Harris, J., Williams, G. C. and Delves-Broughton, J. (1998). The effect of pH on the stability of nisin solutions during autoclaving. *Letts. Appl. Microbiol.* 27, 186–187.

Degnan, A. J., Kaspar, C. W., Otwell, W. S., Tamplin, M. L. and Luchansky, J. B. (1994). Evaluation of lactic acid bacterium fermentation products and food grade chemicals to control *Listeria monocytogenes* in blue crab (*Callinectes sapidus*) meat. *Appl. Environ. Microbiol.* 60, 3198–3203.

Delves-Broughton, J. (1998). Use of nisin in processed and natural cheese. *Bull Int. Dairy Fed.* 329, 13–17.

Delves-Broughton, J., Blackburn, P., Evans, R. J. and Hugenholtz, J. (1996). Applications of the bacteriocin nisin. *Antonie van Leeuwenhoek* 69, 193–202.

Delves-Broughton, J. and Gasson, M. J. (1994). Nisin. In *Natural Antimicrobial Systems and Food Preservation*, edited by V. M. Dillon and R. G. Board. CAB International, UK, pp 99–131.

Delves-Broughton, J., Thomas, L.V., Doan, C.H. and Davidson, P.M. (2005). Natamycin. In *Antimicrobials in Food.* P.M. Davidson, J.N. Sofos, A.L. Branen (eds.), 3rd Edition, CRC Press. pp 275–288

Delves-Broughton, J., Williams, G. C. and Wilkinson, S. (1992). The use of the bacteriocin, nisin, as a preservative in pasteurized liquid whole egg. *Letts. Appl. Microbiol.* 15, 133–136.

De Ruig, W.G., van Oostrom, J.J. and Leenheer, K. (1987). Spectrophotometric and liquid chromatographic determination of natamycin in cheese and cheese rind. *J.Assoc. of. Anal.Chem.*70, 944–948.

De Vuyst, L. and Vandamme, E. J. (1994). Nisin, an antibiotic produced by *Lactococcus lactis* subsp. *lactis*: Properties, biosynthesis, fermentation and applications. In *Bacteriocins of Lactic Acid Bacteria. Microbiology, Genetics and Applications*, edited by de Vuyst and Vandamme, Blackie Academic & Professional. London, pp.151–221.

Duran, L., Hernandez, E. and Flores, J. (1964). Empleo de nisina en la esterilizacion de conservas de pimientos. *Agroquimica y Technologia de Alimentos* 4, 87–92.

El-Samehy, S. K. and Elias, A. N. (1977). The use of nisin in canned okra. *Egypt. J. Food Sci* 5, 78–81.

Eyles, M. J. and Richardson, K. C. (1988). Thermophilic bacteria and food spoilage. *CSIRO Food Research Quarterly* 48, 19–24.

FAO/WHO Expert Committee on Food Additives. (1969). Specifications for the identity and purity of some antibiotics. *12th Report. WHO Technical Report Series*, No 430.

FDA. (1988). *Nisin preparation: Affirmation of GRAS status as a direct human food ingredient.* Fed Reg. 53, 11247–11251.

FDA. (2003a). 21CFR170.30. *Elegibility for classification as generally recognised as safe (GRAS).*

FDA. (2003b). 21CFR 172.55. *Natamycin (pimaricin).*

Fowler, G. G. and Gasson, M. J. (1991). Antibiotics – Nisin. In *Food Preservatives*, edited by. N. J. Russell and G. W. Gould, Blackie Academic and Professional, London, UK, pp 135–153.

Funan, H., Meijun, S., Hebao, D., Linda, L., Jinza, C., Yichi, Y. and Xue, S. (1990). Nisaplin Application Experiment in Sterilisation of Canned Mushrooms. *Report of Scientific Research Institute of Food and Fermentation Industry*, Ministry of Light Industry, Shanghai, Peoples Republic of China, 9 pp.

Gillespy, T. G. (1953). Nisin Trials with Canned Beans in Tomato Sauce and Canned Garden Peas. *Research Association Leaflet No. 3*, Fruit and Vegetable Canning and Quick Freezing Research Association, Chipping Campden, UK.

Glass, K.A. and Johnson, E.A. (2004). Factors that contribute to the botulinal safety of reduced-fat and fat-free processed cheese products, *J. Food. Prot.* 67, 1687–1693.

Gould, G. W. (1964). Effect of food preservatives on the growth of bacteria from spores. In *Microbial Inhibitors in Foods*, edited by M. Molin, Stockholm: Almqvist and Wiksell, pp 17–24.

Gould, G. W. and Hurst, A. (1962). Inhibition of *Bacillus* spore development by nisin and subtilin. *8th International Congress of Microbiology*, Abstract A2–11.

Hamilton-Miller, J.M.T. (1974). Fungal sterols and the mode of action of the polyene antibiotics. *Adv. Appl. Microbiol.* 17, 109–134.

Hammer, P., Lembke, F., Suhren ,G. and Heeschen, W. (1995). Characterization of heat resistant *Bacillus* species affecting quality of UHT milk – A preliminary report. *Kiel Milchwirt Forshungsber* 47, 297–305.

Hernandez, E., Duran, L. and Morelli, J. (1964). Use of nisin in canned asparagus. *Agroquimica y Technologie de Alimentos* 4, 466–470.

Hitchins, A. D., Gould, G. W. and Hurst, A. (1963). The swelling of bacterial spores during germination and outgrowth. *J. Gen. Microbiol.* 30, 445–453.

Jensen, N. (1999). *Alicyclobacillus* – A new challenge for the food industry. *Food Aust.* 51,33–36.

Klima, R.A. and Montville, T.J. (1995). The regulatory and industrial response to listeriosis in the U.S.A: A paradigm for dealing with foodborne pathogens. *Trends Food Sci.Technol.* 6, 87–93.

Knight, K.K., Bartlett, F.M., McKellar, R.C. and Harris, L.J. (1999). Nisin reduces the thermal resistance of *Listeria monocytogenes* Scott A in liquid whole egg. *J.Food. Prot.* 62, 999–1003.

Komitopoulou, E., Boziaris, I. S., Davies, E.A., Delves-Broughton, J. and Adams, M.R. (1999). *Alicyclobacillus acidoterrestris* in fruit juices and its control by nisin. *Int. J. Food Sci. Technol.* 34, 81–85.

Kosikowski, F.V. (1977). *Cheese and Fermented Milk Foods.* 2nd Edition. Edward Brothers, Ann Arbor, MI, USA.

Leistner, L. and Gorris, L. G. M. (1995). Food preservation by hurdle technology. *Trends Food Sci. Technol.* 6, 41–46.

Lewis, M.J. (1994). Heat treatment of milk. In *Modern Dairy Technology* Edited by R.K. Robinson, Chapman and Hall, London pp 1–60.

Lycken, L and Borch, E. (2006). Characterization of *Clostridium* spp. isolated from spoiled processed cheese products. *J. Food Prot.* 69, 1887–1891.

Maslennikova, N.M. and Loshina, P.B. (1968). The use of nisin in canned potatoes. *Konserv. I. Ovoshcheshusi Prom* 23, 12–15 (in Russian).

Maslennikova, N.M., Shundova, Y.U. and Nekhotenova, T. (1968). The effect of nisin in the sterilization procedure of canned whole tomatoes in brine. *Konserv. I Ovoshchesushi Prom.* 23, 7–9 (in Russian).

McClintock, M., Serres, L., Marzolf, J.J., Hirsch, A. and Mocquot, G. (1952). Action inhibitrice des streptocoques producteurs de nisine sur le developpement des sporules anaerobies dans le fromage de Gruyere fondu. *J. Dairy Res.* 19, 187–193.

Meyer, A.M. (1973). *Processed Cheese Manufacture*, Food Trade Press Ltd., London.

Modi, K.D., Chimindas, M.L. and Montville, T.J. (2000). Sensitivity of nisin-resistant *Listeria monocytogenes* to heat and the synergistic action of heat and nisin. *Lett. Appl. Microbiol.* 30, 249–253.

Morris, S.L., Walsh, R.D. and Hansen, J.N. (1984). Identification and characterization of some bacterial membrane sulphydryl groups which are targets of bacteriostatic and antibiotic action. *J. Biol. Chem.* 259, 13590–13594.

Mossel, D.A.A., Corry, J.E.L., Struijk, C.B.,and Baird, R.M. (1995). *Essentials of the Microbiology of Foods. A Textbook for Advanced Studies.* John Wiley and Sons, Chichester, UK.

Nekhotenova, T.I. (1961). The possibility of modifying the sterilization process of green peas by adding nisin. *Konserv i. Ovoshchesushi Prom.* 16, 21–23 (in Russian).

Nursten, H.E. (1997). The flavour of milk and dairy products, Part 1: Milk of different kinds, powder, butter and cream. *Int. J. Dairy Technol.* 50, 48–56.

O'Brien, R.T., Titus, D.S., Devlin, K.A., Stumbo, C.R. and Lewis, J.C. (1956). Antibiotics in food preservation. II. Studies on the influence of subtilin and nisin on the thermal resistance of food spoilage bacteria. *Food Technol.* 10, 352.

Olliver, M. and Rendle,T. (1934). A new problem in fruit preservation. Studies on

Byssochlamys fulva and its effect on the tissues of processed fruit. *J. Soc. Chem. Ind*, London 53, 166–172.
Payne, J., Gooch, J.E.T. and Barnes, E.M. (1979). Heat-resistant bacteria in pasteurised whole egg. *J. Appl. Bacteriol.* 46, 601–613.
Peck, M.W. (2006). *Clostridium botulinum* and the safety of minimally heated, chilled foods: An emerging issue? *J. Appl. Microb.* 101, 556–570.
Pegg, R.B. and Shahidi, F. (1999). Encapsulation and Controlled Release in Food Preservation. In *Handbook of Food Preservation*, edited by M.S. Rahman, Marcel Dekker, New York. pp 611–667.
Pettipher, G.L., Osmundsen, M.E. and Murphy, J.M. (1997). Methods for the detection of *Alicyclobaciius acidoterrestris* and investigations of growth and taint in fruit juice and fruit-containing drinks. *Letts. Appl. Microb.* 24, 185–189.
Pitt, J.I. and Hocking, A.D. (1999). *Fungi and Food Spoilage.* Second Edition. Aspen Publisher, Inc. Maryland, USA.
Plocková, M., Štěpánek, M., Demnerová, K., Čurda, L. and Šviráková (1996). Effect of nisin for improvement in shelf-life and quality of processed cheese. *Adv. Food Sci.* (CMTL) 18, 78–83.
Pohl, I. E. and Smid, E. J. (1999). Combined action of nisin and carvacrol on *Bacillus cereus* and *Listeria monocytogenes. Letts. Appl. Microbiol.* 26, 166–170.
Ponce, E., Pla, R., Sendra, J., Guamis, B. and Mor-Mur, M. (1998). Combined effect of nisin and high hydrostatic pressure on destruction of *Listeria innocua* and *Escherichia coli* in liquid whole egg. *Int. J. Food Microbiol.* 43, 15–19.
Poretta, A., Casolari, A. and Cassara, A. (1968). Impiego della nisina nella reparazione dell succo di pomodoro. *Industria Conserve* 1, 13–14.
Ramesh, M.N.1999). Food Preservation by Heat Treatment. In *Handbook of Food Preservation*, edited by M.S. Rahman, Marcel Dekker, New York. 95–172.
Rice, S.L. (1980). Patulin production by *Byssochlayms* spp. in canned grape juice. *J. Food Science* 45, 485–488,495.
Roland, J.O. and Beuchat, L.R. (1984). Influence of temperature and water activity on growth and patulin production by *Byssochlayms nivea* in apple juice. *Appl. Environ. Microbiol.* 47, 205–207.
Scheldeman, P., Herman, L., Foster, S. and Heyndrickx, M. (2006). *Bacillus sporothermodurans* and other heat-resistant spore formers in milk. *J. Appl. Microb.* 101, 542–555.
Shirk, R.J., Whitelhall, A.R. and Clark, W.L. (1962). The bioassay of pimaricin and its binding effect in orange juice. *J. Food Sci.* 27, 605.
Singh, B., Falahee, M.B. and Adams, M.R. (2001). Synergistic inhibition of *Listeria monocytogenes* by nisin and garlic extract. *Food Microbiol.* 18, 133–139.
Somers, E. and Taylor, S. L. (1987). Antibotulinal effectiveness of nisin in pasteurized processed cheese spreads. *J. Food Prot.* 50, 842.
Splittstoesser, D.F., Churney, J.J. and Lee,Y. (1994). Growth characteristics of aciduric sporeforming *Bacilli* isolated from fruit juices. *J. Food Prot.* 57, 1080–1083.
Stark, J. (1999). Permitted Preservatives. In *Encyclopedia of Food Microbiology*. Edited by Robinson, R.K, Batt, C.A and Patel, P.A., Academic Press, London. 1780–1788.
Struyk, A.P., Hoette, I., Drost, G., Waisvisz, J.M., Van Ekk, T. and Hoogerheide, J.C. (1959). Pimaricin, a new antifungal antibiotic. *Antibiotics Annual 1957–1958*, 878–885.
Thomas, L.V., Clarkson, M.R. and Delves-Broughton, J. (2000). Nisin. In *Natural Food Antimicrobial Systems*, edited by A. S. Naidu, CRC Press, Boca Raton, USA, pp. 463–524.
Thomas, L V., Davies, E A., Delves-Broughton, J. and Wimpenny, J. W. T. (1998). Synergistic effect of sucrose fatty acid esters on nisin inhibition of Gram-positive bacteria. *J. Appl. Microbiol.* 85, 1013–1022.
Thomas, L.V. and Delves-Broughton J. (2001a). Applications of the natural food preservative natamycin. *Res. Adv. Food Sci.* 2, 1–10.

Thomas, L.V. and Delves-Broughton, J. (2001b). New advances in the application of the food preservative nisin. *Res. Adv. Food Sci.* 2, 11–22.
Thomas, L.V. and Delves-Broughton, J. (2005). Nisin. In *Antimicrobials in Food.* P.M. Davidson, J.N. Sofos, A.L. Branen (eds.), 3rd Edition, CRC Press, Boca Raton, pp 237–274.
Thomas, L.V and Isak, T. (2006). Nisin synergy with natural antioxidant extracts of the herb rosemary. A*cta Horticulturae* 709, 109–114
Thomas, M.A. (1977). *The Processed Cheese Industry*, Dept. of Agriculture, New South Wales, Australia.
Thorpe. R.H. (1960). The action of nisin on spoilage bacteria. 1. The effect of nisin on the heat resistance of *Bacillus stearothermophilus* spores. *J. Appl. Bact.* 23, 136–143.
Turtell, A. and Delves-Broughton, J. (1998). International acceptance of nisin as a food preservative. *Bull.Int. Dairy. Fed.* 329, 20–23.
Vas, K. (1963). Use of nisin in the thermal preservation of tomato products. *Fruchtsaft-Industrie ver Confructa* 8, 73–77.
Vas, K., Kiss, I. and Kiss, N. (1967). Use of nisin for shortening the heat treatment in the sterilization of green peas. *Zeitschrift für Lebensmittel-Untersuchung und-Forschung* 133, 141–144.
Walker, E.L and Ferrandini, A.R. (1974). *Process of preserving shellfish meat and product of said process.* U.S. Patent 3 852 486.
Wirjantoro, T I., Lewis, M.J., Grandison, A.S., Williams, G.C. and Delves-Broughton, J. (2001). The effect of nisin on the keeping quality of reduced heat-treated milks. *J. Food Prot.* 64, 213–219.
Wood, S.L. and Waites, W.M. (1988). Factors affecting the occurrence of *Bacillus cereus* in liquid whole egg. *Food Microbiol.* 5,103–107.
WHO. (2002). Natamycin monograph. *WHO Food Additive Series* No.48, 49–76. World Health Organisation, Geneva.
Yamazaki, K., Murakami, M., Kawai, Y., Inoue, N. and Matsuda, T. (2000). Use of nisin for inhibition of *Alicyclobacillus acidoterrestris* in acidic drinks. *Food Microbiol.* 17, 315–320.

17

High pressure processing to optimise the quality of in-pack processed fruit and vegetables

I. Oey, T. Duvetter, D. N. Sila, D. Van Eylen, A. Van Loey and M. Hendrickx, Katholiekie Universiteit Leuven, Belgium

17.1. Introduction

For decades, thermal processing has been widely used as one of the main technologies of processing and preserving fruit and vegetables from small-scale (*home*) up to large-scale (*industrial*) food applications. Numerous researchers have investigated the effect of this technique on food quality and microbial safety. Besides microbial/spore inactivation, exposure to high temperature for a long time may result in changes in nutritional aspects and sensorial/organoleptic attributes of the processed product. The extent of the quality and nutritional losses (e.g. severe losses in heat-labile micronutrients; deterioration in color, texture and flavor properties) depends on the type and the intensity of the thermal process applied.

Consumer demands, especially in developed countries, are pointing to processed food products with '*fresh-like*' characteristics (color, flavor/aroma) and high nutritional value. Therefore, optimization of traditional thermal processing with the aim of improving the balance between food quality and safety remains a challenge. Next to process optimization, different advanced heating technologies such as microwave heating (Alvarez-Alvarez *et al.*, 2005; Cuccurullo *et al.*, 2007) and radio-frequency heating (Zhang *et al.*, 2004) have been introduced to improve the quality of in-pack processed food products. In addition, extensive research on technologies leading to 'minimal processed food', such as high-pressure processing, high electric field processing, etc. have been and are still being explored in order to obtain food products with high microbial safety and minimal food quality alterations (Knorr, 1998, Martin *et al.*, 2002).

In the last two decades, high pressure (HP) processing at room/moderate temperature has been introduced for industrial food processing and preservation. Currently, in countries such as Japan, United States, Spain, France, Czech Republic and United Kingdom, this technique is being applied to produce high-quality foods, including ham, sea food, jam, mixed fruit and vegetables, juices and sauces (e.g. guacamole), etc. (Indrawati *et al.*, 2002a). Typically, foods are pre-packed in flexible containers or in plastic packs before HP treatment.

17.2 High pressure processing: general concept

High pressure technology has been known for more than 100 years, but was adopted and adapted for industrial food applications in the early 1990s (Martin *et al.*, 2002). Pressure has a limited effect on the disruption of covalent bonds (Tauscher, 1998, 1999). Small molecules, such as vitamins, pigments and volatile compounds, are less affected by HP processing compared with proteins/enzymes, which are often characterized by a complex three-dimensional architecture stabilized by various covalent and non-covalent interactions. Consequently, HP processing allows the processing of foods while preserving their '*fresh-like*' properties, due to the limited loss of flavor, color and nutritional value.

Research on HP processing in the pressure domain from 100 to 1000 MPa, combined with subzero to moderate temperatures (–20 °C up to 60 °C) has demonstrated effects on (i) phase transitions (water, lipids) (Lambert *et al.*, 1999; Van Buggenhout *et al.*, 2006); (ii) cell (wall and membrane) disruption processes (Prestamo and Arroyo, 1998; Sila *et al.*, 2007a); (iii) vegetative microorganism inactivation (Ananth *et al.*, 1998; Gervilla *et al.*, 1999; Spilimbergo *et al.*, 2002; Martin *et al.*, 2002), (iv) reduction of toxins (e.g. patulin in apple) (Bruna *et al.*, 1996); (v) modification of biopolymers including enzyme activation or inactivation (e.g. allowing selective knockout of enzymes), protein denaturation and gel formation (Stolt *et al.*, 1999; Fachin *et al.*, 2002; Indrawati *et al.*, 2002b; Verlent *et al.*, 2004a,b; Van der Plancken *et al.*, 2007); (vi) enzyme catalyzed conversion processes (Verlent *et al.*, 2004a,b; Duvetter *et al.*, 2006a; Van Eylen *et al.*, 2006; Sila *et al.*, 2007b) and chemical (e.g. oxidation) reactions (Nguyen *et al.*, 2003; Indrawati *et al.*, 2004a,b; 2005). These observation can be explained by the fact that (bio)chemical reactions and physical changes (e.g. phase transition) resulting in a decrease in total volume (negative activation volume) are enhanced by pressure and *vice versa* (*Le Chatelier* principle). Since the mechanisms of (bio)chemical reactions under pressure cannot always be anticipated by a direct extrapolation of the reaction mechanism occurring at atmospheric pressure, further research is still required to better understand the effect of HP on food products.

From an engineering point of view, HP processing is faced with limitations of heat transfer, although it applies pressure isostatically, i.e. instantaneously and uniformly throughout a mass of food material irrespective of size, shape and

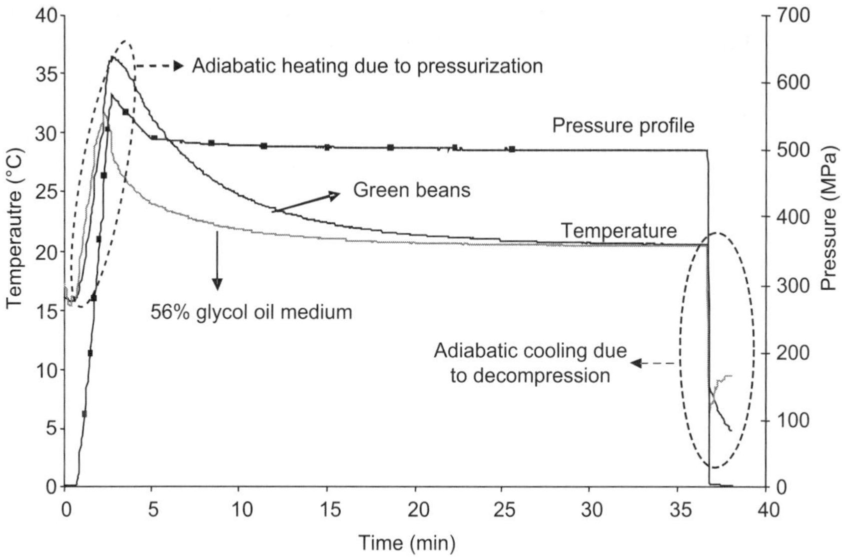

Fig. 17.1 Pressure and temperature profile measured in the center of the vessel of HP equipment (560 mL; 50 mm i.d.; 300 mm height; glycol–oil as pressure transmitting medium) and in the center of young green beans (9 mm o.d.) during HP processing at 500 MPa and 20 °C.

composition. During pressure build-up and decompression, the temperature inside the HP vessel is increased by adiabatic heating and decreased by adiabatic cooling, respectively (Fig. 17.1). The extent of adiabatic heating and cooling depends on various factors such as pressurization rate, pressure level and properties of the pressure transferring medium and the food product. After pressure build-up, the temperature in the vessel achieved due to adiabatic heating evolves to the temperature of the surrounding medium. This temperature evolution influences the homogeneity of the temperature inside the vessel and is largely dependent on the vessel dimension and the capacity of the heating/cooling unit.

For industrial applications, the pressures used range between 100–700 MPa and the processing is done mostly at room temperature for economic reasons. The choice of process parameters (temperature, pressure, time) is determined by the targeted applications and varies, depending on the food product and the expected final result.

In the last ten years, HP processing has been extended, on a laboratory scale, to applications at elevated temperatures (referred to as High Pressure Sterilization/ HPS) to obtain spore inactivation (Meyer 2000; Meyer *et al.*, 2000; Wilson and Baker 2000; Van Schepdael *et al.*, 2002). This application takes advantage of adiabatic heating during pressurization. Initial temperatures higher than 70 °C are used before pressurization, in order to reach temperatures above 100 °C as a result of adiabatic heating. For this application, the sample is pre-heated to the initial temperature before processing. Next to pressure, temperature and treatment time,

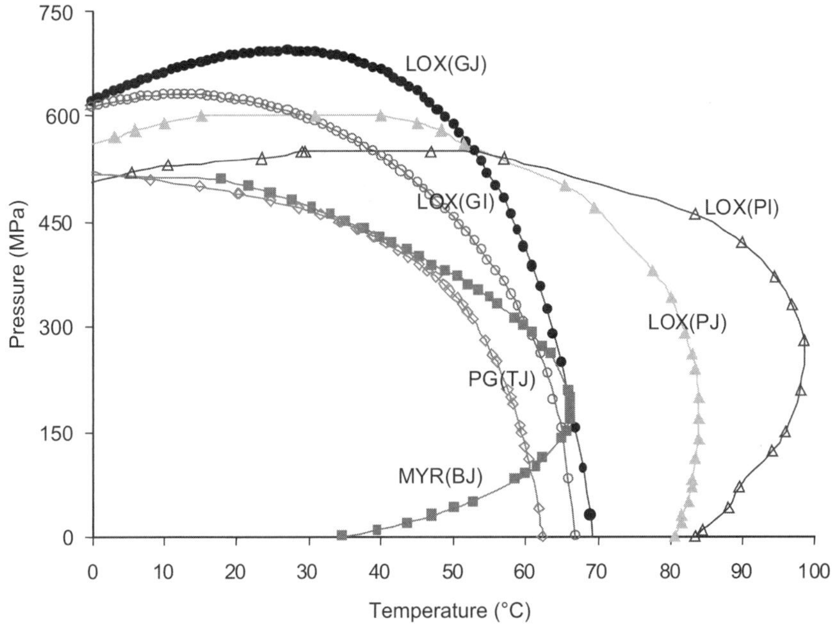

Fig. 17.2 Simulated pressure–temperature diagram of enzyme inactivation up to 1% of enzyme activity retention for total process time of 30 min. based on Indrawati *et al.*, (2000, 2001); Fachin *et al.* (2003) and Van Eylen *et al.* (2007). LOX(PI) = endogenous lipoxygenase/LOX in intact peas (△); LOX(PJ) = endogenous LOX in pea juice (▲); LOX(GJ) = endogenous LOX in green bean juice (●); LOX(GI) = endogenous LOX in intact green beans (○); PG(TJ) = endogenous polygalacturonase in tomato juice (◊) and MYR(BJ) = endogenous myrosinase in broccoli juice (■).

the number of pressure cycles can be an important process parameter for HP. The use of multiple pressurization cycles carried out for a short time (max. 3–5 min) has been introduced for HPS application (Matser *et al.*, 2004).

17.3 Effect of high pressure processing on enzyme activity and stability

Pressure and temperature stabilities of enzymes vary depending on the type of enzyme, the source of enzyme and the intrinsic and extrinsic conditions to which the enzyme is exposed. At constant temperature, increasing pressure can accelerate or decelerate enzyme inactivation; at constant pressure, increasing or decreasing temperature may enhance enzyme inactivation. At elevated temperatures, a moderate pressure increase can act antagonistically towards thermal inactivation (i.e. the inactivation rate constants are lowered by elevating pressure). This antagonistic effect is observed for a number of enzymes, including pectinmethylesterase, polyphenoloxidase, lipoxygenase (Ludikhuyze *et al.*, 2003), and has been recently also reported for mustard seed and broccoli myrosinase (Van Eylen *et al.*, 2006, 2007).

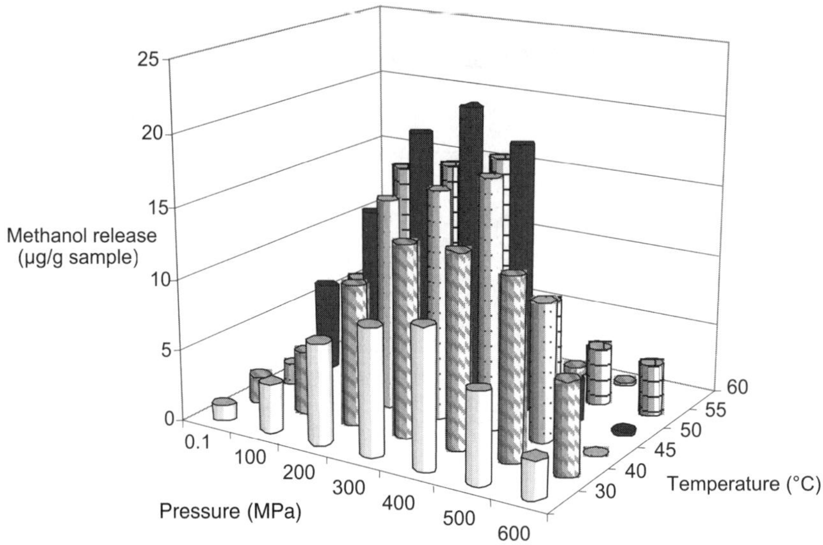

Fig. 17.3 *In situ* PME activity in shredded carrots during HP processing at different pressure and temperature combinations. This illustration is based on the data of Sila *et al.* (2007b).

Pressure and temperature dependence of the inactivation rate constants of various enzymes can be visualized as iso-rate pressure–temperature contour diagrams, i.e. lines connecting different pressure and temperature combinations resulting in the same inactivation rate constant (Fig. 17.2). The elliptically shaped iso-rate contours found for the inactivation of certain enzymes, namely lipoxygenase (LOX), polygalacturonase (PG) and myrosinase (MYR), indicate that these enzymes could be inactivated by low temperature, elevated pressure, high temperature or a combination of these factors.

Studies on enzyme-catalyzed substrate conversions at elevated pressure and temperature combinations, e.g. for pectinases (Verlent *et al.*, 2004a,b; Duvetter *et al.*, 2006; Sila *et al.*, 2007b) and myrosinase (Van Eylen *et al.*, 2006), have been carried out mainly in model systems. Pressure will stimulate substrate conversion in case negative volume changes occur between substrate and reaction products. Figure 17.3 illustrates a case study on pressure-induced enzyme substrate conversion of pectinmethylesterase (PME) (texture-related enzyme) in shredded carrots. The catalytic activity of enzymes can also be influenced by pressure-induced tissue damage, facilitating contact between enzyme and substrate

17.4 Effect of high pressure processing on nutrient stability and bioavailability

In general, HP treatment (at moderate temperatures) can maintain the nutritional value of a food, but this quality aspect could degrade due to the co-existing (bio)chemical reaction during pressure processing. Pressure stability of vitamins in model systems is often different from that in food matrices. Researchers have observed limited direct effect of pressure on the degradation of vitamins in model systems, e.g. for vitamin C and folates (Oey *et al.*, 2006), carotenoids (Tauscher *et al.*, 1998; 1999) and lycopene (Qiu *et al.*, 2006), but the stability of vitamins in food products can be influenced by the coexisting (bio)chemical reactions occurring during or after HP processing (Lambert *et al.*, 1999; Zabetakis *et al.*, 2000; Fernández García *et al.*, 2001a; Krebbers *et al.*, 2002a,b; Polydera *et al.*, 2003; Garcia-Palazon *et al.*, 2004; Kouniaki *et al.*, 2004, Polydera *et al.*, 2005; Suthanthangjai *et al.*, 2005).

The effect of pressure on vitamin C (ascorbic acid) stability has been investigated but less information is available for other vitamins such as vitamins A, B, D, E and K (Van den Broeck *et al.*, 1998; Sancho *et al.*, 1999; Krebbers *et al*, 2002a; Gabrovska *et al*, 2005; Oey *et al.*, 2006; Sánchez-Moreno *et al*, 2006). Studies in model systems show that, in the presence of oxygen, ascorbic acid is degraded during pressure treatment due to oxidation, and the concentration of degraded ascorbic acid is at least twice as high as the molar concentration of the initial oxygen content. The oxidation of ascorbic acid starts to occur at a relatively low pressure level and increases during pressure build-up. When the dissolved oxygen is completely consumed (mostly during pressure build-up) due to aerobic degradation of ascorbic acid, anaerobic degradation will subsequently occur, with the rate of anaerobic degradation being lower than that of aerobic degradation. Therefore, it seems that increasing pressure level or prolonging the pressure holding time (at moderate pressure and temperature combinations for short times) has no/little effect on the ascorbic acid pressure stability (Oey *et al.*, 2006). At extreme temperature and pressure combinations (>60 °C/>700 MPa/>60 min.), ascorbic acid anaerobic degradation is accelerated (Van den Broeck *et al.*, 1998).

Similar phenomena on pressure-enhanced oxidation have been observed for other vitamins which are vulnerable to oxidation, e.g. folates. Folate degradation due to oxidation is enhanced not only by elevating temperature at constant pressure but also by elevating pressure at constant temperature (>40 °C) (Nguyen *et al.*, 2003, 2006; Indrawati *et al.*, 2005). The pressure stability of different folate derivatives differs because it greatly depends on the chemical structure of the folate derivate, and on the intrinsic and extrinsic process factors.

As the molar ratio between oxygen and vitamin may play an important role in the extent of vitamin degradation during HP processing (next to temperature, pressure and treatment time), limiting the oxygen content in the food product (such as by deaeration, application of vacuum, use of antioxidant) before HP processing becomes essential in order to minimize vitamin degradation during processing (Butz *et al.*, 2004; Indrawati *et al.*, 2004a).

As compared to pressure stability of vitamins in model systems, the effect of HP treatment on the stability of ascorbic acid and B vitamins (such as folates/B9 vitamins) in food matrices is more complex because in food products vitamins coexist with other molecules such as proteins, enzymes, anti- and pro-oxidants. Hence, the vitamin stability during pressure treatments may be influenced by the ongoing chemical (such as inter-conversion and oxidation) (Butz *et al.*, 2004; Indrawati *et al.*, 2004a) and biochemical (Melse-Boonstra *et al.*, 2002) reactions. The extent of vitamin degradation in food products also depends on the type of food material and matrix. After pressure treatment at a moderate temperature and pressure combination for a short time, limited degradation of vitamin C (up to 20%) was observed; for example, orange juice after HP treatment at 500 and 800 MPa/25 and 50 °C/1 min (Nienaber and Shellhammer, 2001) or after HP treatment at 400 MPa/25 °C/15 min (Sánchez-Moreno *et al.*, 2005). However, pressure treatment at 500 MPa/room temperature/10 min decreases almost 77% of vitamin C content in sprouted alfalfa seed in citric acid pickle (Gabrovska *et al.*, 2005). At extreme pressure and temperature combinations, HPS processing (2 pulses of 75 °C/1000 MPa/80 s with interval 30 sec/1 atm) can still maintain 75% of the vitamin content in green beans (Krebbers *et al.*, 2002a).

Next to pressure-enhanced oxidation, isomerization can also take place during HP treatment. Recently, it has been reported that the total content of all-*trans* lycopene in hexane is significantly decreased after HP treatments at 500 and 600 MPa and room temperature for 12 min due to isomerization from *trans* to *cis* lycopene but this phenomenon was not observed in food matrices such as in tomato puree (Qiu *et al.*, 2006).

Next to vitamin degradation under pressure, HP treatment can enhance the extraction yield of nutrients, including bioactive compounds, from fruit and vegetables, e.g. carotenoids (Tauscher, 1998; De Ancos *et al.*, 2000; Fernández García *et al.*, 2001a,b; Qiu *et al.*, 2006) and flavanones (naringenin and hesperetin) (Sanchez-Moreno *et al.*, 2005). In addition, HP treatment does not (or only slightly) affect the stability of carotenes in food products. For example, a high retention of carotene in orange–lemon–carrot mixed juice or in tomato puree (observed as lycopene) was obtained after HP treatments at 500 and 800 MPa/ room temperature/5 min (Fernández García *et al.*, 2001a). Under extreme process conditions (e.g. 600 MPa/75 °C/40 min), the carotene content in carrot-based products is slightly decreased (maximally by 5%) after the treatment and remains constant during subsequent storage at 4 °C (Fernández García *et al.*, 2001b).

Since HP treatment influences the stability and the extraction yield of some nutrients, the total antioxidant capacity of fruit and vegetables can be affected by HP treatments depending on the food material. For example, an increase in TEAC (*Trolox Equivalent Antioxidant Capacity*) value is observed for pressure-treated (100 up to 800 MPa/30 up to 65 °C/up to 90 min) carrot juice but not for pressure-treated orange juice (Indrawati *et al.*, 2004b). The increase in TEAC value of pressurized carrot juice is probably due to the pressure-enhanced extraction yield of carotenoids, whereas the decrease in TEAC value of orange juice could be partly explained by ascorbic acid degradation during HP treatment. As pressure can enhance (bio)chemical reactions, changes in total antioxidant capacity may be enhanced during HP treatment. For example, increasing the pressure level (100 up to 800 MPa) at constant temperatures (30 up to 65 °C) enhanced the decrease in total water-soluble antioxidant capacity of orange juice, which is obviously observed for a long HP treatment time, e.g. 90 min (Indrawati *et al.*, 2004b).

However, this phenomenon is not clear when the time duration of the HP treatment is short, e.g. 1 to 5 minutes (Fernandez García *et al.*, 2001a,b; Sánchez-Moreno *et al.* 2005).

Bioavailability studies have shown that vitamins in pressurized fruit and vegetables still possess their bioavailability in the human body after the treatments. Sánchez-Moreno *et al.* (2003) have reported that the consumption of pressure-processed (400 MPa/40 °C/1 min) orange juice increases the plasma vitamin C and decreases inflammatory biomarkers levels in humans. Similar findings have been reported for pressurized vegetable soups, such as gazpacho (Sánchez-Moreno *et al.*, 2004).

In vitro studies have demonstrated that HP (at moderate temperature) processing can increase the vitamin bioavailability of fruit and vegetables due to pressure-enhanced enzymatic conversion reactions. For example, mild pressure treatment (e.g. 200 MPa/room temperature/5 min) promotes the conversion of polyglutamate to monoglutamate endogenous folates (reduced forms of folic acid) (Melse-Boonstra *et al.*, 2002). They found increases in the proportion of monoglutamate folates in different vegetables, such as in leeks (74%), cauliflower (12%) and green beans (82%), after treatments of 200 MPa/room temperature/ 5 min. This implies that HP processing can result in vegetables with a higher folate bioavailability. However, stability of the resulting monoglutamate folates after HP treatment must be guaranteed during frozen and refrigerated storage before consumption.

17.5 Effect of high pressure processing on color and flavor

Color and flavor of food can be maintained immediately after HPT treatment (at moderate temperatures) but these quality aspects can also evolve during and after pressure processing due to (bio)chemical reactions. Combined evaluation by sensory analysis by trained panelists and chemical analysis has shown that HP (at moderate temperature) treated products compare closely in sensorial aspects to untreated ones but that they are superior to thermally-processed products (blanching, pasteurization and sterilization). However, the panelists noticed changes in overall sensorial quality of the HP processed food products during storage. These changes were probably due to the subsequent (bio)chemical reactions, as previously discussed in the context of nutrient stability (Gimenez *et al.*, 2001). As an illustration, examples of pressure effects on color and flavor of certain fruit- and vegetable-based food products are given.

Color

Studies on pressure-stability of color compounds have been conducted in various systems and food matrices, for example, for chlorophyll (Van Loey *et al.*, 1998; Butz *et al.*, 2002; Matser *et al.*, 2004), carotenoid (Tauscher *et al.*, 1998; 1999), and anthocyanin (Garcia-Palazon *et al.*, 2004; Suthanthangjai *et al.*, 2005). In fruit-based food products, for example, pressurized (400 or 500 MPa/2 °C and 400

MPa/40 °C/10 min) white grape juice (Daoudi *et al.*, 2002) and pressure-treated (100-400 MPa/20 °C/15 or 30 min.) mango pulps (Cv Chousa) (Guerrero-Beltran *et al.* 2005), pressure treatments did not result in visual color differences between untreated and pressurized juices immediately after the treatment. During subsequent storage at 4 °C, small changes in color indices can occur (e.g. enzymatic browning due to polyphenoloxidase), which have been observed for pressurized (400 or 500 MPa at 2 °C and 400 MPa at 40 °C/10 min) white grape juice after storage for 60 days (Daoudi *et al.*, 2002), HPT (500 MPa/35 °C/5 min) treated reconstituted orange juice during storage up to 120 days (Polydera *et al.*, 2003) or pressurized (600 MPa/40 °C/4 min) navel orange juice during storage for 64 days (Polydera *et al.*, 2005).

Regarding the color of vegetable-based food products, pressure processing at moderate temperatures resulted in a more intense green color, as observed for HP treated (500 MPa/ambient T/1 min) green beans (Krebbers *et al*, 2002a). This might be due to cell permeabilization during HP treatment resulting in chlorophyll leakage into the intercellular space, yielding a more intense bright green color on the surface of beans. Under extreme conditions such as during HPS processing (2 pulses of 1000 MPa/75 °C/80s), change in the color of green beans from green to olive-green was visibly observed (Krebbers *et al*, 2002a). A similar result was also observed for basil (Krebbers *et al*, 2002b). Another example is pressurized tomato puree. Sánchez-Moreno and co-workers (2006) have found that HP treatment (400 MPa/25 °C/15 min) yields higher color indices compared to thermally treated tomato purees.

Next to the aforementioned applications, HP processing has also been studied as a technique to preserve fermented vegetables such as Korean traditional fermented food (kimchi). Pressure treatments under mild conditions, e.g. 200 MPa/25 °C/10 min do not result in color difference between untreated and treated kimchi after storage at 20 °C. However, higher pressure levels of 400 and 600 MPa induce a color change due to browning (Sohn and Lee, 1998).

Flavor

Effects of HP processing on flavor are dependent on the food matrices and the treatment conditions. HP processing may result in flavor changes of fruit and vegetables due to (bio)chemical reactions, e.g. reaction of incompletely inactivated endogenous enzymes.

A study carried out by Daoudi *et al.* (2002) shows that HP-treated (500 MPa / 2 °C/10 min) white grape juice has similar freshness, aroma and sweetness after storage at 4 °C for one day, as compared with the untreated control juice. Storage for longer time (60 days) causes a reduction in fresh fruit aroma but the product is still acceptable according to sensory panelists.

Regarding strawberry based products, sensory tests carried out by Gimenez and co-workers (2001) indicate that pressurized (400 or 800 MPa/22 °C/5 min) strawberry jams are characterized by a more chemical and rancid flavour than traditional jam. Chemical analysis indicated a change in concentration of volatile compounds in strawberry puree due to elevation of pressure, e.g. HP treatment at

200 up to 800 MPa/18–22 °C/15–20 min (Lambert *et al.*, 1999; Zabetakis *et al.*, 2000). During storage at 4 °C, the volatile composition of strawberry puree changed, presumably due to the activity of incompletely inactivated endogenous enzymes such as lipoxygenase (Navarro *et al.*, 2002).

With citrus-based juices, several authors (Parish 1998; Polydera *et al.* 2003; 2005) have reported that HP (at room temperatures) treated orange juice has better flavor (sensory analysis) than traditional thermally pasteurized juice. Furthermore, the off-flavor usually found in heat-treated juice is not detected in pressure-treated (400 MPa/room T/10 min) juice (Takahashi *et al.* 1993). However, it has also been reported that the taste of orange juice produced by high pressure processing at room temperature (500 MPa/90 s, 500 MPa/5 min, 700 MPa/60 s or 800 MPa/5 min) is not as fresh as untreated juice (Parish 1998; Fernández García *et al.* 2001a). Based on the evaluation of a trained sensory panel and a consumer panel, there is no difference in odour or flavor between high pressure-treated, thermally treated (stored at 4 °C) and frozen orange juices (Baxter *et al.*, 2005).

With regard to herbs, the aroma of HPS (2 pulses/860 MPa/75 °C and 2 pulses/700 MPa/85 °C) treated basil was more intense compared with conventional heat-sterilized, frozen or dried basil, although the content of methylchavicol and linalool (components related to basil aroma) after HPS was not different from fresh basil. During storage at 20 °C for 2 months, the characteristic aroma of pressurized basil is maintained (Krebbers *et al*, 2002b).

For some food products, e.g. pressure-treated (500, 700 or 900 MPa/room temperature/3, 6 or 9 min) tomato juice, HP processing could induce the generation of a strong rancid taste, making the samples inedible and unsuitable for sensory analysis. Based on the chemical analysis of volatile flavour compounds, a remarkable increase in *n*-hexanal content in all pressure-treated samples was observed (Porretta *et al.*, 1995). This could be partly explained by the enzymatic activity of lipoxygenase (Shook *et al.*, 2001). An undesired effect of pressure treatment on flavor was also observed in pressurized (300 MPa/25 and 40 °C/30 min) onions (Butz *et al.* 1994).

17.6 Effect of high pressure processing on texture

Food texture is a broad term that can be defined as 'all the mechanical, geometrical and surface attributes of a product perceptible by means of mechanical, tactile and, where appropriate, visual and auditory receptors' (International Organization for Standardization, 1992). Terms often used to describe the texture of plant-based foods include, amongst others, firmness, hardness and crispness. The texture of plant material depends on the complex interaction between the different levels of structural organization, from the molecular level (regarding cell wall polymers) up to the organization of cells in different tissues and organs (Waldron *et al.*, 1997). Food processing can cause severe changes to the food structure, leading to important changes of textural characteristics.

Thermal processing of plant-based foods generally results in softening. An

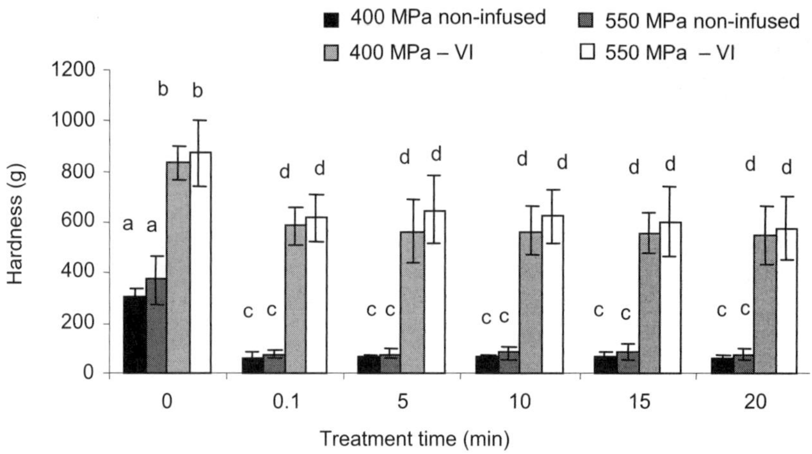

Fig. 17.4 Hardness of non-infused and enzyme vacuum infused strawberries after pressure treatment at 400 MPa and 550 MPa/10 °C for various treatment times (reprinted from Duvetter *et al.* (2005a) with permission from Blackwell Publishing).

initial loss of texture is ascribed to membrane disruptions and the associated turgor pressure loss. Additional softening occurs as a result of cell separation. This is closely linked with the solubilization and degradation, due to the beta-elimination reaction of the pectin polymers in the cell wall that are involved in cell–cell adhesion (Van Buren, 1979; Waldron, 1997). In the case of carrots, the extent and the rate of textural changes upon thermal processing is clearly associated with changes in structure, solubility properties and the rate of beta-elimination of pectin (Sila *et al.*, 2005, 2006; Vu *et al.*, 2006).

Considering novel technologies, high pressure processing induces an immediate texture loss in fruits and vegetables as a result of the pressure pulse, also called instantaneous pressure softening (IPS), the extent depending on the pressure level (Basak and Ramaswamy, 1998; Krebbers *et al.*, 2002a; Duvetter *et al.*, 2005a; Trejo Araya *et al.*, 2007). In some studies, no statistically significant effect of extended high pressure processing time on the texture characteristics was observed (Duvetter *et al.*, 2005a; Trejo Araya *et al.*, 2007). In other cases, the IPS is followed by a gradual change during pressure hold which can include a recovery or further texture loss, depending on the plant material and pressure level (Basak and Ramaswamy, 1998). In general, the effect of high pressure processing on texture is highly dependent on the plant structure and plant type. For plants with a firm structure, such as cauliflower, a firmness level close to the original was obtained after HP treatment (400 MPa, 5 °C, 30 min) (Prestamo and Arroyo, 1998), while for porous fruit such as strawberry, the structure is completely disrupted (400 MPa, 10 °C, 20 min) (Fig. 17.4) (Duvetter *et al.*, 2005a).

At low pressures (100 MPa), the phenomenon of IPS is caused by compression of cellular structures without disruption, while at higher pressures (>200 MPa) severe texture losses occur due to the rupture of cellular membranes and the consequent turgor pressure loss. Studies using scanning electron microscopy revealed losses of turgor pressure due to increased cell permeability and cell collapse for HP treated cauliflower, spinach leaves and cherry tomatoes (Prestamo and Arroyo, 1998, Tangwongchai *et al.*, 2000). Chong *et al.* (1985) have suggested that cell collapse is linked to a pressure-induced inactivation of (Na,K)ATPase which regulates the cell volume. Trejo Arayo *et al.* (2007), correlating effects of HP processing on textural changes of carrots with microstructural responses, concluded that the decrease of carrot hardness was associated mainly with turgidity loss, which was linked with considerable cell wall conformational changes. Another possible contribution to texture loss is the reduced cell adhesion due to the larger intercellular cavities after HP treatment. In HP-treated spinach leaves (Prestamo and Arroyo, 1998) and cherry tomatoes (Tangwongchai *et al.*, 2000), new cavity formation was observed. This was attributed to the rapid expansion of the compressed air upon depressurization.

On the other hand, the texture-firming effect, observed during pressure hold for various plants, is ascribed to the action of PME (Basak and Ramaswamy, 1998). PME catalyzes the demethoxylation of pectin. Pectin with a reduced degree of methoxylation is less sensitive to degradation by beta-elimination. In addition, demethoxylated pectin can be cross-linked by divalent ions such as calcium, resulting in an intermolecular network and gel formation. As mentioned previously (Section 17.3), PME is a pressure stable enzyme, which is hardly or incompletely inactivated by pressure treatments (< 600 MPa) at ambient temperature (Basak and Ramaswamy, 1998; Fachin *et al.*, 2002; Ly Nguyen *et al.*, 2003a,b; Sila *et al.*, 2004, 2007b; Duvetter *et al.*, 2005b). Moreover, the activity of the enzyme can be stimulated at elevated pressures (Verlent *et al.*, 2004b; Duvetter *et al.*, 2006a; Sila *et al.*, 2007b). In addition, the disruption of cellular structures during pressurization allows enhanced interaction between PME and its substrate. For carrots, it was observed that HP processing results in pectin with a decreased degree of methoxlyation and reduced water solubility (Sila *et al.*, 2006). Interestingly, the depolymerization of pectin by beta-elimination, which is accelerated by increasing temperature and has been identified as one of the main reactions explaining texture degradation during thermal processing, does not occur during HP processing at ambient temperature (treatments up to 700 MPa, 20 °C, 1 h) (Kato *et al.*, 1997).

With regard to the differences in pressure/temperature sensitivities of enzymes, HP processing has more potential applications than thermal processing. In the context of obtaining a firm plant tissue, a tight intermolecular pectin network is imperative. This requires controlled activity of PME and inactivation of pectin depolymerizing enzymes such as polygalacturonase (PG). This cannot be achieved by thermal processing since PG is more heat resistant than PME (Fachin *et al.*, 2002, 2004). However, selective knockout of PG while maintaining PME activity is possible by HP processing because of the inverse order of their pressure sensitivity. In tomatoes, in particular, PG is abundantly present. Tangwongchai *et al.* (2000) observed softening due to cell damage after HP treatment (200–600 MPa, *ca.* 20 °C, 20 min) of cherry tomatoes. Both phenomena, however, were less

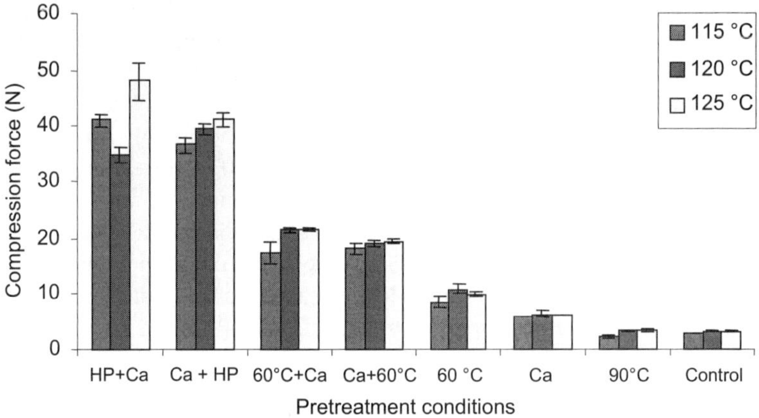

Fig. 17.5 Effect of various pretreatments on the texture of carrot disks after thermal processing at various temperatures for equivalent processes ($F_0 = 6$ min): (HP + Ca: HPT at 400 MPa/60 °C/15 min followed by calcium (0.5%) soaking; Ca + HP: calcium soaking prior to HPT (400 MPa/60 °C/15 min); 60 °C + Ca: preheating followed by calcium soaking; Ca + 60 °C: calcium soaking prior to preheating; 60 °C: preheating for 40 min; Ca: calcium soaking only; 90 °C: preheating at 90 °C for 4 min; control: non-pretreated samples) (reprinted from Sila *et al.*, 2005, with permission from Blackwell Publishing).

obvious after treatments at 500 and 600 MPa compared with 300 and 400 MPa, which was attributed to the inactivation of PG at pressures higher than 400 MPa, while PME activity was maintained.

Various pretreatments enhancing the formation of a calcium pectate network have been developed to protect plant-based foods against thermal softening. These pretreatments aim to increase the level of calcium ions in the plant tissue and/or decrease the pectin degree of methoxylation by activation of the endogenous PME or infusion of exogenous PME (Degraeve *et al.*, 2003; Sila *et al.*, 2005). The activation of endogenous PME is traditionally done by a blanching step prior to cooking (Waldron *et al.*, 1997). However, because of the earlier mentioned activation of PME at elevated pressure, pretreatments involving HP processing have been demonstrated to be a valuable alternative. Pretreatments at elevated pressure reduced the rate and the extent of thermal softening of carrots, which was associated with a decrease in solubility and degree of methoxlyation of the pectins (Fig. 17.5) (Sila *et al.*, 2004, 2005). In some cases, pretreatments have been developed that protect fruits against structural damage during pressure treatment. Vacuum-infusion of strawberries with fungal PME and calcium ions increased the instrumental hardness by more than 100% and limited the texture loss during subsequent HP processing (Fig. 17.4) (Duvetter *et al.*, 2005a). While the structure of non-pretreated strawberries was completely destroyed, the hardness of pretreated fruits after a HP process (550 MPa, 10 °C, 20 min) was higher compared to fresh, untreated samples. The texture improving effect was associated with a decrease in pectin degree of methoxylation and uptake of infusion liquid.

17.7 Conclusions and current research trends

HP is an interesting food processing technology used either as a single treatment or as a complementary treatment. Based on the current know-how, HP processing can provide unique processed food products with other characteristics than those obtained with the currently applied thermal processing techniques. Hence, HP processing opens the horizon of possibilities to create in-pack processed fruit and vegetables having premium/'*novel*' characteristics. However, optimization of the HP process and storage conditions is needed to tap the beneficial effects of high-pressure processing.

The current trend in food technology research is pointing toward the combined use of existing and novel technologies to produce high-quality food products. In this case, the targeted/tailored quality should be compared with the food quality resulting from the existing industrial processing. In parallel, intensive research in packaging materials is still greatly required to guarantee the food quality and microbial/chemical safety of HP processed fruit and vegetables.

Since the temperature domain of high pressure applications is being extended to the elevated temperature region, enabling high pressure sterilization, we need to extend our current know-how on HP (obtained in the low and moderate temperature regions). Research in this area must be broadened, not only to achieve a certain targeted criterion such as spore inactivation but also to have a better understanding of the ongoing (bio)chemical reactions under pressure at elevated temperatures based on a detailed insight into the mechanisms and kinetics of those reactions. Such data will help to formulate criteria for food safety of in-pack pressure sterilized food products, e.g. criteria on microbial and chemical safety.

17.8 Acknowledgements

The authors thank the Research Foundation–Flanders (FWO Vlaanderen), the Commission of the European Communities, Framework 6, Priority 5 'Food Quality and Safety', Integrated Project NovelQ FP6-CT-2006-015710 and the Institute for the Promotion of Innovation through Science and Technology in Flanders (IWT-Vlaanderen, a Ph.D. grant) for their financial support.

17.9 References

Alvarez-Alvarez, J., Guillamon, E., Crespo, J.-F., Cuadrado, C., Burbano, C., Rodríguez, J., Fernandez, C., Muzquiz, M. (2005). Effects of extrusion, boiling, autoclaving, and microwave heating on lupine allergenicity. *Journal of Agricultural and Food Chemistry* 53 (4): 1294–1298.

Ananth, V., Dickson, J.S., Olson, D.G., Murano, E.A. (1998). Shelf-life extension, safety, and quality of fresh pork loin treated with high hydrostatic pressure. *Journal of Food Protection* 61, 1649–1656.

Basak, S., Ramaswamy, H. (1998). Effects of high pressure processing on the texture of selected fruit and vegetables. *Journal of Texture Studies* 29, 587–601.

Baxter, I. A., Easton, K., Schneebeli, K., Whitfield, F. B. (2005). High pressure processing of Australian navel orange juices: Sensory analysis and volatile flavor profiling. *Innovative Food Science and Emerging Technologies* 6, 372–387.

Bruna, D., Voldoich, M., Marek, M., Kamarád, J. (1996). Effect of high-pressure treatment on patulin content in apple concentrate. In Heremans, K. (Ed), *High Pressure Research in the Biosciences and Biotechnology*, Belgium: Leuven University Press.

Butz, P., Koller, W. D., Tauscher, B., Wolf, S. (1994). Ultra-high pressure processing of onions: Chemical and sensory changes. *LWT* 27, 463–467.

Butz, P., Edenharder, R., Fernández García, A., Fister, H.; Merkel, C., Tauscher, B. (2002). Changes in functional properties of vegetables induced by high pressure treatment. *Food Research International* 35, 295–300.

Butz, P., Serfert, Y., Fernández García, A., Dieterich, S., Lindauer, R., Bognar, A., Tauscher, B. (2004). Influence of high-pressure treatment at 25 °C and 80 °C on folates in orange juice and model media. *Journal of Food Science* 69(3), 117–121.

Chong, P.L.G., Fortes, P.A.G., Jameson, D.M. (1985). Mechanisms of inhibition of (Na, K)-ATPase by hydrostatic pressure with fluorescent probe. *Journal of Biological Chemistry* 260, 14484–14490.

Cuccurullo, G., Cinquanta, L., Sorrentino, G. (2007). A procedure to achieve fine control in MW processing of foods. *Infrared Physics and Technology* 49 (3), 292–296.

Daoudi, L., Quevedo, J.M., Trujillo, A.J., Capdevila, F., Bartra, E., Mínguez, S., Guamis, B. (2002). Effects of high-pressure treatment on the sensory quality of white grape juice. *High Pressure Research* 22, 705–709.

De Ancos, B., Gonzalez, E., Pilar Cano, M. (2000). Effect of high pressure treatment on the carotenoid composition and the radical scavenging activity of persimmon fruit purees, *Journal of Food Chemistry* 48, 3542–3548.

Degraeve, P., Saurel, R., Coutel, Y. (2003). Vacuum-impregnation with pectinmethylesterase to improve firmness of pasteurized fruits. *Journal of Food Science* 68, 716–721.

Duvetter, T., Fraeye, I., Van Hoang, T., Van Buggenhout, S., Verlent, I., Smout, C., Van Loey, A., Hendrickx, M. (2005a). Effect of pectinmethylesterase infusion methods and processing techniques on strawberry firmness. *Journal of Food Science* 70(6), S383–S388.

Duvetter, T., Van Loey, A., Smout, C., Verlent, I., Ly Nguyen, B., Hendrickx, M. (2005b). *Aspergillus aculeatus* pectin methylesterase: Study of the inactivation by temperature and pressure and the inhibition by pectin methylesterase inhibitor. *Enzyme and Microbial Technology* 36(4), 383–388.

Duvetter, T., Fraeye, I., Sila, D.N., Verlent, I., Smout, C., Clynen, E., Schoofs, L., Schols, H., Hendrickx, M., Van Loey, A. (2006). Effect of temperature and high pressure on the activity and mode of action of fungal pectin methyl esterase. *Biotechnology Progress* 22, 1313–1320.

Fachin, D., Van Loey, A., Ly Nguyen, B., Verlent, I., Indrawati, Hendrickx M. (2002). Comparative study of the inactivation kinetics of pectinmethylesterase in tomato juice and purified form. *Biotechnology Progress* 18(4), 739–744.

Fachin D., Van Loey A., Ly Nguyen B., Verlent I., Indrawati, Hendrickx M. (2003). Inactivation kinetics of polygalacturonase in tomato juice. *IFSET* 4(2), 135–142.

Fachin, D., Smout, C., Verlent, I., Nguyen, B.L., Van Loey, A.M., Hendrickx, M.E. (2004). Inactivation kinetics of purified tomato polygalacturonase by thermal and high-pressure processing. *Journal of Agricultural and Food Processing* 52, 2697–2703.

Fernández García, A., Butz, P., Bognàr, A., Tauscher, B. (2001a). Antioxidative capacity, nutrient content and sensory quality of orange juice and an orange–lemon–carrot juice product after high pressure treatment and storage in different packaging. *European Food Research and Technology* 213, 290–296.

Fernández García, A., Butz, P., Tauscher, B. (2001b). Effects of high-pressure processing on carotenoid extractability, antioxidant activity, glucose diffusion, and water binding of tomato puree (*Lycopersicon esculentum* Mill). *Journal of Food Science* 66(7): 1033–1038.

Gabrovska, D., Paulickova, I., Maskova, E., Fiedlerova, V., Kocurova, K., Pruchova, J., Strohalm, J., Houska, M. (2005). Changes in selected vitamins, microorganism counts and sensory quality during storage of pressurized sprouted seed of alfalfa (*Medicago sativa* L.), *Czech Journal of Food Sciences* 23(6), 246–250.

Garcia-Palazon, A., Suthanthangjai, W., Kajda, P., Zabetakis, I. (2004). The effects of high hydrostatic pressure on β-glucosidase, peroxidase and polyphenoloxidase in red raspberry (*Rubus idaeus*) and strawberry (*Fragaria* × *ananassa*). *Food Chemistry* 88, 7–10.

Gervilla, R., Mor-Mur, M., Ferragut, V., Guamis, B. (1999). Kinetics of destruction of *Escherichia coli* and *Pseudomonas fluorescens* inoculated in ewe's milk by high hydrostatic pressure. *Food Microbiology* 16, 173–184.

Gimenez, J., Kajda, P., Margomenou, L., Piggott, J.R., Zabetakis, I. (2001). A study on the colour and sensory attributes of high-hydrostatic-pressure jams as compared with traditional jams. *Journal of Science and Food Agriculture* 81, 1228–1234.

Guerrero-Beltran, J.A., Swanson, B.G., Barbosa-Canovas, G.V. (2005). High hydrostatic pressure processing of mango puree containing antibrowning agents. *Food Science and Technology International* 11(4), 261–267.

Indrawati, Ludikhuyze, L.R., Van Loey, A.M., Hendrickx, M.E. (2000). Lipoxygenase inactivation in green beans (*Phaseolus vulgaris* L.) due to high pressure treatment at subzero and elevated temperatures. *Journal of Agricultural and Food Chemistry* 48(5), 1850–1859.

Indrawati, Van Loey, A.M., Ludikhuyze, L.R., Hendrickx, M.E. (2001). Pressure temperature inactivation of lipoxygenase in green peas (*Pisum sativum*): A kinetic study. *Journal Food Science* 66(5), 686–693.

Indrawati, Van Loey, A., Hendrickx, M. (2002a). High pressure processing. In *The Nutrition Handbook for Food Processors*, Henry, C.J.K and Chapman, C. (Eds). Woodhead Publ. Ltd., Cambridge, England. 433–461.

Indrawati, Van Loey A.M., Fachin D., Ly Nguyen B., Verlent I., Hendrickx M. (2002b). Overview: Effect of high pressure on enzymes related to food quality – Kinetics as a basis for process engineering. *High Pressure Research* 22, 613–618.

Indrawati, Arroqui, C., Messagie, I., Nguyen, M.T., Van Loey, A., Hendrickx, M. (2004a). Comparative study on pressure and temperature stability of 5-methyltetrahydrofolic acid in model systems and in food products. *Journal of Agriculture and Food Chemistry* 52, 485–492.

Indrawati, Van Loey, A., Hendrickx, M. (2004b). Pressure and temperature stability of water-soluble antioxidants in orange and carrot juice: A kinetic study. *European Food Research and Technology* 219, 161–166.

Indrawati, Van Loey, A. and Hendrickx, M. (2005). Pressure and temperature stability of 5-methyltetrahydrofolic acid: A kinetic study. *Journal of Agriculture and Food Chemistry* 53(8), 3081–3087.

International Organization for Standardization. (1992). *Sensory Analysis: Vocabulary* (ISI 5492). Available from the ISO Central Secretariat, Case postal 56, 1211 Geneva 20, Swizerland.

Kato, N., Teramoto, A., Fuchigama, M. (1997). Pectic substance degradation and texture of carrots as affected by pressurization. *Journal of Food Science* 62, 359–362, 398.

Knorr, D. (1998). Innovation from progress in food technology. In Stute, R. (ed.) *Food and Science, Wissenschaft in Dienste der Ernahrung: Proceedings of the Symposium 25th Anniversary 'Euro Research and Development', CPC Europe and the 35th anniversary 'Institute for Research & Deveopment'* CPC Deutschland, Heilbronn 5–7 Nov 1997. CPC Deutschland.

Kouniaki, S., Kajda, P., Zabetakis, I. (2004). The effect of high hydrostatic pressure on anthocyanins and ascorbic acid in blackcurrants (*Ribes nigrum*). *Flavour and Fragrance Journal* 19, 281–286.

Krebbers, B., Matser, A.M., Koets, M., Bartels, P., Van den Berg, R. (2002a). Quality and storage-stability of high-pressure preserved green beans. *Journal of Food Engineering* 54, 27–33.

Krebbers, B., Matser, A., Koets, M., Bartels, P., Van den Berg, R. (2002b). High pressure–temperature processing as an alternative for preserving basil. *High Pressure Research* 22, 711–714.

Lambert, Y., Demazeau, G., Largeteau, A., Bouvier, J.-M. (1999). Changes in aromatic volatile composition of strawberry after high pressure treatment. *Food Chemistry* 67, 7–16.

Ludikhuyze, L., Van Loey, A., Smout, C., Indrawati, Hendrickx, M. (2003). Effects of combined pressure and temperature on enzymes related to quality of fruits and vegetables: From kinetic information to process engineering aspects. *CRC Critical Reviews in Food Science and Nutrition* 43(5), 527–586.

Ly Nguyen, B., Van Loey, A., Smout, C., Verlent, I., Duvetter, T., Hendrickx, M. (2003a). Effects of mild heat and high pressure processing on banana pectinmethylesterase: A kinetic study. *Journal of Agricultural and Food Chemistry* 51, 7974–7979.

Ly Nguyen, B., Van Loey, A.M., Smout, C., Ozcan, S.E., Fachin, D., Verlent, I., Truong, S.V., Duvetter, T., Hendrickx, M.E. (2003b). Mild-heat and high-pressure inactivation of carrot pectin methyl esterase: A kinetic study. *Journal of Food Science* 68, 1377–1383.

Martin, S., Barbosa-Carnovas, G., Swanson, B. (2002). Food processing by high hydrostatic pressure. *Critical Review Food Science and Nutrition* 42 (6), 627–645.

Matser, A.M., Krebbers, B., Van den Berg R.W., Bartels, P.V. (2004). Advantages of high pressure sterilization on quality of food products. *Trends in Food Science and Technology* 15, 79–85.

Melse-Boonstra, A., Verhoef, P., Konings, E.J.M., Van Dusseldorp, M., Matser, A., Hollman, P.C.H., Meyboom, S., Kok, F.J., West, C.E. (2002). Influence of processing on total monoglutamate and polyglutamate folate contents of leeks, cauliflower, and green beans. *Journal of Agriculture and Food Chemistry* 50, 3473–3478.

Meyer, R.S. (2000). *Ultra high pressure, high temperature food preservation process.* US Patent 6 017 572.

Meyer, R.S., Cooper, K.L., Knorr, D., Lelieveld, H.L.M. (2000). High pressure sterilization of foods. *Food Technology* 54(11), 67–68, 70, 72.

Navarro, M., Verret, C., Pardon, P., El Moueffak, A. (2002). Changes in volatile aromatic compounds of strawberry puree treated by high-pressure during storage. *High Pressure Research* 22, 693–696.

Nguyen, M.T., Indrawati, Hendrickx, M.E. (2003). Model studies on the stability of folic acid and 5-methyltetrahydrofolic acid degradation during thermal treatment in combination with high hydrostatic pressure. *Journal of Agriculture and Food Chemistry* 51(11), 3352–3357.

Nguyen, M.T., Oey, I., Hendrickx, M., Van Loey, A. (2006). Kinetics for isobaric–isothermal degradation of (6R,S) 5-formyltetrahydrofolic acid in a model system. *European of Food Research and Technology* 223(3), 325–332

Nienaber, U., Shellhammer, T.H. (2001). High-pressure processing of orange juice: Combination treatments and a shelf life study. *Journal of Food Science* 66(2): 332–336.

Oey, I., Verlinde, P., Hendrickx, M., Van Loey, A. (2006). Temperature and pressure stability of L-ascorbic acid and/or [6s] 5-methyltetrahydrofolic acid: A kinetic study. *European Food Research Technology* 223, 71–77.

Parish, M.E. (1998). Orange juice quality after treatment by thermal pasteurization or isostatic high pressure. *LWT* 31, 439–442.

Polydera, A.C., Stoforos, N.G., Taoukis, P.S. (2003) Comparative shelf life study and vitamin C loss kinetics in pasteurized and high pressure processed reconstituted orange juice. *Journal of Food Engineering* 60, 21–29

Polydera, A.C., Stoforos, N.G., Taokis, P.S. (2004). The effect of storage on the antioxidant capacity of reconstituted orange juice which had been pasteurized by high pressure or heat. *International Journal of Food Science and Technology* 39, 789–791.

Polydera, A.C., Stoforos, N.G., Taoukis, P.S. (2005). Quality degradation kinetics of pasteurized and high pressure processed fresh navel orange juice: Nutritional parameters and shelf life. *IFSET* 6, 1–9.

Porretta, S., Birzi, A., Ghizzoni, C., Vicini, E. (1995). Effects of ultra-high hydrostatic pressure treatments on the quality of tomato juice. *Food Chemistry*, 52, 35–41.

Prestamo, G., Arroyo, G. (1998). High hydrostatic pressure effects on vegetable structure. *Journal of Food Science* 63[5], 878–881.

Qiu, W., Jiang, H., Wang, H., Gao, Y. (2006). Effect of high hydrostatic pressure on lycopene stability. *Food Chemistry* 97, 516–523.

Sánchez-Moreno, C., Cano, M.P., de Ancos, B., Plaza, L., Olmedilla, B., Granado, F., Martín, A. (2003). High-pressurized orange juice consumption affects plasma vitamin C, antioxidative status and inflammatory markers in healthy humans. *Journal of Nutrition* 133(7), 2204–2209.

Sánchez-Moreno, C., Cano, M.P., de Ancos, B., Plaza, L., Olmedilla, B., Granado, F., Martín, A. (2004). Consumption of high-pressurized vegetable soup increases plasma vitamin C and decreases oxidative stress and inflammatory biomarkers in healthy humans. *Journal of Nutrition* 134(11), 3021–3025.

Sánchez-Moreno, C., Plaza, L., Elez-Martínez, P., De Ancos, B., Martín-Belloso, O., Cano, M.P. (2005). Impact of high-pressure and pulsed electric fields on bioactive compounds and antioxidant activity of orange juice and comparison with traditional thermal processing. *Journal of Agricultural and Food Chemistry* 53(11), 4403–4409.

Sánchez-Moreno, C., Plaza. L, De Ancos, B., Cano, M.P. (2006). Impact of high-pressure and traditional thermal processing of tomato puree on carotenoids, vitamin C and antioxidant activity. *Journal of Science and Food Agriculture* 86(2), 171–179.

Sancho, F., Lamber, Y., Demazeau, G., Largeteau, A., Bouvier, J. M., Narbonne, J. F. (1999). Effect of ultra-high hydrostatic pressure on hydrosoluble vitamins. *Journal of Food Engineering* 39, 247–253.

Shook, C. M., Shellhammer, T. H., Schwartz, S. J. (2001). Polygalacturonase, pectinesterase, and lipoxygenase activities in high-pressure-processed diced tomatoes. *Journal of Agricultural and Food Chemistry* 49, 664–668.

Sila, D.N., Smout, C., Vu, T.S., Hendrickx, M.E. (2004). Effects of high-pressure pretreatment and calcium soaking on the texture degradation kinetics of carrots during thermal processing. *Journal of Food Science* 69(5), 205–211.

Sila, D.N., Smout, C., Vu, S.T., Van Loey, A., Hendrickx, M. (2005). Influence of pretreatment conditions on the texture and cell wall components of carrots during thermal processing. *Journal of Food Science* 70(2), E85–91.

Sila, D., Doungla, E., Smout, C., Van Loey, A., Hendrickx M. (2006). Pectin fractions interconversions: insight into understanding texture evolution of thermally processed carrot. *Journal of Agricultural and Food Chemistry* 54, 8471–8479.

Sila, D.N., Xu, Y., Van Buggenhout, S., Smout, C., Van Loey, A., Hendrickx M. (2007a). The relation between (bio-)chemical, morphological, and mechanical properties of thermally processed carrots as influenced by high pressure pretreatment condition. *European Food Research and Technology*, 226, 127–135.

Sila, D.N., Smout, C., Satara, Y., Vu, S.T., Van Loey, A., Hendrickx, M. (2007b). Combined thermal and high pressure effect on carrot pectinmethylesterase stability and catalytic activity. *Journal of Food Engineering* 78(3), 755–764.

Smout, C., Sila, D., Truong, V., Van Loey, A., Hendrickx M. (2005). Effect of preheating and calcium pretreatment on pectin structure and thermal texture degradation: A case study on carrots. *Journal of Food Engineering* 67, 419–425.

Sohn, K.H., Lee, H.J. (1998). Effects of high pressure treatment on the quality and storage of kimchi. *International Journal of Food Science and Technology* 33, 359–365.

Spilimbergo, S., Elvassore, N., Bertucco, A. (2002). Microbial inactivation by high-pressure. *Journal of Supercritical Fluids* 22 (1), 55–63

Stolt, M., Stoforos, N., Taoukis P., Autio, K. (1999). Evaluation and modeling of the rheological properties of high pressure treated waxy maize starch dispersion. *Journal of Food Engineering* 40, 293–298.

Suthanthangjai, W., Kajda, P., Zabetakis, I. (2005). The effect of high hydrostatic pressure on the anthocyanins of raspberry (*Rubus idaeus*). *Food Chemistry* 90, 193–197.

Takahashi, Y., Ohta, H., Yonei, H., Ifuku, Y. (1993). Microbicidal effect of hydrostatic pressure on satsuma mandarin juice. *International Journal of Food Science and Technology* 28, 95–102.

Tangwongchai, R., Ledward, D.A., Ames, J.A. (2000). Effect of high-pressure treatment on the texture of cherry tomato. *Journal of Agricultural and Food Chemistry* 48, 1434–1441.

Tauscher, B. (1998). Effect of high pressure treatment to nutritive substances and natural pigments. *Fresh Novel Foods by High Pressure. VTT Symposium 186*. Technical Research Centre of Finland. Helsinki, Finland.

Tauscher, B. (1999). High pressure and chemical reactions: Effects on nutrients and pigments. *Emerging Food Science and Technology*. Tempere, Finland. November 22–24, pp. 58.

Trejo Araya, X.I., Hendrickx, M., Verlinden, B.E., Van Buggenhout, S., Smale, N.J., Stewart, C., Mawson, A.J. (2007). Understanding texture changes of high pressure processed fresh carrots: A microstructural and biochemical approach. *Journal of Food Engineering* 80, 873–884.

Van Buggenhout, S., Messagie, I., Van der Plancken, I., Hendrickx, M. (2006). Influence of high pressure–low temperature treatments on fruit and vegetable quality related enzymes. *European Food Research and Technology* 223(4), 475–485.

Van Buren, J. (1979). The chemistry of texture in fruits and vegetables. *Journal of Texture Studies* 10, 1–23.

Van den Broeck, I., Ludikhuyze, L., Weemaes, C., Van Loey, A., Hendrickx, M. (1998). Kinetics for isobaric–isothermal degradation of L-ascorbic acid, *Journal of Agriculture and Food Chemistry* 46, 2001–2006.

Van der Plancken, I., Van Loey, A., Hendrickx M.E. (2007). Foaming properties of egg white proteins affected by heat or high pressure treatment. *Journal of Food Engineering* 78, 1410–1426.

Van Eylen, D., Indrawati, Hendrickx, M., Van Loey, A. (2006). Temperature and pressure stability of mustard seed (*Sinapis alba* L.) myrosinase. *Food Chemistry* 97 (2), 263–271.

Van Eylen, D., Oey, I., Hendrickx, M., Van Loey, A. (2007). Kinetics of the stability of broccoli (*Brassica oleracea* cv. *Italica*) myrosinase and isothiocyanates in broccoli juice during pressure/temperature treatments. *Journal of Agricultural and Food Chemistry*, 55(6), 2163–2170.

Van Loey, A., Ooms, V., Weemaes, C., Van den Broeck, I., Ludikhuyze, L., Indrawati, Denys, S., Hendrickx, M. (1998). Thermal and pressure–temperature degradation of chlorophyll in broccoli (*Brassica oleracea L. italica*) juice: A kinetic study. *Journal of Agriculture and Food Chemistry* 46(12), 5289–5294.

Van Schepdael, L.J.M.M., De Heij, W.B.C., Hoogland, H. (2002). *Method for high pressure preservation*. European Patent WO 02/45528.

Verlent, I., Van Loey, A., Smout, C., Duvetter, T., Hendrickx, M. (2004a). Purified tomato polygalacturonase activity during thermal and high pressure treatment. *Biotechnology and Bioengineering* 86, 63–71.

Verlent, I., Van Loey, A., Smout, C., Duvetter, T., Ly Nguyen, B., Hendrickx, M. (2004b). Changes in purified tomato pectinmethylesterase activity during thermal and high pressure treatment. *Journal of the Science of Food and Agriculture* 84, 1839 1847.

Vu, T., Smout, C., Sila, D., Ly Nguyen, B., VanLoey, A., Hendrickx, M. (2004). Effect of preheating on thermal degradation kinetics of carrot texture. *Innovative Food Science Emerging Technology* 5, 37–44.

Vu, T.S., Smout, C., Sila, D.N., Van Loey, A.M.L., Hendrickx, M.E.G. (2006). The effect of brine ingredients on carrot texture during thermal processing in relation to pectin depolymerisation due to beta-elimination. *Journal of Food Science* 71, E370–E375.

Waldron, K.W., Smith, A.C., Parr, A.J., Ng, A., Parker, M.L. (1997). New approaches to understanding and controlling cell separation in relation to fruit and vegetable texture. *Trends in Food Science and Technology* 8, 213–221.

Wilson, M.J., Baker, R. (2000). *High pressure/ultra-high pressure sterilization of foods.* US Patent 6 086 936.

Zabetakis, I., Koulentianos, A., Orruño, E., Boyes, I. (2000). The effect of high hydrostatic pressure on strawberry flavour compounds. *Food Chemistry* 71, 51–55.

Zhang, L., Lyng, J.G., Brunton, N.P. (2004). Effect of radio frequency cooking on the texture, colour and sensory properties of a large diameter comminuted meat product. *Meat Science* 68, 257–268.

18

Novel methods to improve the safety and quality of in-pack processed ready-to-eat meat and poultry products

P. L. Dawson, Clemson University, USA

18.1 Introduction

The most influential factors in the development of shelf-stable, non-refrigerated foods; including meats; was probably the need to feed soldiers in the field and explorers during long journeys. Early leaders of fighting men from the Greeks and Romans were well aware of the severe effect on Army morale and physical condition the food supply could have. A description of how well Roman warriors were fed can be found in this excerpt from http://resourcesforhistory.com/Roman_Food_in_Britain.htm.

> The Roman army consumed a healthy combination of simple high-energy food. Bread was their staple food and grain production was increased throughout Britain to meet the demand from the army. Large beehive ovens were positioned all the way around the Legionary Fortress at Caerleon. ... Accounts indicate Roman soldiers ate a lot of bacon. Every group of eight soldiers had a frying pan that folded away in their pack and enabled them to have a fry-up even on campaign. They also ate porridge and stews that would have included meat and vegetables.

Meat was an important part of the diet and was preserved until recent times by drying and salting. In 1932, a 'balanced meat in a can' was developed by a Sanitary Corps Reserve officer in the US Army, consisting of 1 pound of stew containing 12 vegetables and 9 meats. This later officially became known as the 'C ration'. Over the years, army rations have evolved and currently these are known as MREs or meals ready-to-eat.

One of the first commercial in-package processed meats was SPAM, which was first sold in 1937 and is still sold today by the Hormel Corporation™ The name SPAM is speculated to be an acronym for spiced ham, but the name actually came from a contest put on by the Hormel Company with a $100 prize. Hormel developed America's first canned meat in 1926, named Hormel Flavor-Sealed Ham and then, 11 years later, introduced its first canned meat not requiring refrigeration, SPAM.

Early in-pack processed meats were packaged in a metal can. However, today, flexible retort pouches are used to deliver a variety of shelf-stable meats, including tuna fish, poultry and even SPAM. More recently, in-pack processing for ready-to-eat meats has been implemented for refrigerated and cured meat products to address post-processing bacterial contamination.

18.2 The need for ready-to-eat meat in-package processing

There is a new focus on in-package processing for ready-to-eat (RTE) meat and poultry products due to EU and USDA-FSIS rulings and to the presence of *Listeria monocytogenes* in RTE meats (Table 18.1).

There are between 3 and 7.5 cases per 1 million people reported in developing countries such as Australia, European countries, New Zealand and the US. The zero tolerance policy for the presence of *L. monocytogenes* in the US is costing the meat and poultry product industry millions of dollars annually due to recalls. In fact, of all recalls due to *L. monocytogenes* in 2000, around 57% were for RTE meat products. The prevalence of *L. monocytogenes* in RTE meats for the period from 1990 through 2000, averaged from 0.58 for jerky to 4.47% for sliced meats, with most other RTE meat product categories having approximately a 2% prevalence rate. In 2003, a joint risk assessment report from the CDC, FDA and FSIS stated that deli meats were the leading cause of listeriosis among all RTE meat products. The EU instituted a slightly different Listeria policy for all foods, including RTE meats, that requires absence of the bacterium in a 25 g sample before leaving the processing plant but allows up to 100 cfu/g for products in the market if the product is not intended for consumption by infants or for special medical purposes. Due to trade and risk assessment studies, several trade organizations and scientific bodies have recommended that the US change their zero tolerance policy to match that of the EU.

Less than 0.1% of the total amount of meat produced and imported annually in the US is recalled. However, recalls cost the RTE meat industry millions of dollars. Information on recalled products is maintained and updated on the FSIS website at: http://www.fsis.usda.gov/oa/recalls/rec_intr.htm (USDA, 2005). From 1997 through 2002 there has been at least one meat/poultry recall annually greater than 10 million pounds weight, 5 of which involved *Listeria monocytogenes* and 2 involving *Escherichia coli* O157:H7. Over the nine-year period from 1994 through

Table 18.1 Prevalence (%) of *L. monocytogenes* in RTE meat and poultry products, 1990–2000 (modified from USDA-FSIS data)

Year	Cooked, roast, corned beef	Sliced ham and luncheon meats	Small cooked sausages	Large cooked sausages	Jerky	Cooked poultry products	Salads/ Spreads/ pâté	Fermented sausages
1990	6.38	7.69	4.21	5.32	0.00	2.79	5.48	N/A
1991	4.02	5.48	7.24	4.60	0.00	2.62	3.17	N/A
1992	3.86	7.89	6.03	0.42	0.00	2.01	3.32	N/A
1993	3.04	8.05	5.30	2.13	0.00	1.91	2.19	N/A
1994	2.09	5.46	4.81	1.14	2.22	2.37	2.41	N/A
1995	2.68	5.00	4.09	1.14	0.00	2.25	4.69	N/A
1996	3.35	7.69	3.74	0.95	0.00	3.17	2.17	N/A
1997	2.08	4.20	2.74	1.62	0.00	0.95	2.43	9.26
1998	2.15	4.18	3.49	1.19	1.56	2.22	3.11	2.87
1999	2.71	4.58	1.76	0.43	0.00	1.44	1.15	2.09
2000	2.24	3.05	1.26	0.51	0.75	1.24	0.98	1.49
Cumulative	2.95	4.47	2.97	1.09	0.58	1.97	2.83	2.67

Table 18.2 Reasons for meat and poultry recalls in the US (from USDA-FSIS data)

	Number of recalls						
Year	*Listeria*	*E. coli*	*Salmonella*	Other bacteria	Chemical/ physical	Undeclared ingredients	Under-processed
1994	17	3	0	3	16	1	7
1995	11	5	2	2	13	1	7
1996	6	2	1	1	5	3	6
1997	3	6	1	5	8	4	0
1998	7	13	2	2	11	4	5
1999	30	10	6	0	3	8	4
2000	36	20	4	0	5	9	2
2001	25	26	2	0	11	24	6
2002	40	24	4	0	4	36	4

2002, 74% of the recalls were of the Class I category, 74% involved processed meat, 26% raw meat, 23% involved poultry meat, 40% red meat and 69% involved biological hazards (Teratanonat and Hooker, 2004). *Listeria monocytogenes* and *Escherichia coli* O157:H7 were the most frequent causative biological recall agents, despite *Salmonella* spp. and *Campylobacter jejuni/coli* causing many more food-borne illnesses (USDA, 2005) (Table 18.2).

Listeria monocytogenes is widely distributed in nature and is estimated to be present in the feces of about 2–6% of healthy human adults. *Listeria monocytogenes* can grow at refrigeration temperatures, low pH values, high salt concentrations and low water activities. The people most susceptible to serious health effects and death from listeriosis include the elderly, organ transplant patients, unborn fetuses, cancer patients, persons with AIDS and any immuno-compromised individual. Furthermore, *L. monocytogenes* has a 20–25% mortality rate for those becoming infected, and can result in septicemia, meningitis and stillbirths.

The USDA-FSIS has passed an interim rule that requires RTE meat processors implement in order to control *L. monocytogenes*. The rule gives processors three alternatives, which are summarized here. These are listed in order, from least to most extensive, using the USDA numbering sequence.

3. Maintain current plant sanitation practices.
2. In addition to Number 3, add either a post-processing lethal treatment to reduce *L. monocytogenes* numbers or add an antimicrobial agent or process that limits *L. monocytogenes* outgrowth during shipping and storage.
1. In addition to Number 3, add a post-processing lethal treatment to reduce *L. monocytogenes* numbers or add an antimicrobial agent or process that limits *L. monocytogenes* outgrowth during shipping and storage.

As stated, these alternatives are numbered from 3 to 1 since this is the number assigned to these alternatives by the USDA-FSIS who start with the most aggressive (1) and end with the least aggressive (3). The obvious question one might ask is why would a processor implement alternatives (1) or (2) since there will be an additional cost involved. In addition to the motivation of utilizing the best process to prevent human suffering (social consciousness), the processor has two very important economic reasons for implementing a more aggressive *L. monocytogenes* prevention program. First, the more aggressive alternatives (1 and 2) are coupled with less stringent monitoring as required by the USDA-FSIS. A general example of this coupling of testing frequency with monitoring applies to food contact surface testing which is, for alternative (1), about twice per year, alternative (2) about 4 times per year, and for alternative (3) from 1–4 times per month. The second reason for implementing a more aggressive *L. monocytogenes* control strategy is to prevent *L. monocytogenes* product recalls, which, as stated earlier, costs the industry millions of dollars annually. It is likely that the prevention of just one recall would more than pay for the multi-year implementation of one of the more aggressive *L. monocytogenes* alternatives. Antimicrobial packaging is stated in the USDA-FSIS alternatives (1) and (2) as an accepted intervention. The antimicrobial agent must be verified for preventing outgrowth of *L. monocytogenes* during refrigerated storage. To summarize the USDA-FSIS rule, there are new alternatives that employ a two-step intervention program. The first step is a 'kill' process which has the purpose of reducing *L. monocytogenes* presence in the product at the processing plant. This 'kill' step can be heat from steam, a water bath or infra-red sources. High pressure processing and irradiation are also valid processes. The second step is the use of an antimicrobial growth inhibitor which must prevent *L. monocytogenes* outgrowth during storage of the product. If an antimicrobial can both reduce *L. monocytogenes* and prevent outgrowth during storage, the antimicrobial may satisfy the USDA-FSIS designation as a method to reduce *L. monocytogenes* (Step1) and as an inhibitor during storage (Step 2). Antimicrobial packaging may be a very attractive alternative for meat processors since they can request an antimicrobial packaging material from a film supplier which does not require substantial changes to their process.

18.3 Methods to optimize safety and quality

Methods evaluated to reduce or eliminate *L. monocytogenes* or other bacterial contaminants from RTE meats include the addition/use of biopreservatives (Berry *et al.*, 1991; Embarek, 1994; Hugas *et al.*, 1995; Nilsson *et al.*, 1997; Scannell *et al.*, 1997; Winkowski *et al.*, 1993), competitive exclusion (Amezquita and Brashears, 2002), antimicrobial films (Buonocore *et al.*, 2005; Cutter *et al.*, 2001; Dawson *et al.*, 1995, 2002a,b; Hoffman *et al.*, 1997, 2001; Limjaroen *et al.*, 2003; McCormick *et al.*, 2005; Ming *et al.*, 1997; Natrajan and Sheldon, 2000a,b; Padgett *et al.*, 1998), high pressure processing (Ananth *et al.*, 1995; Carlez *et al.*, 1991 ; Hoover *et al.*, 1989; Murano *et al.*, 1999; O'Brien and Marshall, 1996), pasteurization with steam or hot water (Hardin *et al.*, 1993; Muriana *et al.*, 2002; McCormick *et al.*, 2003; McCormick *et al.*, 2005; Murphy *et al.*, 2003a,b; Roering *et al.*, 1998; Selby *et al.*, 2006), and chemical additives (Blom *et al.*, 1997; Juncher *et al.*, 2000; Palumbo and Williams, 1994; Qvist *et al.*, 1994; Samelis *et al.*, 2001; Weaver and Shelef, 1993; Wederquist *et al.*, 1994).

18.3.1 In-package pasteurization

In-package pasteurization is an effective method to reduce post-processing contamination of RTE meats, especially since the packaged product is protected from further contamination until the product package is opened. Hardin *et al.* (1993) reported that the lethality of in-package pasteurization of pre-cooked beef roasts was directly related to the dwell time and temperature of the treatment. *Listeria monocytogenes* populations were reduced by 4 logs on pre-cooked, vacuum-packaged beef roasts using in-package pasteurization at 85 °C for 16 min (Cooksey *et al.*, 1993). Franz and von Holy (1996a) tested a range of in-package thermal treatments for Vienna sausages and found lactic acid bacteria (LAB) were reduced only between 52.9 and 74.6% and total bacteria were reduced only by 84.4% (both less than one log cycle) compared to sausages not exposed to in-package pasteurization and stored at 8 °C for 128 days. In fact, in-package pasteurization did not delay spoilage due to LAB and increased the predominance of *Bacillus* spp. In a second study, Franz and von Holy (1996b) determined the *D* values at 57, 60, and 63 °C of three spoilage LAB to be 52.9, 39.3, and 32.5 s for *Lactobacillus sake*; 34.9, 31.3, and 20.2s for *Leuconostoc mesenteroides*; and 22.5, 15.6, and 14.4s for *Lactobacillus curvatus,* respectively. In-package pasteurization extended the microbiological shelf-life (spoilage level set at 6.3 log cfu/g) of Vienna sausages from 2 to 8 weeks compared with samples not exposed to the in-package treatment (Dykes *et al.*, 1996). Roering *et al.* (1998) reduced *L. monocytogenes* 3 log cfu/g in vacuum-sealed summer sausages with a hot water submersion, using time/temperature treatments of 30, 60, and 90s at 99, 88, and 77 °C, respectively. Roering *et al.* (1998) reported the corresponding *D* values as 2.08, 0.84, 0.37, and 0.28 min for 66, 77, 88, and 99 °C, respectively with only a 2 log cfu/g reduction of *L. monocytogenes* after 240 s at 66 °C. Murphy and Berrang (2002) found that pasteurization of fully cooked, vacuum-packaged chicken breast strips in steam and hot water at 88 °C for 25 and 35 min lowered *Listeria innocua* populations by 2 and 7 log cfu/g, respectively, with no difference in the effectiveness of hot water *vs* steam. Muriana *et al.* (2002) tested submersion in hot water for the reduction of a four-strain *L. monocytogenes* cocktail resuspended in product purge before inoculation into packaged RTE deli meats. Post-pasteurization at 90.6, 93.3, and 96.1 °C for 2 to 10 min resulted in a 2 to 4 log cfu/g reduction in population; however, *L. monocytogenes* reductions in product challenge studies were lower, a difference attributed to product surface imperfections that shield bacteria. A pilot-scale, 100 °C post-packaging hot water system for RTE beef snack sticks and natural-case wieners reduced *L. monocytogenes* by 2 or more log cfu/g after 1 min for individually packaged sticks and 4 min for packages of 4 or 7 sticks (Ingham *et al.*, 2005). These researchers reported a 7 min treatment using the small pilot system was needed to achieve a 1 log reduction of *L. monocytogenes* on packages of four natural-casing Vienna sausages.

18.3.2 Modeling in-package pasteurization

Heat resistance, expressed as *D* and *z*-values, has been determined for several pathogenic bacteria in RTE meats exposed to in-package pasteurization. Murphy *et al.* (2003a) found significant differences in the *D* and *z* values for *Salmonella* spp., *L. innocua*, and *L. monocytogenes* among turkey, duck and chicken products, indicating product-to-product variation and the need to determine lethality of pathogens for each specific product. McCormick *et al.* (2003) reported that *D* values of low-fat turkey bologna slices were 124 and 16.2 s at 61 and 65 °C for *L. monocytogenes* and 278 and 57 s at 57 and 60 °C for *S. typhimurium*. *z* Values of 4.44 and 5.56 C were calculated for single slices of turkey bologna for *L. monocytogenes* and *S. typhimurium*, respectively. Bacterial survivor curves often display non-linearity; however, the most common approach used to describe bacterial inactivation is first-order kinetics (*D*- and *z*-values), which is linear. First-order kinetics is not compatible with tailing, curvature, or a shoulder of the thermal survival curve (Virto *et al.*, 2006). A first-order relationship to thermobacteriology assumes each bacterial cell has the same probability of being killed (van Boekel, 2002). This is probably most often not the case; that is, bacterial cells have varying resistance to inhibitory treatments and acquire some heat resistance quickly. Mangalassary *et al.* (2007) applied nisin and/or lysozyme to bologna to lower the heat tolerance of *L. monocytogenes* and found a non-linear thermal death curve during in-package pasteurization by testing for linearity using orthogonal polynomials. Two models often used to describe non-linear thermobacteriology are the Weibull distribution and the log-logistic model:

- *Weibull model*: The cumulative form of the Weibull distribution suggested by Peleg & Cole (1998) is given by

$$\text{Log}\ \frac{N}{N_0} = -\ bt^n$$

where N_0 = the initial number of cells after come-up time, N = the number of survival cells after an exposure time t, b is the scale factor which is a characteristic of time, and n is the shape factor. An $n < 1$ corresponds to a concave upward survival curve, $n > 1$ to a concave downward curve and n equal to 1 to a straight line. In the Mangalassary *et al.* (2007) study, all of the treatments yielded a value of $n < 1$, indicating an upward concavity of the curve. But values of the estimate of n in all cases were close to 1 (0.75–0.95), indicating that the upward concavity was minimal. Upward concavity can be interpreted as evidence that weak or sensitive members of the population are destroyed at a relatively fast rate, leaving behind survivors of higher resistance

- *Log-logistic model*: A modified version of the original log-logistic model (Cole *et al.*, 1993) with the following equation was used by Mangalassary *et al.* (2007) to fit the survivor *L. monocytogenes* curves obtained for in-package bologna processed at 60 °C;

$$\text{Log}\ \frac{N}{N_0} = -\ bt^n$$

where A is the difference between lower and upper asymptotes, σ is the maximum rate of inactivation (maximum slope of inactivation curve) and τ is the log time to the maximum rate of inactivation.

While the Weibull model fits data that shoulders or tails, it does not fit data that displays both upper and lower asymptotes and, in the same in-package bologna pasteurization study (Mangalassary *et al.*, 2007), the log-logistic was the best model at 60 °C, while the Weibull model was the best fit at 62.5 and 65 °C due to the shape of the inactivation curves (Mangalassary *et al.*, 2007). In this study, the Weibull parameter b did not show any systematic pattern among treatments or between the pasteurization temperatures tested, and according to Mafart *et al.* (2002) parameter b has little significance. The non-linear models can accurately predict the shape of thermal inactivation curves, and some of the parameters such as the Weibull shape parameter (n) may be useful for determining similarities between pasteurization treatments or bacterial population inactivation kinetics but their usefulness may be limited to processors. The Institute of Food Technologists' second research summit suggested that food preservation process performance should be communicated based on log cycle reductions.

18.3.3 Factors affecting quality

Some factors affecting the rate of surface heating include meat surface roughness, product composition, packaging film, and product size (thickness). For example, Murphy *et al.* (2003b) reported that the 'roughness' of 4 kg RTE turkey breast surfaces (varying in depth by >15 mm) required 50 min at 96 °C to achieve a 7 log cfu/cm^2 reduction of *L. monocytogenes*, while surfaces with depth variation of less

Table 18.3 In-package pasteurization *D* values for *Listeria monocytogenes* in selected ready-to-eat meat products

T (C)	*D* (min)	Product	Reference
96	2.33	4 kg turkey breast	Murphy *et al.* (2003)
93	0.5	1.8–5.9 kg ham	Gande and Muriana (2003)
93	0.16	0.9 kg bologna chubs	Gande and Muriana (2003)
62.8	6.9	2–5 kg whole-formed turkey	Muriana *et al.* (2002)
65.6	1.23	2–5 kg whole-formed turkey	Muriana *et al.* (2002)
68.3	0.42	2–5 kg whole-formed turkey	Muriana *et al.* (2002)
71	0.16	2–5 kg whole-formed turkey	Muriana *et al.* (2002)
62.8	1.12	2–5 kg ham	Muriana *et al.* (2002)
65.6	0.5	2–5 kg ham	Muriana *et al.* (2002)
68.3	0.25	2–5 kg ham	Muriana *et al.* (2002)
71	0.08	2–5 kg ham	Muriana *et al.* (2002)
88	2.86	227 g packs of chicken strips	Murphy and Berrang (2002)
88	4.14	454 g packs of chicken strips	Murphy and Berrang (2002)
61	2.1	Turkey bologna slice	McCormick *et al.* (2003)
65	0.26	Turkey bologna slice	McCormick *et al.* (2003)

than 15 cm required significantly less time to achieve the same log reduction. Package film thicknesses ranging from 0.08 to 0.33 mm had a significant effect on the heating rate of vacuum-packaged turkey breast during hot water in-package pasteurization at 68 °C for 10 to 120 s (Murphy *et al.*, 2003, 2003b). Murphy *et al.* (2003a) developed a model to predict lethality of *Salmonella* spp. and *L. innocua* for different thicknesses of fully cooked, vacuum-packaged chicken breast exposed to in-package pasteurization. The model was verified using an inoculation study and fell within a 95% confidence level to achieve a 7-log cfu/g reduction using a hot-water cooker set at 90 °C.

Post-process pasteurization of RTE meat is a surface pasteurization; therefore product thickness or size might be expected to have a minimal effect on surface heating rate. Based on the large range of D values for *L. monocytogenes* on RTE meat (Table 18.3) and from discussions with equipment industrial representatives, product thickness effects on surface heating rate can have a large effect on processing times.

Mangalassary *et al.* (2004) investigated the surface heating rates for 1, 3 and 5 bologna slices representing 4, 12 and 20 cm thick samples and two types of bologna having 13 and 18% fat. Four pasteurization temperatures were used (60, 70, 80, and 90 °C) and the researchers found that surface heating rates were significantly slower for 12 compared to 4 cm thick and slower for 20 compared to 12 cm thick bologna stacks (Fig. 18.1). The difference in surface heating rate was so dramatic that, to achieve a 5-log reduction of *L. monocytogenes* at 80 °C, the hold time would be 0.72, 2.56, and 4.12 min for each 8 mm increase in thickness for 4, 12, and 20 mm sample thickness, respectively.

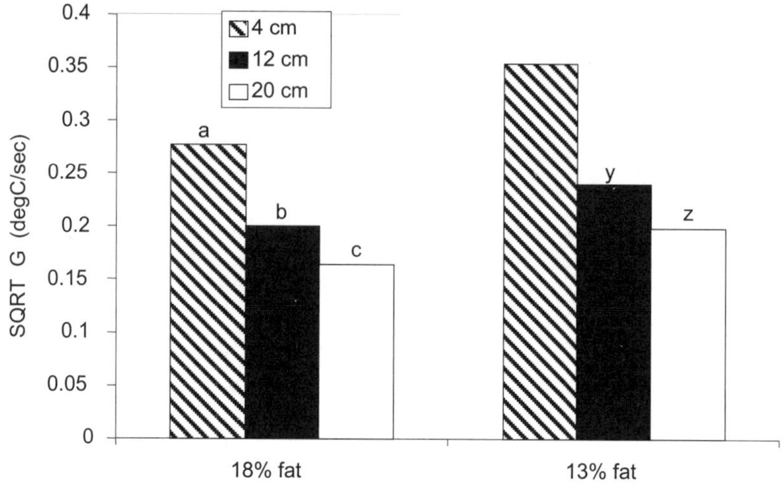

Fig. 18.1 Surface heating rates for different thicknesses and for two types of bologna.

18.4 Use of antimicrobials

18.4.1 Approaches to delivering antimicrobials

There are several possible approaches to delivering antimicrobials to RTE meat surfaces. These include:

(i) direct application to the meat surface,
(ii) application to a film surface,
(iii) incorporation into the meat product,
(iv) incorporation into a packaging film, and
(v) incorporation into a carrier or encapsulation which is subsequently placed on the meat or packaging film surface.

There are many antimicrobials available for possible application to meat, including commercially available formulations directed at limiting *L. monocytogenes* on RTE meats. For instance, Danisco™ markets a variety of natural antimicrobials with several targeted for fresh and ready-to-eat meat products. Their nisin-based formulations include Nisaplin® and Novasin™ which are effective against *L. monocytogenes.* Danisco™ also sells what they call antimicrobial systems Guardian™ and Novagard™ that also inhibit Gram-positive bacteria.

Some examples of applying an antimicrobial directly to food surfaces include the use of ozonated water (Fabrizio *et al.*, 2002), electrolyzed water (Fabrizio and Cutter, 2005), lactic acid bacteria (Amezquita and Brashears, 2002), organic acids (Dykes *et al.* 1996; Samelis *et al.*, 2001; Geornaras *et al.*, 2005), bacteriocins (Geornaras *et al.*, 2005), and silver ions (Nobile *et al.*, 2004). Some specific results of these direct application studies found that a combination of organic acids

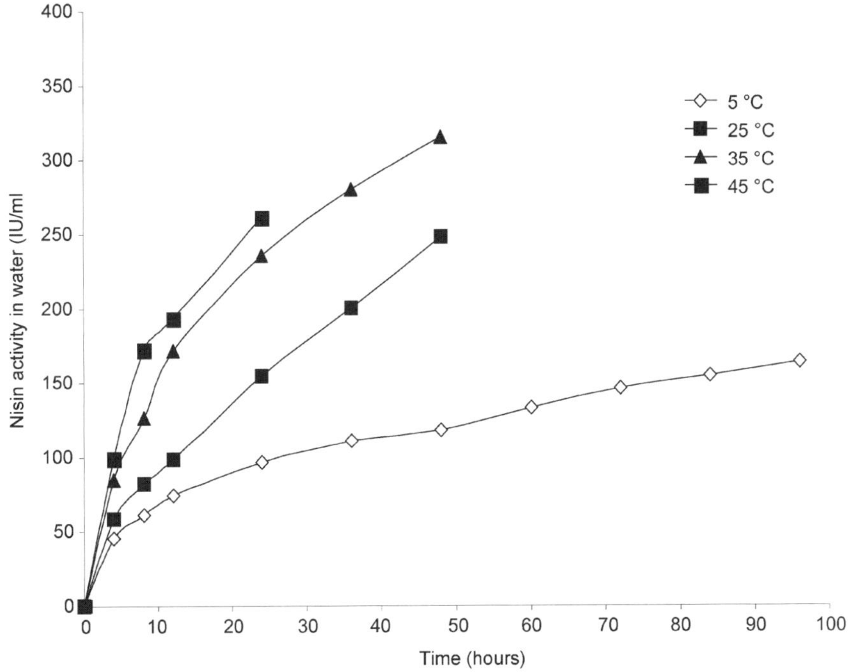

Fig. 18.2 Temperature dependence of nisin activity release from bio-based films (Teerakarn *et al.*, 2000).

(acetic, ascorbic, citric, and lactic) added to vacuum-packaged Vienna sausages, actually decreased shelf-life by about 1.8 days since the predominant spoilage organisms were lactic acid bacteria (Dykes *et al.*, 1996). Geornaras *et al.* (2005) reported on dipping bologna or ham before vacuum-packaging in solutions containing 2.5% acetic acid, 2.5% lactic acid, 5% potassium benzoate, 0.5% nisin (as Nisaplin™ equivalent to 5000 IU/ml) and nisin in single combination with the other preservatives. The acid treatments alone were effective in preventing *L. monocytogenes* outgrowth through 48 days when stored at 10° C. The nisin dip ion combination with (followed by) acid dipping reduced *L. monocytogenes* by 2–3 logs and prevented outgrowth during 10 °C, 48-day storage.

18.4.2 Antimicrobial films

Incorporation of antimicrobials into 'plastic' or 'bio' polymer films has been studied using a variety of antimicrobials and polymer materials. Antimicrobials tested include silver ions, triclosan, chlorine dioxide, peptides (such as lactoferrin), bacteriocins (primarily nisin), organic acids, enzymes, chelating agents, essential oils, seed extracts, plant skin extracts, plant pigments, and other plant extracts. One advantage of incorporating an antimicrobial into a film is that gradual release that can be achieved from the film to the food surface during storage, which in some cases is temperature dependent (Fig. 18.2).

Table 18.4 Heat/seal temperatures for selected polymer films

Packaging resin	Melt temp (°C)
Polyethylene	104–171
Polyvinyl chloride	138–171
Polyvinylidene chloride	121–149
Ethylene/vinyl alcohol	177–204
Ethylene/vinyl acetate	66–177

'Zeolites' have been a commonly tested carrier for antimicrobials in petroleum-based polymer packaging. Zeolite is an inorganic porous material having a highly regular structure of pores and there are hundreds of different zeolites found in nature and made by man. Chemically, zeolites are a group of natural or synthetic hydrated aluminosilicate minerals which contain alkali and alkaline metals. Several commercially available packaging materials utilize zeolite as a carrier for antimicrobials (Oliveira and Oliveira, 2004) such as Alpha San™ (triclosan), Iraguard™ (chlorine dioxide), Johnson-Matthey™ and Agion™ (silver ions). Addition directly to petroleum-based polymer packaging has also been tested; for instance, Limjaroen *et al.* (2003) added nisin, sorbic acid or potassium sorbate to Saran™ (polyvinylidene chloride) to inhibit *L. monocytogenes*, while film material with lactoferrin or sodium diacetate added to Saran™ did not inhibit *L. monocytogenes*. The lowest concentrations to show *L. monocytogenes* inhibition for nisin, sorbic acid and potassium sorbate were 1, 1.5 and 2%, respectively. Petrikova *et al.* (2003) reported antimicrobial activity against a broad spectrum of bacteria and fungi associated with foods for films having silver zeolite incorporated into polyethylene. Another interesting polymer film with silver incorporation was reported by Nobile *et al.* (2004) who produced a silver-containing nano-composite film on polyethylene oxide by plasma deposition of silver particles. The silver composite film was effective in inhibiting the growth of spoilage bacteria of acidic foods but the level of inhibition was directly related to the release of silver from the film.

With the exception of some ethylene/vinyl acetates, the temperatures used to form petroleum-based polymer films are greater than 100 °C (Table 18.4), thus limiting the incorporation of heat labile antimicrobials into commercial polymer-based films.

Cast films and coatings can reduce the loss of activity when incorporating antimicrobials since less heat is used to form the film, but casting films adds several other issues to be addressed, particularly in large-scale production. A cast food-coating would likely alter the appearance and taste of meat product as many edible biopolymers carry characteristic colors and flavors. In addition, casting large quantities of film would require a large, flat surface area for film formation since most cast films require several hours to solidify. Furthermore, some solvents used in cast films require adequate venting and storage capabilities.

A wide variety of biopolymer films have been tested as antimicrobial carriers,

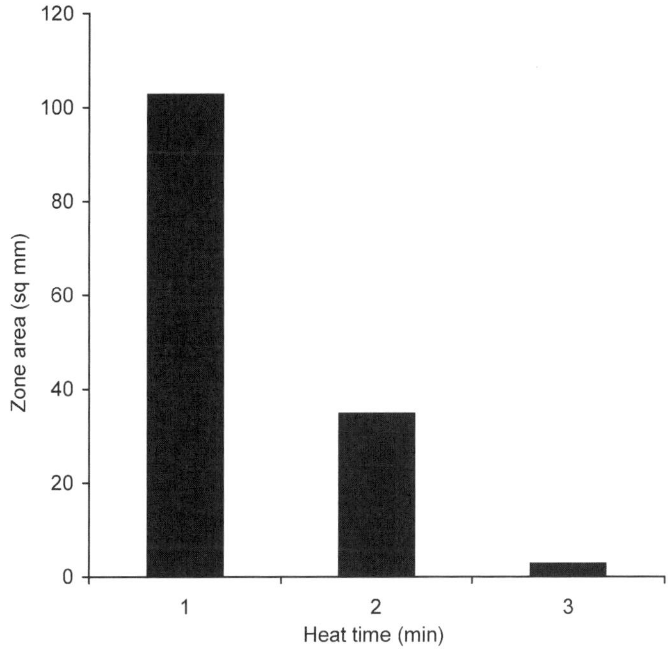

Fig. 18.3 Effect of heat extrusion time on nisin activity in polyethylene films heated at 149 °C (300 °F) for 1, 2 and 3 minutes.

being mostly polysaccharide (starches, cellulose, alginates, gums, chitosan, agar) or protein (corn zein, wheat gluten, soy, milk [whey], peanut, collagen, gelatin) based. Chitosan (a deacetylated form a chitin, derived from the exoskeleton of shellfish), for instance, is a polysaccharide-based material that possesses antimicrobial properties and forms flexible films and coatings. Park *et al.* (2003) found that lysozyme added to chitosan films at a 20 and 60% basis (w/w) resulted in a 3–4 log reduction of *Escherichia coli* and *Streptococcus faecalis* in selected growth media over 25 hours. Park *et al.* (2002) had earlier added chitosan to polyethylene to form inhibitory films against *Lactobacillus plantarum* and determined that the degree of acetylation was an important factor in the antimicrobial activity of chitosan. The antimicrobial activity of edible and biopolymer films was demonstrated by Dawson *et al.* (1995) for nisin and lysozyme added to soy and corn protein-based films. It was later determined that nisin activity could be retained in heat-extruded polyethylene films but that extending the heat-extrusion times would reduce nisin activity in the resulting film (Fig. 18.3) (Hoffman *et al.*, 1997). However, nisin-impregnated cast films generally had a more bacteriocidal effect than heat-set films containing nisin (Padgett *et al.*, 1998). When Nisaplin™ (commercial grade of nisin, ~2.5% pure nisin) was added to either heat-set or cast films, a trend was observed that, at approximately 4–5% and higher levels of addition to the films, little or no additional antimicrobial affect was gained (Hoffman *et al.*, 2001; Orr *et al.*, 1997; Dawson *et al.*, 1995; Padgett *et al.*, 1998).

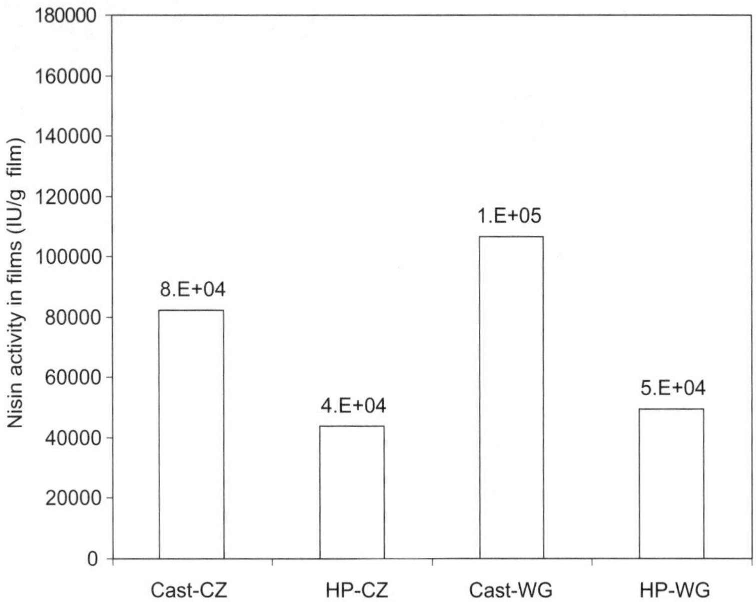

Fig. 18.4 Nisin activity retention after film formation for cast or heat-extruded (HP) wheat gluten (WG) and corn zein (CZ) films.

Thermally extruded soy films containing 5% Nisaplin® reduced cell populations suspended in peptone water by 1 $\log_{10}$ cycle, while films containing both Nisaplin® and 8% lauric acid reduced *L. monocytogenes* populations to below detection levels (> 6.5 $\log_{10}$ reduction) (Dawson *et al.*, 2002a). Dawson *et al.* (2002b) reported that thermally extruded soy films containing nisin applied to inoculated bologna surfaces reduced *L. monocytogenes* populations by 1 $\log_{10}$ over 21 days of refrigerated storage.

Numerous studies using bacteriocins (primarily nisin) with meat packaging have been published, with several specific findings described here. Ming *et al.* (1997) found 5000 IU nisin/g was ineffective while 640 000 AU pediocin/g completely inhibited *L. monocytogenes* on the surfaces of beef and turkey when dialyzed in 10% solutions onto meat casings. Retention of antimicrobial activity was reported with collagen and natural meat casings dipped into solutions containing various levels of Nisaplin® (Kassaify, 1998). Nisaplin® was added to wheat gluten, soy protein, egg albumen, and whey protein film-forming solutions prior to casting the films to yield film inhibitory to *L. monocytogenes* in inoculated media (Ko *et al.*, 2001). Natrajan and Sheldon (2000a) found polyvinyl chloride to be more inhibitory against *Salmonella typhimurium* than linear low density polyethylene and nylon when treated with a nisin formulation including 100 μg/ml nisin, and varying concentrations of EDTA (0., 5.0, 7.5 mM), citric acid (3.1, 3.0, 3.0) and Tween 80 (0, 0.01, 0.5%). Natrajan and Sheldon (2000b) also coated chicken skin samples with calcium alginate containing either 0, 200, 500, 600, or

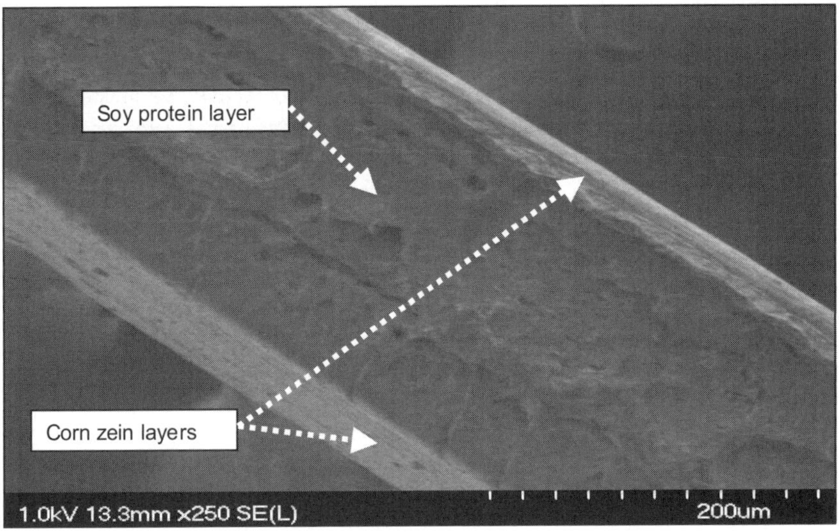

Fig. 18.5 Micrograph of corn zein and soy protein laminated film

1000 μg/ml nisin with 5.0 mM EDTA, 3.0% citric acid, and 0.5% Tween 80 and found the level of kill was proportional to increasing exposure time and produced a 1.8 to 4.6 $\log_{10}$ reduction in *S. typhimurium* populations after 96 hours of exposure. Cutter *et al.* (2001) achieved a 1.7 to 3.5 $\log_{10}$ cycle reduction of *Brochothrix thermosphacta* over 21 days of refrigerated storage on vacuum-packaged beef with polyethylene and polyethylene oxide films having 0.1% Nisaplin® and EDTA incorporated into them.

Dawson *et al.* (2003) reported that nisin activity retained after formation of a cast or heat-extruded film (when identical bio-based materials were used) was about twice that for cast films as for the heat-extruded films. The loss of nisin activity in the heat-extruded films was probably related to heat denaturation of the antimicrobial peptide nisin during film formation (Fig. 18.4).

Incorporation of antimicrobials into packaging films can result in substantial loss of antimicrobial activity. This is partly due to binding of the antimicrobial to film components, denaturation of antimicrobial peptides due to solvents or heat used in the film-forming process, and retention of antimicrobial molecules in the film bulk. Buonocore *et al.* (2005) demonstrated the effect of film as a barrier to antimicrobial release in a study using multi-layer films with the middle layer containing lysozyme and two outer layers devoid of antimicrobial. This multi-layer film was compared to a mono-layer film containing lysozyme which resulted in greater release of lysozyme activity from the mono-layer compared to the multilayer film. Several strategies have been evaluated to remedy the loss of antimicrobial activity to the film bulk. One strategy is the incorporation of the antimicrobial into the food contact layer by laminating the antimicrobial film layer to another layer (Pol *et al.* 2002). This can be seen in corn zein layers laminated to a soy protein core in Fig. 18.5. The lighter-colored corn zein layers carry the

Table 18.5 Nisin activity retained and released from biopolymer films and food-grade powders

Bio-polymer films	Activity retained/released
Cast corn zein	12.1 %
Cast wheat gluten	15.8 %
Heat-extruded wheat gluten	7.4 %
Polyethylene	<6.5 %
Food-grade powders	**Activity retained/released**
Calcined diatomaceous earth	63.4 %
Calcium silicate hydrate	16.7 %
Diatomaceous earth, Type A	57.2 %
Diatomaceous earth, Type B	76.0 %
Corn starch	54.2 %

antimicrobial allowing the antimicrobial to be concentrated on the film surface rather than in the film bulk. A second strategy to avoid loss of activity is to incorporate the antimicrobial into a carrier that can be used to coat either the film or the meat itself. By incorporation or encapsulation of nisin into food grade particles, 4–5 times the retention and release of antimicrobial activity was achieved compared with when nisin was incorporated into polymer films (Dawson *et al.*, 2004) (Table 18.5).

Therefore, the latter approach of carrying the antimicrobial in a particle or powder may be a more efficient approach for delivering bactericidal activity to the packaged meat surface. This approach minimizes the loss of antimicrobial activity while also controlling the release of the activity. Several food-grade powders have demonstrated the capability to adsorb nisin and then release nisin's listericidal activity into an aqueous solution. Some of these nisin-activated powders were able to reduce *L. monocytogenes* populations from 8 log cfu/ml to below detection levels within hours (Fig. 18.6.)

18.5 Combining in-package pasteurization with antimicrobials

Antimicrobials can reduce the heat resistance of bacteria, thereby reducing the thermal treatment required to eliminate harmful bacteria while minimizing negative quality affects due to heat. This novel approach has been examined by several researchers, including Murphy *et al.* (2004) who reported no synergistic effect of using sodium lactate with in-package pasteurization at temperatures between 55 and 70 °C to inactivate *Salmonella* spp. and *Listeria monocytogenes*. Previously, Dykes *et al.* (1996) found no advantage for shelf-life extension of Vienna sausages by combining organic acids with in-package pasteurization. Chen *et al.* (2004) used the surface application of pediocin (ALTA 2341) with hot-water post-packaging treatments at 71, 81, or 96 °C for 30, 60 or 120 s to reduce *Listeria*

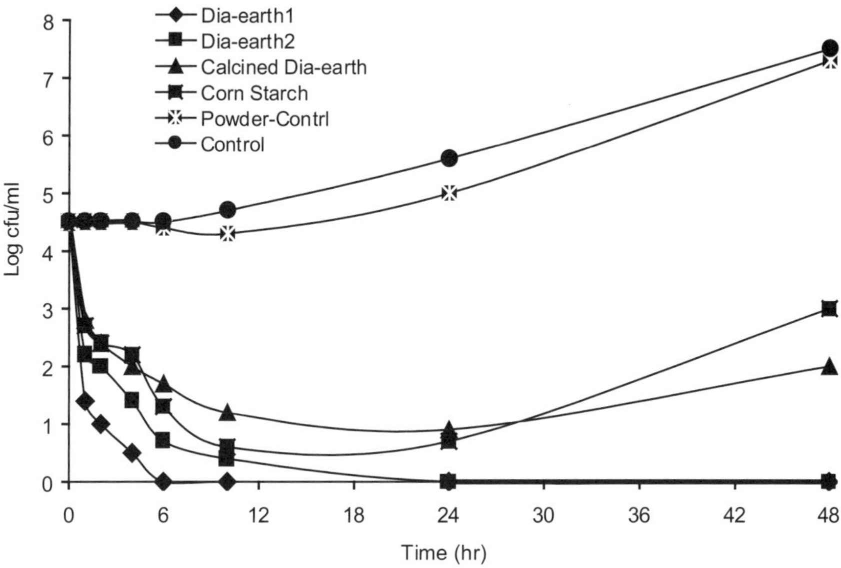

Fig. 18.6 Reduction of Listeria monocytogenes populations in solution when exposed to nisin-adsorbed powders

monocytogenes on frankfurters; however, at least 81 °C for 60 s was required to attain a 50% reduction of initial populations. The pediocin/heat treatments did prevent *Listeria* growth on frankfurters for 12 weeks at 4 or 10 °C, and for 12 days at 25 °C. Combination of heat with wheat gluten antimicrobial films was also effective in reducing and preventing the outgrowth of *L. monocytogenes* on refrigerated turkey bologna (McCormick *et al.*, 2005). In this study, slices of turkey bologna were packaged in conventional polymer films with a thin layer of heat-extruded wheat gluten film in contact with the bologna surface (Fig. 18.7a). Bologna slices subjected to in-package pasteurization reduced *L. monocytogenes* populations of between 3.8 and 7.0 log cfu/g with the surviving populations fluctuating between 1.2 and 3.8 log cfu/g during 2 months of refrigerated storage (Fig. 18.7b). McCormick *et al.* (2005) also found that wheat-gluten/in-package pasteurization reduced *Salmonella typhimurium* on inoculated bologna by 5.7–7.3 log cfu/g, with the remaining populations decreasing to less than 100 cfu/g during 2 months under refrigeration. The combination of antimicrobial films with pasteurization was synergistic against *L. monocytogenes* but gave no added effect for *S. typhimurium.*

Nisin has shown synergistic antilisterial activity with heat in cultures (Modi *et al.*, 2000; Ueckert *et al.*, 1998), lobster meat (Budu-Amoako *et al.*, 1999), and liquid egg (Knight *et al.*, 1999). Budu-Amaoko *et al.* (1999) found that the addition of 25 mg/kg of nisin to lobster brine prior to in-package thermal processing at 60 °C for 5 min or 65 °C for 2 min resulted in 3 to 5 log reductions of *L. monocytogenes*, while the use of nisin or heat alone resulted in 1 to 3 log reductions. Chung and Hancock (2000) found an additional 1 log reduction of

(a)

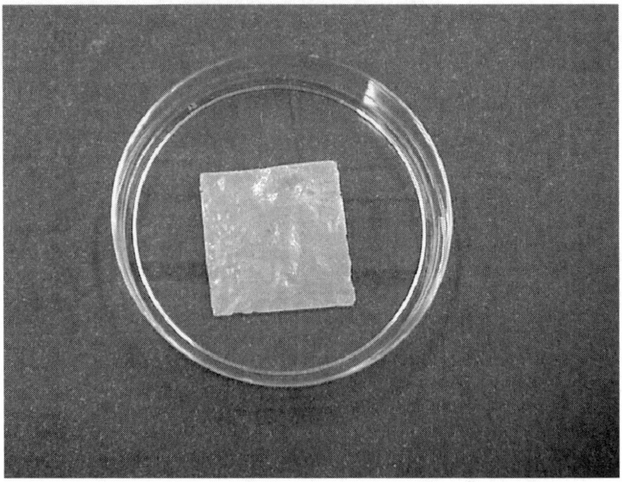

(b)

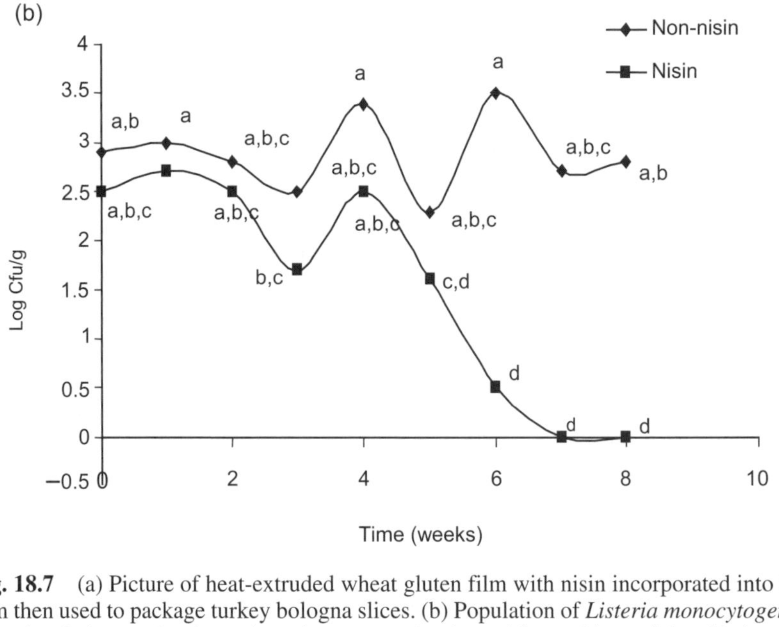

Fig. 18.7 (a) Picture of heat-extruded wheat gluten film with nisin incorporated into the film then used to package turkey bologna slices. (b) Population of *Listeria monocytogenes* on turkey bologna after in-package heat pasteurization, with or without nisin added to the wheat gluten food contact film, during 8 weeks of refrigerated storage.

L. monocytogenes during heating due to the addition of a nisin–lysozyme mixture, and attributed the added reduction to increased cell membrane damage or the inhibition of cell membrane repair by nisin–lysozyme. Mangalassary *et al.* (2007) evaluated nisin and lysozyme, singly and combined as heat sensitizers for *L. monocytogenes* for the in-package pasteurization of turkey bologna. No synergism for either antimicrobial was found at 60 °C; however, nisin and nisin–lysozyme reduced *L. monocytogenes* heat resistance at 62.5 °C, and lysozyme alone and nisin alone or with lysozyme enhanced the heat inactivation of *L. monocytogenes*.

18.6 High-pressure processing

An emerging process for RTE meats is high pressure processing (HPP) which kills bacteria with little or no change in meat quality. Pressure is often applied hydrostatically using a water-filled vessel surrounding the packaged product. This method applies pressure equally to all sides of the packaged product, which can temporarily reduce the package volume up to 15%. The package must be able to withstand this 15% volume change, after which the package will return to its original volume. For HPP, food products are usually packaged then placed in a high-pressure vessel after which the vessel is sealed and filled with water. The pressure is raised to a set point by pumping water into the sealed vessel, then the pressure is held constant for a certain amount of time. Pressures used for foods range from about 100 MPa (14 700 lb/in^2) to 1000 MPa (147 000 lb/in^2); while typical pressures for meat are in the 300–700 MPa range. Atmospheric pressure is about 1 kg/cm^2 or 14.7 lb/in^2 therefore100 MPa is equivalent to about 14 500 lb/in^2 or 1000 Xs atmospheric pressure and 1000 MPA is about 145 000 lb/in^2 or 10 000 Xs atmospheric pressure. Some of the capabilities of HPP are:

(i) bacterial inactivation
(ii) spore germination or inactivation
(iii) enzyme denaturation
(iv) meat marination

18.7 Future trends

18.7.1 Bacteriophages

Bacteriophages are essentially viruses that infect, replicate and kill bacteria; and upon killing the bacteria, release from 10 to 200 active bacteriophages that can infect other bacterial cells. These 'lytic' bacteriophages are very specific to the type of bacteria they infect which is good since they do not kill 'good' bacteria or other cells. Phages were discovered just before 1900 in India, but were not isolated until 1915. Bacteriophages were used to treat human skin infections in the 1920s and have been utilized to successfully treat bacterial infections for over 50 years in Russia. Bacteriophages were used in the US during the 1930s but fell out of favor with the advent of antibiotics. Antibiotics had the advantage (and disadvantage) of being effective against multiple types of bacteria and recently have been under scrutiny for the development of bacteria with resistance to multiple antibiotics.

The use of bacteriophages on ready-to-eat meat was approved by the US FDA in mid-2006. The combination of six viruses is designed to be sprayed onto ready-to-eat meat and poultry products, including sliced ham and turkey, according to a representative from the producer of the bacteriophage treatment, Intralytix Inc. The mixture of bacteriophages is designed to kill strains of *Listeria monocytogenes*.

18.7.2 Gas plasmas

Plasma is produced when gas between two electrodes is energized by a radio-frequency power source, similar to a fluorescent light. Atmospheric, non-thermal plasma (NTP) kills bacteria on surfaces or wherever the ions come into contact with bacterial or viral cells. NTP requires minimal power for operation and heating effects can be minimized such that the food experiences no temperature increase during treatment, thus providing a non-thermal treatment option. Numerous studies have explored NTP (Zhao *et al.*, 2005a,b,c,d; Jampala *et al.*, 2005; Johnson, 2004; Laroussi and Leipold, 2004; Laroussi, 2002; Gao *et al.*, 2001). Over the last several years, researchers investigating the efficacy of non-thermal atmospheric pressure plasma have demonstrated its inhibitory activity against several bacterial pathogens and other organisms including *E. coli, Listeria monocytogenes, Bacillus subtilis*, and *Salmonella* spp. (Shore *et al.*, 2005; Oldham *et al.*, 2004; Maeda *et al.*, 2004; Trompeter *et al.*, 2002; Birmingham *et al.*, 2000). Thus, NTP may be an alternative to existing technologies, such as dry heat, steam, high pressure, chemical/bio-preservatives, and gamma irradiation, which are the predominant processing technologies used for RTE meat preservation.

18.8 Sources of further information and advice

Antimicrobials in Foods. 2nd edition. Davidson, P.M. and Branen, A.L., editors, 1993. Marcel Dekker, Inc., New York, NY.

Edible Coatings and Films to Improve Food Quality. 1994. J.M. Krochta, E.A. Baldwin, and M.O. Nisperos-Carreido, editors. Technomic Publishing, Lancaster, PA.

Protein-based Films and Coatings. 2002. A. Gennadios, editor. Technomic Publishing, Lancaster, PA.

Food Biopreservatives of Microbial Origin. Ray B. and Daeschel, M., 1992. CRC Press, Inc. Boca Raton, FL.

Natural Food Antimicrobial Systems. Naidu, A.S., editor, 2000. CRC Press. Boca Raton, FL 33431.

Natural Antimicrobial Systems and Food Preservation. Dillon V.M and Board R.G. editors, 1994. CAB International. Wallingford, UK.

Thermal Food Processing: Modeling, Quality Assurance, and Innovations. Da-Wen Sun, editor. Marcel Dekker, NY.

18.9 References

Amezquita, A. and Brashears, M.M. 2002. Competitive inhibition of *Listeria monocytogenes* in ready-to-eat meat products by lactic acid bacteria. *Journal of Food Protection* 65(2), 316–325.

Ananth, V., Murano, E., and Dickson, J.S. 1995. *Shelf-life extension and safety of fresh pork treated by high hydrostatic pressure*, Pittsburgh, PA: International Association of Milk, Food and Environmental Sanitarians. Abstract.

Berry, E.D., Hutkins, R.W., and Mandingo, R.W. 1991. The use of bacteriocin-producing *Pediococcus acidilactici* to control postprocessing *Listeria monocytogenes* contamination of frankfurters. *J. Food Protection* 54, 681–686.

Birmingham, J.G. and Hammerstrom, D.J. 2000. Bacterial decontamination using ambient pressure nonthermal discharges. *IEEE Transactions on Plasma Science* 28, 51–55.

Blom, H., Nerbrink, E., Dainty, R., Hagvedt, T., Borch, E., and Nissen, H. 1997. Addition of 2.5% lactate and 0.25% acetate controls growth of *Listeria monocytogenes* in vacuum-packed, sensory-acceptable servelat sausage and cooked ham stored at 4 °C. *Int. J. Food Microbiology* 82, 567–577.

Budo-Amoako, E., Ablett, R.F., Harris, J., and Delves-Broughton, J. 1999. Combined effect of nisin and moderate heat on destruction of *Listeria monocytogenes* in cold-pack lobster meat. *J. Food Prot.* 62, 46–50.

Buonocore, G.G., A. Conte, M.R. Corbo, M. Sinigaglia, and M.A. Del Nobile. 2005. Mono- and multi-layer active films containing lysozyme as antimicrobial agent. *Innovative Food Science and Emerging Technologies*. 6, 459–464.

Carlez, A., Cheftel, J-P., Rosec, J.P., Richard, N., Saldana, J.L., and Balny, C. 1991. Effects of high pressure and bacteriostatic agents on destruction of *Citrobacter freundii* in minced beef muscle. In *High Pressure and Biotechnology*, C. Balny, K. Hayashi, and P. Masson, editors. Vol. 224, pp. 365–368. John Libbey Eurotext, Ltd., Montrouge, France.

Chen, C.M., Sebranek, J.G., Dickson, J.S., and Mendonca, A.F. 2004. Combining pediocin with post-packaging thermal pasteurization for control of *Listeria monocytogenes* on frankfurters. *Journal of Food Protection* 67, 1855–1865.

Chung, W., and Hancock, R.E. 2000. Action of lysozyme and nisin mixtures against lactic acid bacteria. *International Journal of Food Microbiology* 60, 25–32.

Cole, M.B., K.W. Davies, G. Munro, C.D. Holyoak, and D.C. Kilsby. 1993. A vitalistic model to describe the thermal inactivation of *Listeria monocytogenes*. *Journal of Industrial Microbiology* 12, 232–239.

Cooksey, D.K., Klein, H.P., McKeith, F.K., and Blaschek, A.A 1993. Reduction of *Listeria monocytogenes* in pre-cooked vacuum-packaged beef using post packaging pasteurization. *Journal of Food Protection* 56, 1034–1038.

Cutter, C.N., Willett, J.L., and Siragusa, G.R. 2001. Improved antimicrobial activity of nisin-incorporated polymer films by formulation change and addition of food grade chelator. *Letters in Applied Microbiology*, 33(4), 325–328.

Dawson, P.L., J.C. Acton, I.Y. Han. T. Padgett, R. Orr, and T. Larsen. 1995. *Incorporation of antibacterial compounds into edible and biodegradable packaging films*. Research and Development Activities Report for Military Food and Packaging Systems 47(2), 203–210.

Dawson, P.L., Carl, G.D., Acton, J.C., and Han, I.Y. 2002a. Effect of lauric acid and nisin-impregnated soy-based films on the growth of *Listeria monocytogenes* on turkey bologna. *Poultry Science* 81, 721–726.

Dawson, P. L., G.D. Carl, J.C. Acton, and I.Y. Han. 2002b. Effect of lauric acid and nisin-impregnated soy-based films on the growth of *Listeria monocytogenes* on turkey bologna. Poultry Sci. 81, 721–726.

Dawson, P.L., Hirt, D.E., Rieck, J.R., Acton, J.C., and Sotthibandhu, A. 2003. Nisin release from films is affected by both protein type and film-forming method. *Food Research International*. 36(9–10), 959–968.

Dawson, P.L., Harmon, L., Sotthibandhu, A., and Han, I.Y. 2004. Antimicrobial activity of nisin adsorbed silica and corn starch powders. *Food Microbiology* 22, 93–99.

Dykes, G.A., Marshall, L.A., Meissner, D., and von Holy, A. 1996. Acid treatment and pasteurization affect the shelf life and spoilage ecology of vacuum-packaged Vienna sausages. *Food Microbiology* (1), 69–74.

Embarek, P.K. 1994. Presence, detection and growth of *Listeria monocytogenes* in seafoods: A review. *International J. Food Microbiology* 23, 17–34.

Fabrizio, K. A., Sharma, R. R., Demirci, A., and N. 2002. Comparison of electrolyzed oxidizing water with various antimicrobial interventions to reduce *Salmonella* species on poultry. *Poultry Science* 81(10), 1598–1605.

Fabrizio, K. A. and Cutter, C.N. 2005. Application of electrolyzed oxidizing water to reduce *Listeria monocytogenes* on ready-to-eat meats. *Meat Science* 71(2), 327–333.

Franz, C.M., and von Holy, A. 1996a. Bacterial populations associated with pasteurized vacuum-packaged Vienna sausages. *Food Microbiology* 13, 165–174.

Franz, C.M., and von Holy, A. 1996b. Thermotolerance of meat spoilage lactic acid bacteria and their inactivation in vacuum-packaged Vienna sausages. *International Journal of Food Microbiology* 29, 59–73.

Gande, N., and Muriana, P. 2003. Prepackage surface pasteurization of ready-to-eat meats with a radiant heat oven for reduction of *Listeria monocytogenes. J. Food Prot.* 66(9), 1623–1630.

Gao, C.H., Herald, T.J., and Muino, P.L. 2001. Application of a plasma reactor to modify egg ovalbumin for improved solubility. *J. Food Science* 66(1), 89–94.

Geornaras, I., Bek, K., Scanga, J., Kendall, P., Smith, G., and Sofos, J. 2005. Post-processing antimicrobial treatments to control *Listeria monocytogenes* in commercial vacuum-packaged bologna and ham stored at 10 °C. *Journal of Food Protection* 68(5), 991–998.

Hardin, M.D., Williams, S.E., and Harrison, M.A. 1993. Survival of *Listeria monocytogenes* in post pasteurized precooked beef roasts. *Journal of Food Protection* 56, 655–660.

Hoffman, K.L., Dawson, P.L., J.C. Acton, I.Y. Han. and A.A. Ogale 1997. *Film formation effects on nisin activity in corn zein and polyethylene films.* Research and Development Activities Report for Military Food and Packaging Systems 47(2), 203–210.

Hoffman, K.L., Han, I.Y., and Dawson, P.L. 2001. Antimicrobial effects of corn zein films impregnated with nisin, lauric acid, and EDTA. *Journal of Food Protection* 64, 885–889.

Hoover, D.G., Metrick, C., Papineau, A.M., Farkas, D.F., and Knorr, D. 1989. Biologicical effects of high hydrostatic pressure on food microorganisms. *Food Technology* 3, 99–106.

Hugas, M., Garriga, M., Aymerich, .T. and Monfort, B. 1995. Inhibition of *Listeria* in dry fermented sausages by the bacteriocino-genic *Lactobacillus sake* CTC494. *Journal of Applied Bacteriology* 79, 322–330.

Ingham, S.C., D'Evita, M.D., Wadhera, R.K., Fanslau, M.A., and Buege, D.R. 2005. Evaluation of small-scale hot-water post-packaging pasteurization treatments for destruction of *Listeria monocytogenes* on ready-to-eat beef snack sticks and natural casing wieners. *Journal of Food Protection* 68, 2059–2067.

Jampala, S.N., Manolache, S., Gunasekaran, S., and Denes, F.S. 2005. Plasma-enhanced modification of xanthan gum and its effect on rheological properties. *J. Agricultural Food Chemistry* 53(9), 3618–3625.

Johnson, F.M. 2004. *Atmospheric plasma inactivation of foodborne pathogens on fresh produce surfaces.* Masters Thesis, The University of Tennessee, Knoxville.

Juncher, D., Vastergaard, C.S., Soltolft-Jensen, J., Weber, C.J., Bertelsen, G., and Skibsted, L.H. 2000. Effect of chemical hurdles on microbiological and oxidative stability of cooked cured emulsion type meat product. *Meat Science* 55, 483–491.

Kassaify, Z.G. 1998. *The potential use of nisin on fresh sausages and onto sausage casings.* M. Sc. Thesis. Oxford University, Oxford, UK.

Knight, K.P., Bartlett, F.M., McKellar, R.C., and Harris, L.J. 1999. Nisin reduces the thermal resistance of *Listeria monocytogenes* Scott A in liquid whole egg. *Journal of Food Protection* 62, 999–1003.

Ko, S., Janes, M.E., Heittiarachchy, N.S., and Johnson, M.G. 2001. Physical and chemical properties of edible films containing nisin and their action against *Listeria monocytogenes. Journal of Food Science* 66, 1006–1011.

Laroussi, M. 2002. Nonthermal decontamination of biological media by atmospheric-pressure plasmas: Review, analysis, and prospects, *IEEE Transactions on Plasma Science.* 30(4), 1409–1415.

Laroussi, M. and F. Leipold. 2004. Evaluation of the roles of reactive species, heat, and UV radiation in the inactivation of bacterial cells by air plasmas at atmospheric pressure, *International Journal of Mass Spectrometry*. 233, 81–86.

Limjaroen, P., Ryser, E., Lockhart, H., and Harte, B. 2003. Development of a food packaging coating material with antimicrobial properties. *Journal of Plastic Film and Sheeting* 19(2), 95–09.

Maeda, Y., N. Igura, M. Shimoda, and I. Hayakawa. 2003. Bacterial effect of atmospheric gas plasma on *Escherichia coli* K12, *International journal of Food Science and Technology*. 38, 889–892.

Mafart, P., Couvert, O., Gaillard, S., and Leguerinel, I. 2002. On calculating sterility in thermal preservation methods: Application of the Weibull frequency distribution model. *International Journal of Food Microbiology* 72, 107–113.

Mangalassary, S., Dawson, P.L. Rieck, J.R., and Han, I.Y. 2004. Thickness and compositional effects on surface heating rate of bologna during in-package pasteurization. *Poultry Science* 83, 1456–1461.

Mangalassary, S., Han, I.Y., Rieck, J.R., Acton, J.C., Jiang, X., Sheldon, B., and Dawson, P.L. 2007. Effect of combining nisin and/or lysozyme with in-package pasteurization on thermal inactivation of *Listeria monocytogenes* in ready-to-eat turkey bologna. *Journal of Food Protection*. In press.

McCormick, K., Han, I.Y., Sheldon, B.W., Acton, J.C., and Dawson, P.L. 2003. D- and Z-values for *Listeria monocytogenes and Salmonella* Typhimurium in packaged low-fat ready-to-eat turkey bologna subjected to a surface pasteurization treatment. *Poultry Science* 82, 1337–1342.

McCormick, K.E., Han, I.Y., Acton, J.C., Sheldon, B.W., and Dawson, P.L. 2005. In-package pasteurization combined with biocide-impregnated films to inhibit *Listeria monocytogenes* and *Salmonella* Typhimurium in turkey bologna. *Journal of Food Science* 70(1), M52–7.

Ming, X., Weber, G.H., Ayers, J.W., and Sandine, W.E. 1997. Bacteriocins applied to food packaging materials to inhibit *Listeria monocytogenes* on meats. *Journal of Food Science* 62, 413–415.

Modi, K.D., Chikindas, M.L., and Montville, T.J. 2000. Sensitivity of nisin-resistant *Listeria monocytogenes* to heat and the synergistic action of heat and nisin. *Letters in Applied Microbiology* 30(3), 249–253.

Murano, E.A., Murano, P.S., Brennan, R.E., Shenoy, K., and Moreira, R.G. 1999. Application of high hydrostatic pressure to eliminate *Listeria monocytogenes* from fresh pork sausage. *Journal of Food Protection* 62, 480–483.

Muriana, P.M., Quimby, W., Davidson, C.A., and Grooms, J. 2002. Post-package pasteurization of RTE deli meats by submersion heating for reduction of *Listeria monocytogenes*. *J. Food Prot*. 65, 963–969.

Murphy, R.Y., and Berrang, M.E. 2002. Effect of steam and hot water post-process pasteurization on microbial and physical property measures of fully-cooked vacuum-packaged chicken breast strips. *Journal of Food Science* 67, 2325–2329.

Murphy, R.Y., Duncan, L.K., Driscoll, K.H., and Marcy, J.A. 2003a. Lethality of *Salmonella* and *Listeria innocua* in fully cooked chicken breast meat products during post-cook in-package pasteurization. *Journal of Food Science* 66, 242–248.

Murphy, R.Y., Duncan, L.K., Driscoll, K.H., Marcy, J.A., and Beard, B.L. 2003b. Thermal inactivation of *Listeria monocytogenes* on ready-to-eat turkey breast meat products during post-cook in-package pasteurization with hot water. *Journal of Food Protection* 66, 1618–1622.

Murphy, R.Y., Osaili, T., Duncan, L.K., and Marcy, J.A. 2004. Effect of sodium lactate on thermal inactivation of *Listeria monocytogenes* and *Salmonella* in ground chicken thigh and leg meat. *Journal of Food Protection* 67, 1403–1407.

Natrajan, N., and Sheldon, B.W. 2000a. Efficacy of nisin-coated polymer films to inactivate *Salmonella* Typhimurium on fresh broiler skin. *Journal of Food Protection* 63, 1189–1196.

Natrajan, N., and Sheldon, B.W. 2000b. Inhibition of *Salmonella* on fresh poultry skin using protein- and polysaccharide-based films containing nisin formulations. *Journal of Food Protection* 63, 1268–1272.

Nilsson, L., Huss, H., and Gram, L. 1997. Inhibition of *Listeria monocytogenes* on cold-smoked salmon by nisin and carbon dioxide atmosphere. *International Journal of Food Microbiology* 38, 217–227.

Nobile, M.A. del., Cannarsi, M., Altieri, C., Sinigaglia, M., Favia, P., Iacoviello, G., and D'Agostino, R. 2004. Effect of Ag-containing nano-composite active packaging system on survival of *Alicyclobacillus acidoterrestris*. *Journal of Food Science* 69(8), E379–E383.

O'Brien, J.K., and Marshall, R.T. 1996. Microbiological quality of raw ground chicken processed at high isostatic pressure. *Journal of Food Protection* 59, 146–150.

Oldham, C.J., S.A. Shore, B.A. Hartman, X.Li, B.W. Sheldon, K.M. Keener, and J.J. Cuomo. 2004. North Carolina State University, unpublished work.

Oliveira, L.M. de, and Oliveira, P.A. de. 2004. Main antimicrobial agents used in plastic packaging. *Brazilian Journal of Food Technology* 7(2), 161–165.

Orr, R.V., I.Y. Han, J.C. Acton, and P.L. Dawson. 1997. Comparison of the zone of inhibition and enumeration assay for evaluating the effectiveness of antimicrobial packaging films. In: *Proceedings of the 42nd International Congress of Meat Science and Technology*. American Meat Science Association, pp 26–28.

Padgett, T., Han, I.Y., and Dawson, P.L. 1998. Incorporation of food-grade antimicrobial compounds into biodegradable packaging films. *Journal of Food Protection* 61, 1330–1335.

Palumbo, S.A. and Williams, A.C. 1994. Control of *Listeria monocytogenes* on the surface of frankfurters by acid treatments. *Food Microbiology* 11, 293–300.

Park, S.I., Marsh, K.S., Dawson, P.L., Axton, J.C. and Han, I. 2002. Antimicrobial activity of chitosan incorporated into polyethylene films. *Institute of Food Technologists Annual Meeting*, June 15–19.

Park, S.I. 2003. *Antimicrobial Effect of Films Developed from Chitosan*. Ph.D. dissertation. Clemson University, Clemson, SC.

Park, S.I., Daeschel, M.A., and Zhao, Y. 2004. Functional properties of antimicrobial lysozyme–chitosan composite films. *Journal of Food Science* 69(8), M215–M221.

Peleg, M., and M.B. Cole. 1998. Reinterpretation of microbial survival curves. *Critical Reviews in Food Science and Nutrition* 38, 353–380.

Petrikova, J., Ruzickova, A., Janyi, I., Korenova, J., Simko, P.l and Khunva, V. 2003. Influence of the packaging material on the growth of microorganisms in foodstuff. *Agriculutre Pol'nohospodarstvo* 49(8), 412–421.

Pol, H., P. Dawson, J. Acton, and A. Ogale. 2002. Soy protein isolate/corn zein laminated films: Transport and mechanical properties. *Journal of Food Science.* 67, 212–217.

Qvist, S., Sehested, K., and Zeuthen, P. 1994. Growth suppression of *Listeria monocytogenes* in a meat product. *International Journal of Food Microbiology* 24, 283–293.

Roering, A.M., Wierzba, R.K., Ihnot, A.M., and Luchansky, J.B. 1998. Pasteurization of vacuum-sealed packages of summer sausage inoculated with *Listeria monocytogenes*. *Journal of Food Safety* 18, 49–56.

Samelis, J., Sofos, J. N., Kain, M. L., Scanga, J.A., Belk, K. E., and Smith, G. C. 2001. Organic acids and their salts as dipping solutions to control *Listeria monocytogenes* inoculated following processing of sliced pork bologna stored at 4 °C in vacuum packages. *Journal of Food Protection* 64(11), 1722–1729.

Scannell, A.G., Hill, C., Buckley, D., and Arendt, E.K. 1997. Determination of the influence of organic acids and nisin on shelf-life and microbiological safety aspects of fresh pork sausage. *Journal of Applied Microbiology* 83, 407–412.

Selby, T.L., Berzins, A., Gerrard, D.E., Corvalan, C.M., Grant, A.L., and Linton, R.H. 2006. Microbial heat resistance of *Listeria monocytogenes* and the impact on ready-to-eat meat quality after post-package pasteurization. *Meat Science* 74, 425–434.

Shore, S.A., K.M. Keener B.W. Sheldon, and J.J.Cuomo. 2005. Atmospheric plasma treatment to eliminate food pathogens. Undergraduate Poster Session. *ASAE Annual International Meeting*. July 17–20. Tampa, FL.

Teerakarn, A., D.E. Hirt, J.R. Rieck, J.C. Acton, and P.L. Dawson. 2002. Nisin diffusion in protein films: Effects of time and temperature. *J. Food Science* 67, 3019–3025.

Teratanonat, R. and N. Hooker. 2004. Understanding the characteristics of US meat and Poultry recalls:1994–2002, *Food Control* 15, 359–367.

Trompeter, F., W.J. Neff, O. Franken, M. Heise, M. Neiger, S. Liu, G.J. Pietsch, A.B. Saveljew. 2002. Reduction of *Bacillus Subtilis* and *Aspergillus Niger* Spores Using Nonthermal Atmospheric Gas Discharges, *IEEE Transactions on Plasma Science.* 30, 1416–1423.

Ueckert, J.E., ter Steeg, P.F., and Coote, P.J. 1998. Synergistic antibacterial action of heat in combination with nisin and magainin II amide. *Journal of Applied Microbiology* 85, 487–494.

USDA, 2005. *Recalls*. Available: http://www.fsis.usda.gov/oa/recalls/rec_intr.htm. Accessed: March 28, 2005.

Van Boekel, M.A. 2002. On the use of Weibull model to describe thermal inactivation of microbial vegetative cells. *International Journal of Food Microbiology* 74, 139–159.

Virto, R., Sanz, D., Alvarez, I., Condon, S., and Raso, J. 2006. Application of the Weibull model to describe inactivation of *Listeria monocytogenes* and *Escherichia coli* by citric acid and lactic acid at different temperatures. *Journal of the Science of Food and Agriculture* 86, 865–870.

Weaver, R.A., and Shelef, L.A. 1993. Antilisterial activity of sodium, potassium, or calcium lactate in pork liver sausage. *Journal of Food Safety* 13, 133–146.

Wederquist, H.J., Sofos, J.N., and Schmidt, G.R. 1994. *Listeria monocytogenes* inhibition in refrigerated vacuum packaged turkey bologna by chemical additives. *Journal of Food Science* 59, 498–500, 516.

Winkowski, K., Crandall, A.D., and Montville, T.J. 1993. Inhibition of *Listeria monocytogenes* and *Lactobacillus bavaricus* MN in beef systems at refrigeration temperatures. *Applied and Environmental Microbiology* 59, 2552–2557.

Zhao, G.B., Hu, X., Argyle, M.D., Radosz., M. 2005a. Effect of CO_2 on Nonthermal Plasma Reactions of Nitrogen Oxides in N_2. 1. PPM-level Concentrations, *Industrial & Engineering Chemistry Research*, 44, 3925–3934.

Zhao, G.B., Hu, X., Argyle, M.D., Radosz., M. 2005b.Effect of CO_2 on Nonthermal Plasma Reactions of Nitrogen Oxides in N_2. 2. Percent-level Concentrations, *Industrial & Engineering Chemistry Research*, 44, 3935–3946.

Zhao, G.B., Garikipati, S.V.B J., Hu, X., Argyle, M.D., Radosz., M. 2005c.The Effect of Oxygen Addition on Nonthermal Plasma Reactions of Dilute Nitrogen Oxide Mixtures in N_2. *AIChE Journal*, 51, 1813–1821.

Zhao, G.B., Garikipati, S.V.B J., Hu, X., Argyle, M.D., Radosz., M. 2005d. The Effect of Reactor Configuration on NO Conversion and Energy Consumption in Non thermal Plasma Parallel Tube Reactors. *AIChE Journal* 51, 1800–1812.

19

Novel methods to optimise the nutritional and sensory quality of in-pack processed fish products

D. Skipnes, Norconserv AS, Norway and M. Hendrickx, Katholieke Universiteit Leuven, Belgium

19.1 Introduction: the range of in-pack thermally-processed fish products

Preservation by heat remains one of the major methods for extending the shelf-life of packed fish, and will always be the first choice for products which are to be cooked, before purchase by the end consumer. In-pack thermally processed fish products have been produced for almost two centuries. While Nicolas Appert started the first cannery for meats and vegetables in 1803, the first production of canned sardines took place in France in 1830. As one of the results of the ongoing globalization process, the exported volume of canned seafood products has almost tripled over the last 30 years. Although Thailand remains the largest producer, the largest volume increase in exports over the last 10 years is seen in China and Ecuador, whereas the largest volume increase in imports is seen in Spain, Italy and the USA. The export of canned tuna still constitutes a major category, but shrimp and prawns represent the largest export value. A number of products have achieved a specific status, not only as a canned, preserved version of a fresh product, but something special in itself. Examples include canned tuna, canned kippered herring and canned brisling sardines in olive oil (Fig. 19.1).

Healthy eating trends have given rise to increased demand for fish and seafood, which is acknowledged to offer many health benefits. In addition, heart health has become a major concern among modern consumers, with heart disease being

Fig. 19.1 A traditional sterilised product: Sardines in olive oil packed in a ¼ Dingley can and wrapped in decorated cellophane (Photo: King Oscar AS).

responsible for a large number of deaths in the developed world. The benefits of fish oil over animal fats is becoming more widely recognised among consumers and this has also led to a shift away from red meat towards white meat and more importantly fish and fish products.

According to Euromonitor International, the ready-meals market in Western Europe was worth €19.9bn in 2006 and will rise to €21.8bn by 2010. The European market for chilled ready-meals amounted to €1.775m in 2000, with the UK having over 70% of sales. The most important factor driving growth in the ready-meals market is the increasing demand for convenience. Many consumers do not like to cook fish themselves, even if they like the fish and believe it is a nutritional alternative to meat products. Several reasons may be mentioned for this and some of them are the smell during cooking, bones in the fish and fish sticking to the pan. Another problem may be the relatively short shelf-life of fresh fish, which means it has to be cooked shortly after purchase. An example of a ready-meal product with fish as a main ingredient is shown in Fig. 19.2.

19.1.1 Parameters affecting quality and limitations of traditional methods for producing in-pack processed fish products

Thermal processing can be subdivided into several more or less overlapping groups, based on temperature regime, method or equipment for thermal processing, fish species, packaging method or the microbial target of the process. In Table 19.1, the methods are classified by the heat load on the product, which is directly related to the microbial lethality achieved.

Sterilisation is the classical method. The products are undergoing a process aiming for inactivation of all pathogenic bacteria and their spores. The temperature regime during processing may vary from 110 °C to 135 °C. For low acid foods, the

Fig. 19.2 A 'sous vide' ready meal available in the Norwegian retail market. The cardboard box contains three plastic pouches, one with salmon in sauce, one with potatoes and one with carrots, which can be reheated in a kettle with boiling water while still in the plastic pouch (Photo: Lisa Westgaard by courtesy of Fjordland AS).

Table 19.1 Classification of thermally processed products containing fish sorted by heat load (pH > 4.5, a_w < 0.85)

Level of processing	Effect on micro-organisms	Reference micro-organism	Heat load
Commercial sterilisation	Inactivation of all pathogen spore formers	*Clostridium botulinum* A/B	F_0 >3.0min
Pasteurisation	Inactivation of all vegetative cells Inactivation of pathogen spore formers with ability to grow above 10 °C	Non-proteolytic *Clostridium botulinum* B/E *Bacillus cereus*	$P_{90°C}^{10°C}$ > 10 min (T > 90 °C) $P_{90°C}^{7.5°C}$ > 10 min ($T \leq$ 90 °C) $P_{100°C}^{9°C}$ > 48 min
Mild pasteurisation	Inactivation of all vegetative cells, survival of spores	*Enterococcus faecalis*	$P_{70°C}^{10°C}$ 17.5 min
Minimal processing	Inactivation of some pathogens, survival of others	*Listeria monocytogenes*	$P_{90°C}^{7.5°C}$ > 2 min

process is aimed to inactivate the spores of *Clostridium botulinum* type A. This is sometimes referred to as commercial sterility, as some spore-forming non-pathogenic strains may survive this heat load. Other sub-groups of sterilisation are also in use but are not presented here.

Pasteurisation is intended to inactivate vegetative cells but is not intended to inactivate the spores of all pathogenic bacteria. The term is often related to the heat treatment of acid foods or refrigerated foods where growth of surviving spores is prevented by a pH below 4.5, a low temperature, or by other means. A variant of pasteurisation is sous vide processing, i.e. mild thermal processing of vacuum-packed products.

19.1.2 Microbial constraints and legislative aspects

The quality of thermally processed products is highly dependent on the heat load, which is determined by the requirements for inactivation of microorganisms. Legislation on inactivation of pathogenic microorganisms often includes a safety margin which may lead to over-processing of some goods. In the USA, the Food and Drug Administration (FDA) previously had an restrictive view regarding pasteurized products (Rhodehamel, 1992) and it has been common to distribute pasteurised fish products frozen.

Safety of hermetically packed and thermally processed foods has been an issue since the invention of the technology, and detailed legislation is in use. National legislation is often very different from country to country, but there are also several things in common. For commercially sterilised products, the rules have been more focused on the safety of the end product. Sterilised foods have been used world-wide for several decades and one of the requirements most countries have in common is that the least sterilising value F_0 should be 3.0 for a low acid canned food. This general requirement has also been applied to fish. Internationally recognised guidelines are published by the United Nations Codex Alimentarius Commission for canned foods (Anon., 2001). In most legislation for hermetically packed, heat-preserved foods, the following topics (in addition to more general issues such as hygiene) are of major concern:

- Determining of a safe heating procedure, i.e. requirements for sterilisation or pasteurisation values
- Achieving the required sterilisation/pasteurisation, i.e. heat penetration tests or other measuring techniques, and determining a scheduled heating process (e.g. sterilising time and temperature)
- Reproducing a scheduled process, i.e. control of heat distribution and constant heat transfer conditions in the product
- Validating procedures and equipment (at least calibration of thermometers) and record keeping
- Controlling end product. For sterilised products this includes incubation and microbial sample testing.
- Ensuring integrity of packaging

European Guidelines for canning have been published by the Campden and Chorleywood Food Research Association (CCFRA) (May, 1997). Volunteer organisations, such as the National Food Processors Association (NFPA) and the Institute for Thermal Processing Specialists (IFTPS) have issued several publications which are widely used (IFTPS, 1992, 1995, 2002). All these guidelines describe how to perform heat penetration tests in canned products and heat distribution tests in autoclaves. These are issues also important to milder heat preservation techniques, and several elements can be transferred from the canning guidelines. The guidelines for heat penetration tests of canned foods could be successfully used for almost any heating regime for packed foods.

During the development of any new heat-treated product, it is essential to assess the combined effects of the total system, consisting of heat process, preservatives, packaging and storage conditions, in order to ensure that the product is of good microbiological quality and does not present any food safety hazard.

Several reviews have focused on heat resistance of spores (Gerhardt, 1988;Gerhardt and Marquis, 1989; Gombas, 1983; Setlow and Johnson, 1997). Spores formed by the genera *Bacillus, Clostridium, Desulfotomaculum* and *Sporolactobacillus* are hot topics in food microbiology. Several guidelines and codes of practice have been published with respect to safe production of refrigerated processed foods of extended durability (REPFEDs) (ACMSF, 1992, 1995; Betts, 1996; ECFF, 1996). Most of these are targeted at preventing growth and toxin production by non-proteolytic *C. botulinum*. A general recommendation in the guidelines mentioned above is that the heat treatments or combination of processes utilized, should reduce the number of viable spores of non-proteolytic *C. botulinum* by a factor of 10^6 (6D). Accordingly, a minimum heat treatment of 90 °C for 10 min or equivalent lethality in the slowest heating point of the product has been recommended by ACMSF (1992, 1995). This is based on a D_{90} of 1.6 min and a z-value of 7.5 °C when the temperature in the product is below 90 °C, and a z-value of 10 °C at higher temperatures.

Listeria monocytogenes is a Gram-positive bacterium, mobile by means of flagella. As has been mentioned, the 6D concept is also applicable for *Listeria monocytogenes*. Accordingly, a minimum heat treatment of 70 °C for 2 min or equivalent lethality in the slowest heating point of the product has been recommended by ACMSF (1992, 1995). This is based on a D_{70} of 0.33 min and a z-value of 7.5 °C. There is, however, a wide range of kinetic data reported for *L. monocytogenes* depending on the strain and the model system used for determining the heat resistance (Ben Embarek and Huss, 1993). For cod, Ben Embarek and Huss (1993), investigated the heat resistance of *L. monocytogenes* O62 and found a D_{70} of 0.03 min and a z-value of 5.7, while he found a D_{70} of 0.05 min and a z-value of 6.1 for *L. monocytogenes* O57. This indicates that a 6 log inactivation would require a pasteurisation value P_{70} in the range of 0.18 to 0.30 min, but further studies of heat resistance of *L. monocytogenes* in fish are needed to draw a conclusion. In recent years, the commonly used first-order inactivation models described above have been challenged by more sophisticated non-log-linear modelling (Peleg, 2006). This might also give opportunities for more accurate optimisation of thermal processes in the future if this can be adopted in legislation and standards, and more detailed knowledge is gained.

In conclusion, two levels of heat load can be used as thresholds for refrigerated vacuum packed fish preserved by heat:

(i) A mild thermal process designed to inactivate *Listeria monocytogenes*: $P_{70}^{7.5} > 2$ min resulting in a shelf-life of 10 days max.
(ii) A thermal process designed to inactivate spores of non-proteolytic *Clostridium botulinum*: $P_{90}^{7.5} > 10$ min, resulting in a shelf-life of 21 days max.

19.1.3 Changes during heat processing

Aitken and Connell (1979) reviewed the effects of heating on fish and reported that the cooking losses varied greatly with the fish species, the method of heating, and the heating regime (i.e. sterilisation, pasteurisation, etc.). There are, however, few publications discussing the effects of heat on fish, the differences between thermally processed farmed and wild fish and the effects of filleting pre- and post-rigor on the quality of the heat-processed product. To the consumer, the most obvious quality parameters are flaking (whether the fish is falling apart on the plate or not), tenderness and juiciness. From this perspective, the two main issues are the heat denaturation of proteins and the water-holding capacity (WHC).

Heating converts the translucent, jelly-like cellular mass into an opaque, friable, slightly firm and springy form. The muscle is shrinking during heating, resulting in release of liquid. The proteins in this liquid may coagulate on the surface of the solid fish as a curd. The connective tissue holding the cells together is easily damaged and thus, cooked fish easily falls apart and becomes palatable on mild heating (Parry, 1970). Post-rigor processed fish does not fall apart as easily as pre-rigor processed, but it is more likely to break across the myomera. Even at temperatures as low as 37 °C, the tensile strength is reduced to zero for codfish after 30 minutes (Aitken and Connell, 1979), and a visible softening of the connective tissue occurrs after 15 min at 35 °C. Below this, no temperature effects were found by Aitken and Connell. However, Howgate and Ahmed (1972) showed that thermal effects on proteins were the most important aspects for texture in their study of drying of cod and hilsa at 30 °C.

Approximately 95% of the water in muscle is mechanically immobilised water, i.e. held by capillary and surface tension forces and is often referred to as 'free water'. This water is not tightly associated to proteins, e.g. by hydrogen bonds, and is free to migrate throughout the muscle structure when the muscle is subjected to mechanical forces. The fat content of cod muscle is in the order of 0.3%. It has been reported that the average total liquid loss for non-fatty fish species is 18.6% (Aitken and Connell, 1979). Thus, the liquid loss must consist mainly of water and dissolved proteins. The nuance between water and liquid-holding capacity has therefore no practical relevance for non-fatty fish species and all liquid removed can be regarded as water.

It is generally accepted that the forces that immobilise 'free water' within the muscle are generated by surface tension (Hamm, 1986). More specifically, the water is trapped within the muscle by capillary action generated by small pores or 'capillaries' (Trout, 1988). The pores producing the capillary forces are located between myosin and actin, and measure up to approximately 10 nm under normal conditions (Hermansson, 1983). Changes in the volume of the myofibrils will induce changes in water held by the muscle. In raw meat, absorption of water occurs by the entry of water into the myofibrils. Conversely, loss of water occurs by expulsion of water from the myofibrils as they shrink when filaments approach. WHC depends on heat-induced structural changes, sarcomere length, pH, ionic strength, osmotic pressure and state of *rigor mortis* (Ofstad *et al.*, 1996a).

The WHC of cod during heating has been studied by Ofstad *et al.* (1993), who showed that the main structural changes occur in the connective tissue at low temperatures (<40 °C) for both cod and salmon, and they concluded that water loss at these temperatures is mainly due to denaturation and melting of collagen. These authors used a heating rate of 1 °C/min to the required temperature and a 10 min holding time. The solid material was placed on a net in a sample holder and centrifuged at 210 × g for 15 min. It was found that maximum water loss was attained when the muscle cell had shrunk due to denaturation of myosin and the extra-cellular spaces were widened. The centrifugation loss decreased as a function of temperature when extra-cellular, granulated material became visible. Ofstad *et al.* (1993) found the centrifugation loss to be at maximum between 40 °C and 60 °C, while it was lower at temperatures above 60 °C.

The properties of comminuted fish are affected both by the structure of the biological tissue and the new structures formed during processing (Ofstad *et al.*, 1996b). However, as the changes in microstructure showed a close relation to WHC of the tissue during heating for coarsely chopped muscle, it was claimed that this would also be the case for comminuted fish. As expected, it was also found that increasing temperature from 30 °C to 60 °C increased the liquid loss. Ofstad *et al.* (1993) showed that liquid loss at 60 °C was much higher when the muscle was comminuted before heating. Heating at higher temperatures results in further cook loss for both minced and whole cod muscle samples when isothermally heated at 60 °C, 75 °C and 90 °C for 10 min (Skipnes *et al.*, 2007). When heating salmon at 121.1 °C, the cook loss increased rapidly to 14% after 5 min heating and then increased more slowly to 20% after two hours (Kong *et al.*, 2007). Freezing, followed by thawing, will result in rapid increase in cook loss and higher loss in WHC at lower temperatures than for fresh fish (Skipnes *et al.*, 2007).

For a solid portion of fish, e.g. a fillet-cut of 140 g, during a conventional heating process in a steam cabinet, a temperature gradient from the surface to the core of the fish will exist. This has been studied for transport of moisture in tuna muscle. Bell *et al.* (2001) found that moisture transport out of the tuna muscle primarily resulted from the denaturation of muscle proteins and the resulting pressure gradient. The changes in mass transfer properties in the near-surface region reduced the mass loss rates but enforced the pressure gradient. The ratio of surface area to total volume is not expected to be of importance for a vacuum-packed fish portion.

The most labile muscle proteins of fish are collagen and α-actin. Collagen denaturation starts at about 30 °C and α-actin becomes insoluble at 50 °C. Any temperature sufficient to cook cod is sufficient to disrupt collagen so that it is not a factor in the toughness of the cooked material. Myosin becomes insoluble at about 55 °C and actin at 70–80 °C. Tropomyosin and troponin are the most heat-resistant and become insoluble at about 80 °C (Hultin, 1996).

Muscle proteins are generally classified as sarcoplasmic (water-soluble), myofibrillar (salt-soluble) or stromal (insoluble) (Skaara and Regenștein, 1990), of which the myofibrillar constitute the largest fraction of the total protein in cod (76%), and the stromal (extracellular), fraction, which mainly consists of collagen, the smallest (3%) (Suzuki, 1981).

The sarcoplasmic proteins, as such, do not contribute much to texture changes because they have a very low capacity of immobilisation of water in their structure (Dunajski, 1979). The enzymes present in the sarcoplasmic fraction, however, may influence gelation of intact muscle.

Myosin and actin are the myofibrillar proteins directly involved in contraction–relaxation. Due to the low collagen content of fish, their role in gelation and texture is even more important than in meat (Brown, 1987). Much work has been done on thermal gelation of myofibrillar proteins, especially related to surimi (Arai, 2002; Gill *et al.*, 1992; Lanier, 1986). The special mixture with salt, however, and the fact that the studies of surimi focus on low temperature gelation or 'setting' caused by transglutaminase activity (Tsukamasa and Shimizu, 1990, 1991), reduces their relevance for regular thermal processing of intact muscle.

The effects of proteolytic enzyme activity in fish products have been reviewed by Haard (1994) and Skåra and Olsen (2000). The problems related to residual enzyme activity in fish products after thermal processing are mainly quality changes in the texture. Deng (1981) studied textural changes in mullet during heating and found different profiles in shear force when changing the heating process. He suggested that the texture of cooked fish is tougher due to physical protein denaturation only, while a slower and/or stepwise heating process gives a tender/softer texture which he explained by alkaline protease activity. Most enzymes are inactivated at temperatures above 50 °C (Svensson, 1977), although some are known to be heat stable (with remaining and stable activity over a certain range of time and temperature) and therefore have the potential to change the product quality. In fact, some heat-stable alkaline proteases are more or less inactive at physiological temperatures and activated at high temperatures only (Dahlmann *et al.*, 1985; Toyahara *et al.*, 1987).

Warmed over flavour (WOF) sometimes develops 1–2 days after thermal processing and is characterised as the taste of carton or paint. WOF is caused by oxidation of lipids. Thermal processing releases iron and other ions from the fish meat and increases the rate of oxidation in the temperature range 60–70 °C. Above 80 °C, and especially around 100–110 °C, Maillard reactions will slow down the reaction as reduction of sugar and proteins have an anti-oxidative effect. Addition of antioxidant to the fish feed (e.g. E-vitamin, alfa-tocopherol) will reduce the rate of oxidation.

19.2 Novel methods of optimising the quality of in-pack processed fish products

As for any refined food, the quality of cooked fish depends on raw material quality. There has been a long tradition in the Scandinavian canning industry of using frozen fish, mostly post-rigor filleted. Wild fish has to some extent been frozen or gutted and filleted offshore, pre-rigor, while the rest have had to be landed before processing, i.e. post-rigor. Due to lack of a stable raw material supply, industrial production has been based on frozen fish. This has put some limits on the possible end quality of the finished products. Today, several farmed species are available, and there are already two decades of experience on large-scale farming of salmon and trout. Closed life-cycles have been developed for haddock, pollock and hake, but production of these species has been rather modest while other species, e.g. cod and halibut have been farmed on an industrial scale since 2006 and are expected to grow fast in the coming years (Anon., 2006; Rosenlund and Skretting, 2006). This has also made a more stable supply of pre-rigor fish possible for those packing and heat-processing fish.

19.2.1 Pre-processing steps

All kind of pre-processing may be of importance to the end quality of the product, but some are of special importance to thermal processing. In the case of thawing of frozen products and pre-cooking, the initial temperature of the product when starting the autoclave is very important and must be kept above the scheduled minimum process temperature. Products that are still partly or completely frozen will result in a process that cannot be controlled and must be prohibited.

Cutting, grinding, slicing, etc. must be specified in order to avoid pieces with larger sizes than used during the heat penetration studies. During weighing and dosing, the amounts must be within specified limits established during the heat penetration studies. Any changes in the recipe of the product may change the heat transfer conditions, and must be verified with a heat penetration study or other means.

The pre-processing must also be designed to avoid degradation of quality. Trimethylamine-n-oxide (TMAO) is found in high concentrations in several marine fish (Oetjen and Karl, 1999). It is reduced to trimethylamine (TMA) by enzymes or spoilage bacteria (Malle *et al.*, 1986), resulting in the characteristic smell of iced fish (Pedrosa-Menabrito and Regenstein, 1990; Simeonidou *et al.*, 1997).

19.2.2 Equipment for thermal processing

Sterilisation requires temperatures above the boiling point of water at atmospheric pressure, and an overpressure is required. In the autoclave, the pressure must be carefully controlled during the process to avoid damage to the packaging and the autoclave circulation pump (cavitations if the pressure is to low). The forces on the packaging should be reduced to a minimum by determining the correct counter

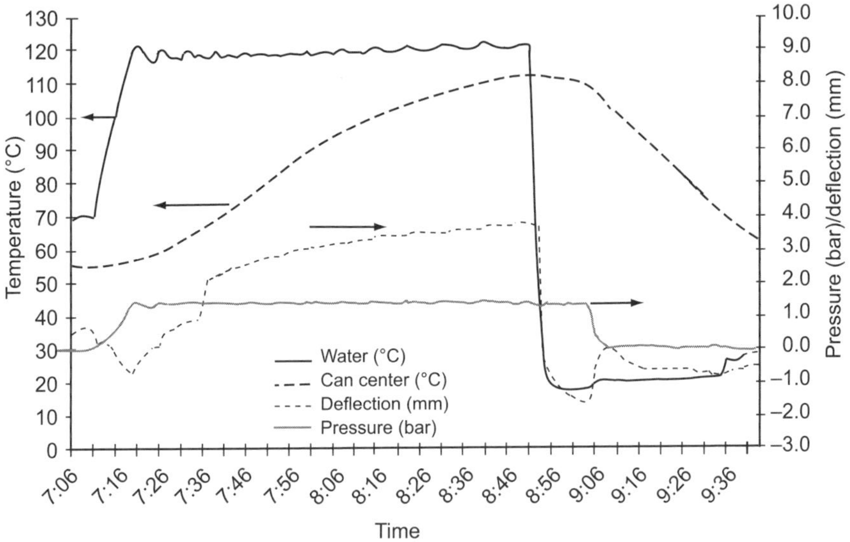

Fig. 19.3 Temperatures in autoclave and product, deflection of can, and pressure of autoclave during a real-life sterilisation process.

pressure. This can be done by using the average temperature inside the can during the process for calculating the expansion of the food and the headspace gas. Instruments measuring the deformation of the packaging are an expensive but more exact way of doing this, especially for foods releasing gas during processing and flexible packaging. Release of gas or formation of new gas may result from chemical reactions during thermal processing. An example of this is canned mackerel in tomato. In Fig. 19.3, curves for deflection and pressure in a can are shown. During the first phase of cooling, pressurised air is used to control the pressure and results in some deflection of the can, but the lid of the can returns to the initial position after processing. The pressure may also be of importance to the heat transfer and the sterility of the product (Skipnes *et al.*, 2002). A low pressure may result in a dead space between the food and the packaging, and insulate the food. A sudden pressure change at the start of cooling may result in ebullition and an unexpected rapid temperature fall inside the product.

Rotary retorts increase the heat transfer considerably, even for particulate foodstuffs. For liquid products, rotation may reduce the sterilisation time (and the heat load) dramatically (Eisner, 1988). This compensates, in many cases, for the lower capacity of rotary autoclaves compared to still autoclaves. For fish products, agitation during heating is unusual because the fish is vulnerable and easily falls apart. This problem is a challenge for processing involving rotation and rocking. More recently, the so called Shaka™ system has been introduced, with vibrating retort baskets. In these machines, a reciprocal movement of the products with 100 to 150 strokes per minute results in a dramatic improvement in heat transfer for convection-heated products, but examples with fish products are not known.

For pasteurisation at temperatures of about 90 °C and above, a counter-pressure may be desirable for flexible packaging materials and in some cases (e.g. easy peel top film) even necessary. As for sterilisation, this results in the need for an autoclave, but even at temperatures below 90 °C the autoclave may be the preferred solution because of the possibility of counter pressure and a temperature distribution that normally is much better than the alternatives.

Alternative equipment for mild heat treatment involves water immersion and steam cabinets. As for autoclaves, these solutions have their continuous variants with steam tunnels and water baths with conveyors. For a water bath with sufficient circulation (at least 50% exchange of water per min) and spreading system, a temperature distribution comparable to a modern autoclave should be possible. For cabinets, the performance depends on the mixture of air and steam in the cabinet and the fan system. Large temperature deviations must be expected in cabinets (Sheard and Rodger, 1995), and the variations in steam/air ratio may also result in uneven heat distribution, even for an otherwise seemingly acceptable temperature distribution. Measurements reported by Nicolaï (1994) revealed oven temperature differences of up to 15 °C between different locations for a pre-set temperature of 70 °C in a combi-steamer, using a hot air/steam mixture. There are usually also larger deviations from the set point temperature in cabinets than in autoclaves, but there are examples of improvements by advanced control strategies being used in spite of an inhomogeneous temperature distribution (Ryckaert *et al.* 1999; Verboven *et al.*, 2000a,b). The large production of cabinets for kitchens has resulted in moderate prices, but for industrial fish processing, cabinets of high capacity and temperature uniformity within 1 °C are required.

Rapid heating methods are more suitable for continuous processing. Microwave heating still suffers from problems with uneven heating and limited penetration depth (a few millimetres) (Ryynänen, 2002). An example of these problems is shown in Fig. 19.4. Here, a fish pudding formed as a sausage has been heated in a microwave oven at 2450 MHz and cut by a string immediately after heating. Thermal images of the cross-sections have then been recorded with a thermal imaging camera (FLIR Systems AB, Stockholm, Sweden). The picture illustrates how the limited penetration depth results, combined with the focusing effect of the cylindrical shaped pudding, results in a pattern where the temperature is highest in the middle between the surface and the core. Radio-frequency (RF) heating gives a better penetration into the product and can be stabilised by a water immersion system, but RF heating also suffers from uneven heating and is, so far, rarely used for pasteurisation of foods (McKenna *et al.*, 2006).

When the required F-value has been determined, the calculation has to be linked to the heat transfer properties to determine the required processing time and temperature. The basics of mathematical methods for calculation of a sterilisation process were developed in the 1920s. Ball's mathematical method, after C. Olin Ball, was published in 1923 and has been refined several times. Other mathematical methods have been issued later and also adapted to pasteurisation processes, like the Stumbo calculation method (Stumbo, 1973). Modified versions of these methods are today found in computer programs, side by side with numerical

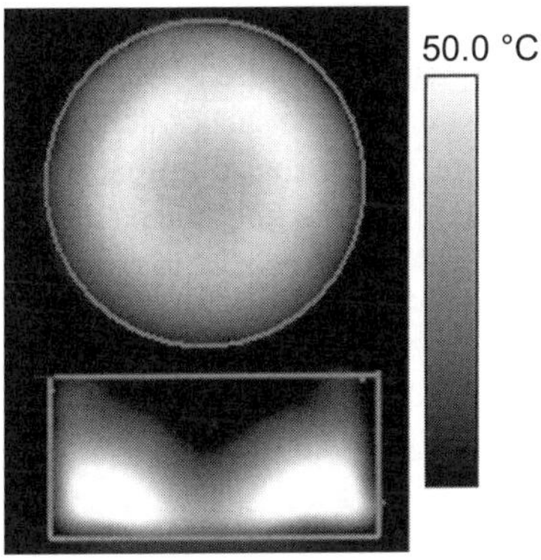

Fig. 19.4 Cross-section of fish pudding cut longitudinally (bottom) and cut at half height (top).

simulation. One example of this is CTemp (Tucker and Holdsworth, 1991; Tucker *et al.*, 1996) issued by Campden and Chorleywood Food Research Association. Among the numerous publications in this field, Pflug (1988) has presented several of the most central papers up to 1978. A review of the mathematical methods have been presented by Stoforos *et al.* (1997) and there are also several publications on how to bring these methods to the next step: on-line control of autoclaves based on thermal inactivation modelling (Teixeira *et al.*, 1999).

The use of time–temperature integrators has gained popularity over the last decade. One practical result of this is the use of small beads, placed within the food and analysed after thermal processing, indicating either the maximum temperature reached or even the achieved lethality as given by the F value. Such time–temperature integrators can also be used for studying factors of quality. For an overview of these very interesting and promising techniques, refer to Hendrickx *et al.* (1995) and Van Loey *et al.* (2004).

For pasteurised products, the thermal processing must also take into account the desired shelf-life of the product. The shelf-life depends on:

- The time for surviving micro-organisms to germinate and reach undesired levels or produce toxin.
- The time for enzyme activity or other chemical reactions to degrade the product to an unacceptable level. For mild heat treatment the processing might even accelerate the activity of some enzymes.
- Physical factors, such as discolouring by light.

A shelf-life study must be done to validate the shelf-life of the products. This study should include an evaluation of sensory, chemical and microbiological parameters of a product stored at or above expected realistic temperatures for a period longer than the expected shelf-life. Microbial modelling may be used in addition to a shelf-life study and is often a good tool in the early stages of product development. Software for microbial modelling is available today both as commercial and free software, and should be carefully evaluated as there always will be an uncertainty related to such models. Therefore, Dalgaard and co-workers (2002) suggested a bias factor for seafood spoilage micro-organisms that should be between 0.75 and 1.25 for a microbial spoilage model to be successfully validated, but no generally accepted criteria for successful validation of predictive models are available at present. However, modelling of specific micro-organisms and growth media, e.g. *Listeria monocytogenes* in cold smoked salmon, has successfully been done (Gimenez and Dalgaard, 2004).

19.2.3 Quality optimisation by calculations and other methods

Ever since the invention of thermal processing, the focus has been on how to minimise the thermal damage to the product (with the combination of microbial inactivation). An obvious approach to optimise the quality is sensory evaluation of a range of time–temperature combinations that result in the desired safety (*F* value). This is time-consuming and expensive, but gives the opportunity to optimise directly in terms of consumers of the product and as long as the only concern is to find the optimal time and temperature combination, this might be a feasible approach. The quality changes during storage are not to be forgotten when optimizing the thermal processing of a product.

Sophisticated methods for optimisation, including TTIs and computational fluid dynamics have been introduced, as summarised in Richardson (2004). One of the easiest ways of performing optimisation is to calculate the cook value (*C* value), as described in the previous paragraph, for several time–temperature combinations that result in the same *F* value; the time–temperature combination resulting in the lowest *C* value correlates to the best quality retention (Richardson, 2004). Quality changes are often too complex to be described in terms of a *z*-value, but there are exceptions. For instance, the denaturation enthalpy of fish proteins during heating has been studied by differential scanning calorimetry. Denaturation of cod muscle proteins has been compared to cook loss and loss of water-holding capacity, and it has been shown that the protein denaturation occurs in a lower temperature range (35–66 °C) than the appearance of major cook loss (above 80 °C) when cod muscle is heated (Skipnes *et al.*, 2008). Other mechanisms for release of water than protein denaturation should therefore also be considered, but as this is in addition to liquid loss due to denaturation of several proteins, it does not fit into a first-order model and cannot easily be quantified by a *C* value. However, kinetic parameters for a first-order inactivation of separate proteins are available, i.e. a *D*- and *z*-value for actin have been presented (Skipnes *et al.*, 2008) and may perhaps be used for *C* value calculations if they can be related to sensory acceptance of a cooked appearance.

Another limitation to the end quality of the product is given by the package geometry. The 'cold spot' of a product is the place where the temperature is lowest during heating. This is not necessarily the geometric centre of the product and has to be determined for each product. The distance to the cold spot is crucial for the time necessary to achieve the desired sterilisation/pasteurisation through the whole product. The thermal conductivity of fish products depends on several parameters,

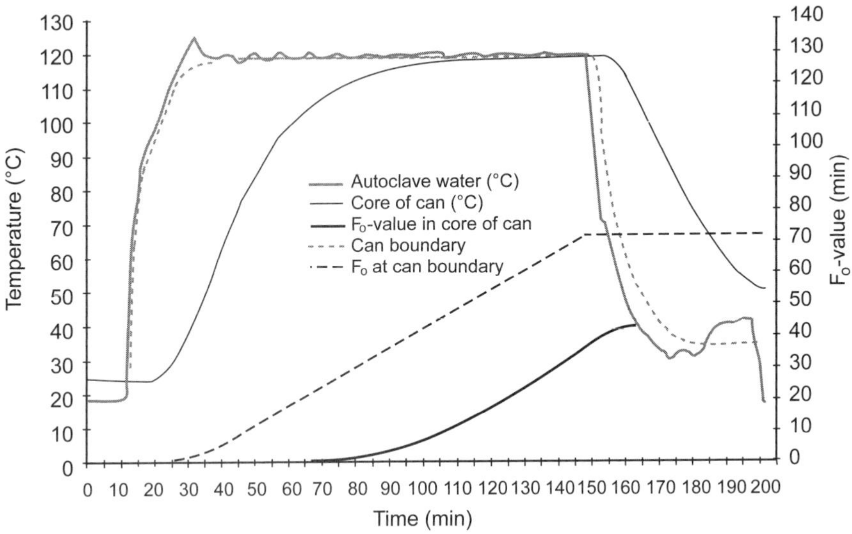

Fig. 19.5 Temperature and sterilisation value (F_0) in core of can (continuous lines), temperature and sterilisation value (F_0) at can boundary (dashed lines) and autoclave temperature (thin, continuous line).

the most important being the water content. The order of magnitude of thermal conductivity of fresh fish is 0.5 W/m K. For a solid product, heat is transferred by conduction. At the boundaries of the product the temperature will be close to the water temperature during the sterilisation, resulting in a much higher heat load than at the cold spot. This is illustrated in Fig. 19.5. The heat load is, in this case, nearly doubled at the boundaries compared with the core, resulting in severe over-processing for most of the contents of the can.

Producers desiring a reproducible quality should keep in mind that continuous changes in the fish as raw material cannot be avoided, and therefore optimisation should be regarded as a dynamic process to be updated on a regular basis or even for each product batch. Thus, the optimisation tools need to be easy to use and give rapid results, or even on-line process control.

19.2.4 Quality and heating rate – effects of rapid heating

Thermal processing of cylindrical cans with conventional methods results in a much higher heat load on the surface compared with that on the centre of the product, and a number of methods may be used to reduce the difference between surface and cold spot. It is often desired to achieve a similar temperature curve in the centre as on the surface as this would reduce the total heat load on the volume average of the product. This effect has been demonstrated by using cans of different diameters and it has been shown that the optimum processing temperature is decreasing when the can diameter is increasing (Ohlsson, 1980).

Rapid heating can be achieved in many ways other than reducing the size of the food product. For liquid or semi-liquid products, rotation of the package during conventional heating is very effective (Eisner, 1988). Conventional heating can also be speeded up by vibrations or by employing ultrasound to the product. However, for some fish products, shaking, rotating, etc. will make the fish fall apart and the use is therefore limited.

Rapid volumetric heating has been suggested as a method for reducing the heat load on the volume average of the product. Ohmic heating is a method that can generate temperature though an electrically homogeneous product, but the need for electrodes in contact with the food limits the method to aseptic packaging.

Heating with electric fields is more convenient for packed products, as long as appropriate packaging is used, such as plastics; use of metal containers is limited. Microwaves and radio-frequency is more commonly used for thawing or tempering of foods than for pasteurisation or sterilisation in the food industry. For radio-frequency heating, water immersion of the product can be used to stabilise the temperature at the product boundaries and may result in a more widespread industrial use in the future (Skipnes *et al.*, 2003). Reduced cook loss and better controlled texture attributes are proposed in the ongoing research (Lyng *et al.*, 2007; Skipnes and Pfeiffer, 2005). There might, however, be some limitations to the effect of rapid heating. Protein denaturation will typically be a process of low z-value, i.e. a small change in temperature will result in a great change in the time required for denaturation of proteins. Thus, changes in cook loss and water-holding capacity will be highly influenced by process temperature, while process time becomes less important. Most of the target microorganisms will show a higher z-value and therefore a low temperature and relatively long process time will be favourable. For a process where this is the case, there is little to gain by rapid heating. An example of such a situation is shown by Kong (2007).

19.3 Future trends

The limited fish resources and the predicted collapse in the fisheries (Worm *et al.*, 2006) will eventually result in increased use of farmed fish and underutilised fish species. This will result in better control of raw material quality but, from a consumer viewpoint, not necessarily a better final quality. The change in raw material properties such as pH and texture will also affect the possibilities for optimisation of thermal processing. Furthermore, the fish may be aqua-cultured and fed to obtain desired product properties with respect to further processing. With increasing raw material prices, better utilisation of fish may be expected and this will also affect the quality of the products.

In addition to the raw material situation, the fish meat is very sensitive to heat compared to vegetables and meat of mammals. As a conclusion of the discussion above, some future trends can be expected and are listed below:

- Raw material quality will change.
  - Limited resources of wild fish may result in higher prices and increasing focus on quality.

- Farming of fish will increase and a better controlled raw material quality and a more stable delivery can be achieved.
- Pre-rigor filleting of fish will reduce lead time from slaughter to packaging and thermal processing.

- Processing technology will be enhanced
  - Rapid heating methods will be more widespread in industrial production, e.g.
    - volumetric heating with microwaves and/or radio frequency, and
    - improvement of conventional processing by agitation during heating and cooling, either by rotation, rocking, shaking or ultrasound for products where it is not so important to maintain the fish in one piece (e.g. Bacalao).
  - Combination of thermal processing and other techniques, e.g. high pressure processing will be adapted

- Enhanced optimisation techniques will be found, e.g.
  - modelling tools for calculating inactivation of micro-organisms and enzymes, combined with calculation of quality retention, available to everyone working with optimisation of thermal processing
  - new knowledge on inactivating kinetics and increased knowledge of quality changes during processing
  - Calculation of shelf-life, assisted by modelling of microbial growth during storage

- Better process control
  - New validation tools (e.g. TTIs and wireless temperature loggers) will be more widespread
  - Reduced deviations in process and raw materials and reduced safety margins will be attained.

Finally, novel packaging solutions may also present opportunities for new products. The metal can (steel or aluminium) is still common for sterilised fish products but most products available in Europe now come with easy-opening. In recent years, a number of flexible packaging solutions have also become available to the producers of sterilised products; retort trays and retortable pouches are increasingly used for fish products.

19.4 Sources of further information and advice

Numerous textbooks and articles have been published on thermal processing and there are also textbooks dedicated to quality aspects of sterilised and sous vide processed foods. Among them the following can be recommended: Ghazala, 1998; Lopez, 1987; Richardson, 2001.

Market reports and literature

The World Market for Chilled Food, *Euromonitor Report*, Jan 2004.
Ready Meals in the United States, *Euromonitor Report*, Nov 2001.

The Chilled Ready Meals Market in North America. 2004. Food for Thought Strategic Information Services.

Anon. 2004. Meal preparation time. *Food Processing* 65(2), 20.

Anon. 2004. A global wave. (US and European microwave ownership, meal preparation time and chilled ready meal market.). *Food and Drink Technology* 3(4), 28–29

Rhodehamel, E. J. 1992, FDA's Concerns with Sous Vide Processing, *Food Technology*, Vol. 46, No. 12, pp. 73–76.

US 21 Code of Federal Regulations, Federal Register, National Archives and Records Administration, Washington DC, USA.

Websites

Selected organisations

www.iftps.org
www.fao.org
www.onefish.org
www.norconserv.no
www.campden.co.uk

Producers/Products

www.kraftfoods.com
www.conagrafoods.com
www.hormel.com
www.pcpizza.com
www.landolakes.com
www.lloydsbbq.com
www.lsg-skychefs.com
www.jimmydean.com
www.smithfieldfoods.com
www.tyson.com

Statistics/Trends

http://www.natick.army.mil/soldier/media/fact/food/atp.htm
www.intrafish.com
http://www.ers.usda.gov/publications/FoodReview/DEC2002/frvol25i3a.pdf (Food Review)
www.nfi.com (National Fisheries Institute)
www.noaa.gov (NOAA)
www.nmfs.noaa (NOAA Fisheries)
www.fao.org/docrep/003/R6918E/R6918E00.HTM

Regulations

www.cfsan.fda.gov/~dms/reg-2.html
ec.europa.eu/food/food/foodlaw/index_en.htm

19.5 Acknowledgement

This work has been partly supported by the Research Council of Norway (158929/I30).

19.6 References

ACMSF (Advisory Committee on the Microbiological Safety of Food) 1995, *Annual Report*, Her Majesty's Stationery Office, London, UK.

ACMSF (Advisory Committee on the Microbiological Safety of Food) 1992, *Report on Vacuum Packaging and Associated Processes*, HMSO, London, UK.

Aitken, A. and Connell, J. J. 1979, Fish, in *Effects of Heating of Foodstuffs*, R. J. Priestley, ed., Applied Science Publishers Ltd., Barking, Essex, UK; pp. 219–254.

Anon. 2001, Fish and fishery products, *Codex Alimentarius*, Vol., pp. 9A, viii.

Anon. 2006, Aquaculture – *Production of aquatic organisms (2000–2005) – Aquaculture research: From sea to table [HAVBRUK – Produksjon av akvatiske organismer (2000–2005) – Havbruksforskning: Fra merd til mat]*. Edited by Thomassen, M., Gudding, R., Nortvedt, B., and Jørgensen, L. 1–373.Oslo, Norway, Norwegian Research Council. Ref Type: Serial (Book, Monograph)

Arai, K. 2002, Denaturation of muscular proteins from marine animals and its control, *Nippon Suisan Gakkaishi*, 68(2) pp. 137–143.

Bell, J. W., Farkas, B. E., Hale, S. A., and Lanier, T. C. 2001, Effect of thermal treatment on moisture transport during steam cooking of skipjack tuna (*Katsuwonas pelamis*), *Journal of Food Science*, 66(2), pp. 307–313.

Ben Embarek, P. K. and Huss, H. H. 1993, Heat resistance of *Listeria monocytogenes* in vacuum packaged pasteurized fish fillets, *Int. J Food Microbiol.*, 20(2), pp. 85–95.

Betts, G. D. 1996, *Code of Practice for the Manufacture of Vacuum and Modified Atmosphere Packaged Chilled Foods with Particular Regard to the Risk of Botulism*, Campden and Chorleywood Food Research Association, Chipping Campden, UK.

Brown, W. D. 1987, Fish muscle as food, in *Muscle as Food*, P. J. Bechtel, ed., Academic Press Inc., Orlando, FL, pp. 405–451.

Dahlmann, B., Rutschmann, M., Kuehn, L., and Reinauer, H. 1985, Activation of the multicatalytic proteinase from rat skeletal muscle by fatty acids or sodium dodecyl sulphate., *Biochem.J.*, 228, pp. 171–177.

Dalgaard, P., Buch, P., and Silberg, S. 2002, Seafood Spoilage Predictor – Development and distribution of a product specific application software, *International Journal of Food Microbiology*, 73(2–3), pp. 343–349.

Deng, J. C. 1981, Effect of Temperatures on Fish Alkaline Protease, Protein Interaction and Texture Quality, *Journal of Food Science*, 46, pp. 62–65.

Dunajski, E. 1979, Texture of fish muscle, *Journal of Texture Studies*, 10, pp. 301–318.

ECFF, 1996, *Guideline for the Hygienic Manufacture of Chilled Foods* Helsinki, Finland.

Eisner, M. 1988, *Introduction into the Technique and Technology of Rotary Sterilization*, Second edn, Private Author's Edition.

Gerhardt, P. 1988, The refractory homeostatsis of bacterial spores, in *Homeostatic Mechanisms in Microorganisms*, R. Wittenbury *et al.*, eds., Bath University Press, Bath, United Kingdom, p. 41.

Gerhardt, P. and Marquis, R. E. 1989, Spore thermoresistance mechanisms, in *Regulation of Procaryotic Development*, I. Smith, R. A. Slepecky, and P. Setlow, eds., American Society for Microbiology, Washington D.C., pp. 43.

Ghazala, S. 1998, *Sous Vide and Cook–Chill Processing for the Food Industry* Aspen Publishers Inc, Maryland.

Gill, T. A., Chan, J. K., Phonchareon, K. F., and Paulson, A. T. 1992, Effect of Salt

Concentration and Temperature on Heat-Induced Aggregation and Gelation of Fish Myosin, *Food Research International*, 25(5), pp. 333–341.

Gimenez, B. and Dalgaard, P. 2004, Modelling and predicting the simultaneous growth of *Listeria monocytogenes* and spoilage micro-organisms in cold-smoked salmon, *Journal of Applied Microbiology*, 96(1), pp. 96–109.

Gombas, D. E. 1983, Bacterial spore resistance to heat, *Food Technology*, 37(11), p. 105.

Haard, N. F. 1994, Protein hydrolysis in seafood, in *Seafoods: Chemistry, Processing, Technology and Quality*, F. Shahidi and J. R. Botta, eds., Blackie Academic and Professional, London, pp. 11–33.

Hamm, R. 1986, *Muscle as Food*, Academic Press, New York.

Hendrickx, M., Maesmans, G., De Cordt, S., Noronha, J., Van Loey, A., and Tobback, P. 1995, Evaluation of the integrated time–temperature effect in thermal processing of foods', *Critical Reviews in Food Science and Nutrition*, 35(3), pp. 231–262.

Hermansson, A. M. 1983, Quality of plant foods, *Human Nutrition*, 32, p. 369.

Howgate, P. F. and Ahmed, S. F. 1972, Chemical and Bacteriological Changes in Fish Muscle during Heating and Drying at 30 °C, *Journal of the Science of Food and Agriculture*, 23, pp. 615–627.

Hultin, H. O. 1996, Amino Acids, Peptides, and Proteins, in *Food Chemistry*, 3rd edn, O. R. Fennema, ed., Marcel Dekker.

IFTPS 1992, *Temperature Distribution Protocol for Processing in Steam Still Retorts, excluding Crateless Retorts*, Institute for Thermal Processing Specialists (IFTPS), Washington D.C.

IFTPS 1995, *Protocol for Carrying out Heat Penetration Studies*, Institute for Thermal Processing Specialists (IFTPS), Washington DC.

IFTPS 2002, *Temperature Distribution Protocol For Processing In Still Water Immersion Retorts, including Agitating Systems Operated in a Still Mode* Washington D.C.

Kong, F., Tang, J., Rasco, B., Crapo, C., and Smiley, S. 2007, Quality Changes of Salmon (*Oncorhynchus gorbuscha*) Muscle during Thermal Processing, *Journal of Food Science*, 72(2), p. S103–S111.

Lanier, T. C. 1986, Functional Properties of Surimi, *Food Technology*, pp. 107–114.

Lopez, A. 1987, *A Complete Course in Canning*, 12 edn, The Canning Trade, Baltimore.

Lyng, J. G., Cronin, D. A., Brunton, N. P., Li, W. Q., and Gu, X. H. 2007, An examination of factors affecting radio frequency heating of an encased meat emulsion, *Meat Science*, 75(3), pp. 470–479.

Malle, P., Eb, P., and Tailliez, R. 1986, Determination of the quality of fish by measuring trimethylamine oxide reduction, *International Journal of Food Microbiology*, 3, pp. 225–235.

May, N. 1997, *Guidelines for Batch Retort Systems – full water immersion – raining/spray water – steam/air*, Campden and Chorleywood Food Research Association, Chipping Campden, Gloucestershire, GL55 6LD, U.K., Guideline No. 13.

McKenna, B. M., Lyng, J., Brunton, N., and Shirsat, N. 2006, Advances in radio frequency and ohmic heating of meats, *Journal of Food Engineering*, 77(2), pp. 215–229.

Nicolaï, B. M. 1994, *Modeling and uncertainty propagation analysis of thermal food processes,* Ph.D. thesis, Katholieke Universiteit Leuven.

Oetjen, K. and Karl, H. 1999, Improvement of gas chromatographic determination methods of volatile amines in fish and fishery products, *Deutsche Lebensmittel-Rundschau*, 95(10), pp. 403–407.

Ofstad, R., Egelandsdal, B., Kidman, S., Myklebust, R., Olsen, R. L., and Hermansson, A. M. 1996a, Liquid loss as effected by post mortem ultrastructural changes in fish muscle: Cod (Gadus morhua L) and salmon (Salmo salar), *Journal of the Science of Food and Agriculture*, 71(3), pp. 301–312.

Ofstad, R., Kidman, S., Myklebust, R., and Hermansson, A. M. 1993, Liquid Holding Capacity and Structural-changes During Heating of Fish Muscle – Cod (*Gadus-Morhua* L) and Salmon (*Salmo-Salar*)', *Food Structure*, 12(2), pp. 163–174.

Ofstad, R., Kidman, S., Myklebust, R., Olsen, R. L., and Hermansson, A. M. 1996b, Factors influencing liquid-holding capacity and structural changes during heating of comminuted cod (*Gadus morhua* L) muscle, *Food Science and Technology–Lebensmittel-Wissenschaft and Technologie*, 29(1–2), pp. 173–183.

Ohlsson, T. 1980, Optimal Sterilization Temperatures for Sensory Quality in Cylindrical Containers, *Journal of Food Science*, 45(6), pp. 1517–1521.

Parry, D. A. 1970, Fish as food, in *Proteins as Human Foods*, R. A. Lawrie, ed., Butterworths, London, p. 365.

Pedrosa-Menabrito, A. and Regenstein, J. M. 1990, Shelf-life Extension of Fresh Fish – A review. Part III Fish Quality and Methods of Assessment, *Journal of Food Quality*, 13, pp. 209–223.

Peleg, M. 2006, It's time to revise thermal processing theories, *Food Technology*, 60(7), p. 92.

Pflug, I. F. 1988, *Selected Papers on the Microbiology and Engineering of Sterilization Process*, Environmental Sterili, Minneapolis.

Rhodehamel, E. J. 1992, FDA Concerns with Sous Vide Processing, *Food Technology*, 46(12), pp. 73–76.

Richardson, P. 2001, *Thermal Technologies in Food Processing,* Woodhead Publishing Ltd., Cambridge, England.

Richardson, P. 2004, *Improving the Thermal Processing of Foods*, 1 edn, Woodhead Publishing Ltd., Cambridge, England.

Rosenlund, G. and Skretting, M. 2006, Worldwide status and perspective on gadoid culture, *ICES Journal of Marine Science*, 63(2), pp. 194–197.

Ryckaert, V. G., Claes, J. E., and Van Impe, J. F. 1999, Model-based temperature control in ovens, *Journal of Food Engineering*, 39(1), pp. 47–58.

Ryynänen, S. 2002, *Microwave heating uniformity of multicomponent prepared foods,* PhD EKT series 1260, University of Helsinki, Department of Food Technology (Dissertation).

Setlow, P. and Johnson, E. A. 1997, Spores and their significance, in *Food Microbiology. Fundamentals and Frontiers*, 1 edn, M. P. Doyle, L. R. Beuchat, and T. J. Montville, eds., ASM Press, Washington D.C., pp. 30–65.

Sheard, M. A. and Rodger, C. 1995, Optimum Heat Treatments for 'Sous Vide' Cook–Chill Products, *Food Control*, 6(1), pp. 53–56.

Simeonidou, S., Govaris, A., and Vareltzis, K. 1997, Effect of frozen storage on the quality of whole fish and fillets of horse mackerel (*Trachurus trachurus*) and Mediterranean hake (*Merluccius mediterraneus*)', *Zeitschrift fur Lebensmittel-Untersuchung Und-Forschung A-Food Research and Technology*, 204(6), pp. 405–410.

Skaara, T. and Regenstein, J. M. 1990, The structure and properties of myofibrillar proteins in beef, poultry and fish, *Journal of Muscle Foods*, 1, pp. 269–291.

Skåra, T. and Olsen, S. O. 2000, *Proteolytic enzyme activity during thermal processing*, S. A. Georgakis, ed., Greek Society of Food Hygienists and Technologists, Leptocarya – Pieria, Greece, pp. 268–286.

Skipnes, D., Øines, S., Rosnes, J. T., and Skåra, T. 2002, Heat transfer in vacuum packed mussels (*Mytilus edulis*) during thermal processing, *Journal of Aquatic Food Product Technology*, 11(3/4).

Skipnes, D. and Pfeiffer, T. 2005, Radio-Frequency Heating of Food in a Water Bath. Building a FEM Model of the Process, *The Third International Symposium on Applications of Modelling as an Innovative Technology in the Agri-food Chain*, Leuven BE.

Skipnes, D., Pfeiffer, T., Skåra, T., Folkvord, L., Øines, S., Sveinsdottir, K., and Palsson, P. G. 2003, *Radio-Frequency Heating of Minimally Processed Fish Products – Possible Quality Improvements by Rapid Heating*, Icelandic Fisheries Laboratory, Reykjavik, Iceland, pp. 36–38.

Skipnes, D., van der Plancken, I., Van Loey, A., and Hendrickx, M. 2008, Kinetics of heat denaturation of proteins from farmed Atlantic cod (*Gadus mohua*), *Journal of Food Engineering*, 85(1), pp. 51–58.

Skipnes, D., Ostby, M. L., and Hendrickx, M. E. 2007, A method for characterising cook loss and water holding capacity in heat treated cod (*Gadus morhua*) muscle, *Journal of Food Engineering*, 80, pp. 1078–1085.

Stoforos, N. G., Noronha, J., Hendrickx, M., and Tobback, P. 1997, A Critical Analysis of Mathematical Procedures for the Evaluation and Design of In-container Thermal Processes for Foods', *Critical Reviews in Food Science and Nutrition*, 37(5), pp. 411–441.

Stumbo, C. R. 1973, *Thermobacteriology in Food Processing*, 2nd. edn, Academic Press Inc..

Suzuki, T. 1981, *Fish and Krill Processing Technology*, Applied Science Publishers, London.

Svensson, S. 1977, Inactivation of Enzymes during Thermal Processing, in *Physical, Chemical and Biological Changes in Food Caused by Thermal Processing*, T. Høyem and O. Kvåle, eds., Applied Science Publishers Ltd., London, pp. 202–218.

Teixeira, A. A., Balaban, M. O., Germer, S. P. M., Sadahira, M. S., Teixeira-Neto, R. O., and Vitali, A. A. 1999, Heat transfer model performance in simulation of process deviations, *Journal of Food Science*, 64(3), pp. 488–493.

Toyahara, H., Nomata, H., Makinodan, Y., and Shimimizu, Y. 1987, High molecular weight heat stable alkaline proteinase from white croaker and chum salmon muscle: Comparison of the activating effects by heating and urea., *Comp.Biochem Physiol.*, 68B(99), p. 102.

Trout, G. R. 1988, Techniques for measuring water-binding capacity in muscle foods – A review of methodology, *Meat Science*, 23(4), pp. 235–252.

Tsukamasa, Y. and Shimizu, Y. 1990, Setting Property of Sardine and Pacific Mackerel Meat, *Nippon Suisan Gakkaishi*, 56(7), pp. 1105–1112.

Tsukamasa, Y. and Shimizu, Y. 1991, Factors Affecting the Transglutaminase-associated Setting Phenomenon in Fish Meat Sol, *Nippon Suisan Gakkaishi*, 57(3), pp. 535–540.

Tucker, G. S. and Holdsworth, S. D. 1991, *Optimisation of Quality Factors for Foods Thermally Processed in Rectangular Containers*, Technical Memorandum 627, CCFRA, Chipping Campden, UK.

Tucker, G. S., Noronha, J. F., and Heydon, C. J. 1996, Experimental validation of mathematical procedures for the evaluation of thermal processes and process deviations during the sterilization of canned foods, *Food and Bioproducts Processing*, 74, pp. 140–148.

Van Loey, A., Guiavarac'h, Y., Claeys, W., and Hendrickx, M. 2004, The use of time–temperature integrators (TTIs) to validate thermal processes, in *Improving the Thermal Processing of Foods*, 1st edn, P. Richardson, ed., Woodhead publishing Ltd., Cambridge, UK, pp. 365–379.

Verboven, P., Scheerlinck, N., De Baerdemaeker, J., and Nicolai, B. M. 2000a, Computational fluid dynamics modelling and validation of the isothermal airflow in a forced convection oven, *Journal of Food Engineering*, 43(1), pp. 41–53.

Verboven, P., Scheerlinck, N., De Baerdemaeker, J., and Nicolai, B. M. 2000b, Computational fluid dynamics modelling and validation of the temperature distribution in a forced convection oven, *Journal of Food Engineering*, 43(2), pp. 61–73.

Worm, B., Barbier, E. B., Beaumont, N., Duffy, J. E., Folke, C., Halpern, B. S., Jackson, J. B. C., Lotze, H. K., Micheli, F., Palumbi, S. R., Sala, E., Selkoe, K. A., Stachowicz, J. J., and Watson, R. 2006, Impacts of biodiversity loss on ocean ecosystem services, *Science*, 314(5800), pp. 787–790.

Index